# Beyond Biology

# Beyond Biology

## METAPHYSICAL BRAIN SCIENCE

Charles S. Yanofsky

**Outskirts Press, Inc,**
**Denver, Colorado**

Beyond Biology
Metaphysical Brain Science

V2.0



Outskirts Press, Inc.
http://www.outskirtspress.com

ISBN: 978-1-4327-5401-3



PRINTED IN THE UNITED STATES OF AMERICA

“Friends, Hold my arms! For in the mere act of penning my thoughts of this Leviathan, they weary me, and make me faint with their outreaching comprehensiveness of sweep, as if to include the whole circle of the sciences, and all the generations of whales, and men, and mastodons, past, present, and to come, with all the revolving panoramas of empire on earth, and throughout the whole universe, not excluding its suburbs. Such, and so magnifying, is the virtue of large and liberal theme! No great enduring volume can ever be written on the flea, though many there be who have tried it.”

-Herman Melville: Moby Dick

# Table of Contents

# Introduction

Science is a seductress. 'Scientism' holds that by answering an infinite series of small questions we will uncover the final riddle of The Big Question in an epiphany of resplendent beauty. The scientist waits in breathless patience as each small layer is slowly peeled away, devises elaborate schemes to gain a better view of the object of his passion, maintains abiding faith in the rightness of his mission. Sadly, as layer after layer is stripped away, he is frustrated by unfulfilled desire.

Yet all of us believe in the scientific method, which is at once seduction and religion. If religion is, roughly speaking, a belief in something or a set of principles larger than ourselves, then all of us, even professed atheists and agnostics, have got religion. Our new faith, deeply held, is the belief in science, scientism. No matter whether a person believes in any god, he or she still shares awe in the power of science to unveil the deepest mysteries of life and death. And even those who believe in god, in our day, accept scientific principles. Science is a bedeviling nymph wooing us, intoxicated, to the brink of deep enduring truth. We know about the beginnings of the universe up until a limbus of but a tiny fraction of a second. But for this knowledge we would know how it is the world began. But we'll never find the ultimate cause or the ultimate fate of our universe. We know about the mechanism of motor function in the brain, but may never be able to discern where exactly the slightest command, even to move a finger, originates in the brain. Science brings us to the brink. She seduces and tantalizes us, but delivers us to a certain point. We're condemned never to answer the Big Question, to know naked truth.

In the not so distant past, truth was the province of the seer or prophet. Truth was revealed by selected visionaries and then filtered down to the masses. In our time, truth is democratized and can be experienced and tested by anyone. What is true can be reproduced in anyone's laboratory. No one claims to have a special vision. If something isn't there for everyone to see, it isn't true. Today's secrets are meticulously teased out by scientific method from a penurious universe that does not easily give up its secrets. Truth is discovered, not revealed, for the vast majority of us today, and is available for all to examine. A corollary of an acceptance of scientific principle is belief in matter, as opposed to spirit, and a mechanistic view of nature. Nature is manifest in readily perceptible material objects having no divine purpose, design or enterprise.

But this is not the end of the story. Science is wrong a lot of the time. Indeed the major advantage of the scientific method is that our most cherished beliefs and theories are overthrown by new facts. The greatest part of science is admitting that we are continuously in a state of some ignorance. It follows then that science is not capable of telling us everything about reality or even the mind of man. Science falls short. This volume is largely about this deficiency in science and what it all means, the consequences of admitting that there is far more to human experience than can be explained by science.

The Big Bang theory is highly touted, yet fundamentally flawed. The Big Bang either does not strictly conform with fact, or fails to explain certain phenomena. Observations that remain to be explained include: 1. Abundances of Hydrogen and Helium may not be as predicted by theory. 2. Estimates of the age of the universe and the Hubble constant do not jive with empirical estimates of the age of the universe or the speed of expansion. 3. The ultimate origin of everything over the first fraction of a second still remains to be explained because physical laws as currently understood do not apply to these initial conditions. 4. The survival of matter over anti-matter. Is the all the stuff of our entire cosmos really made of the miniscule excess of matter over anti-matter after the explosion and why this tiny inequality? Both matter and anti matter very likely came to be in almost equal amounts. 5. Sufficient inhomogeneity (bumps or bubbles) in the original explosion to account for coalescence of galaxies as we observe them. 6. Failure to give even the grossest predictions about the ultimate abundance of matter. This applies especially to so-called “dark matter”. The Big Bang theory does not predict the quantity of matter in the universe. The result is that scientists don’t know whether the universe is closed, that is if it ultimately may end up collapsing upon itself in a big crunch, or if is open and will continue to expand forever. 7. The origin of everything, absolutely everything, which came to be, perhaps from nothing, in a *singularity* smaller than the size of a pinhead is extremely counter-intuitive, even if this might be possible in theory as represented by equations on some physicist's blackboard. In other words as has been remarked many times, our universe may have been created in a space smaller than our physicist’s laboratory! The idea that the universe began as a tiny point, a singularity is reductio ad scientum, which taxes the limits of credulity, at least common sense, but alas, it may be found in the end to be right.

Explanatory “modifications” or correctives have had to be derived which begin to sound far-fetched. These include certain “super-string”, and “inflationary” models, dark matter which holds local galaxies together, dark energy or a force pulling far-flung galaxies apart at a faster rate than can have occurred from a big bang. No one knows if they are true; they are not verifiable in and of themselves. Furthermore only an elite cadre of highly skilled mathematicians knows their deep meaning. Since these persons have some intellectual tools unavailable to most of us, it is reasonable to ask whether they are the new prophets or if they will one day be able to explain their discoveries in a form that can be objectified for the rest of us. At any rate we all know that whatever anyone says today about the really big questions, the origin of the universe and what is to become of everything in the end, only a fraction of current scientific doctrine will be found to be true.

The pangs of doubt that we are experiencing are similar to the Ptolomaic epicycles in the Middle Ages. In medieval times, elite scholars, at that time primarily churchmen, were the only persons felt to be qualified to criticize a complex system later found with the help of outsiders, Kepler, Copernicus and Galileo, to be wrong. Men in those times at first tried to fit empirical data about planetary motion into an erroneous model of an earth-centered planetary system. Only later when the theory crossed a certain threshold of inaccuracy, was the Ptolomaic earth-centered system finally jettisoned in favor of our current heliocentric model.

The real meaning of the Big Bang theory is that matter, all the material of our entire universe, came to be suddenly out of nothingness. No one knows the ultimate origin of all things of course, because physical principles do not apply to conditions at the very beginning of time. It is tempting to speculate that all matter derived out of an idea, that everything as we know it is an emanation of the mind of the Creator. .

Living creatures function on the basis of chemical and physical properties. While in physics the pure mechanical theories of Newton have been replaced with the quantum then relativistic theories, this has yet to trickle down into applied biology and cognitive science. Biological descriptions are on the human or animal scale macro-world where Newton's principles are a good approximation, but new physics raises a doubt at least, about whether cognition, emotion and experience are reducible to biophysical mechanistic constructs. Is all feeling and experience reducible to mechanical events within brain substance, or will some overarching theory someday be invoked? One may speculate that some principle other than electromagnetism may carry information within neuronal systems, perhaps even the strong or weak force on the sub-cellular level or another more encompassing physical principle yet to be discovered that is capable of storing information and experience. In the not too distant future we may be shocked to discover a whole different methodology in biology, that not all can be explained by a mechanistic or current scientific methods.

Scientism has the same defects as religious doctrine. The wide world is seen from the vantagepoint of a single model. In every age theories are born, and some are overthrown, but all die hard. Those who have spent their lives developing or working within the confines of a scientific theory are not disinterested parties. They are loath to give it up. They espouse untenable points of view well after one would have thought on a scientific basis they should have given these theories up. There are, in fact, too many examples to cite in scientific history. Not only church, but scientific authorities stuck to the Ptolomaic earth-centered view of the universe well after observations of planetary motions should have overthrown it. Even today some persons stubbornly insist that the age of the universe is only some six thousand or so years refusing to be influenced by an enormous quantity of preponderant evidence to the contrary. So much must rests on some cherished beliefs, perhaps a whole world view. Semmelweis, Lister and Pasteur struggled to convince their medical world to accept the germ theory of disease that we now take for granted. In our own time, it seems, the Big Bang theory of the creation of the universe, though only accepted in recent decades will, at the least, require major revisions, but these will come slowly because so many careers have been built on physical principles deriving from the Big Bang.

Just when scientific medicine is beginning to make important inroads upon disease, to have a demographic impact and prolong life expectancy and improve the quality of life, many persons, some of them not ignorant in the classical sense of the term, are running enmasse to pre-scientific alternatives. Medical academicians are petrified about criticizing obvious mountebank and charlatan champions of “alternative medicine”, despite the proven efficacy of scientific medicine in our time. We hear almost daily about therapies and surgeries developed and championed. Yet when they are not found to be effective they fall by the wayside though way too slowly. Examples are surgeries such as were used for thoracic outlet syndrome, cranial artery bypass procedures, and physical, chiropractic and oriental therapies, craniofacial therapy, and therapeutic touch to name just a few that persist without having a shred of theoretical or empirical evidence for efficacy. People invest in a system of thought. Indeed whole scientific careers are embedded in individual theories that become belief systems as real as any religion. Scientists are even more susceptible than non-scientists to pseudo science. Some examples here the IQ and eugenics movement in the beginning of our century which resulted in turning away and death of immigrants who would have made our great country even greater and even worse racist movements in Germany which rationalized euthanasia of undesirables and discrimination and genocide. We see pseudo-science in creationism and animal rights and environmentalist movements. The point is how easily myth persists for relatively long periods of time beneath a scientific mantle.

As far as a belief in science, I have to count myself a co-religionist, though I believe also a bit of renegade. I've spent my professional career not as a laboratory researcher, but as a practitioner in the real world, seeing patients. Over this time I have been making my own observations. Though my practice is thoroughly based on scientific principles, I have needed to treat whole human persons. In so doing I have had daily to entertain other modes of thought, some emotional, pre-logical even illogical and contradictory. I have the greatest deference for science, but I do not uncritically accept all scientific dogma.

Science is extremely powerful and has proven to be great tool for explaining the world. Science has apparently its own set of limitations as well and it seems to me their are other useful ways we have of seeing how the world works and our own place within the world. There is a mode of scientific reasoning that is different from other forms of human perception and insight.

| SCIENCE | OTHER |
| --- | --- |
| mechanism | purpose (teleology) |
| analysis (divide and conquer) | Holism |
| lesion experiment | intact function |
| Analysis, mathematics | Intuition |
| Sensory data | theory or faith |
| Observation | Invention |

Tables, such as the one above, help to highlight some of the differences experienced in various modes of thought. My own point of view, I should point out, is really much more in the scientific column. This is not to completely exclude some elements in the second column. Just examining this table which compares scientific thinking with alternative modes of thought, it would seem we would want to choose science rather than alternatives. Yet these other modes not only define a legitimate and real form of human experience, a personal way of seeing the world that completes a picture of the universe that just may be excessively constrained by science. After science there must be something more.

My overall strategy in this book is to point out what the brain and individual neuron can do and then to take stock of man's capacities, which go considerably farther. We will thus discover that human means go well beyond biological endowments. This is a fundamental departure in strategy compared with other discussions by neurologists and brain scientists who, in the main, discuss the brain from the point of view of various brain lesions. There is nothing intrinsically wrong with looking at the brain by analyzing the effects of lesions that impair the function of various parts of the nervous system. It's just that in doing so we emphasize dysfunction, not intact function. If a person has a problem we try to find a lesion in a particular part of the brain. This could be an injury or tumor or stroke that impairs the work of a certain part. Through this method we have been able to define the function of various parts of the brain. So often though, we are enmeshed in this method and lose sight of our real purpose, and that is to explain the totality of human experience. We should try not to lose sight of how the whole nervous system functions as an intact entity, the repository of consciousness and experience. While the lesion model has been extremely fruitful it is now possible to go further in explaining brain function and thus the totality of human experience. This will have profound implications philosophically and practically.

Apraxia is a case in point. A person should by rights be able to perform a certain complex task, by virtue of muscle strength, sensation, coordination, even motivation but just can't. Presumably his brain is just unable to figure out how to proceed. A man stares in the mirror, the water running, razor in hand, but can't figure out how to shave. This is apraxia.

What we end up considering less, because it doesn't usually tell us all that much about the function of specific parts of the brain, is intact human function and in particular we neglect the situations where people perform much better than you expect them to. Instead of a-praxia, lack of ability to do what one should be able to do, we might call this hyper-praxia, performing better than expected. Given our biological means we have traveled well beyond our wildest expectations, far beyond anyone's ability to explain them, yet this is barely noticed. Medical texts which look only at disease states, descriptions of nervous function we have at hand, ignore the well-functioning organisms and have nothing to say about better than normal function. Yet this positive side would be a fruitful area of inquiry, what I would designate as the other side of understanding of human mental function.

The study of dysfunction from the vantagepoint of physical lesions, diseases and alterations in mental function due to abnormalities of the body and effects of chemicals and drugs, has been a very fruitful one and I don't disparage that approach one bit. But it seems to me there are just as many examples of superlative function of an intact human brain and mind and that we can also learn a lot from these examples, one point of view that has not been stressed up until now. In the lofty realm of intact and even superlative ability, we will find it much more difficult to find a ready mechanistic or biological explanation. Indeed, study of this area of function may be outside of or beyond biology.

What is there to learn? Here are just some of the questions we can ask by looking at this other side and the useful answers that may be derived. How does the record breaking runner or long jumper accomplish perform in his athletic event? How can we make his performance better? What is involved in superlative performance of an athletic function in a physiologically healthy human? What roles do motivation and training play? Can these same questions be asked for examples of superlative mental performance? From what source does the Theory of Relativity spring in an intact human brain, or a Beethoven symphony? It seems to me one might thereby uncover combinations of factors that contribute to superlative human performance in an intact organism and that by examining this, we might learn even more about what mental function is about.

Taken from the point of departure of the lesion experiment, the best of all possible worlds is merely undiseased output. The very best that can be expected is an organism and brain that is merely functioning properly. Hence the arguments as to human function are highly deterministic as if to say, "Such and such is expected to occur as long as everything works. We know this because we can tell you what happens when it doesn't work." Everything in this negative case is neatly explained by knowing biology. But what if I could show you examples of what sometimes albeit rarely occurs when everything does work, examples of high performance, athletic feats and genius. Then the explanations tend to be less biological on one side, and also much less deterministic. Along the way we might discover what the human experience, at least for some, is all about. It is for the first time a conception of man as more than a machine. One major conclusion from all of this: Intact biological function is a necessary condition, a springboard from which athletic prowess, genius, hyper-praxis, even free will, develop. All in all we have something far less deterministic and predictable, but the other side of the story of human experience something beyond biology. It's just fine to pat ourselves on the back and brag about what we do know. All of those strides in the area of basic biological science have provided us with a truly wonderful understanding. But as miraculous as they are, they are only part of the story.

We have long ago left our meager biological endowment in the dust. How? With machines that carry us faster than our legs could ever take us and with athletic training that extends physical abilities. We sense light frequencies well beyond the visible spectrum, peering out of the confines of our galaxy into the macro-world over distances signifying the very beginning of time, also into the micro-world of viruses and even the atomic nucleus. We see in other words further and deeper than by rights we should be able to, given the meager biological endowment that is the eye. We see this far and this deep, by using our mind. And we build libraries, computers and other data storage devices to expand memory and reason and do mathematical calculations. Humans inhabit grand buildings, not mere shelters from the cold and dark, write symphonies and stories that transport us from the mundane. As we master physical and genetic handicaps we are heir to, influence our own biology with our minds in other words, it becomes abundantly clear that our biological organism no longer defines us. No longer are we confined to a limiting body. This explosion of technical and perceptive ability suggests not a reductio ad materium of the person but the re-enchantment of life. Examples of how we outwit biological limits are numerous. Why is this is a useful observation?

First of all this makes us unique among all living things on earth; other organisms seem determined, imprisoned by their biology. Perhaps they can adapt rapidly as bacteria and insects do, but they are only able to do whatever their organism is capable of. But that is not the main point. In our scientific materialistic society many persons safely assume that all that we are has some biological or physical explanation. Emotions and deepest thoughts have some physical substrate even if this often remains undefined. Most of our higher function takes place in the brain and we would have a complete explanation for all behavior, if we understood everything about our physiology which is incompletely understood. From the philosophical standpoint this is unpalatable because we then have to accept that knowledge is a mere product of our physiology and bears no relationship to any concept of abstract objective truth. Using and manipulating what we know has merely helped us in the practical sense to survive among other animals, but there are no abstract or fundamental considerations only practical ones. That idea in itself is most disturbing but that aside, what if could be shown that in many ways human belies biology goes well beyond it? Most reasonable people would admit that we do not have a good understanding of all mental phenomena and capacities. Most would say that what we don't understand can be more deeply scrutinized with more scientific research. I would counter that biological explanations for human capacities are so inadequate it is just as likely that not everything is explainable using a biological or purely mechanistic scientific method. Unexpected human ingenuity and ability, make the purely biological explanations for human thought and behavior all the weaker. There might possibly be there is another factor that we have failed to take into account, some will or soul (these are some words used for this nebulous substance in the past) that is not explained biologically. Or perhaps this just gives us reason to admit that we just don't know everything.

Looking at things this way, from the standpoint of whole intact function rather than only by analysis paves the way for a new path of introspection, finding out about ourselves. The end point of relentless biological analysis is by definition ultimate reductionism, that human function can be nothing more, than a sum total of biochemical and electrical mechanisms that occur on the cellular level. What I am saying is maybe, there is more to us than that. Our dignity and purpose is holding on here perhaps by little more than a thread, when we look at man from the scientific perspective here at the end of the Millennium. Possibly there is hope, there is more than meets the eye.

In the next decades we will succeed in altering our very biology by our own design, the very behavior, thought and feelings that are felt to reside and which are in some circles completely explained by our biology. This is a paradox when considered fully. We are on the verge of genetic discoveries that will give us the wherewithal at some time to alter human form and ability. At that time the human organism will be a moving target. Persons will not have a standard heritable form. Instead, that form, if there is one, will be altered at will. At that time we can truly raise questions about where this human will resides. Consider the fact that there will no longer be a constancy of form that we will be able to depend on. Some time in the not too distant future, an Olympian may wish to alter the genes that determine muscular power or strength in order to better compete in an event, or given the genetic polymorphisms that we know about at least choose to have in his own genetic complement a particular heritable version of a protein that maximizes muscular function. Understand that there are thousands of such heritable protein types or polymorphisms that all of us inherit many of which do a creditable job, but which don't maximize function. A prospective parent may choose to pass on the type which appears to function the best. Or undoubtedly such genetic polymorphisms exist for mental function as well.

A simple example being studied at present is a genetic variable for apolipoprotein E. This comes in three forms inherited by all of us designated E2, E3 and E4. We each inherit two copies with our two copies of genes so that a person may be described to have genes E2/E2, E2/E3, E2/E4... etc. It seems that persons who have one or more copies of E4 don't do nearly as well on a statistical basis in older age. There is a much higher incidence of such chronic diseases of Alzheimer disease and heart disease. This Apolipoprotein in all of its variant forms, is implicated in the transport of certain fats in the blood and variations of function may affect, loosely fat cleanup and storage. In order to maximize personal life expectancy and function, many of us would, quite reasonably, want to eliminate the effect of the E4 gene. But this is only one of what probably what probably will run into the thousands of examples of a desire to change genetic endowment and even our own form and function. Putting aside ethical implications or even desirability of doing any of this, this kind of manipulation is inevitable as knowledge expands. Actually, we are used to altering biology and our own relations with other humans and organisms every day. Psychiatric drugs change by design something basic about how a person responds to his environment if only by eliminating troubling hallucinations or making him falsely optimistic enough to do his work. Even an antibiotic alters the patient's relation with commensal and pathologic bacteria and fungi. Perhaps these changes are subtler and less permanent than those that would be wrought by a changing inheritance, but they do alter basic biology. The whole notion of describing what a person is by referring to his organism is now all the more slippery, less tangible. Why? Because what he is biologically has changed and it's changed by his own or someone else's design. The argument that nature, physiology or biological processes determine what he is or what he does from there in on is much weakened, and so the deeper meaning of it all is that as we acquire the means to alter our own biology, perhaps even heritable characteristics that define our own human species, we lose all sense of what exactly constitutes a stable biological self. This is what I mean when I refer to our organism as a moving target. It is all kind of a thought experiment which proves and I think once and for all, that human essence and motivation, intellect, feeling, are not entirely based in biology. When we look at our organism we have nothing stable to refer to anymore. We have broken out of our original biocapsule or cocoon and this further weakens the arguments of those who believe that biology can give us the total picture of human behavior. Is it possible that after biology is considered there is some residual essence or principle?

To many writers science looks at the material nature. Biological studies prove that thought arises out of matter and is a product of physical laws, chemistry and electronics. I believe science shows us more than that. Science, properly considered, does not diminish man. It does not take away his soul. On the contrary, only with a true understanding of science can we grasp what is truly noble, feel our soul, the metaphysical.

This book is organized into five sections. Each will build upon arguments raised in previous sections. Neurological descriptions will capitalize on nervous system's organization into three limbs: *afferent, efferent*, and *associative*. Thus is the structure of the book like that of the nervous system.

1. "Inside the Neuron- Consciousness, Unconsciousness, Split Consciousness" Begins with a consideration of brain death as ultimate lesion experiment. The concept of lesion is taken to the point of remotest absurdity. From there this unit will look at basic neurology as we know it, alternating between micro or cellular functions and macro considerations, reviewing understandings about brain *modules* and how these tie together to yield the basic output of the brain which is consciousness and states of consciousness.

2. "Vision": The major premise is that we are able to see much more than basic design would predict. Vision is a model of sensory or *afferent* function. The major argument is that scientific analysis cannot fully explain how we experience a visual image. We see a great deal more than we ought to be able to given our meager biological endowment. The chapter begins with the design and development of the visual system, then describes how impulses ramify through the brain, mingle with visual and multimodal sensory data, emotion, and memory to form a percept or transforming experience. The physiology of vision has vast implications for reality and esthetics.

3. "Beginnings": Memory is the simplest cognitive or *associative* function but we will see that it is far from simple or trivial. How does memory function fit into the puzzle of

experience? The chapter starts with the physiology of memory. It emphasizes the important fact that persons develop embryologically. We are not made whole as a Ford car or other object. This development must somehow reside in memory and have ultimately an effect on who we have become as does post-natal experience. Libraries and computers are "brain tools" that have expanded the capacities of the mind just in terms of storage. What we are is no longer confined to the space of one physical body or a single brain. A huge store of information is accessed an engram at a time, like a ribosome reads RNA one nucleotide triplet at a time. This is a model for focussing attention. We have a collective consciousness that can be accessed at any time be each of us. Persons living together are as human microprocessors working in parallel. The idea is that collective memory is now far more than is present in our head. It is a great store recorded in print magnetic, optical, and other media.

4. "Iconoclasm": Section on executive or *efferent* function - organization of effort and labor to accomplish long term goals. Asks and answers questions such as: Why do persons act in a way contrary to their own selfish short-term goals? Here I explore the physiology of motor planning, and delayed gratification as well as how we make use of instruments outside the brain that enhance executive function. There is much more here than can be predicted given knowledge of the brain.

5. "Beyond Biology": Our abilities have expanded outside biological boundaries. Indeed, we have acquired so much cognitive ability that it can no longer be housed within the confines of head or brain. We are already well into changing and improving our genetic endowment, implanting of electronic and other devices inside our bodies and brain. What does "The "Marriage of Carbon and Silicon" imply about our view of ourselves? I explore how implantable

> Silicon computer devices and alterations of function and form that will inevitably come from genetic engineering will redefine the person. Indeed, human form and function is no longer fixed but a moving target. As technology allows us to step well beyond our biological endowments this will prove once and for all that there is far more to a person than can be encompassed by any pure biological or materialistic description. The general structure of the argument is to go from lesion and hypofunction, to intact function to look to the future of expanded function, a dilated human consciousness.

Science illuminates and helps define reality. Science is a tantalizing seductress, coldly incapable of satisfying our deepest desires. The scientific method leads us on by answering some consequential questions, but somehow, it seems, we will never get to what matters to us most. We may dream of one day crossing a great chasm and entering into a different world. We will have to conclude that material and mechanical models are insufficient. On that day we'll view an ultimate truth that lies beyond material considerations, beyond biology.

# Chapter 1

## INSIDE THE NEURON:
## Consciousness, Unconsciousness, Split Consciousness

Their statues are made of silver and Gold
The product of the hands of man
Nostrils they have, but smell not;
They have hands, but feel not;
Feet they have, but walk not,
Neither do they utter a sound with their throats.
Those who make them, shall become like them,
Also those who confide in them.
-Psalm 115

### Entrée: Consciousness: A Chorus of Choruses:

You can make the argument that brain science is the most powerful science of them all. The line of reasoning goes something like this: All experience funnels through our brain, the instrument of consciousness. Thought, emotion, behavior are expressed as physical events inside neural circuits. If one day we master our own biology we will know everything. When it comes down to it, all knowledge is registered in the physical brain.

Mind and brain comprise one machine conforming to the laws of chemistry and physics. Human action is the output of brain and body, the result of physical processes, of automatic cause and effect. The reason persons seem to act as free agents, is that scientists have yet to uncover all details of the working brain. If all experience results from electrophysical processes, then every mental event is determined by its antecedents. Thus the idea of free will is illusory and conflicts with biological knowledge.

For those not preoccupied with these issues, science has encroached on their self-concept very little. They are deluded into thinking they are acting of their own free will. But make no mistake. Scientific knowledge will engulf them. New discoveries and scientific techniques will one day alter their cherished beliefs. We do well to learn all we can about our physical selves, which is the ultimate source of self-knowledge.

Another view is that the brain merely brings experience to fruition. In that case, biology is slave to another process, an outside will, soul or essence. Electrical and chemical processes express mental events, but aren't their initiators, or authors. One major support for this point of view, is that no one can tell us where the desire to perform even the simplest act, begins in the brain. The anatomist cannot say where in the brain the command even to move one little finger, starts. The philosophy separating the physical brain from some numinous agent controlling it, is the apostasy of dualism, refuted by scientific knowledge.

Scientific determinism on the one hand and dualism on the other, are insufficient models of human behavior. Both are too simplistic. The reason, I think, is that people have very long ago learned to leverage their mental abilities. We have formed complex social, linguistic, artistic, technological structures. That is the human enterprise. All of this is extra-cerebral, outside of our heads, certainly outside the purview of an individual's head, and it is the key to our free will in the sense that these structures are no longer determined by biological events. You can make the argument that all of these endeavors began as some kind of biological or adaptive initiative, but they have attained a life of their own by this time and no longer have biological determinants. Indeed the reason for a good part of human endeavor is to somehow escape natural events. It's why we create shelter, plant crops, fight disease, form groups, design machines, create art, study the world around us. Biology does give us basic information about our innards, but cannot tell us everything about ourselves.

Alzheimer's disease is a case in point. In Alzheimer patients we witness a mysterious disintegration of cognitive function and then of the entire personality. Very simply, it now seems that beta-amyloid accumulates in the brain and causes the disease. Thus the dissolution of the personality reduces to the accumulation of a chemical in the brain. The slow destruction of cognitive capacity and personality is a process to be followed under the microscope. Alzheimer's develops in Down's patients who have three copies of the twenty-first chromosome coding for extra amyloid and thus make amyloid in excess. In four types of familial Alzheimer disease that so far discovered the common feature is Beta Amyloid accumulation.[1] Amyloid accumulates in the senile plaque, a structure recognized under the microscope. The senile plaque is the major pathological element in Alzheimer disease, a probable first cause. It means little that in some studies the absolute quantity of senile plaque does not correlate as directly with dementia as much as some other microscopic features such as neurofibrillary tangles or that proteins, one designated "tau" may also be involved. This latter is an effect whereas Amyloid is the cause.

All at once we have a simple mechanism of causation of this mysterious disease. One day soon doctors will beat Alzheimer's disease by blocking the accumulation of beta-amyloid in senile plaque. More to the point, here is an example of the gross disintegration of the personality explained by events on a microscopic and biochemical level

Amyloid, and so-called paired helical filaments and other Alzheimer changes also accumulate in muscles of aged individuals. All of these changes are clearly visible in the muscle disease common after age 50 termed “inclusion body myositis[2]“. Here we find the very same pathological elements as are present in brain right in peripheral muscle!. Inflammation seems to be part of the problem, at least in many cases where the lymphocyte, a type of inflammatory white blood cell, infiltrates muscle. Changes in blood elements such as platelets reproducibly appear in Alzheimer disease suggesting that certain blood cells may be instrumental in transporting and depositing injurious substances in Alzheimer disease and inclusion body myositis, a form of senile muscle disease. Inclusion body myositis is partly ameliorated with anti-inflammatory medication and theoretically anti-inflammatory medicine may help prevent Alzheimer disease as well.

Still, what is visible to family friends and physicians too in Alzheimer’s, is the disintegration of the person. Little by little former interests fall away as inclinations, motivation an inner fire extinguishes. This gives insight into normal function as well. It’s obvious that some persons have more intellectual fire and motivation and intense level of interest than others, a natural continuum of the level of interest in the world. Persons in the process of losing intellect often resemble those of lower capacity who never had it, so that you get an idea that intellect and the quality of experience may be determined by some mathematical representation of the total cognitive power of a brain. The decline in the cognitive power of the brain in Alzheimer disease relates to accumulation of a toxic substance and cell death. Persons possessing a certain gene, ApoE4, are much more likely to have Alzheimer's because beta-amyloid is more easily transported into their neurons. That is an adequate mechanism for the disease.

So Alzheimer’s is an example of how a relatively simple process, here accumulation of a toxic chemical, may have profound consequences. Fine you may say. A mystery has been solved. Having done so you open up a Pandora’s box of further issues.

Suppose modern medicine were able to prevent Alzheimer's disease, as it seems likely to the case soon, by coming up with chemicals that simply block the accumulation or transport of amyloid? That would have profound social effects. As upwards of 50% of nursing home admissions are blamed on dementia, productive life would extend well into advanced years, so long as mental function could be preserved. The discovery of a treatment for Alzheimer's would undoubtedly extend life. On the surface of it, this would be wonderful. Cognitive function could be preserved in one's last years. But that would lead to other forms of social upheaval.

Industrialized countries already face a sharp increase in the proportion of elderly to young individuals who compete for economic resources. One wonders if over the long run medical advances that extend the lives of the very old are in fact a blessing. With age comes the accumulation of debility, but even more important, continued consumption of resources and wealth that could be used to improve lives of the young, the infusion of new life and ideas contributing to the well-being of a society as a whole. Always when we find a solution to a problem such as this, we create a host of other difficulties. Which is more adaptive – to extend the life and productivity of the very old or experience the springtime of renewal that can be attained only by the young? Successful treatment of Alzheimer's and other diseases of the aged would alter our society in fundamental ways. Robust grandparents and great grandparents might be the primary caregivers of the very young, freeing persons in their middle years for other forms of productive labor. Even so lengthened life expectancy would place persons of all ages in competition for scarce resources.

Given that it is natural for the old to become less functional and die off, a planned obsolescence of sorts, one wonders if this in fact isn't for the best. Life is finite so as to clear the way for the trial of the new and very young. This is the end result of natural selection, a trial of many possibilities. The old must make way for the new for a group to survive and adapt.

It will not be easy to find a sufficient biological explanation for another complex and poorly understood phenomenon, male homosexuality. A biological explanation for homosexuality would at one blow, negate almost all philosophical, religious and political rhetoric as those who maintain that male homosexuality is a lifestyle choice may be proven wrong. Male sexual preference is established very early in life and is likely to be biologically determined. In many primate species relatively few males have the pleasure of performing the majority of heterosexual copulations. This means that not all males contribute sperm to the gene pool, even when the ratio of males to females is close to 50:50. The presence of more males than is needed to procreate may add to the survival of a tribe, for example to increase cohesion or provide defense, so that homosexuality may be adaptive for a group whose members share many genes. Families with male homosexuals might even have some genetic advantage in that dominance hierarchies and division of labor among males may be firmly established or, alternatively, limiting numbers of offspring. New work implicates certain anatomical differences in the homosexual brain that may be heritable, especially in the hypothalamus, a tiny but powerful structure governing appetites sexual and endocrine function. These differences have not been replicated by others and are somewhat conjectural.

Consider the possibility that a homosexual tendency resides on the X chromosome. This putative gene could change the hypothalamus that controls basic emotions or work through another mechanism. If this work pans out, making homosexuality a model of biological causation, many other complex behaviors are sure to be explained as well. But it is just as reasonable to suppose that the entire answer is not in physiology or anatomy[3]. Identical twin studies among homosexuals find only an impressive concordance rate of between 50 and 67%, but still a non-genetic element is part of the story. Homosexuality must have a much more complex mechanism that the simple accumulation of a chemical substance in the brain as described for Alzheimer's disease but if we succeed in describing the biology of homosexuality, this would have far-reaching consequences. I suspect that we may never be able to do that, that homosexuality is in part biological and part motivational. If so, or if this proves to be the case for other complex sets of behaviors, this would underscore the limits of biological causation.

Obesity is falling under the sway of brain research. Is it motivational or genetic? Pima Indians[4] who live in Arizona are obese compared with their close relatives in who reside in Mexico. That would argue against a genetic mechanism for obesity. There are also significant differences in identical twins reared apart, and a marked increase in obesity of immigrants who move into economically advantaged regions within just one or two generations. Thus obesity appears to be caused by the availability of food. Most Americans have food whenever we want it and obesity is on the rise. Persons, who have inherited efficient use of calories that has enabled survival under conditions of scarcity and starvation, suffer when things go well. They are the first to get fat and to gain weight. This implies that in conditions of scarcity of food, some obese persons who we consider to be unfit, might be the ones who survive, while those who are thin and fit in our own environment of relative abundance of food would be the ones to die off.

What then should be the approach to dealing with obesity in the economically advantaged parts of the world? In the USA obesity is a major public health problem. Policymakers have suggested that unhealthful behavior such as eating a lot of hamburgers and fats ought to be discouraged through a change in our tax structure or that insurance should be more expensive and the like. Fat people should be made to pay for their gluttonous habits. But do people truly choose to be fat? No one knows.

A combination of appetite-suppressing drugs, that simulate the action of Serotonin perhaps, or a host of newly discovered proteins that signal satiety within the brain, plus false nutrients, low calorie or non-absorbed fats and sugars, drugs that cause fat malabsorption are proposed remedies. Ironic that rich persons with access to all the food they could possibly want and more, are so bloated they need to starve themselves to escape obesity, or they need to do heavy exercise just to get rid of the excess calories they can't stop from consuming. Other folks who are less fortunate die from malnutrition and poverty. Gluttony vs. poverty. Does it seem as if we will one day find a biological or pharmacological solution to these problems? Antibodies against hunger hormones - grelins, or satiety hormone strengthers - leptins, or blockers of inner marijuana "munchies" brain receptors, might hold a key to obesity. Here I have grave doubts. This too underscores the limitations of proposing a biological solution to what seems to be an non-biological problem. But more importantly it underlines a question about where pure biology ends and where more slippery, or at less scientifically definable concepts such as voluntary behavior, motivation, start.

The three examples above progress gradually from Alzheimer's disease whose cause is likely to be biological, to male homosexuality which is likely to be multifactorial, to obesity which is most likely to be primarily behavioral except exceptional situations. It is clear that a biological model has sharp limitations in all of these instances.

Alcohol and drug addiction, aggression and other personality characteristics may have a biological cause. Scientists just don't know the role of brain chemistry and physiology in these disorders. When we find changes in the brain of an addicted person are they cause or effect? An addictive tendency does tend to run in families and could be inherited. A drink of alcohol has less effect on person who is prone to become alcoholic while a single beverage provides a more dramatic quick buzz to someone who is less prone to drink, so one argument goes. A future alcoholic drinks more to attain the same effect. This is before he has habituated to the effects of the drug. Or with alcohol as with other drug-seekers, the addiction-prone may be seeking a thrill, due to a genetic character that increases activity in the thrill seeking chemical Dopamine in the brain. This mechanism would seem more applicable to stimulant type drugs, rather than sedatives such as alcohol. In any case, other sets of behaviors that seem now to be a matter of choice, will fall under the sway of biological causation. Biological research will demystify these behaviors as vague concepts such as choice and human will acquire mechanical explanations. Complex behaviors and personality characteristics that today stimulate moral arguments may be reduced to simple biological processes or turn into definable disorders treatable by medical intervention. The trend is for "psychiatric" diseases turn into "neurological" disorders as soon as problems are found within the brain, though significantly we have yet to see any very dramatic examples that have fundamentally changed our view of behavior.

Schizophrenia exemplifies this trend. Schizophrenia, or severe thought disorder with a poor prognosis, was at one time related to upbringing. Certain theorists championed concepts such as a so-called schizophrenogenic mother, who through her outrageous behavior caused her children to become schizophrenic. This theory quickly lost popularity as data about schizophrenia accumulated. The effectiveness of certain medications in schizophrenia, allowing patients with this chronic, and terrible disease to function outside the hospital, made a powerful argument for a physical cause. Certain anatomical abnormalities in the brains of schizophrenic patients relate tantalizingly to the disease. These include enlarged ventricles, indicating a loss of brain volume and anatomic changes within the temporal lobes of schizophrenics. Psychotic depression as well, is surely a neurological disease. Again the major argument at this point is pharmacological. Psychotic depression relates to disordered neural transmission, and the most efficient treatments are pharmacological. The effect of medications on these disorders has been considerable, allowing severely ill persons to function, yet the effects hasn't exactly been staggering. Persons afflicted with psychosis very much still have their disorders even when functioning very well.

Even the approach to neuroses that traditionally have been felt to have non-physical causation, most particularly one of the most common, panic disorder, is also chemical. Insurance carriers have noticed that psychotherapy is ineffective and expensive. Holding the purse strings for medical treatments insurance payers strongly "encourage" drug treatment.

You can argue that panic disorder is caused by psychological mechanisms, even though it certainly responds to drug treatment over the short term. The brain is in the middle here, and short-term results can be gotten with drugs even though the basic cause is really an extreme fear brought about by thought processes having to do with unfulfilling life circumstances and pressures. Perhaps this process goes out of control and the fear response assumes a life of its own. Efficient short-term treatment is in the form of a quick acting sedative, more potent long-term solutions may be connected with brainstem transmitters such as Serotonin. But the basic process begins with certain ennui or life-threatening non-fulfillment that eventually needs to be addressed. One strong possibility is that the psychological process uses the brain as a substrate, a stage on which to act out its drama, and temporary treatment regimens may utilize brain chemistry even if the root cause lies elsewhere. At the very least, no one has been able to prove or disprove any of this and it is a mechanism worth remembering, because the same kind of faulty logic pervades understandings about the brain and psychological mechanisms, which affect it. Motivational processes, heretofore somewhat mysterious, are reduced to biological tendencies even disease states, removing all concepts of culpability and punishment, even considerations of motivation and free will. Consider the implication of discovering the biology of alcohol, drug addiction even kleptomania or even of finding the changes in the brains of murderers and violent criminals. When biological explanations for deviant and normal behavior too are deemed sufficient, philosophical, religious and political vision thoroughly changes. What we have then is a whole spectrum of behaviors and personality characteristics, some of which are more, some less, proven to have a biological mechanism. Yet there is the sense at least that all behavior can be reduced to a consideration of biology, mechanistic even automatic, not at all controlled by free will.

If matter and physiology is all, the real question is whether someone will one day be able to put together a mechanical device, most likely made of Silicon, that will recreate or even go beyond human experience. Brain scientists are on the verge of concluding that consciousness is no more than the mere interaction of modules or pieces brain real estate, anatomical sections of brain, now more precisely described than ever, that communicate well enough to give the illusion of experience. Even at our current level of knowledge, more data is accumulating indicating the failure of this model or at least showing that it requires fine-tuning. Consciousness produced by interacting modules yes, which include the reticular activating system, limbic structures, thalamus, frontal, parietal lobes, but these and other smaller and larger sections of brain function simultaneously and in parallel. Connections and interactions don't necessarily function in the rarefied realm of computer logic either. Some aspects of the connectivity may be illogical. These issues will be described in greater detail in the pages that follow.

The product of this interaction, we perceive as human experience, is more like a chorus composed of individual voices singing together or even a chorus of choruses. My model here is Gustav Mahler's huge 8th Symphony otherwise known as "The Symphony of a Thousand" a chorus of separate choruses (plus orchestra). A neuron is the individual voice coordinated in a group or smaller chorus, or nucleus of cells, Nuclei, sometimes silent and sometimes active, are the smaller choruses. The harmonious (or cacophonous) output, the chorus of choruses, is the symphony of consciousness. Note that the connection between these elements is not logic alone, but involves quite a bit more, emotional, musical and other relationships variously amalgamated into a cohesive conceptualization, just the kind of relationships it would be near impossible for the computer scientist to reproduce. Nervous system functional anatomy is also simultaneously hierarchical and interactive, even more a chorus of separate voices not a full linear logical progression in a top-down or interactive design. These are concepts that will be dealt with in the pages that follow.

Looking at a gross picture of the human brain and one thing is striking. There is far more white than gray matter. The white mater is the wiring, the communication pathway between gray matter, neuron cell structures, and cortex and brain nuclei composed of cell bodies. You have to come to the conclusion that the brain must be more involved in communication between parts of the structure than initiation of messages. Messages ramify and resonate to a far greater extent than they are actually created. If experience is a chorus of different parts then it is logical to ask where is the chorus master or conductor? The simple answer is that we do not know. It is fascinating that we have come tantalizingly close to a physiological understanding of human experience but have yet to define the most important element anatomically, physiologically, a central organizing executive. Some may say that this is somewhere in the frontal lobe of the brain, the part that has most lately developed in evolution. But this is at best a vague concept. It is far more likely that what we actually perceive as a unitary experience is organized in much the same way as a choral piece or a symphony. We have a group of voices, tied together by some unitary conceptualization that comes from some outside agent, such as a composer of the work. The problem is that it's not possible to find this organizing principle anywhere in brain anatomy.

The upshot of all of this is that we shan' t expect any computer scientist to reproduce in Silicon, human brain function any time soon. Material, physiological descriptions fall far short of reality.

Considered from the vantage point of abnormalities and deficiencies, from a disease model, you are almost forced into a mechanistic materialistic version of brain function. The brain, even the personality misfires like a car lacking some of its pieces. But the brain can function optimally, which means without disease with no lesion or disorder. The absence of disease the brain, like the rest of the body, will *allow* optimal normal function, just as the car with everything working will put out an optimal performance. In other words the absence of disease has nothing more than a *permissive* effect on behavior. Absent a certain amount of brain matter a person will not be able to learn well and he will have mental retardation or dementia. Lacking the chemical Serotonin in some areas of the brain, a person will be depressed, and so forth. A normal brain will *permit* someone to learn, and not to be depressed. At the same time you get the feeling that slow learning and depression will always have a physically definable cause. The question is whether these physical processes account for anything more than function in the absence of disease, whether we can extrapolate from consideration of deficits that impair function, into all aspects of normal or even better than normal human behavior. What is responsible for our going beyond our basic level of performance and is that physical or intangible? Will or will not biology eventually account for everything or will there always be some residuum that is beyond biology?

## Death in the ICU or What Coma Teaches Us about Consciousness

"I don't think anyone unconscious is dead inside. They may be only more difficult to access."

-Oliver Sacks

The favorite method of neuroscience is the lesion experiment. To determine the function of a part of brain, you destroy it. Then you examine the subject animal for deficits. You conclude that the area of brain lesioned helps do the function that is lacking when it is gone. Cut out the right frontal lobe and an animal is paralyzed on its left side. The right frontal lobe must control movement on the animal's left. A simple deduction.

Brain death is the ultimate lesion. The lesion is all of the intracranial contents. So brain death gives us a chance to start with nothing. Nothing is assumed. It gives me an opportunity to start from the beginning and to build upon my arguments from scratch. The hospital Intensive Care Unit is the place where people encounter Brain Death. Because of nominal survival of the rest of the body enormous cost of continued care, it is a much harder issue to deal with than ordinary total body death. People tend to treat brain dead persons as living, even if, according to statutes in most states, they are legally dead. Even professionals have a terrible time talking to relatives of the brain dead patient because as the heart is still beating, and machines give breaths, the patient looks alive. Hence they tend to speak in equivocal terms often bewildering to the family.

The brain defines the individual. This is a given. Brain death is the very model of this teaching. When the brain is dead life is over. The quality of experience reduces to the function of the brain. An attempt to measure the quality of life reduces to assessment of cerebral function. This is what we assume, but is it true? Is it possible that brain only an apparatus of awareness, or is awareness nothing more than the sum of complex biological processes?

## The Brain Is Not Different than Other Organs:

Having spent my professional life treating diseases of the nervous system, I've acquired a slightly different perspective on consciousness and the brain. As a medical doctor you learn to use the very same strategies to preserve cerebral function as you use for other organs. Daily experience shows the brain to be just like all the other organs in the body. It causes you to question the idea that the brain is initiator or author of experience rather a medium for experience. Just like other organs the brain depends on a specific molecular milieu for survival and optimal function, only perhaps more so; the brain is a little more particular than other organs are. Body temperature, serum sodium, potassium, oxygen, and acid-base balance must be right. This is just what the kidneys or the gut or the liver require. True, the brain is semi-isolated by a blood-brain barrier, but the brain is an organ, and behaves the same as other organs do.

Compared with the liver, kidney, I tend to put the brain on a pedestal, yet paradoxically, I have to come to the conclusion that all the great scientists who write about the brain and talk about it as the repository of consciousness and everything human, are overglorifying the structure. I see myself as the proverbial soldier experienced enough point out that all political problems can't be solved by war. My daily experience has humbled me. I see how little I can accomplish. The brain aids in the enterprise of consciousness but is not responsible for the consciousness in toto at least it seems to me given the explanations below. Most everyone I've read who writes about the brain wants to find the totality of human experience in biology. But it seems to me with our state of knowledge now and in the foreseeable future, that's an impossible aim. As much as the brain is involved with consciousness, it is largely peripheral to the process of consciousness, perhaps not as peripheral as the gut or kidney which by the way are also involved, but peripheral. I am aware this is a minority view and that it even sounds preposterous but it will be proved as we go along. At the very least we will see that by far the biggest bulk of the brain, and very probably by extension all of the brain, is peripheral to the process of consciousness, initiation of the totality of thought and feeling.

An example is a brain operation that is popular for Parkinson's disease, the pallidotomy. The procedure is used with advanced Parkinsonians whose level of function fluctuates greatly. These poor folks, because of long periods spent with their disease, alternate between total immobility and being overdosed with their medicines, which causes them to writhe and twist about. This operation lesions out a tiny area of the Globus Pallidus, (literally gray ball because area containing primarily neuron cell bodies look gray) deep in the motor controlling area of the brain and is done with advanced physiologic monitoring with the patient awake. The movement areas of the brain that are destroyed by Parkinson's disease, connect extensively with the globus pallidus, but even so, why exactly these patients improve has not been worked out. What you see after the surgery is done can be considered in some cases, a miraculous improvement. This is an operation on the brain an one of the higher areas of the brain yielding striking improvement in gait and other functions especially on the side of the body opposite the surgery. Improvement is so dramatic it can transform a person from an invalid to an active person in one fell swoop.

How is this different than the replacement of a limb by an artificial device, a replacement that helps a person to walk anew, or a kidney or heart transplant that transforms a medical cripple into a functional human being? The answer is that this surgery on the brain is not different. The brain acts as an organ like any other whose normal function enables a person to have a normal and fulfilling existence. As such the brain acts as a tool or enabler of normal function. In the future, we may be able to enhance cerebral capacities beyond the norm. This will bring about a truly revolutionary enhancement of human capacities and a redefinition of ourselves. We would not be the same, after all but have enhanced capacity. But that is some time in the future. We will then have broken at that instant the biological ties that limit us and our potential may one day truly be limitless.

## Behavioral Output Isn't Everything:

We have the two groups of phenomena, the output of the brain i.e. human experience and behavior, and the biological brain itself. We do not know how they relate precisely. We may agree with the consensus that all thought can be reduced to an understanding of biophysical processes. One problem is that we rarely examine anything except the physical representation of consciousness, behavior. Some of this is observed with instruments such as the electroencephalogram that looks at electrical brain wave activity, but is behavior nonetheless. A lot of people have tried to simplify the debate by maintaining that manifest behavior is all there is because behavior, the output of the brain, is all that can be examined and compared objectively. In looking at the brain, behavior is a convenient starting point as the output of the organism but there are so many more phenomena that can't be ignored by wishing them away or agreeing that you are going to ignore them. Those who look at behavior alone, such as psychologists, rarely take into account the anatomy and pathology of the brain. They may ignore internal mental states simply because they cannot be observed by anyone but the person experiencing them i.e. these states are felt to be "subjective" and therefore not applicable to scientific objective scrutiny. But because internal states are not observable doesn't mean they aren't there or can be ignored. What we can most obviously see simply doesn't tell the whole story. It is only the part of human experience that is directly observable.

Brain death is our strongest argument that all behavior is caused by biological events because you can't show that there is any residue of the personality or behavior after the brain is dead. Thus the lack of any demonstrable behavioral output in a brain dead individual argues forcefully that the sum total of the personality is expressed in the brain. The other side of this is the distinct possibility that brain death means only the destruction of the *instrument of expression* not the end of that consciousness.

There is another interpretation. There could be some other agent controlling brain function in which case the brain is merely used as a physical tool of expression of this outside executive. What I'm talking about is some other entity akin to the soul or human will, an executive hovering over and controlling brain function. After the brain is gone, there is no way to show that any such entity exists, not only that such an executive, it is frequently argued, does not exist, because there is nothing, no phenomenon, that it would help to explain. We have no need to invoke, an extra-cerebral executive to explain any currently observed phenomena so why talk about it? Francis Crick, co-discoverer or the structure of DNA, says as much in his book THE ASTONISHING HYPOTHESIS: The Scientific Search for the Soul. His "astonishing hypothesis" seems to be, that the soul does not exist.. The hypothesis is not so astonishing in that it seems certainly consonant with mainstream science today. Certainly scientific principles will eventually explain everything. Mental activity, emotions are epiphenomena of neural and physical events. By the end of the book Crick seems almost to trivialize philosophical concerns over free will, in localizing it to a little talked about structure in the brain, the anterior cingulate sulcus[5].

There is no denying that once you have killed the brain, you observe no further behavior at all. But on the other hand, the same cannot be said about subtotal but very extensive brain lesions where the outcome varies. Given a giant stroke that obliterates one half of the brain, a lot of people never come back as an intact entity. They never regain their personality or responsiveness. But once in a while you may witness a surprising preservation of a personality. The affected person may have had a very strong personality structure before his injury. Then despite a seemingly devastating large brain lesion, his basic personality will be preserved. A medical colleague of mine had a huge right hemisphere embolic stroke. You could see on his head scans that fully half of his brain was gone and there was, at first, tremendous brain edema or swelling that affected the other side of the brain as well. It took him a long time to come around, months before he was any good for anything, yet his struggle was heroic and today he is a very impaired but he is very much the same person. Examining him, you will find certain "deficits", basic difficulties in much the same way that you would if he had lost an arm or a leg. Vladamir Lenin suffered a whole series of strokes near the end of his life, yet there are adequate records to show that he continued to function in much the same way almost until the very end, continued with his former perceptions and assessments. Indeed near the end in his final testaments he issued perceptive pronouncements about Stalin proving that he was more clairvoyant than his contemporaries. Boris Yeltzin's extra-cerebral problem after his after his series of heart attacks and alcohol addiction is qualitatively not different than Lenin's. The loss of a part of the brain brings us closer to basic cognitive function but it is very difficult to change what is basic about a person.

We are still left with the proposition that whenever we see a fellow human being, we are evaluating him on the basis of his behavior only, what is manifest, what we can see. Various neurological conditions produce degrees of inadequacy of behavioral output. A stroke may stop a person from moving his right side or from speaking even though he wants to move and may have a lot to say.

The difference between manifest behavior and internal processing can be seen in a condition known as locked-in syndrome. This terrible naturally occurring condition leaves a patient virtually unable to move at all even though his level of awareness is there. The subject is awake and aware and thinking but is unable to move, in a state of perfect impotence, completely dependent and unable to change his environment. Though he has basic bodily functions, his heart beats and he breathes. On the surface the pitiable patient with locked-in syndrome seems to be in coma, since he does not respond to his environment. Only in the past few decades was the true nature of the condition, a form of "pseudo-coma", appreciated, for the subject is awake and aware, only he is unable to move. Such persons were thought to be in a coma until someone noticed a level of awareness in their eyes which can be made to move and even to express thoughts via an agreed upon code with vertical up and down eye movements. Gazing into an affected person's eyes gives the first clue that a level of awareness is maintained.

The commonest lesion causing a person to be locked-in is a stroke in the pons, part of the brainstem where in a small area all of the motor output fibers course. The reason why vertical eye movements are spared is that these emerge from a slightly higher area of the brainstem, the midbrain, which is unaffected. Level of function and sophistication, sapience is still judged on the basis of behavioral output in the end. Only here, our breadth of observation is restricted to eye movements which is the only preserved voluntary motor output. Certain patients with other neuromuscular conditions causing near complete paralysis are also locked-in. Since the voluntary motor system is the major behavioral output that one can observe, and the motor system is efferent, outgoing from the brain to the environment, we say that such an unfortunate person is *de-efferented,* disconnected from his environment so far as motor output is concerned.

Now this is an interesting condition of complete and utter helplessness, an inability to do almost anything, to influence anything in one's environment, particularly if one's condition goes unrecognized as it sometimes does and life is just simply preserved without anyone noticing that there is still an awake thinking person inside. This is a sort of eerie "brain-in-a-bottle" experiment. By this I mean that supposing for a moment that after you died one could store your brain, in an awake and aware state, away from your body, keep it alive, with you thinking and dreaming, in a sort of nutrient fluid, floating in a jar on a shelf. How would that be for a person? Possibly one of the world's most horrifying experiences. But there are motor feedback loops as well as sensory ones, a constant communication and interplay with motor outputs that keeps the brain active in much the same way that you stay awake better even when you are tired while you are actually engaged in some activity much more than if you're sitting around just reading or listening and so very likely the de-efferented subject is somewhat numbed, anesthetized, during his long experience being cut-off from his motor output. The experience may not be quite as horrible as it seems to be. I know of cases where persons have recovered from severe paralysis, in myasthenic crises or after Guillian-Barre syndrome, an inflammatory condition of nerves causing a paralysis that can last weeks or months. Such persons aren't comfortable but don't describe a horror as bad as I would imagine under the circumstances. Perhaps some of us, being somewhat more kinetic, would not do as well under the same circumstances.

## Brain Isolation:

Along these lines it is possible to come up with another means of disconnection from one's environment which is so far is not well documented. A person may be unable to hear, see, taste, or feel. All sensory input could conceivably be interrupted in which case all sensory data from the environment would be cut off so a relevant response would also be impossible. Such a subject would be cut off from sensing his environment. He would retain all simple reflexes such as a simple withdrawal, eye movements, breathing swallowing and voluntary motor activity as well except that such activity would be irrelevant to environmental context because no data would reach consciousness and no voluntary motor output could be connected contextually to environmental stimuli. He would superficially seem to be awake and aware, but being

that he receives and processes no sensory input, he would not be responding to his environment except via automatic reflexes. Therefore he would seem to be awake but unaware.

This is similar to persistent vegetative state due to severe brain injury. The subject seems to have normal sleep and awake stages and may be looking about the room, but not fixating visually and while all the basic reflexes are there, even some rudimentary motor output, there is no evidence whatever of any awareness, awake but not aware. The eminent neurologist, Fred Plum, MD, famous as co-author of the classic text on Coma, presented a collection of patients with persistent vegetative state with complex organized motor activity even irrelevant speech. The key was that motor output, however complex, was not related to environmental stimuli[6]. But it is at least theoretically possible that even though a subject fails to show any contextual motor response, still he is awake and retains a level of awareness that is not apparent behaviorally due to the fact that all sensation is cut off. Since we speak of sensation as being afferent or traveling toward the brain, we have the obverse of the de-efferented state here what can be termed a *de-afferented* or *locked-out* syndrome. The disconnection with the environment is on the sensory side[Φ]. One way to think about the thalamus is that New Testament expression, "No one enters the kingdom of heaven but through me." Sensory data, as a general rule doesn't make it to the brain, except through the thalamus.

This very situation was brought to the fore when the brain of Karen Quinlan was examined. This young woman' s plight was widely publicized one of the first "right-to-die" cases and she had been in a persistent vegetative state for many years before she finally died. Her parents petitioned for the right to withdraw medical care in her case. An examination of her brain disclosed that the brunt of brain damage was in the thalamus, the major sensory way station of the brain, through which pass tactile, visual, auditory, gustatory inputs. Such information at least raises the conjecture that a persistent vegetative state may result from ablation of sensory inputs to the brain without destroying large areas of cortex, which leaves a person both awake and aware yet behaviorally vegetative. That being said, the operational definition for a persistent vegetative state in medical parlance is that the person is allowed to be awake but is unaware, that is there is no contextual content to consciousness. This occurs in most instances with widespread destruction and dysfunction of most of the entire cerebral cortex leaving a good part of the brainstem, the lower areas of the brain, intact. The basic problem is that we are forced to make these judgments on the basis of behavioral output. The philosophical conundrum is that by these simple discreet lesions taken together, pontine destruction causing the locked-in syndrome, plus thalamic destruction causing a locked-out syndrome, that a person may attain a state of perfect untrammeled awareness, being awake, and conscious yet be completely disconnected from the environment. Further, an outside observer would be likewise unaware of the pure thought processes still occurring in that person's brain because he would be unable to get at them to make behavioral measurements. Tantamount to a soul floating bodiless in another world unable totally to communicate with our own, but while theoretically possible, this is entirely conjectural at this point[7]!! In Eastern religious circles, this perfect isolation of thought might be considered a state of nirvana. To most of us conventional types, it is more like hell.

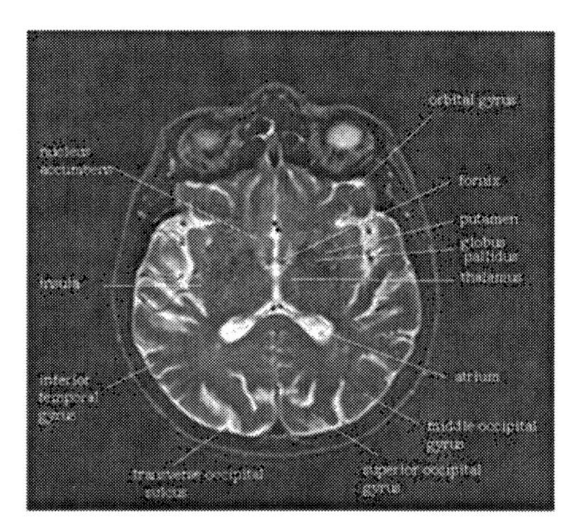


**Figure 1: An MRI brain slice. The thalamus is centrally located here, bordering the fluid-filled third ventricle. Eyes are on top, occipital lobe or back of the brain at the bottom.**

I'm not saying that having pure isolated functioning cerebral cortex ever exists, even in an intensive care unit. Such a situation where a discreet lesion is made in the thalamus and also the pons would never and likely has never occurred naturally. Deafferentation in particular, is bound to have many other effects on consciousness so that a total isolation of a thinking brain from its environment is probably impossible. For one thing, the thalamus also helps to keep the cortex in a waking state. I only describe it here to illustrate the existence within the brain of afferent, incoming sensory vs. efferent, out-going, motor limbs or divisions of function.

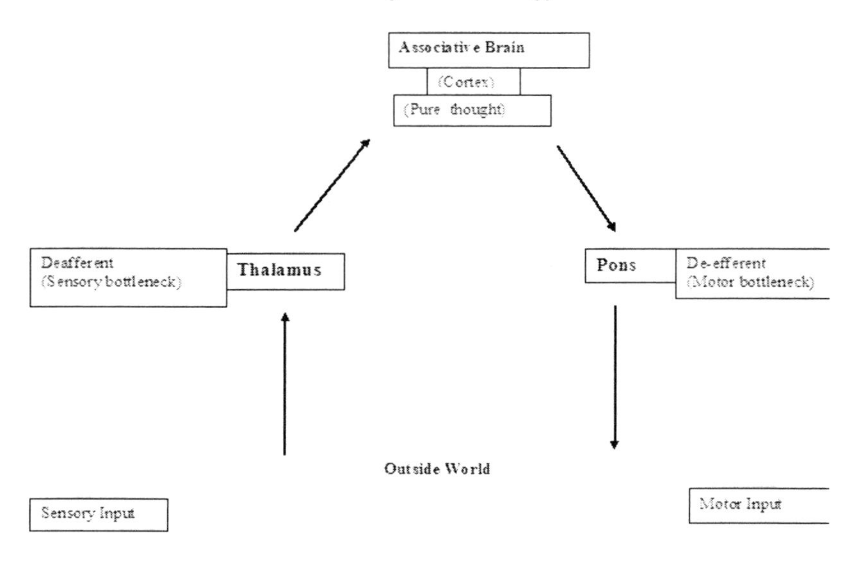


**Figure 2: Isolating lesions of the brain. The brain is separated from the outside world through the bottlenecks of thalamus and pons. Even with Afferent or Efferent limbs amputated it may think and emote.**

One possible way to look at de-afferented state specifically would be through the sense of smell. If we had a patient with the specific thalamic destructive lesion, we could test the hypothesis of Deafferentation. Our putative thalamic lesion would not deafferent our subject entirely because olfaction (sense of smell) does not course through the thalamus, but goes instead directly to the frontal lobe. If a person in a persistent vegetative state is conscious but disconnected from all other sensory modalities, then it might be possible to communicate with that person using the sense of smell. This would have to be in a case that must be exceedingly rare, if it exists at all, of discreet thalamic damage causing a vegetative state.

Deafferentation undoubtedly has other clinical correlates. It is akin to sensory deprivation, only the complete anatomically pure case of sensory deprivation that is bound to have extreme consequences for awareness. In prisoners and other subjects who are sensory deprived, there are profound psychological changes. Lacking stimuli, many subjects will start to hallucinate. This may be due to the physiological mechanism of denervation hypersensitivity, described later. Neurons lacking input are hungry for it and tend to over-respond. Hence a person subjected to sensory deprivation may begin to come up with his own stimuli or hallucinate sometimes extremely.

## Afferent, Efferent, Associative Limbs:

From the foregoing and on the basis of all current data, about brain function, we conclude that neural function has three components or "limbs". These are **Afferent, Efferent,** and **Associative.** Afferent refers to the sensory side or processing of incoming information. Efferent, the motor or doing side, by which the nervous system manipulates one's own body and the outside world. Associative or relating parts of the brain, refers to the sum total of internal processing of thoughts and feelings. This is what I refer to as the NeuroWeltanschuung or neurologist's world view. For the neurologist the nervous system is the center of the world and he breaks nervous system function into afferent, efferent, and associative limbs

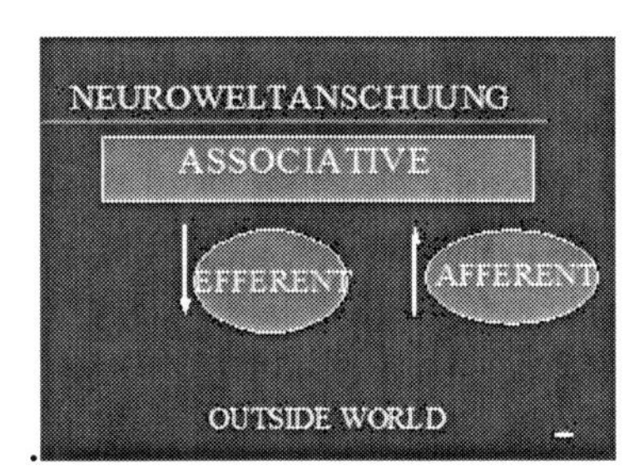


**Figure 3: The neuro worldview with afferent, efferent and associative components.**

A closer look at this simple construct will at the same time help explain and clarify while showing how useful these ideas are. We've all noticed that certain personality types have one or the other of these limbs de-emphasized or atrophied and others hypertrophied. A facile example is the artist who, stereotypically at least, is hypersensitive. The artist has a hypertrophied or overdeveloped afferent limb. Artists are supposed to feel, to receive, more profoundly than the rest of us, so much so that they are often nonproductive. They lack at times, motivation and output, and are easily blocked. What does the artist do when he needs to put out his product? He retreats to his loft, to his workshop, or better, to the mountains, where he can finally cut himself off from overwhelming sensory stimuli and begin to produce, to finally paint, or to write or compose. He thereby cuts off his troubling afferent limb and finally begins to put out his oeuvre. Of course all of us know that many of the most successful artists are primarily doers rather than feelers and that most of them have a prodigious output under even adverse circumstances.

Contrary to this stereotype are both the athlete and the entrepreneur. They don't think or feel much, not to let too many thoughts or feelings get in the way, but are always producing. They have an hypertrophied efferent limb. Then it is easy to appreciate the armchair philosophers, or mathematicians. Divorced from connections with the outside world, these folks are paid to think and associate. For the athlete his efferent limb is huge, but he doesn't feel too much and the afferent area is shriveled. Contrariwise the artist has a hypertrophied afferent area and maybe not such a muscular efferent limb. Ronald Reagan was stereotypically described not to think or feel to much but much more of a do-er. They are "neurostereotypes" meant to present a simple model of behavior.

The brain lesion model provides a very useful platform for describing the necessary components for creating a conscious output. The parts of the brain necessary for conscious response have been well described using this methodology. The brain scientist breaks consciousness into two operational components, the maintenance of arousal and content of consciousness. To be conscious a person needs to be awake. Later we'll get into the argument about whether sleep and other states are really altered states of awareness, but for the time being it simplifies things to require our conscious subject to be awake and aware.

The part of the brain that maintains arousal is a collection of anatomical structures in the brainstem designated the reticular activating system. These brainstem structures are under the cerebral cortex or you may say behind the cortex, closer to the tail of the animal. There are groups of cells that maintain arousal by virtue of connection to the cortex that essentially excites it, I like to say revs the cerebral cortex, keeps the engine of consciousness humming. When a large portion of the central brainstem is affected by such processes as trauma, tumor, infection stroke or any destructive process, a person will be unable to maintain wakefulness. On the other hand as we have seen, if the cortex is affected, a person may be awake but unaware. He may lack any substance or content to consciousness, even though seeming, superficially, to be awake. Thus the brainstem ascending reticular activating system or ARAS keeps the person awake and a good deal of its input is by way of the thalamus that connects extensively with cortex. The ARAS is ascending because the influences are going up or toward the highest areas of the brain. The thalamus is the rostral or most forward part of the ARAS.

## Conditions Correlate with Specific Brain Lesions:

The table below is not fully understandable at this point. It is meant to apply the modular or component or mechanistic point of view of consciousness I have been discussing above. In the hospital ICU we see various conditions that affect behavioral output. With modern understandings we are able to relate these various conditions to specific lesions inside the brain with a fairly high degree of precision.

| CONDITION | AWAKE | AWARE | LESION | EEG | EVR | REFLEXES |
|---|---|---|---|---|---|---|
| PVS | S/W CYCLES | NO | BILAT. CORTEX, OR THALA MUS | LOW & SLOW OR UNRE-ACT. ALPHA | SER IS OUT BAER OK | YES |
| LOCKED IN | AWAKE (& altered sleep) | YES | PONS OR nerve or muscle | Near NORMAL | BAER ABN SER NL | YES |
| COMA | CAN'T BE AROUSED | NO | BRAIN-STEM OR BILAT CORTEX | LOW & SLOW OR UN-REACT. ALPHA | VARI-ABLE | YES |
| DEGEN DIS | YES (S/W Cycles) | REDUCED | VARI-ABLE (severe and diffuse) | NL TO SLOW | VARIES ACC'G TO CAUSE | YES |
| BRAIN DEATH | NO | NO | ENTIRE BRAIN | FLAT | ABSENT | SPINAL ONLY |
| AKINETIC MUTISM | YES | probable | MID-BRAIN OR FRONTO -MEDIAL (cingulate) | VIRTU-ALLY NORMAL | NOR-MAL | YES |

**Table 1 : DISTINCTIONS IN CLINICAL SYNDROMES PRESENTING WITH ALTERED CONSCIOUSNESS.(PVS= the persistent vegetative state.)**

**NOTES:** 1. "CONSCIOUSNESS" = WAKEFULLNESS AND AWARENESS

2. PERSISTENT VEGETATIVE STATE (PVS): Complete unawareness of the self and the environment accompanied by sleep-wake cycles with either partial or complete preservation of hypothalamic and brainstem autonomic functions. In add'n: No sustained,

reproducible, purposeful, or voluntary behavioral response to sensory or noxious stimuli. No language comprehension or expression. Bowel and Bladder incontinence. Full chew and swallow not present.

## You Need to Be Conscious to have Pleasure or Pain:

A lot of times I'm asked if a comatose patient is suffering. How much does he know? Does he experience any pain? My stock answer is that you have to be awake and aware in order to truly experience pain and suffering. This is exactly the same argument that was used by surgeons forgoing anesthesia in newborns and prematures. The brain isn't well developed or myelinated with all connections intact. The baby is preverbal in his development. Does the infant suffer? A newborn subjected to pain or surgery will withdraw, move, grimace and cry but what is going on in his cerebral cortex and how much does he feel? You can apply the same reasoning to lower animals. What happens when you vivisect and insect or boil a lobster? Again you may observe certain behaviors that make such activity look cruel. The lobster struggles to get out of the scalding pot, but how much gray matter is there to appreciate pain and pleasure and how much suffering takes place? Tough questions. A person in coma has no awareness that we can gauge behaviorally. If he survives and rearouses and he was truly in a coma, he will never tell you about details of his experience or complain about how much he suffered, this is assuming that he was not awake at the time. So my answer has to be that there was no suffering or pain. But to be honest, I'm still not sure. What if he does suffer but doesn't remember it? You can use a drug to affect recall but has no analgesic properties, such as atropine, which is used in anesthesia. Very rarely patients, inadequately anesthetized, will recall pain. This creates an interesting philosophical conundrum. You don't use a narcotic or anesthetic that actually reduces pain or lowers the level of consciousness to dull or obliterate the experience. You merely give atropine and the person is unable to recall the experience, even though he may have experienced intense pain. He cannot recall it[Φ]. Is it cruel to inflict pain that the person is later unable to recall? As far as our subject is concerned, it never happened, yet during our procedure he may have been in agony. It's a little like the tree falling in the forest with no one to hear it. Did the sound happen with no one to recall it.? On even a more metaphysical level, I do know so far as my ability to measure it, that a comatose or brain dead person does not suffer. This is related to the question of whether brain function defines the person. We have no knowledge of experience apart from the assessment of brain function that is measured behaviorally. Some religionists say that there is a soul that is separate from the brain, perhaps hovering over the bed. But we cannot corroborate this belief phenomenologically. If we bring our comatose patient back, he will not recall suffering. Brain death and coma are a metaphor for the position that the brain makes the person, the argument made on the basis of the most extreme case. Brain function is gone, ergo what is human is gone also.

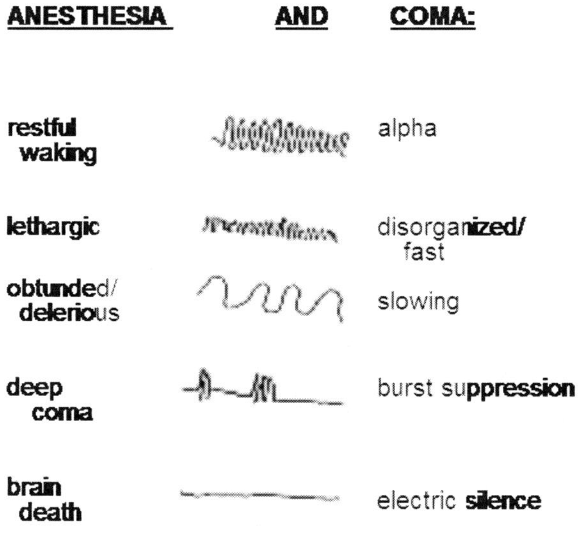

| **ANESTHESIA** | **AND** | **COMA:** |
|---|---|---|
| restful waking | | alpha |
| lethargic | | disorganized/ fast |
| obtunded/ delerious | | slowing |
| deep coma | | burst suppression |
| brain death | | electric silence |

**Figure 4: BEHAVIORAL AND ELECTRICAL CORRELATES OF LEVELS OF CONSCIOUSNESS[8]**

You can follow the brain through its course of decline as global function deteriorates through the various decreasing levels of consciousness through coma, which may be considered brain failure, and then finally to brain death. The electroencephalogram (EEG) is a test that looks at the brain wave activity of the cerebral cortex and makes it convenient for us to follow the various levels of consciousness which are the same whether the person is undergoing deeper levels of anesthesia or sinking deeper and deeper into coma. Giving larger doses of inhalation anesthetics induces levels of anesthesia. The technique of anesthesia is really just a refinement of the use of volatile gases like chloroform used in the nineteenth century. Today we measure the anesthetic as a percent of inhaled and exhaled gas applied to the alveoli, the diffusing area deep in the lungs that has access to the blood. Various volatile gases such as isoflorane and analogues are breathed in with a tube and a ventilator. Levels of anesthesia can be correlated with percentages of inhaled and exhaled gas.

When you are awake, brain activity appears somewhat disorganized with rapid changes and unexpected electrical events. The tracing looks almost chaotic. The person is moving about and thinking about various things using different parts of the brain unpredictably. Mostly fast activity tells us that the brain is electrically active. But let our subject sit down and close his eyes in a relaxed or sometimes focused manner, and suddenly the electrical activity becomes very organized and regular. There is almost a perfect sine wave pattern a lot of times, usually between the frequency of 8-13 cycles per second that is called alpha wave activity. It is the EEG concomitant of restful waking. Oftentimes you see a regular pattern with a gently modulated gradually waxing and waning amplitude.

Not too long ago some psychologists were enamored with controlling brain alpha activity. They found it remarkable that people could be "conditioned", actually "learn", to produce alpha activity, something that was considered to be involuntary. Maharishis and patients wowed the psychologists with how much alpha they could produce at will. But anyone who looks at a lot of EEG's finds the conditioning or learning how to make alpha activity absurd. Almost all of us do it naturally and effortlessly and if we are relaxed we can bring it out any time. You only have a problem making alpha if you can't relax or focus but you certainly don't need to be any kind of a guru.

As the level of consciousness becomes a little depressed, regular organized restful waking is again not possible and the EEG begins to become fast and disorganized. Perhaps the cerebral cortex is no longer under the tight control of lower levels of the brain that produce the alpha rhythms in the cortex. This is in the thalamus, a group of neurons that connects to wide areas of the cortex. There is fast (beta) and very fast activity of varying amplitude. This pattern occurs in persons given certain sedative especially Valium like drugs and alcohol, but also in anxious individuals. In the early stages when you start to administer anesthesia, you see this varying amplitude fast activity. Perhaps the cortex is actually more active or is less well modulated or controlled in an organized fashion by the thalamus. As a general rule on the EEG, faster activity signifies that brain cells are more easily stimulated but in a less controlled manner with more chaotic activity again. Slow activity means the cells are hyperpolarized in other words stimulated with difficulty and, conversely, fast activity implies the cells are partly depolarized. If this sounds electrical, it is. We can look at the electrical activity in individual brain cells. This will be explained below. Fast or disorganized activity seems to occur with some minimal level of impairment then, light anesthesia of low doses of some other sedative (Valium-like) drugs. What will our subject look like? You may be able to appreciate very little. He may seem mildly or subtly impaired perhaps with slurred speech. He may be somewhat irascible or uninhibited or even a little delirious.

As you go on administering more anesthesia behavior will change by degrees. As the person is even more impaired and this can either be by disease processes that injure the brain or by anesthesia where the changes are very reversible, the person's level of arousal is ever more impaired. It will become hard enough to arouse our subject to call him lethargic or even obtunded or there may still be unfocused attention, the impossibility of staying on task in an organized fashion for anything more than an instant, the impaired excited response to the environment that is delirium. At this point with even more anesthetic effect, the EEG slows, indicating that neurons are hyperpolarized and harder to stimulate. The activity is called theta, which is 3-7 cycles per second. When the EEG looks like this the impaired awareness is less subtle. A diffuse slow pattern and the behavior that goes with it are what is meant by "encephalopathy" a word that simply means the brain as a total entity is affected or diseased. The anesthesia will affect brain function reversibly. Indeed the whole job of anesthesia is to temporarily and reversibly impair brain function. Disease processes are less reversible.

Coma is an even more extreme impairment of cerebral function. As opposed to being lethargic or obtunded where our subject is still arousable, a coma is unarousable unresponsiveness. Here the EEG becomes very slow. In the extreme case the EEG is absolutely flat for a while but then some electrical activity escapes from whatever is suppressing it. It almost looks like a seizure pattern as high amplitude electrical activity suddenly appears out of a flat pattern. This is burst suppression. When a person is in a coma due to a medical illness and has this pattern, the prognosis is poor. Very few people will ever reawaken. Most of the time they are not breathing at all or breathing is too inefficient or unreliable to leave without some kind of mechanical support. The limbs may be flaccid. This shows that something extreme is affecting the brain. However, depending on what exactly is wrong, recovery might still be possible. We know this because we can reproduce this exact pattern with a high enough dose of anesthesia, in other words anesthetize a person to this very deep level and the person will make an uneventful recovery. The EEG can also be completely flat or isoelectric. Here the brain shows no electrical activity at all that is detected by our instruments. In an intensive care unit, with a coma from a medical illness, almost no one will recover from this EEG pattern. Still we know that under certain conditions this pattern is reversible, for example if the body is extremely cold (hypothermia) or with very high doses of drugs, especially Phenobarbital that suppress brain activity. In fact Phenobarbital and other barbiturates, produce exactly this sequence of EEG and clinical pattern as the dosage is gradually increased and these drugs serve almost as models for levels of consciousness, anesthesia and EEG patterns. Other anesthetics used are a little more quirky and don't exactly reproduce this sequence of patterns. Some have a little more in the way of stimulant properties as well as anesthetic properties but in general the sequence is as presented above and in the figure.

To review I have presented a thumbnail sketch of levels of arousal from being normally awake and responsive to coma, and brain death, all of which may be produced by medical illness or reversibly induced by anesthesia. Anesthesia depth provides insight into medical coma and brain failure because we can appreciate that under the right circumstances severe brain impairment may turn out to be reversible. A lot depends on the total context of brain impairment. The EEG gives us a physiological way to correlate (albeit loosely) physiology with behavior seen. Anesthesia is a nearly perfect model for levels of awareness that occur in nature..

Levels or arousal or wakefulness may implicate as we have seen, a set of structures in the brain designated as the *reticular formation,* much talked about in popular discussions of the brain. Anesthetics tend to affect the function of the cortical neurons directly not the reticular formation. Roger Penrose in his popular discussion *The Emperor's New Mind*[9] seems to confuse the idea of consciousness in his discussions, with the much simpler concept of level of arousal. Arousal is a simpler concept than consciousness implicating these deeper brainstem reticular formation structures which keep the cortex awake. The discussion of consciousness in general and why biological organisms have it and machines, even computing machines appear to lack it, has little really to do with the reticular formation but involves more the whole functioning brain in a complete organism.

## The Reticular Formation Controls Sleep Wake Cycles:

At one level the reticular formation and brainstem structures that regulate arousal could be viewed as a switch which in a binary manner is either off or on. Arousal, loosely speaking whether or not a person is awake and has any self-awareness (something which as far as we can discern only animals and not machines have capacity for) is an electrical process. Judging the state of arousal is much the same as viewing a city skyscraper and observing whether or not more or less lights are on or not on, or looking at a complex computer control room with silent vs. Blinking or non-blinking lights. Activity is or is not taking place on a high level. This general electrical activity or signs of life is what determines both arousal and self-awareness and it is regulated by brainstem structures whose connections ramify widely over the entire cortex or surface of the brain. This is a switching system that is high coordinated, disorders of this coordination mechanism cause untold disturbances in arousal cycles.

The reticular formation helps to coordinate sleep *stages* that are somewhat the same but also differ in some ways with levels of consciousness. Sleep is broken into four stages of slow wave sleep and REM or dream sleep also called paradoxical sleep. A person in REM sleep has some physiological responses that we observe in him when he is awake, more rapid heart rate, respirations and eye movements and is mentally very active, that is actually dreaming. There may be more brain oxygen consumption than during waking. Brain EEG activity actually does resemble an awake individual with his eyes open, that is desynchronized, low voltage irregular activity. Yet in REM sleep, the subject is hardest to arouse and muscles are completely flaccid, give no resistance at all to movement.

Narcolepsy is a disease where elements of this peculiar REM sleep stage intrude into a person's normal waking day. Normally you don't go into REM sleep until you've descended through the four slow wave sleep stages. The brainstem precisely orchestrates this response. Narcoleptics typically have an irrepressible urge to fall asleep during the day and when they do, they go right off into REM sleep and may feel satisfied after just 10 or 15 minutes. When they feel strong emotions they may crumple to the floor as their muscles suddenly totally weaken and are unable to support them, acquiring on of the characteristics of REM, total flaccidity, a symptom termed cataplexy or falling down.

Cataplexy is a feature of Narcolepsy is derived from an adaptive feature of REM sleep. Imagine what would happen if during REM or dream sleep you could more than experience your dream, you could act it out! Then all of the movement inside a dream, the movement perhaps involved in imagined sex or violence in a struggle with an imagined intruder would cause a person to actually act out. The bedroom would be a shambles, one's bed partner might be beaten up and you would be exhausted in the morning after a night's exercise. This is just what happens in REM behavioral disorder, a rare condition in which subjects fail to lose the ability to move intrinsic to normal REM sleep. Subjects with this disorder have indeed committed violent acts in their sleep and this bizarre disorder has been invoked in criminal trials with sleep experts brought in on the side of the criminal defendant.

In narcolepsy REM sleep happens where and when it should not, as part of the waking day. Cataplexy is tied to high emotion because of close anatomic relationship of the upper brainstem with the limbic system whose product is emotion. Narcoleptics have other periods of weakness and paralysis connected with REM stages of sleep, which happen during their day. Also some have vivid dreams during waking or the earliest or latest stages of sleep, hypnagogic hallucinations, so called because dreams with a person awake become vividly realistic sensory experiences. What is lost in this condition is the ordinarily tight coordination of sleep and wake cycles administered by the reticular formation. Φ

The normal awake state, slow wave sleep and REM sleep are three normal, well-coordinated states of consciousness. Transitions between these three states are controlled by the Reticular activating system. The highest or anatomically the most forward or rostral part of the RAS is the thalamus. The thalamus is a group of nuclei, or clumps of gray matter, groups of neurons. Just one thalamic function is to drive the cerebral cortex. Inside the thalamus are some nuclei that project, connect to, wide areas of cerebral cortex. These nuclei constitute a revving apparatus for higher brain centers. The easiest way to see their effect is with the electroencephalogram. On the EEG various states of consciousness are easily distinguished by waves that have a certain rhythms which mark out, delineate, these states.

For example another beautiful sine-wave type pattern other than "alpha" described above is the sleep spindle. At 7-14 cycles per second (Hz) spindles wax and wane in amplitude and are recorded on the EEG over wide areas of cortex. The sleep spindle defines stage II of slow wave sleep and is generated or made by the reticular nucleus of the thalamus. The reticular nucleus is a group of cells that connects with a lot of cortical real estate and is thus a rhythm making area of the brain. One function of the sleep spindle may be to block the flow sensory impulses from periphery to the upper levels of the brain i.e. consciousness. Remember that if you are sleeping, as a general rule you don't wish to be affected by stimuli around you, you prefer to be disconnected from your environment (deafferented).

20-40 Hz rhythms termed gamma activity most probably generated by intralaminar thalamocortical cells characterizes waking and REM sleep10. So thalamic nuclei drive or rev upper brain levels. Not all cerebral EEG rhythms are driven by the thalamus, some like the slow waves of slow wave sleep and high amplitude waves termed K-complexes seem to be made autominously by networks of cortical neurons.

We are talking here about levels of consciousness self-imposed. What evolutionary advantage there is for tightly controlled sleep wake system is a mystery. Very likely we've evolved a sleep system to regulate alternations of high performance and rest necessary for the function of every machine not excluding living creatures. When you exercise heavily lactic acid and other waste products accumulate in muscle. You are able to sustain a certain amount of activity for a while, but accumulate a debt that has to be repaid and is done so with rest that regenerates the organism and gives one time to get rid of metabolic end products. This is supported by the fact that we all feel fatigued and even much more of an urge to sleep induced by heavy exercise, not just after you exercise which is why you may find it hard to fall asleep if you work out very late in the day, but after a day of heavy exercise.

Sleep stages do not precisely mirror levels of arousal as above described. What sleep and impaired arousal levels do have in common is lessening of electrical cerebral activity as seen on the EEG. When electrical frequencies are slow, it is harder to excite neurons. Slow activity signifies hyperpolarization. The individual neuron is harder to stimulate. When a lot of cortical neurons participate in this process the person is much harder to rouse and the EEG with electrodes placed over the scalp, is looking at the brain on a macro or holistic level, showing us what is happening in many neurons at once. Many different deeply situated, widely projecting cells make up the reticular formation. These influence wide sections of brain altering excitability.

In sleep we profit by looking at three parameters: EEG, muscle activity and eye movements. Recall that a person restfully awake has alpha activity on the EEG, with frequencies ranging from 8 to 12 per second. In stage I sleep the alpha diminishes and activity becomes more disorganized and is lower in amplitude. As the alpha pattern diminishes and more random activity appears over the scalp, the nice controlled sine wave pattern goes away and different regions of the brain appear to be less controlled and to function autonomously. We say activity is *desynchronized*ψ. It is as if the part of the brain controlling brain activity has turned off, which it has. Stage II sleep is remarkable. Instead of seeing some simple EEG slowing, Brain activity is wilder best appreciated over the scalp vertex, high amplitude waves that look almost as if the person is having an epileptic seizure, so called, vertex sharp waves. It is very like an uncontrolled seizure and many persons will actually have a muscle jerk or twitch or sudden movements of their legs, awakening them or impaired sleep at this stage. A slight jerk in the legs may occasionally occur in any one of us at this stage. But it is not a true seizure. What looks wild or out of control is precisely controlled. Coordinated rhythmic electrical activity is seen in sleep spindles also in stage II sleep. These appear superficially similar to alpha activity in their patterned waxing and waning amplitude and sine wave like frequencies in about the 15 cycle per second range. The rest of the EEG activity is slow. In stages III and IV of slow wave sleep, the sleep spindles and vertex activity virtually disappear and we are left with just slow activity as the person achieves deeper levels of more classical sleep. I say this because of the relative loss of patterned electrical activity and inactivity of the brain. This serves as a prelude to REM sleep. REM happens precisely after stage IV sleep 60 to 90 minutes into the night and again and again in later cycles of slow wave then REM sleep for 4 to 6 complete cycles during a normal average night. We may spend, depending on our age, younger people spend more, older less time, babies spend the bulk of their time in REM sleep, about 25% of our nighttime sleep in REM. If you awaken a normal subject in REM sleep, he may recall a dream very vividly. Given time, he will not remember a good part of the dream. Deprive him of REM and he will not feel refreshed after a night's sleep. Under extreme circumstances, he will go into REM sleep more easily, trying to make up for his deficit, hallucinate or even die.

One of the cardinal signs of narcolepsy REM sleep out of control, is sleep-onset REM in which REM occurs not after stage IV sleep as it is supposed to, but just after falling asleep. Dreams can happen abnormally and appear quite vivid. Some of this disordered sleeping may cause hallucinations in the elderly who lose some control of their sleep-wake cycles and explain why in the elderly as in other degenerative processes, hallucinations occur readily. Some hallucinations may actually be dreams. REM in animals, especially cats which are the best studied, is heralded with so-called PGO (for ponto-geniculo-occipital) waves. These are discharges initiated by the pons where REM is initiated to visual areas of the brain. REM is present in almost all mammals, which may imply that most of them dream, and some elements of REM can even be seen in birds. REM is present in newborns in greater quantity than inadults. Perhaps this means that young folks dream more than old people do. Eye movement has more of a scanning quality during REM which may imply the person is looking around at the scene of his own making. Muscle activity is nil and the person is flaccid, all except for the breathing muscles, but pure voluntary activity is absent.

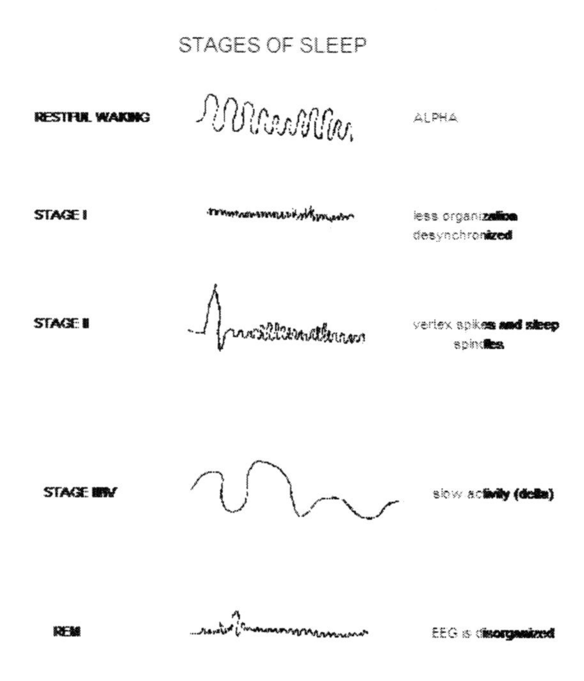


**Figure 5: Stages of Sleep**

In summary we all have a well-orchestrated sleep-wake cycle. Levels of consciousness are controlled by a rather vague physiological construct, the reticular activating system of the brainstem. This concept developed partly out of work done in the 1930's by Frederich Bremer. He cut the brainstem of a cat at different levels and observed that if the brainstem is cut high enough, the cat will remain permanently asleep. Many other researchers since have contributed to the concept of an activating center of the brain that effects the excitability of the cortex. Nowadays we have defined the effects of specific structures that project to higher brain areas as well as transmitters that seem to be involved in waking and the various levels of sleep. Some of these brain areas show up on the diagram[11]. In sleep you are aroused with varying difficulty that depends on your sleep stage. In that one respect we can see superficial resemblances with pathological alterations of consciousness and with coma induced by the anesthetist. The most important difference between sleep and coma is that sleep is an active process in which mental activity continues. In coma, as far as we can measure, there is no mental activity at all and none is recollected.

Different nuclei or volumes of nerve cells each play their own role in controlling states of arousal and most of them connect diffusely to the cortex. For example, the Raphe area of the midbrain uses the neurotransmitter Serotonin and seems to be involved in deep slow wave sleep. The Locus Ceruleus of the Pons, slightly lower down, in the brainstem secretes norepinephrine and is implicated in REM or dream sleep. From this we derive a lot of clinical information. For example drugs that counteract Serotonin or which destroy the midbrain Raphe can cause sleep disturbances or insomnia. Patients with strokes in the midbrain sometimes have vivid visual hallucinations and behave wildly. Loosely speaking, the serotoninergic Raphe may oppose the norepinephrinergic locus Ceruleus. Does the unopposed Locus Ceruleus take over leaving these poor folks in a state of dream sleep (from which hallucinations are derived)? This is a matter of some conjecture. In the hospital we see this a lot with older folks when sleep wake cycles are disturbed by night day confusion and drugs. It is common for a delirious person not to be able to separate sleep and awake states such that they are partially awake and asleep at the same time. They might vividly hallucinate, still being in a partial state of dream sleep.

## Sleep Is an Active Mental State

Two important points about sleep are: (1) The stages are precisely controlled. (2) Sleep is not synonymous with reduced or absent cerebral electrical activity. You can't compare sleep to brain death or even deep coma because in sleep there is mental activity. This fact makes you think twice about the purpose of sleep. Is it only for restoration or is something else involved? Pure rest could be accomplished much more efficiently. Why go to the bother of having all that carefully planned alteration of brain activity? Why do we need paradoxical REM sleep?

Man is not the unique in possessing REM sleep. As mentioned some of the seminal early work on sleep was done on cats. Chances are that they imagine in their sleep as well. Moses Maimonides, a Twelfth century physician who did not, of course, have an EEG at his disposal, remarked that animals have imaginations. Imagination he said is not what separates men from animals. What distinguishes man from animals according to Maimonides is intellect by which he seems to mean logical scientific thought processes.[12]

From ancient times men have drawn parallels between sleep and death, positing that sleep as a bridge from the waking conscious world to a different world to come. [13] The evidence is to the contrary. Even in coma, as we have seen, and I am talking here of deep coma, chances are there is no thought content there. Patients I have spoken to do not recollect mental activity and EEG changes are not organized and not like those or REM sleep. We only see a slowing or decrease in brain activity, no orchestration as I like to call it, rather a slowing and disintegration of activity.

People have thought about dreams for thousands of years. Aristotle wrote a short treatise, "On Dreams, and On Divination in Sleep." And the basic notion of dreams as being revelatory of the intent of the God's and thus somehow prophetic or of the soul's being disinhibited in sleep and exercising clairvoyant power by virtue of dreaming, was widespread. Oracles used the technique of Incubation, the practice of seeking significant dreams by sleeping in a sacred area or for the purpose of healing, was in widespread use.[14] The Bible treats dreaming pretty monolithically. Dreams are prophesy telling of future events. Deuteronomy warns of false prophets and dream interpreters but throughout the Bible the dream is a vessel conveying deep truths. Probably the first great interpreter of dreams was Joseph who was a prototypical dreamer in his own right. Joseph explained the Pharaoh's troubling dream, which turned out to be an accurate foretelling of the future, rising to a level of great power for a foreigner as a kind of first minister. Pharaoh altered his name into something that can be translated as "the explainer of hidden things". In the Bible, you can find many examples of dreams conveying deep meanings and prophecies. Daniel gained great fame for his deep learning and wisdom, but especially as a dream interpreter and diviner. King Nebuchadnezzar of Babylon was a second monarch disturbed by a dream that foretold the future. Daniel gains his deep understanding directly from heaven, "He revealeth the deep and secret things: He knoweth what is in the darkness and light dwelleth with him." Because of his wisdom, and studiousness and attention to deeper matters, Daniel, and others like him are more or less susceptible to the knowledge that dreams provide. The Bible is full of deep insights basically symbolic visions of a future time gained through dreams, perhaps none more spectacular than the visions of Ezekiel. Apparently prophets who communicate directly by verbal conversation such as Abraham and Moses, may be on a higher level than those who do their work through dreams. Dreams and visions in the first five books of the Bible predict what will happen to the kingdom of Judah many hundreds of years later. This is good evidence of course, that these documents were actually written, or at least revised, at that much later time. Even so, what a great literary technique this is, what greater way to make your point and at the same time give your message a profound theistic spin, a bold stroke, when you consider that the oldest books of the Bible were at least revised about twenty five hundred years ago. The best example of all is found in the book of Numbers where the non-Jewish prophet Balaam, predicts Israel's destruction by the Assyrians and even the Romans who rose to prominence over 1000 years later! But witchcraft and false prophesy are strongly condemned even necromancy, the art of predicting the future with dead spirits of the past, common, at the time in Mesopotamia[15] but the Bible is resolute in condemnation of such practices, something which ironically enough, resulted in mass persecutions in Medieval times of accused witches, such persecution betraying the superstition of the inquisitors.

Ironic as it may seem today psychologists are still telling us that our deepest personal insights come to us in sleep. Freud to Jung are our Century's greatest dream interpreters. But dreams for us may are a window to a Subconscious or Unconscious, not a prophetic or mystical visions or view of a future world.

The Jungian anthropologist Joseph Campbell who specialized in the role of myth in culture and many others maintain that modern men have lost contact with a subconscious that defines us. Living in our technological world we have become adept reality testers and have subjected our mythology religious and otherwise, to skeptical scientific and intellectual inquiry that has, in general, rejected the veracity and utility of our basic yearnings. We are alienated from our mythic past, our basic biological and social origins living in a more rarified and sterile environment which we have to find unfulfilling. We have lost contact with our basic identity, best defined in mythic and dream symbols. The symbol in our dreams is a polestar reorienting us and pointing the way home. Some symbols are discoverable in literature and art but the majority are even more elemental. That they continuously reappear even to the unsophisticated and unread among us is evidence that their meaning is simply universal. Some psychic symbols are personal as well, an idiosyncratic part of our own life experience. But others are archetypes whose presence and use owes to universal experiences such as birth, initiation, mate selection, child rearing, and death, life cycle events. These symbols emerge psychically with illogical behaviors, irrational fears in dreams and myth. They are felt by all of us as long as they are not denied or rejected, a part or the human journey. They need not be taught and are understood innately because they are part of all of us if we pay attention. They are thus part of a *collective unconscious,* which is almost an inborn or genetic defining part of humanity, a useful concept.

## Sleep is Evidence For a Composite Consciousness, Two or More Simultaneous Streams of Awareness:

What is the role of the concept of conscious or unconsciousness in this conceptualization? The meaning is that these symbols work on ourselves beneath the level of the strict scrutiny and editorial processes of our stream of waking consciousness. That is why they come to the fore, emerge and are operated upon but not by the light of day. Better to appreciate them, when they can be recalled at all, at night. If we can only rediscover this myth or subliminal level of awareness, then some of our psychic angst, or alienation from part of our basic self, our biological struggle, should, so the theory goes, be reduced. This is the work of psychoanalysis. As everyone knows, the style of dream thought is fundamentally different from the logic or waking thought. Unreasonable things happen in our dreams. Our logical thought police or editors are put to rest nightly as we enter a dream world unfettered by logic or rational linguistic methods of thought. In dreams, words have many meanings. Logic and physics of the real world takes a rest. We are allowed to enter a different alter world, unburdened by the constraints of logic and formal language. The evidence is that all of us, no matter how unimaginative we are while awake, live simultaneously in two worlds or (at least) two separate states of consciousness[16]. During the day we inhabit our workaday mundane conscious world. Then there is a more imaginative dream-like state of consciousness that we journey to at night. Therefore, there are at least two parts to our mental fabric, probably best called the conscious which is language based, Euclidean, Aristotelian (our mundane waking existence) and alter-conscious for alternative conscious state rather than "sub" or "un" conscious. Jung gives the example of a recurrent dream in which he sees his house, a symbol for himself, and constantly discovers other rooms which he had no awareness of, presumably another part of his inner self and new avenue of interests and research So it is with all of us. Rather than being "un" or not conscious, or the slightly more palatable "sub" or beneath the table type of consciousness, this paradigm is an admission that there are two or more alternative streams of mental awareness that co-exist within an individual. In honor physics where the universe supposedly started with symmetry of matter and anti-matter, both components of the real stuff that makes up our universe, we may have consciousness and anti-consciousness but this is defective since it implies an adversarial relationship in which one component of awareness will annihilate the other. All of us (though some may deny it) are intrinsically conscious and alter-conscious, but it matters little what names you use as long as you recognize that your total being is broken up into at least two parts. And these are not entirely separate from one another either, though as we have seen they one or the other predominates at certain times and altered states constantly intrude on one another. Where this is most visible is in our memories of perceptions that never reach consciousness but are stored away nevertheless and many are forgotten totally except we know they are present the brain because they are always unexpectedly dredged up in our dreams. Things we saw or heard of no significance in our waking conscious life assume central importance in our dreams most likely because they are indeed important to our anti-conscious existence[17]. All of us live in at least two simultaneous worlds whether we want to admit it or not, a real practical world and a world of our imagination.

Popular psychology has made a great deal of Left and right brain functions. The left hemisphere, responsible for language in most of us, is thought to be the seat of logic, the right hemisphere, perhaps something akin to dream states and intuition. Musical function and art are deemed to be right brain functions. It is true that word production, syntax and processing are primarily performed by the left brain or "dominant hemisphere". However artistic and musical functions particularly among those gifted enough to have significant output, involve both the left and right hemispheres. The old neurology literature cites many examples of composers and performers who lost their abilities due to left hemisphere lesions. Why? Musically gifted individuals also tend to be musically literate and logical, as they are "artistic". And as any musician can tell you music, as is every other advanced human endeavor, is a highly composite function, involving large volumes of both cerebral hemispheres. The right hemisphere is even more involved with simple speech production than we'd bargained for. It turns out large right frontal lesions cause loss of *prosody* or musicality and melodiousness of speech. A more posterior right hemisphere lesion may cause there to be a problem interpreting sentences with different inflections such as a question which we typically ask by raising the pitch of our voice at the end. There does not exist a right hemisphere *artistic* vs. A left hemisphere *logical* brain. Things aren't that simple. Φ

In sleep our minds are mentating but differently than when we are awake. We have much less direct contact with our environment and most thoughts, not all certainly, come from inside the brain but yet not completely disconnected. A man dreamt he was guillotined when a bedpost accidentally fell over his neck. You may dream of being hosed down with water when actually drenched with sweat etc, but for the most part dream stuff is made of memories and a different kind of internal mental processing.

## Hypnosis is not Sleep

Many people put the phenomenon of hypnosis in the same category as sleep. Indeed the word comes from Greek root for sleep and hypnosis resembles sleep superficially. Hypnosis had questionable beginnings in the 18th Century. It is generally accepted that the discoverer of hypnosis was one Anton Mesmer remembered today as a medical charlatan who promulgated such ideas as animal and planetary magnetism but it is a real phenomenon, which despite vigorous investigation over the years remains largely unexplained. For example, a good level of anesthesia can be achieved for some persons undergoing invasive procedures, even amputations, with hypnosis and behavioral effects are difficult to deny. The subject is not asleep though. Most persons on the EEG are in a relaxed state of waking as evidenced by a normal alpha pattern. Some persons in a deep hypnotic state have slower activity. The subject's attention is intensely focused on a boring repetitive object or stimulus. Some modern studies utilize PET scanners that look at active brain metabolism show that there can be an increase in glucose and oxygen utilization in hypnosis, in other words an intensity of attention, arousal and brain activity. At that time he somehow eschews or casts off his normal regular awake personality and ordinary logic. He is intensely relaxed but will not drop an object out of his hand. He may even relax all muscle groups yet maintain an uncomfortable cataplectic position for extended periods of time. He will have a reduced or altered response to ordinary sensory stimuli and as mentioned, a person may even be operated on and women have had vaginal deliveries without having any pain or recollection of pain. E.R. Hilgard who studied hypnosis found that a subject who denied having any pain when asked verbally nevertheless reported pain when induced with a suggestion for automatic writing but he was not aware of this writing exercise.[18]

Using evoked responses we can electrically track sensory impulses arriving at the brain. In this way we can see that hypnosis in no way interferes with the arrival of sensory impulses to the brain. The subject is made oblivious to these stimuli. How is this achieved? Apparently through unusual concentration and focusing by relegating his attention to another area of awareness. We've all had the experience of having an injury which should by rights have been painful, yet because of concentration on something else, we feel nothing, until we look down and note dripping blood. All of this is taken to mean that there is achieved through hypnosis, not only an intense awake focussing, but also a kind of dissociation akin to what occurs in hysteria especially dissociate type hysteria where there exists more than one personality or consciousness within the same person. For hysterics awareness on the part of one personality for the other is essentially nil.

The comparison of hypnosis to hysteria was long ago made by Pierre Janet and it seems to be an apt one. Dissociated hysterics are those subjects made quite a lot about in fiction and by the press who have multiple personalities, a phenomenon usually attributed to childhood sexual abuse or severe trauma as in war suffered as an adult. Alternatively they may experience periods of amnesia or what is termed a fugue or fleeing state. The theory is that some aspect of one's real life was so traumatic that person needs to amputate it or at least to flee from this part of his personality. Using hypnosis a person can toggle between personalities yet, supposedly, one personality has little to no knowledge of the other. Many instances of this phenomenon remain poorly documented enough for the skeptic to ask, who is pulling whose chain? Is the therapist being fooled by a good actor? Or maybe the therapist is trying to fool a gullible public, yet there is little doubt that states of altered knowledge or awareness of sensory stimuli, especially painful stimuli can be induced by hypnosis.

Dr. Dorothy Otnow Lewis, as psychiatrist has studied perpetrators of brutal murders and has found multiple personalities in a number of them. Her interview bring to light unspeakable ordeals of sexual and physical abuse in early childhood in a number of her subjects. The stories of abuse at least appear be substantiated by other evidence. At her beckoning she has spoken with a number of totally separate personalities residing in the same individual murderers and rapists. Questions remain about the effect of her own belief system and suggestion and whether these inmates on death row are merely play acting in a last ditch effort to try to save their own skins. Courts and juries have certainly tended to ignore her evidence and pleas for mercy. Deep unresolved questions remain about the possible veracity of her observations, whether it is ever possible for personalities to split off in this way within a single body only one of which is actually responsible for committing a heinous act! For society to continue the unspeakable abuse of a young person tortured and sexually abused as a child and end that life in an electric chair seems a cruel and unusual, certainly an irrational and uncivilized response[19].

No matter how big a skeptic you are, from a purely physiological standpoint then it is fair to ask what hypnosis is. What hypnosis isn't is a kind of sleep or in any way an altered state of awareness. The hypnotized subject is awake relaxed and intensely focused. However what he has shown is the ability to split consciousness and compartmentalize sensory stimuli and responses to these in different simultaneous personalities. Between these personalities the bridge of memory or a stream of consciousness is affected. Notice that in dream sleep, a memory bridge between the illogical state of dreaming and logical waking is also broken so that it is possible to say that at least two different simultaneous personalities live within the same person!! Notice that I don't maintain they are separate persons, only separate personalities (for want of a better word) because even though the *manifest* connection of waking memory and any obvious stream of consciousness between these states is not apparent, still, we can appreciate these parts interact with one another and effect the manifest waking person subtly. (See discussion above on dreams and sleep) I shall have more to say about all of this in succeeding discussions in the chapter on memory.

A lot has been written about hypnotized subjects being made to recall childhood memories and being regressed to distant times as a child to rework experiences. Subjects have been made to relive traumatic events and there is lively debate about what can be admitted in a court of law. Quite commonly, subjects "remember" forgotten traumatic events especially connected with sexual abuse in childhood. How much is remembered and how much is fabricated, we do not know. There are a lot of examples of spectacular accusations initiated by hypnosis. In February 1994 Cardinal Joseph Bernadin of Chicago, a man who championed the cause of sexually abused children, was exonerated after being wrongfully accused on the basis of data obtained through hypnosis of sexual abuse, illustrating the dangers of a court of law relying on information obtained during a supposed trance. The perpetrators of these false accusations might have been complete fakes, of course. Others have even been regressed to former lives, and it seems there are simple parsimonious ways to explain such clinical regressions, not the least of which is the expectation on the part of the subject and his examiner, each trying their best to either fool or live up to the expectation of the other. Can remote or pathological inaccessible memories be dredged up using the technique of hypnosis? This is hard to prove and personally remain unconvinced. In the few situations I have seen, I have been convinced that the subject is fooling his psychiatrist.

The celebrated case of Rachmaninov illustrates how hypnosis may be used as psychotherapy, a means of liberating the creative spirit. Rachmaninov, after the disaster of his first symphony, which he wrote for a forbidden and unrequited love, hit a dry spell, characterized by a deep melancholy that put an end to his creative output. He couldn't compose. A certain Dr. Dahl supposedly helped Rachmaninov out of his deep depression. Of course, we all know that depressions of this type have a good record of spontaneous recovery, so it is debatable how much value this therapy had. The first movement of his second piano concerto, it seems to me, is an allegory for the return of his creative spirit. The orchestra is an enabler for the solo piano, as the work begins in a state of dark melancholy but gradually moves into the Springtime of hope. You can actually see a delicate flower bloom as the paralyzing melancholy ends but leaves behind a residue of deep emotion. This is as good an illustration as I know, of hopeful recovery from the effects of depression, a recovery thought to have been accomplished through hypnosis and psychotherapy.

What I am after here though is that true hypnosis as well as dreaming as an indicator of the existence of multiple simultaneous conscious selves within the same individual. In clinical hysteria, we may be seeing the very same thing. If it occurs pathologically it also is part of our normal existence and we have to ask why or how this occurs. Not two sides or hemispheres of the brain separate out but two or more multiple strands of awareness seem to be working simultaneously. We have our most direct connection with the conscious logical critical waking self. It sometimes seems possible to compartmentalize consciousness. There may be techniques that affect the a single stream of memory traces, in fact a way to break the bridge that ties together consciousness into a single whole, in other words, memory. Some of these ways include dreams, which are rarely remembered but give a view into a simultaneous but active irrational consciousness that seems to be functioning alongside a rational brain, and hypnosis and hysteria and even multiple personalities when they truly exist.

Our waking behavior is unaffected or is at least not explicitly affected by memories of these parallel awarenesses, memories forgotten, dreams, illogical alternate personalities, alter or anti-egos. Yet there is a whole phenomenology attesting to their existence.

## Alternative Modes of Thought (Not Just Logic)

Most of the time we insist on logic and material proof. But we suspend logical judgment at other times when it could be said we are waxing more poetic or artistic. Thus it could be said that we have at least two sides one having limited access to the other but they are far from being mutually exclusive. There is all manner of evidence that one part interacts extensively with the other. Poetry is full of reasoning by similes and makes the most of symbolic logic yet the best poems are constructed with the utmost care and conscious calculation. Even pure music, good music is part magic and emotion but logical even purely linguistic processes are not thrown out. The best music, music that really challenges, is the product of meticulous design, working and reworking as is the best poetry. In the same way many leading scientists describe how they are helped along with a pathmaking insight when they listen to their non-logical selves in other words when they suspend logic for a while and work by a powerful insight. Some researchers describe how vexing problems are literally worked out in their sleep and in dreaming. This is undoubtedly because we are mentally active while sleeping and obsessed night and day when trying to work a problem out.

What we have then, is an interactive process that constantly occurs between all of these sides, aspects of consciousness. In truth, there is no split, but a culling and coordination of separate processes within the brain that merely contribute to the whole product that we call awareness or consciousness, if everything is working the way it should. Doubtless, as we have seen, we focus our attentions on one aspect or the other depending on our state. We attend to the dream world that is less than logical in our sleep and are more logical when awake. Still, as we have also seen, dream processing is active in an awake person and is constantly gathering fodder, material, from our waking life. Similarly, scientific insight comes from inductive and deductive logic and even from the imagination of the dream. For an individual person at a specific time, these aspects of consciousness, or splits, either are, or are not, integrated well. We may talk about an ego structure as the integration of the whole, or prefer to call this holistic principle by another name. By whatever appellation, it refers to the integration of the whole, selective focus as certain appropriate times yes, but at the end, integration, that we call the whole person. The schizophrenic seems to be permanently severely even fatally split, perhaps do to a defect in the brain, he has failed to develop this integrating principle, failing to organize and utilize these splits to his advantage.

| LEFT HEMISPERE | RIGHT HEMISHPERE |
|---|---|
| LOGIC | EMOTION |
| CONSCIOUSNESS | UNCONSCIOUSNESS |
| INDIVIDUAL | COLLECTIVE CONSCIOUSNESS (JUNG) |
| OBJECTIVE | SUBJECTIVE (SELF) CON-SCIOUSNESS |
| LANGUAGE BRAIN | NON-VERBAL |
| AWAKE | ASLEEP |
| DEDUCTION | INDUCTION |

**Table 2: Introducing some ways we traditionally split consciousness. Do we flip-flop between splits or integrate them into a whole?**

Sometimes an accident or stroke destroys brain regions connected with the arousal mechanisms. Relatively small lesions in the right area of the brain can devastate a person. Clinically we see all permutations of arousal. Some persons have a relatively small brain lesion, usually involving an area in the upper brainstem, yet remain comatose, even though the cortex is not doing that badly. One of the most heartbreaking problems is when a person seems to be awake but there is no content to consciousness. This is possible when widespread areas of the cortex are destroyed. These brain cells are much more vulnerable to the effects of oxygen deprivation or the interruption of the blood supply to the brain. In that case the person is awake, as he appears to be. In fact, he may have normal sleep and wake cycles and be rousable with little other function. There is no evidence that he processes thoughts, moves with true volition or even follows any commands. He can be awake and even look around. This the persistent vegetative state, mentioned above. Akinetic mutism on the surface, almost looks like a persistent vegetative state. Akinetic mutism occurs where there is destruction of a relatively small area either in the frontal lobe or at the top of the brainstem that controls responsiveness. In the case of frontal lobe destruction as when an aneurysm bursts into medial frontal lobe structures akinetic mutism can result, presumably because this area of the frontal lobe, is responsible for motivation and drive. Specifically the area of frontal lobe involved is usually the Anterior Cingulate gyrus. A few patients who recover from akinetic mutism in this region of the brain, recalled an extreme motivational deficit, not that they can't talk, but that nothing is worth saying. Without this basic motivational component, responsiveness is colorless, zombielike. It is almost as if all personal motivation is amputated. The patient appears to be almost unresponsive, though awake, their speech, if it exists at all, expressionless. Akinetic mutism, unmoving silence, has also been associated with damage to the midbrain and thalamus. Here, the problem may be due immobility, due to damage of certain motor areas, a little like the immobility we see in Parkinson's disease except in a more extreme form so that the problem is entirely different. In either case, arousal is preserved, because the brainstem reticular activating system is relatively intact, but the subject stares ahead, virtually immobile, apparently without volition.

## Arousal Vs. Content of Consciousness

This dichotomy in anatomy, brainstem vs. cortex, allows us to operationally distinguish consciousness as a combination of *arousal* and the *content* or quality of arousal connected with the brainstem and cortex respectively. It alters the basic way that we look at human consciousness, superseding the many purely philosophical points of view that have appeared over the years. Akin to the level of arousal is the focusing of attention. Where brain function is slightly impaired, attention mechanisms are also. For example we notice that a slight impairment of brain function occurs with such insults as lowering oxygen concentration in the blood, with a high fever, slight change in blood constituents such as Sodium concentration that affects the brain more or less diffusely. This is what we mean by the term *encephalopathy.* A person can be awake but has trouble staying on a mental target. He cannot follow complex ideas and is easily distracted and becomes unfocussed and upset abandoning a difficult task at hand. By contrast, a highly functioning individual has no problem staying with a programme and taking it to completion. In delirium distractibility increases mental function and staying on task is a problem and there is often excessive movement, disinhibition and irascibility.

Certain persons function in this way habitually. This is the concept of hyperactivity disorder of childhood, aptly termed "minimal brain dysfunction". What we see is abnormal disruptive behavior, learning disabilities some of which derive from an inability to focus attention, disinihibition and increased motor activity. Some recent work has found decreased metabolic function which can be seen on certain function measuring PET scanners in kindreds of children identified with Attention Deficit disorder of Childhood. By some accounts the disorder has proven to be a matter of hypofunction affecting a good portion of both frontal lobes which is perhaps genetic. We see problems not unlike this among many psychiatric patients especially chronic schizophrenics. What we notice is that brain function is diffusely impaired. In schizophrenia we find a high incidence of brain atrophy, enlarged intracranial spaces or ventricles within the brain.

All of this is brain stuff. We take it for granted that consciousness and levels of awareness have to do with the brain and chances are you didn't give this a second thought. It was not always this way. Only recently has the brain been placed in the exalted position as the central arbiter of consciousness. In the Bible the heart is the seat of intellect, not emotion. You should love God with all your heart (read intellect), soul and might. The soul was given and taken away through the nostrils and symbolically represented in the breath of life. Thus Adam was given life when he was given breath. When Aaron's sons disobeyed who came too close to God were killed by having their very breath sucked out of them through their nostrils. Air is an invisible essence, which life is, and at the same time, breathing defines life, at least for animals like man.

All of these romantic notions aside, we know today that the brain mediates emotion and intellect. The heart and lungs exist to supply the all-important brain with oxygenated blood. For Aristotle the purpose of the soft mushy remote brain, was to cool the body. I wonder if there was some person of that day, or doctors that came after him and held him in high regard, even suspected that the brain was the seat of intellect. We are a lot wiser today but only in the last 20 years or so has death been legally defined and only in some states, as an irreversible cessation of brain function. Before that death happened when the heart and lungs stopped working. Then we discovered that we could make the lungs work artificially with a ventilator and we had machines and drugs to keep the heart functioning as well and when we did we soon found out the hard way that life was over anyway because the brain ceased to function and was not viable, we had to change our definition of death.

## Brain vs. Kidneys:

No one doubts today the brain's role in making us human. But the brain is at the same time an organ in the body like all the others, composed of tissues and cells. It's interesting to ponder on what makes the brain different from other organs, and in what ways it is the same. The brain is obviously different in that it holds that key to all of our thoughts, perceptions and emotions. It has an exalted position. You may see the kidney as a filtering machine. When you really study it you find it is a whole lot more than that. The kidneys remove toxins from blood and play a crucial role in homeostasis, keeping concentrations of ions and other substances within very narrow limits of concentration. The kidneys also control blood volume and pressure and even secreting hormones that prevent drops in blood pressure when you stand and the controls red blood cell production. OK, the kidney is very complex, but still, its function is a whole lot easier to grasp than brain function. For the kidney you need only define what it does. study its methods and hopefully manipulate function to improve the patient who may have problems. We have no trouble looking at the kidney purely as a machine. When it comes to the brain there has been quite a bit more debate.

It's not unusual that the kidneys suddenly fail. Urine output stops and waste products, expressed as BUN, and Creatinine, increase in the blood. This is renal shutdown. At such times whether or not the kidneys will come back, and what you should do to try to make their function return is much debated. Arguments turn on various questions. What was the level of function before the insult? What was the nature of the process that caused the kidneys to stop functioning, for example was it heart failure, fluid restriction, shock, or sepsis? Given this information one can make an intelligent guess as to whether kidney function will return and what one should do to ensure recovery of the organ. The interesting thing for me is seeing that the best prognostications of specialists are often wrong. The best ones know that they don't know. The attitude is that they don't give up but give the kidneys all the help they can and then hope for the best. Under the best of circumstances, they can't prognosticate with 100% accuracy. I've also noticed a different attitude among less experienced medical residents in training. Seeing a severe level of dysfunction they tend to be overly pessimistic and often are out “hanging crepe” for already anxious family members. At best the poor victim may survive but require a lifetime of dialysis or at best a kidney transplant if he's lucky enough, they maintain. But experience teaches that some patients start peeing when you least expect them to. Kidney function may in fact improve unexpectedly, miraculously.

Deliberately I'm using so mundane and mechanical an organ as the kidney to make a point. First, in a lot of ways the kidney is not that different from the brain. You may make the argument that since the kidney exists to serve the brain, kidneys function in essence only to provide the nervous system with the exactly controlled and comfortable milieu, so the brain can do what is really important, so kidneys are different than brains. The truth is kidney cells are subject to just same tasks to ensure survival and threats to function as brain cells. First of all the brain too is vulnerable to the above threats of heart failure, fluid imbalance, shock, and sepsis. Indeed the brain is even more vulnerable to these very insults. But think of how much more complex prognostication must be for the organ that is the seat of the soul. One important obstacle is that the brain, far more than the kidney, is broken into individual parts allowing specific functions, for example the cerebellum, the cerebral cortex, basal ganglia etc. each of which may be affected to a different degree and present with different vulnerabilities. For the brain we can determine severe damage to large parts and imaging with CT or MRI scans may show in black and white severe damage, swelling which affects the whole brain compartment or an EEG can show a severely disorganized or even flat graph. A lot of factors may conspire against the brain to severely impair brain function and we may see the patient in a deep coma. Our resident will, even more, go out and hang crepe with the family. But again, once in a while we may witness a miraculous save, so much so that our approach in prognostication has to be as a general rule statistical. We prognosticate on the basis of a group of patients in coma under approximately the same conditions, for example, after an out of hospital cardiac arrest and can say statistically on the basis of a large group of similar patients, that there may still be a tiny percentage of persons who will regain consciousness. Indeed patients in a persistent vegetative state have to be observed for three or more months before we can state with confidence that they will never again enter a sapient, a knowing state[20]. This is for a large group of patients in a gray area. For other subsets we can predict confidently that there is no chance for recovery or that the chances from the brain's standpoint are very good. A lot of times of course just looking at the brain, things don't seem to be all that bad but the poor patient is tottering on the brink of multiple organ failures and will die for that reason.

In critically ill persons the brain is frequently an item of intense concern. A change in brain function may be picked up many times by the intensive care nurse who is with the patient. Most of the time there is abundant sedation and the patient can be agitated but cognitively impaired, in other words delirious, or he may be somnolent or comatose, an impairment of level of consciousness, less often some specific or focal problem such as weakness on one side, hemiparesis. On one level the brain is merely the frailest organ but even more critically, the brain is the whole point to the exercise. If it can be determined that the brain will not recover then why continue doing anything? So it is most frequent that I am there to make this determination. What is the chance that mental function or so-called sapience, will return? If there is to be a brain impairment, how bad will it be? Are we devoting all of our best treatment and resources only to produce a totally dependent nursing home candidate?

So here we are again. We are asked to make some assessment based on both a physical and a mental essence. At base the answer to these questions resides in the individual brain cell. For the patient laying in an intensive care unit, you have to decide whether these individual cells are dead in which case no one can ever bring them back, or if there may be something impairing cell functions so much so that you see this translated into a gross impairment of behavior that is an alteration of consciousness. Brain cells may merely be placed in an inhospitable milieu that simply impairs their function in which case are likely to function again. More and more nowadays, questions about the brain bear upon the function of individual brain cells.

## Brain Death, Brain Cell Death

In recent years brain lesioning especially as caused by diseases and other natural processes has been more and more conceived to be a cellular process. As regions of the brain die, a chain, actually multiple chains of events have been described. We still have a lot to learn about this complex cascade of events which has been called *neuroptosis* literally, a drooping neuron a type of programmed cell death. We do know that organisms need to have a way to get rid of some of their own cells. This happens in the embryo. As certain structures develop other cells are eliminated in order to form normal structure. With invasion of a disease organism, certain infected cells are eliminated and cancers may occur when abnormal cells are not adequately killed off. In these cases and others cell death is an adaptive and normal event. As a cell dies it goes through a whole sequence of events which is pre-planned and certain specific chemicals, some destructive are released and processes unleashed that implicate even simple intracellular substances such as Calcium and Nitric Oxide. If a chain of events can be precisely described, then we may someday discover a way to intervene and interrupt the process of cell death, saving our patient.

## LOOK AT THE BRAIN FROM THE CELL'S PERSPECTIVE:

Can decisions be taken, strategy made, in as small a unit as the individual cell? The above is a picture of an ameba, a single celled animal predator, engulfing another ameba, a single cell prey. A struggle appears to ensue, very similar to the predator-prey conflicts familiar in larger animals, in which the engulfed organism, tries to escape and actually does. A drama unfolds between two single-celled protagonists, each expressing a biological imperative or will to live, the very same as the human struggle for life and limb proving that such struggles take place daily in the life of the single cell. We do not know if such decisions can be taken by even smaller units of biology, at the level of the organelle, DNA or individual protein elements. Clearly, specialized nervous system tissue is not necessary for the struggle of life, but is a unique component of living tissue. How then did the nervous system develop and why is it now considered to be the primary repository or at least mediator of awareness?

How did it all begin? What adaptive function does the nervous system serve? At first blush you might think that any complex organism containing large numbers of cells would require specialists in communication and control, but complex organisms survive quite well without nervous tissue including plants and lower animals. Tiny animals, protozoa and microscopic creatures with just a few cells exhibit complex behavior in order to survive in their specialized microscopic world. Single celled Paramecia display incredible specialization of function. The Paramecium is one cell with organelles (the diminutive form for little organs) which performs many of the same vital functions that larger communities of cells perform through specialization of cells as tissues and organs. Paramecia have a gullet, cilia for locomotion, light sensitive spot, excretory and sexual organelles and emit complex behaviors of avoidance, predation and courtship. Even single cells have a certain executive capacity. Just one thing that distinguishes them from non-living machines, which can only be vaguely described from an objective vantage point, is a will to live. You may say that machines are not alive and possess no will to live, though it's of course possible to build into them procedures that help ensure their own survival. For example, a computer may control an air conditioning unit which keeps the temperature at a certain range necessary for its circuits to function and you may ask, how is this different than the hypothalamus of the brain which also controls an inner milieu. Animals have built into them emotional responses and seem to strategize in order to avoid their own demise, to react with fear and depression at the prospect of death, but it theoretically possible to build these responses into a machine which would at least make it appear to respond in an emotional way, to mourn, exhibit a fear or rage response etc. You may counter that the machine would simply be following a procedure designed to make it appear to emote by exhibiting emotional behaviors and there would be no inner content there, only some automatic response which is not what apparently occurs in a human or an animal and you would be right. See how the ameba victim struggles against its predator in order to get free. One has the impression right or wrong that even this tiny creature strategizes to get free, that is his response in automatic only to an extent but that if conditions were only slightly different, say the was surrounded by more predator ameba protoplasm on one side than the other, he would try to break free on the thinnest side and would thus alter his response in specific ways given slight changes in his situation. All of this he does without a nervous system (and much of this behavior al response can also be reproduced by a suitably programmed machine.)

In bigger animals cells become specialists, and are more than willing to sacrifice themselves for the greater good. This is because the genetic information handed down is identical for each constituent cell which is a major determinant for such altruism, a topic we'll return to later on. As larger aggregations of cells acquire specialists, constituent cells lose something of themselves. What illustrates this better than anything else is that a good amount of genetic information is suppressed in cells in order to make special tissues and organs. A single celled protozoa expresses pretty nearly its full genetic potential. Nervous tissue expresses only a small part of its genetic potential and is no different from other types of specialized cells as nervous tissue evolved in just the same way other tissues did. A neuron or nerve cell is no smarter than any other cell specialist. We have an obvious analogy today in our medical system composed of "too many specialists", with the criticism that they forget about general medical training and are no longer able to handle basic medical problems. Specialist doctors have lost a lot of what makes them what they are, but presumably this allows the whole structure to achieve a higher level of organization. The question is whether or not this is true. Are we giant humans achieving more than tiny Amebas and Parameciums??

## The Brain Represents Design, Biology, the Triumph of Randomness, Trial and Error

It is really questionable whether having a brain confers any real biological advantage. Indeed the jury is still out with regard to this question. Having a brain gives an organism a chance to strategize, for command and control and planning. A good advanced brain provides plasticity of response. Lots of people maintain that due to the development of an advances human brain the rest of biology will be vanquished that man does not have to worry about being crowded out of the evolutionary world. Surely many human weaknesses that would have meant certain death and impaired survival now persist our gene pool, that is persons with these have been able through medical to survive and reproduce. But it is also true that we are daily fighting for our lives as a species mostly against organisms whose mode of operation is to mutate very fast. For example it may very well be the case according to a variety reports, that more people not less, in the U.S. and worldwide, are dying of infectious disease, this in spite of our enormous industry in development of hundreds of second and third generation antibiotic compounds and sophisticated means of drug delivery. In an earlier age of antibiotic use, researches boasted that we would vanquish infectious diseases caused by bacteria, viruses, Ricettsiae, mycoplasmas and other organisms[21]. Now we know this isn't true.

We are experiencing the very same type of competition for food from the insect world and against fungi and other microorganisms in agriculture. We are competing with these varied primitive organisms directly and it isn't at all clear which side will win, we humans or bacteria and viruses or plant eating insects. The conflict is one of design based competition that comes from a central command and control unit of logical output which is the human brain, designer of the advanced third generation antibiotic and insecticide and of the tractor, vs. The non-logical not centrally designed biological organism which continues to depend on natural selection and random biological variation and recombination, the brain and logic against the random solutions of biology. Again the victor is not at all clear. This is analogous to a centrally planned Soviet economy vs capitalism in which the most effective and useful endeavors carve out a niche and survive.

In honor of the realization that bio-evolutionary strategy may over the long run be proved more effective than a central planning agency of the brain, design theorists and computer scientists have started to capitalize on successful biological paradigms. How is biology so successful as opposed to human planning? Biology has "reckless and random" ways. "Because humans rely on logical processes, they consider a fairly narrow range of solutions. Nature, on the other hand, takes a sprawling trial-and -error approach that tests many more potential solutions."[22] The basic conflict in strategy is preplanned logical design versus a wider trial and error strategy not limited by logic. Even computer theoreticians seem to be coming around to the idea that a random trial and error strategy may win. A trick is to design computer programs whose lines of instructions will mutate and recombine similar to the way genes and chromosomes combine and reproduce in an effort to achieve an optimal fitness. Next you employ computational systems of sufficient speed to test these recombinant strategies at great speed, simulating many generations of evolution.

Large multi-celled organisms easily get along without a nervous system. All plants do of course. They use chemical signals sent between cells that take over part of this function. Plant's reactions to environmental stimuli tend to be stereotyped simple and slow by our human standards. Plants make extensive use of chemical messengers such as Auxins and root hormones to communicate between cells and promote growth. Indoleacetic acid utilized as an Auxin, is a chemical very similar to a mammalian transmitter involved in the control of sleep and depression, Serotonin, and its amino acid precursor, Tryptophan, something that illustrates the universality of chemical messengers that exert powerful organic effects. All organisms are currently felt to have a single ancestor. Different lines are not thought to have arisen independently in different places. The most potent argument for a common line of descent is the similarity of organelles in all cellular organisms and especially the homology between strips of nucleic acids and amino acids that make up DNA and proteins. No one has discovered lines of organisms with fundamentally different strategies for survival or totally novel subcellular components that make up their structure but this is not to say that this is impossible. You may also argue that organisms may have arisen independently but on the subcellular level at least converged so as to utilize the same proteins, nucleic acids and organelles given some general homogeneity of life on our planet, i.e. there is only one optimal biological strategy for survival. But our planet is highly diverse just looking at differences in climate as regards temperature, availability of water, oxygen etc, a consideration multiplied even further when you take into account the profound geological changes that have occurred over the eons of time since the earth's existence and the dawn of life some 3.5 billions of years ago. Just as an example the atmosphere at that time contained almost no oxygen. Initial conditions even on our own planet were not friendly to life. Our earth was down right hostile.

But consider the striking homology of organelles such as chloroplasts and mitochondria. It is generally accepted that these structures may be remnants of early bacterial invaders of cells, early symbionts or commensals with primitive early single cells that ended up helping these cells handle their energy needs. Chloroplasts and mitochondria still survive as symbionts sharing their own DNA with that of the cell's nucleus. They contain a remnant of their own DNA to this day but perhaps started out with their own DNA when they originally invaded other cells as parasites.

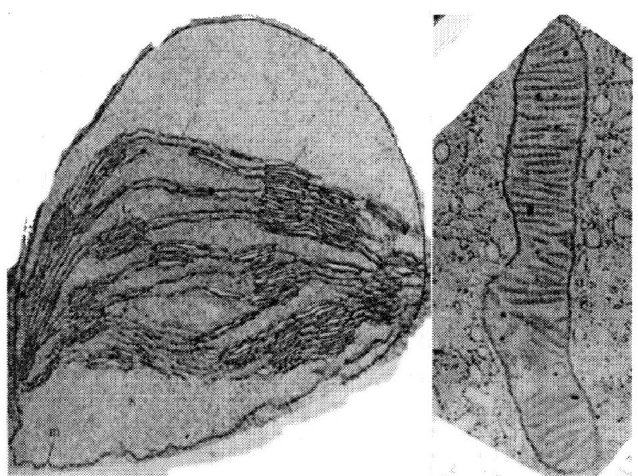

**Figure 6: The mitochondrion (L) and Chloroplast (R). Note similarity in structure. Both organelles are enery converters.[23].**

So our best understanding is that all living cells and organisms are derived from a single cell and the best evidence for this is that animals and plants have a limited repertoire of chemical and physiological responses that they use over again. Proteins, nucleic acids and organelles are highly preserved. The chloroplast, which began as a remnant of a symbiont, is the single best example of a subcellular organelle that gives evidence of early divergence of groups into animals and plants.

## Slow and Fast Communication – Nervous System Phylogeny:

Plants make abundant chemicals that exert profound effects on humans. These undoubtedly evolved for far different purposes, not to be used as drugs by man. A minority of these chemicals did evolve to influence other animals, but most affected other plants, some as with antibiotics as poisons against other competitors. Biology has a limited repertoire of chemical signals which with slight deviations affect a wide array of cell types. Just some examples that come to mind include digoxin, psilosibin, curare, Caffeine, cyclosporine, and penicillin. One of the reasonable fears of environmentalists is that some of these substances that may have potential medicinal value may be lost as large numbers of as yet undiscovered species are lost.

Getting back to Auxins, they help control and give direction to growth toward or away from the earth and light that nourishes the plant. Plant growth is something simpler and slower than a reflex termed a tropism. Tropism means literally going toward something nourishing such as the sun. Auxins simply give the order for plants to grow. In phototropism a plant grows toward light; in geotropism, toward the earth. The slow response to a chemical messenger controlling growth is the way plants respond to their environment.

Growing toward or away from something is an awfully slow way to respond to one's environment but fitting for an organism stuck in the soil. Plants have reasonably adapted to relatively unchanging conditions. Something's to be said for constancy when you our kids who flip from one to another station on their TV sets and end up seeing nothing, or the fellows who are trying to time the stock market, are trying to move their money in and out and never make any headway. In more instances than we allow for, inertia proves a very reasonable way to make a living. Alternate strategies do exist as in animals, which evolved efficient patterns of locomotion. It was animal motility that led to the need for nervous tissue specialized in conveying messages between cells at a rapid rate. The most primitive example is to be found in the Hydra, a tiny carnivore. A specialized nerve net coordinates movement of a whole colony of cells helping the organism to quickly seize the moment in pursuit of its prey.

This primitive nerve net functions orders of magnitude faster than can chemical signals and sets the Hydra apart from sedentary plants responding only slowly to environmental change. The Hydra's nerve net adapts rapidly, sensing change with an afferent limb coordinating sensory inputs connected to and intimately involved with an efferent component effectuating controlled movement. Nerve cells are necessary when animal cells specialize and the whole complex organism needs to respond quickly to environmental change. The first nervous tissue like other organ systems, therefore was a revolutionary solution for this need.

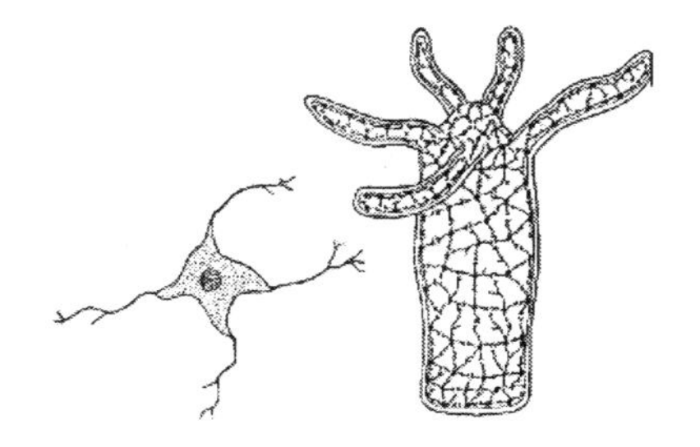

**Figure 7 Single nerver cell and primitive "nerve net nervous system. How it all began.**[24]

The most primitive nervous systems respond stereotypically. They tend to be hard-wired, meaning that they are designed merely to bring about a highly reproducible response to a specific stimulus. For the Hydra, sensing prey will be made to fire a tiny hook from his nematocyst attached to a thin thread to reel the victim in. Or his tentacles will extend when neurons in some way sense movement or the presence of prey. These are simple stereotyped responses hard-wired to occur when sensing a certain stimulus. In lower animals and in lower levels of the human nervous system too, we find a specific nexus between a sensory stimulus and a motor response coordinating pre-ordained motor responses.

An animal may respond to simple components abstracted from a large complex sensory field connected to an equally simple motor response. The responsive neurons need to latch onto specific elements of the stimulus in a large stimulus field that serve to distinguish it from among all other stimuli and at a certain specific time. Thus an attack by a male stickleback fish may be induced by a model with a red underbelly, the appearance of a male fish in breeding season. Perhaps he will respond if the red is on a certain specific shape on the underside of an object and it moves. And the female stickleback is attracted preferentially to the male with a brighter red underbelly. Ducklings will follow a moving object even as large as a human, bonding to this object as a mother a long as they are stimulated within a critical period, a phenomenon called imprinting. On some level, humans too respond to a single simple stimulus extracted from what may be a complex stimulus field, in a stereotyped fashion, a male to the woman's breast or buttocks is an obvious example. Yet we cannot over-generalize about all human or even animal behavior on the basis of this. This simple response to a releaser, the red underbelly, the moving object, the breast, is only part of a response repertoire.

Nervous systems allow two different modes of response, one a coordinated response to a simple stimulus stereotyped, always reliable, but not plastic. To change such a response, would require an alteration in genetic material since such stereotypy is ordinarily inherited, reflected in anatomical change in nervous system structure. Examples of such heritable characteristics include changes in cricket and bird songs that allow females to find and mate with eligible males of their own species. Both the song and the nervous system's pattern recognition system are inherited in the same way that a body pattern or a color would be passed down. Behavioral characteristics of animals are inherited in just the same way that physical characters are.

## The Hard-Wired vs. Plastic Response

Nervous system anatomy which brings about song production and reception are inherited. The architecture of neural networks involved in song production in fact, mimics reception, that is if a male cricket of a species produces a patterned sound say ···- , then there is a group of neurons in the receptor circuits of the receiving female, which fires or responds to this same exact pattern, ···-,so that the patterns fit together like lock and key. Songs are thus designed to stimulate neurons of animals of the same species. Animal song is critical to the mating process of crickets, katydids and songbirds. In birds though their song sounds inspired by the male desire to copulate, the basic form of the song composition is inherited. Then certain other details, what may be termed “dialects” are learned by the young male when he hears the song of more seasoned adults. Females are the recipients or objects of song while males are the makers of song[Ψ]. If you listen for it at any given time, there is a veritable concert of insect and bird mating calls, female receptor neurons responding to exact patterned firing or stridulating or vocalizing male motor neuron patterned firings. Male songbird production of song patterns will not occur if his brain is not exposed to sex hormones early in life and there are sexual differences in brain structure that result from such exposure[25].

Behavioral patterns reflect patterns of inherited nervous system structure. The bloodsucking leech, for instance, goes through a complex coordinated sequence of behaviors that ends in its successfully finding, swimming toward, biting, and finally sucking the blood of its mammalian (often human) host. When satiated, alternate behaviors take over. There are programmed anatomical patterns in the leech, which owns a relatively simple nervous system consisting of just 32 ganglia containing an average of 400 nerve cells, which making this simple nervous system good subject for research. Nerve cells respond predominantly for the purpose of feeding, to the chemical messenger, Serotonin. While there are some superficial similarities in nervous systems, complex behaviors in mammals have less of a tendency to be a function of hard-wired inherited neuronal networks.

The contrasting strategy in nervous tissue design is to build in plasticity of behavioral response, not just a simple reflex. With such a design the animal need not await the passage of generations to be passed down as no permanent anatomical change is specified by genetic machinery. Indeed, if a change in the nervous system is reflected in anatomy at all, this is hard to demonstrate. Such a change is pretty subtle even under the microscope. Plasticity, whatever the mechanism for bringing it about, is an advance in neural strategy. Adaptation can take place rapidly. The down side is the response is much less reliable, so it is of less use where you need a specific response you can count on. Such a situation happens with learning. Only certain neural circuits have the capacity to learn and they are probably the minority. Even systems that learn need to have a lot of hard wired responses built in. In the most advanced animals learning alters not only a motor response, but may change a whole internal set. This change may not be very obvious even easily tested for. For example a person changes his point of view after reading a book. You can't see him change his mind unless you know to ask the right questions. Again, one does not wait through generations to change. Change is accomplished rapidly. The whole process of change is more efficient.

While simple stereotyped behavioral responses are easily correlated with neuroanatomic change, more subtle complex adaptations involved in learning are more difficult to correlate with anatomy. Given the current level of knowledge, such behavioral changes are difficult if not impossible to see anatomically or chemically though presumably these changes do take place. We have to depend on what we see from the macroscopic behavioral perspective. Simple behaviors in animals with few neurons have often been used as a model for human psychology but these models fall short of expectations. They do give some rough idea of what must take place in more complex neuronal systems. Still we suspect learned behavioral changes may someday be better correlated with anatomic findings, which are bound to be microscopic or chemical.

Quite a bit of debate has been generated on the subject of brain size, neural plasticity and intelligence. On the surface it would appear that that the total size of the nervous system should tell you the most about an animal's intelligence. Men have larger brains than women do, or certain races seem to have larger brains than others, and so are more intelligent, or so the argument goes. Theoretically, having more volume of neurons should allow for greater processing power. But consider that an animal such as the elephant may have a brain as large as 6000 grams in comparison to ours averaging about 1300 grams (about 2.9 pounds) or that of a porpoise whose brain weighs in at about 1600 grams or a blue whale's at 9 Kg!! Are these animals exceptionally bright? Perhaps they have deep imaginative powers that are not apparent to human observers.

All this must be tempered by the obvious fact noted by numerous observers that brain size does seem to correlate *loosely* with mental ability when body size is taken into account. In other words, the elephant's or blue whale's brain size when compared as a ratio of its body weight compares unfavorably with the primates or human brain. One measure is the so-called Jerrison's encephalization quotient, but the principle is to compare body size to brain weight and when that is done, the dolphin particularly, fares very well in relation to other primates though still below man[26]. Our intelligence resides in the large area of the cerebral cortex and the amount and number of brain cortical convolutions. The surface area of the brain across many species is proportionate to the brain weight. Still it would seem that the total computing power of the brain should be governed simply by the total mass of neuronal material.

The problem at hand is that except for some neat little tricks, the regard of a mother for its young, a not-too-large repertoire of vocal sounds and responses and some anecdotes about heroic behavior, dolphins do not seem to give us all that much evidence in their behavior of any kind of advanced intelligence even comparable to more evolved primates. But the dolphin brain's size is impressive, the amount and number of convolutions and the area of its cerebral cortex. The answer to this enigma can be found under the microscope. The human cerebral cortex is composed of six layers whose architecture determines anatomic regions called Brodmann areas. In the human cortex the cellular architecture is distinct so that numerous areas can be distinguished. These are broken loosely into more primitive archi and paleocortex and more evolved neocortex with six distinct layers. Primary sensory areas for vision, hearing and surface sensation tend to have large numbers of distinct granule cells in layer iV, while primary motor areas have large pyramidal cells in layer V. But the dolphin brain has few of these distinctive cells and regions of the cortex are not so well developed at all. Dolphin cortex, though much larger than primitive placental mammals, shares characteristics with more primitive land mammals forms from which they descended before returning from land to sea.

The human brain has on its side precision in its architecture with well-developed and differentiated sensory (granule) and motor (pyramidal) cells that were utilized more extensively later in mammalian evolution and with distinct specialized areas of cortex which developed later in land mammals. Early workers were much taken with the size of the dolphin brain which in itself proved to be a deception. Size isn't everything.

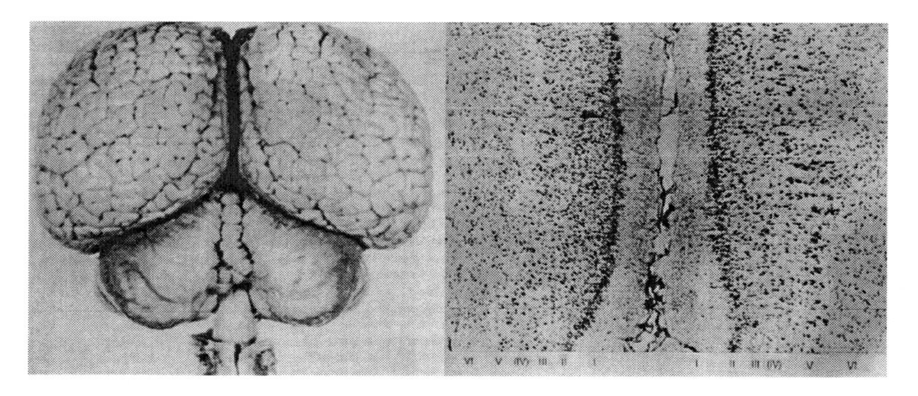


**Figure 8: The dolphin brain with its large complement of convoluted cerebral cortex above with microscopic cell layering below. Layers are indistinct[27]. [28].**

We've gone from the consideration of the nervous system as a reason for being in the beginning of this chapter to a concern for a biological imperative as a strategy for adaptation. I have talked about neuronal systems as a method to bring about a rapid motor response and as a vehicle for plasticity of response in a rapidly changing stimulus field. The nervous system is more than a reason for being in biology. It has evolved into an executive, controlling nearly every aspect of biological function in higher animals and all other organ systems are subservient. This is not obvious or even true when you look at more primitive living things which have some reason other than cognition for their being alive. In animals with advanced nervous systems other organ systems are more and more controlled by the nervous system. Indeed it can be said these other organ systems need not exist at all except to serve the brain, whose highest representative is the cerebral cortex.

## Neuron as Executive

As we see the nervous system as a whole in its executive function, a central command and control station, then we conceive of the individual neuron as a single miniature unit in this decision making process. The neuron or single nerve cell is a mini-executive, a single soldier in an army of neurons. Nerve cells have different rank, depending on how exactly they connect with other neurons, how many other cells they connect with, and their specific location. But the analogy goes only so far, since there is no one cell generalissimo in the brain, rather a complex feltwork of cells. Each cell is subject to a complex of inputs. In most cases the individual neuron makes a binary decision, that is, at a given moment it is either firing or not, similar to a series of 0's and 1's in a computer. The computer analogy is even more obvious but even more imperfect, as we shall see.

If the nervous system is a complex of individual cells each of which at any moment is in one of two states, excited or not, the brain is nothing more than a large scale information processor of complex binary digits (bits). Understanding nervous function, hence thought and feeling, comes from data about the either on or off state of each neuron, and a knowledge of how these neurons are wired together (their anatomy). The oft used highly seductive brain computer analogy has, until recently, been very compelling and has a number of attractions, not the least of which is the resemblance of a logic machine to its maker which is the brain. In this computer logic machine, the brain sees a reflection of itself. A computer may think and if it can its thought processes must be much like the men who created it and may even be used to model for human thought processes.

In the computer, you have the human brain creating a functional machine in its own image. Therefore brain scientists may learn a great deal from looking at computers and logic circuits. Many aspects of neural function can be understood from the vantage point of computer design. To mention just one of many examples, we tend to look at brain function today as a composite of function involving modules, circuit boards essentially. Consciousness, as we have seen, involves various groups of cells, each group responsible for a specific function. The ARAS awakens the brain, the cortex gives content to consciousness, the emotional or limbic centers color consciousness, the frontal lobes as motivation and will, and so forth. Take out a module, say the limbic system, and consciousness will be altered, perhaps the subject will be awake and aware, but will lack emotion, will be a Mr. Spock or Data. The more complex the brain and the specific function you are examining, the more useful the concept of modularity. A group of neurons is a circuit board or even a chip.

Computer engineers have learned a great deal from looking at the brain as well. They are constantly trying to emulate brain function and are hungry for knowledge about brain circuitry. Parallelism in brain function was not appreciated until fairly recently. Your ordinary computer has a single microprocessor. All information must flow through this single unit which functions at great speed, but still, every piece of data must flow through the microprocessor in sequence, one bit at a time. Logic is thus 100% sequential. The brain however, is capable of handling a lots pieces of information in parallel, through many pathways simultaneously. Each individual datum in the brain is processed by computational standards, fairly slowly. In order to walk we must simultaneously and in parallel process three dimensional visual data, input from vestibular balance centers, proprioceptive input from our peripheral nerves to mention just a few information sources, then compute a whole program of muscle activations and then vary all of this over time, all this just to walk. No wonder computers have not up until this point been able to turn out anything resembling a full human gait pattern. Sensory input in the brain and motor output, are "massively parallel". Innumerable microprocessors, that is, neurons and groups of neurons, are working in the brain, in parallel. In recent years, computer scientists, realizing the advantages of this major divergence in design, have built it into computers, which now boast vast arrays of microprocessors working in parallel.

Computers are growing more and more powerful, partly owing to designs that emulate brain function, but far outstripping human capacities in many areas especially data storage and retrieval and in calculating ability. It makes you wonder whether computers will one day be able to "think" and whether output of advanced computers, say computer speech, literature, or musical composition will always be as inferior as it is today from the output of talented humans. Do we expect computers to become high-speed initiators of perception, thought, feeling, and action, be conscious in other words? Or computers may even best people in some intellectual endeavors. Is it possible to build into a computer self-awareness, anxiety about death, an idea of personal boundaries and space, emotion, all the things that define awareness of biological Carbon based beings? Time will tell. Superficially computers, mechanical objects based on Silicon, resemble Carbon based biological machines. (Of course, we have created a cognitive tool in our own image!!)

As neurons exist in two states only, there is an electrical action potential or there is no action potential, so computer function can be reduced to a series of 0's and 1's that is translated on higher and higher abstract levels until with all the things computers do, word processing, mathematical reasoning, switching and other tasks this series of 0's and 1's is invisible. Brains and computers are both machines, so the argument goes, and the basic structure of each is not relevant to the similarities in their behavioral output. Carbon and Silicon are enough alike in any event and each machine, brain and computer utilizes arrays of binary elements that at any given moment in time are in one of just two states. The precise configuration of binary elements, invisible when you are looking at a specific task or output, determines content. This brain-computer analogy is at once very seductive and false, seductive because it works and is actually a good description to a limited extent.

First, the binary analogy is not strictly true for neurons. Though at any given moment a neuron is either firing or not, and hence may mathematically be assigned a '1' or a '0' to describe its state, this is nearly always an incomplete description of the neuron's state at a given time. The '1' for the firing neuron is straightforward enough but after firing almost all neurons will be unable to fire again for a certain length of time, which is the refractory period for that cell. The nerve cell has an absolute refractory period over which time it cannot be made to fire under any circumstances, and a relative refractory period when it is merely more difficult to excite the neuron. A given person, is only able to achieve an orgasm at a certain frequency. For a time you may try to excite him and get no response at all. Over a longer period of time, it is difficult to excite him though that can be done. The neural action potential is similar to an orgasm. There are the absolute and relative refractory periods. The refractory period for neurons is different for each cell.

Further, at a given moment, a neuron is at a certain state of excitation. A mathematical description that assigns a single value to a neuron dependent on whether it is firing is not completely describing that cell. Superficially neurons resemble Silicon based machines whose states may be represented as a series of "all or none", firing or not firing, '0's' and '1's', but mathematical models may not take into account that at any given moment a neuron is more or less easy to get into its firing state, it is more or less excitable. We shall see that excitability can partly or wholly described by the current state of depolarization of the neuron. The neuron responds by adding up in some way all of its inputs, typically synapses, excitatory and inhibitory, which can run into the thousands for an individual neuron. If it has only just fired, it can't fire again. In addition, there are global factors that modify neuronal excitability including among many things, fatigue, hormone and drug effects, availability of energy, and Sleep/wake State, appetite, and satiety. For example, an animal that hasn't eaten in a long time is hungry and we may say accurately though in a simple-minded fashion, that neurons in a putative hunger center are poised for action.

Not all neurons are locked into a binary paradigm anyway. Some make a graded electrical response. Some of the light receptor cells within the retina of the eye function in this way, but also many other nerve cells particularly sensory cells. Nerve cells store data in a number of ways. Neurons that we know about are affected by electric charge in individual cells, but another possibility is to change in electrochemical relationships with neighboring cells. There is good evidence that changes within the nervous system are induced internally not only via the acquisition of new information (learning or change in software) expressed in changes in electric and chemical signals, but also in changes in structure. Nerve cells and synaptic connections between them form and break down over time. There are also permanent changes within nerve cells that occur with environmental change and in learning. For example, it is thought that learning may involve synthesis of intracellular chemicals such as RNA. Unlike the array of binary electrical elements within a computer, neurons are influenced by supporting cells especially changes in other numerous cells called glia (for "glue", the cells that help bind neurons together and perform myriad other functions).

## EVEN ON A CHESSBOARD:

Nothing illustrates the differences between the human brain and computer as well as the game of chess. It should be emphasized that for most fields of endeavor the computer, even for modern machines, is so much more primitive than the brain there is no means for comparison. But chess is an artificial situation in which moves are confined to an eight by eight square board and pieces having restricted and well-defined geometric moves in only 2 dimensions, a task perfect, or so it would seem, to pit a calculating machine against a human brain. In real life we work in three dimensions and consider many divergent inputs at once, but even in the limited field of chess, profound differences can be seen between the workings of a conscious machine which is the human brain and a computer.

You don't have to know much about the game of chess to appreciate a difference but it helps. Chess is a game of strategy that involves analyzing one's position and a specific sequence of moves that needs to be accomplished to maximize future positions. It is meant to simulate a battlefield except that it is, eminently sequential. A real general needs to manage a number of changes occurring at the same time, Not so on a chessboard where each move occurs in sequence, hence chess should theoretically be a again a perfect task for a computer since moves are, by nature, sequential. The chessboard has become the standard field to pit brain power against computer power. It's great too because it shows fundamental differences in action between brain and machine.

A human plays the game by developing a strategy and pursuing it. The human chessplayer is a schemer capable of acting on a number of hunches at once. He looks at his position on the board, sees an opportunity and makes a plan to accomplish a perceived goal. In short, he strategizes. He may very well know his opponent and his weaknesses but the most important element is that a person keeps a specific goal in mind and makes plans on how to attain it. He may covet control of a specific space or region, say at the center or the board or may need to capture a certain piece. However a computer is a machine and so has no goals or plans. For a computer chess is a particular situation translated into numbers and values. To plan its next move it will have to evaluate its position and have some means of comparing the relative advantages and disadvantages or outcomes of all of the possibilities for subsequent moves. Each of the possibilities is assigned a relative value which and these numbers are compared, determining the next move, so that a souped up calculation is made to simulate human behavior. The computer is a calculating machine that compares relative values nothing more, but has to look like it is "deciding" on a move, which it isn't, what it is really doing is picking an alternative based on a numerical comparison of an outcome. The computer looks at the possibilities for a next move, comparing, depending on its size and power, what may happen one, two, three or more moves into the future. The bigger the computer the more calculations it can do in one second, the more moves into the future it can compare and it can look at an enormous number of alternatives which no human can do.. The bigger the computer, the faster its microprocessors, the more microprocessors there are, the more calculations it can do, the more moves it may compare, and the farther it can look into the future, the better human chess player it can beat. You can build a huge computation machine that will have all the advantage against a human player. Put in a load of microcircuits and perfect the software with appropriate input from the best players. The idea of using a big machine with rapid as possible calculation is called brute force and as things now stand it is the major method by which calculating computing machines compete against humans. And the machine won't make any stupid human errors either.

The interesting thing is that while computers can easily defeat human chess players of average ability, at this point they are not better than the best human players, the grand masters. In February of 1996 there was a well-publicized tournament between “Deep Blue” an IBM machine and the grand-master Garry Kasparov. Kasparov lost the first game, but he later won the tournament not through his incredible calculating ability, the machine was far faster than he was, but by strategizing, perceiving the weaknesses of his computer opponent, learning about the methods used and where they fell through. It is significant that he defeated the machine only later on and not in the first game that he had to learn about the weaknesses of play, that is that there was no conscious strategy at all for the machine only numerical comparisons, hence no design, no goal in mind and method to attain it. This concept is akin to motor planning, scheming and goal directed behavior which behaviors in neurology are attributed to the frontal lobes of the brain, one of the major factors that makes us human. You can't talk about a computer having tactics, or plans, not at least at this stage of the game, even on the limited eight by eight field of chess let alone in real life.

The final denouement is that in May1997 Kasparov was narrowly defeated by Deep Blue after each side won a single game (there were also three draws). This is a very close record. Some say that Kasparov at one point lost his nerve and conceded in one game before he was sure he’d lose. The computer had no nerve at all. In fact the major advantage the machine has, is the lack of psychological weakness and doubt that we are all subject to. The computer doesn't make silly mistakes and isn't influenced by sickness, or mood. It will capitalize relentlessly on the mistakes of its human opponent.

Deep blue ended up winning its 6 game rematch with Kasparov, not by much, with a score of 2.5 for Kasparov, vs. 3.5 for the machine. There were three draws with Kasparov winning the first game of the set, the computer the second and sixth. The design of the IBM computer, the RS/6000 was 'massively parallel' meaning there was an array of microprocessors capable of analyzing bits of information simultaneously, not only sequentially as would have to occur if there were a single processor. Another thing the computer had in its favor was brute force, deploying tremendous computational power. The RS/6000 IBM machine could examine 200 million movers per second, which is a little faster than Kasparov's brain. Moreover the machine was able to make up for its lack of strategy because built into its design were algorithms specifically made to defeat Kasparov himself, to capitalize on his weakness. Could the same machine defeat any other chess grand master?

The human vs. the computer in chess brings up the same arguments as were raised in man's competition with bacteria and insects. Does the human brain which can learn and perform acts by volition, design and the deliberate strategy, have the advantage over a machine capable of examining millions of possibilities but without an aim or a goal? Bacteria and insects can mutate many orders of magnitude faster than man, exploring myriad possibilities, though quite stupidly, along the way. Man doesn't muddle as much. He designs, but at the end, one wonders if the action by design is any better than the brute force exploration of all the myriad possibilities. Which method is ultimately more adaptive, strategy, design, intelligence or brute force? Will man win against arch biorivals bacteria and insects? Or you can even generalize and ask whether our world or the cosmos developed merely from the exploration of myriad possibilities or is everything the culmination of some design. The jury is still out.

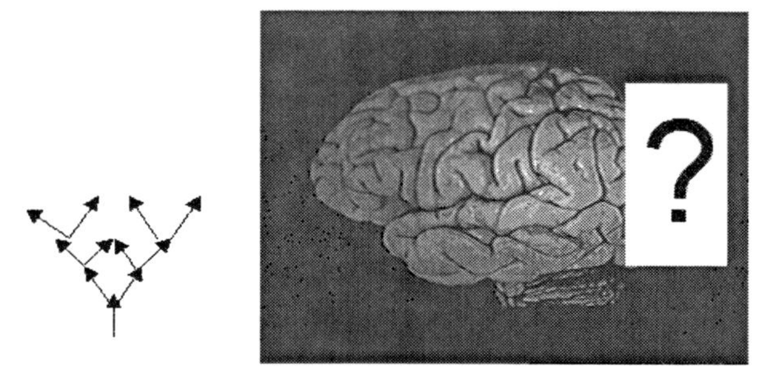


**Figure 9: Which will win out, the brute force examination of myriad possibilities, or the grand design?**

In a sense the computer is a high-speed exploiter of possibilities though its process is aimless. A computer has no goal of its own, unless a human gives it a goal, no strategy. The computer's inability to function as competently as the human brain in less structured areas that is as its own expert has been a major disappointment. Artificial intelligence has been discussed for a very long time, and for a long time seemed within reach. There were high hopes within the field just a few years ago. Now the term is anathema as the fondest hopes and dreams have failed to become to fruition. Originally computers promised to replace humans in vast areas of intellectual endeavor. Developers of systems of artificial intelligence would, it was thought, create human reason machines that would replace highly skilled people in various fields. In medicine computers would replace experienced clinicians, and make unerring diagnoses. At the very least they would have access to huge banks of data and would not be subject to human error. Machines could take the drudgery out of diagnosis and make it more precise. Engineers, architects, and attorneys would become similarly obsolete.

As early as 1963 Joseph Weizenbaum at M.I.T., introduced a program called Eliza, designed to simulate questions of a psychiatrist. Eliza was not about to replace your analyst but was tongue in cheek, more of an intellectual sleight of hand. The program merely re-questioned the human subject utilizing key verbs and nouns culled from his previous response. "How are you feeling today."; "I'm feeling down."; "Can you tell me more about why you're feeling down today." This program had been able to fool human subjects who did not recognize that there was no cognition driving the programs responses, only the return of phrases in the form of questions. The quest for computerized simulation of human thought processes has turned out to be a giant disappointment, as early workers in the field made a series of promises that could not be kept. Major strategies involved what are known as rule based systems. The idea was to compile a set of rules used in reasoning within certain fields. Next you mix in a large data base to utilize these rules (computers are particularly excellent at storing and retrieving huge quanta of information) and create an algorithm or set procedure for following these rules to recreate the mind of an expert. In medicine a matching system may attach various numerical weights to symptoms and physical findings using programmed rules to determine a diagnosis. One can even alter the numerical weights assigned to these rules in order first to reproduce the competence of human experts and later to outperform them adjusting these values according to the success the program experiences when confronted with certain specific situations. Modifying the numerical value assignments of various characteristics is but one way a program can be made to "learn" or alter itself in order to be made to perform as well as or better than an expert.

However, while they have been useful as educational tools such systems have mostly failed to replace expert opinion. One famous example called "Mycin" attempted to diagnose bacterial infections on the basis of data presented. Unbeknownst to the creators of such systems, human experts don't usually follow a given set of rules. We may teach our medical students to follow rules, and there is a certain basis of factual knowledge that is necessary if we are to perform an expert function. But there is also a stage if a human is to function on anything but the most rudimentary level, where rules become expendable and are no longer followed by practitioners. Experts function differently than beginners in that they depart from the program when called to do so. A student masters a subject by learning rules only to discover later that his older colleagues function at an even higher level by not following these rules precisely. Indeed, the more advanced practitioner may have cast away many rules that he depended on to learn the ropes at an earlier time in his career.

I often find medical students and residents baffled by a therapeutic course set by an experienced staff member. The experienced practitioner may not follow the rules in any precise manner but does better by the patient. In teaching examination of the patient we always have students go through a certain sequence which they initially follow closely so as not to "miss" any specific part. However all experienced clinicians find themselves looking at many aspects of the patients simultaneously, and if we're worth our salt, honing in on the particular problem(s) confronting us. In this way one misses very little, but an awful lot of irrelevant information is relegated to the scrap heap and may not be worth mentioning. You need to know what is and what is not important. In outlining for the novice precise rules utilized in recognizing disease, certain findings invariably mean more while other critical elements while usually noted, are not fully appreciated. In many instances an experienced person can make a diagnosis almost instantly without resorting to rules at all. A certain pattern of speech may make to diagnosis of amyotrophic lateral sclerosis almost unmistakable. Similarly an experienced mechanic knows before he opens an engine that a certain tapping sound comes from a loose valve cover. This is the facility of recognition which computers are not as competent at as humans. Computers follow rules exquisitely well, much better than medical students, but recognize faces and sound patterns not as well. Recognition is just one of the cognitive facilities that humans have and can call upon at any time, that computers first have to be designed to have. Then computers or any similar cognitive device would of course be expected to call upon that facility on an as needed basis as humans do, a tall order.

Prosopagnosia is a curious brain disorder that says a lot about how the brain works. In prosopagnosia there is an inability to recognize faces, those of acquaintances, relatives, even oneself, a simple function we take for granted. Recognition is accomplished holistically, not through analysis of individual elements in a picture. You don't recognize a friend by noting that his eyes are spaced a certain number of inches apart, that he has brown hair or a certain shaped mouth, you simply see his face and know it. Prosopagnosia can affect the ability to distinguish objects within a general class such as a farmer recognizing each individual cow, or picking out one's car in a parking lot. Such objects are not distinguished on the 'conscious' level by analyzing individual characteristics. However, a victim of prosopagnosia has to depend on a conscious search among distinguishing characteristics, such as the letters on the license plate of his car or the mailbox address of this house, or even nonvisual cues. This disorder celebrated in the book, The *Man Who Mistook His Wife For A Hat* and in a certain opera by the same name, illustrates certain principles of brain function. Firstly recognition is immediate and is not performed analytically, that is as a sequential feature-by- feature task. An entire object is not broken into its components. Secondly, a simple brain lesion, in this case a disconnection between the visual or occipital area of the brain and the recognition areas in the temporal lobes on both sides of the brain, frayed wiring, if you will, may interrupt this recognition process. For over a century there has been a debate among brain scientists between those who sought to localize particular cerebral functions in precise brain regions (the phrenologist's approach -Phrenologists attached great significance to the study of bumps on the head) and those who took a more holistic approach.

Paradigmatic among disorders in which localization of function is significant are the aphasias. These are disorders of language function localized to the dominant (usually the left) hemisphere. After long years of examining patients with localized brain lesions, strokes, head injuries, tumors, abscesses etc., neurologists discovered that disorders of language function occur when an area on the left side of the brain has been affected. Moreover they found out that destructive lesions in the frontal lobe cause weakness or paralysis, while a lesion in the parietal lobe causes numbness or a problem with sensation. Because we are built in such a way that nerve fibers cross to the opposite side of the body, destruction of the left side of the brain causes paralysis or numbness on the right or opposite side of the body. Aphasias, dysfunctions of language, follow this same general scheme. Destruction of the left posterior frontal lobe will cause a problem with movement on the right side of the body and also trouble with the motor aspect of making speech and with writing i.e. an expressive aphasia. A lesion of the parietal lobe and the temporal lobe which lie behind the frontal lobe, will cause a problem both with sensation on the right body, sometimes even visual loss on the right side, and also a problem receiving understanding and interpreting speech and written language. This can now be appreciated in the living patient using various brain scans and electrophysiologic techniques that weren't available when these discoveries were initially made. The recognition and classification of Aphasias provides one of the most reliable localizing techniques in clinical neurology, lending strong support to those who maintain that there are particular functions that reside in certain brain regions. Among the persons who believed in such localization was Paul Broca. Sigmund Freud was a no less distinguished spokesman for those who maintained a more holistic approach. But the argument for precise localization in the brain is not as simple as the mechanists would have us believe. After localized destruction occurs, other areas begin to assume a good part of the affected function. Part of the recovery process seen in patients is a goal directed. They try to recapture lost function by whatever means become available. Various supplementary areas not ordinarily used for specific functions, begin to take over function of damaged areas not only in the opposite hemisphere, as was once thought, but also on the same side as the neurological event. That brain areas initially uninvolved in a certain activity can take over this function in a pinch has profound meaning. Brain functions are controlled locally with certain brain areas destined to perform specific functions. On the other hand, most brain areas are pluri-potent, while in the natural state performing a fixed function, are able to assume other functions if necessary. This is partly how the brain repairs itself to preserve the organism. This is also not something generally observed in machines even Silicon-based machines. You don't find a hard disk taking over the function of the CPU or the printer doubling as a random access memory device. But the brain is different. In the repair process and in order to preserve function, we see such seemingly disparate areas of brain suddenly becoming active and assuming function. In stroke patients who are trying to perform a function impaired by their stroke, say trying to moved a paralyzed right arm, you can watch as far-flung brain regions are recruited to accomplish a task. One method is to use the PET scan, which looks at localized glucose utilization that in the intact person are not used to perform the task. Whereas in an undamaged person you may have seen the motor strip of the left frontal lobe become active, in a stroke patient whose motor strip is non-functional you see other areas become involved. The supplementary motor area on both sides of the brain and even the cerebellum pitch in with their effort. These areas are all recruited in order to accomplish a certain task which is second nature to an intact person but accomplished with greater effort when impaired. Now other areas of the brain have to become involved presumably as this damaged individual works harder to accomplish even a simple task. The brain may be the only machine that is able to jury rig itself in a pinch. It uses what is available. The impetus may be strong motivation to accomplish something. Motivation is not something you can observe in man made machines.

You can also see this when a less adept person who is not damaged performs as task that is more difficult for him. For example girls for certain math tasks, boys for language tasks. In order to perform the same function more widespread areas of the brain need to be called in any time a task is more difficult. A minor difference of opinion may be settled adeptly by your diplomats. If your leaders are just a little less competent you may have to call out the army.

Such considerations have now become a basic feature of rehabilitation for example after stroke and head injury. The brain is at the same moment, *localized* and non-localized, holisitic. These contrary elements, localization vs. holism, in consideration of how the brain works are as basic to neurobiology as particle vs. Wave models are to physics. They are different aspects of the same phenomenon. The brain is a whole structure composed of modules or elements. Depending on how you ask a question you will see one or either side of its nature.

The brain is centered about performing a certain function. Whether it be writing, or reading or moving an arm, throwing a ball, it will work until the job is done. In order to accomplish a task we may have to recruit brain areas not ordinarily used to perform a given function. It's the same when you break a leg, the leg is casted and you try to walk. You will recruit the opposite leg, your arms on crutches but you know you have to walk and get from place to place and you do it. A machine is different. It's designed to accomplish a task and an algorithm or sequence of moves is incorporated into its design, in order to accomplish the task. If one part of the sequence fails the work will not get done. There is no goal direction, only a series of instructions.

The patient's frustration at his lack of function seems to speed his recovery. The human patient is goal directed something we don't appreciate in a Silicon based machine. Because the brain is plastic, it recovers function even after cell death. An inanimate machine designed for a deliberate purpose different than an organism that develops through the biological process of evolution, trial and error, in a real environment. The designed machine suffers from the same shortcomings as a planned economy in a communist country as opposed to a capitalist or natural unplanned economy. Economic planning does not work as well as a naturally derived economy. The latter, is primarily goal and task rather then design directed and is bound to be more plastic. We continuously discover how much we tend to underestimate human and animal plasticity. Plasticity is part of what defines the brain as a living tissue. Reading and letter recognition, which comes easily for the human brain, is very difficult to design into a machine. A program may possibly recognize a precise written figure on a specified background as long as parameters of shape and size are specified with mathematical precision. Even then there are problems in recognizing these patterns in a slightly different orientation or in different form for example, the same letter in a different person's script or in a different slant or orientation. The computer's "perceptions" work again via analysis, a non-holistic approach. Images are mostly analyzed into tiny boxes (pixels or picture elements) hundreds or thousands of these making up a final form. The position of each box is mathematically described. The brain performs recognition functions easily because it does not work through analysis but rather performs its function more holistically. Computers bear little resemblance to human brains. Computers perform logical processes sequentially their responses being hard wired and determined. By contrast the human mind usually functions in a non-sequential manner. Thus a computer with one loose connection will probably cease to function. The brain loses components, nerve cells and other elements almost continuously yet still keeps on all the while improving learning and increasing function, primarily because it does not depend on sequential but instead mostly parallel and overlapping processes. As a living organ it is constantly working and repairing itself. a malfunctioning or sick brain most of the time doesn't lose its oneness, personality and basic method of coping. A human with prosopagnosia or aphasia is still the same human.

## Comparing Brain to Computer

Computer scientists, realizing how the brain works as a parallel machine, have tried to emulate brain function with their electronic silicon based machines. They have realized the advantages of designing machines that reflect biological methods. This means using a strategy of parallel instead of sequential processing. In a computer everything must be processed by a centralized processing unit (termed the CPU) which is a microchip. All operations need to wait their turn and go through this tight bottleneck one step at a time, which is why machines emphasize the speed of this microprocessor (for example the Intel 486 or Pentium Chip.) Experts have designed parallel distributive processors (used even for the case of the eminently sequential game of chess, by the way). One of the tasks they have set out to mimic is recognition. In massively parallel machines with overlapping function the loss of a few elements does not shut down the function of the whole machine. For a serial processor the likes of which are our own home computer, the loss of any particular element, especially the central processing unit, would end everything. Newer machines employ parallel arrays of CPU's instead of just one the CPU being Silicon based analog of the individual neuron. It is common knowledge that as we age, we lose tens of thousands of neurons daily, yet our performance in certain tasks, especially in our 20's, 30's and 40's actually improves as learning takes place and we see perhaps what even amounts to an increase in synaptic connections. Sooner or later the loss of neuron's effect, overtakes the offsetting process of learning and we see the ravages of aging on cognitive function. Only in recent years has this offsetting effect of learning on aging been fully recognized. It is one potent method to delay aging effects.

| | COMPUTER (MACHINE) | BRAIN |
|---|---|---|
| **PROCESSING** | SEQUENTIAL | PARALLEL |
| **ATTACK** | ALGORITHM | STRATEGY |
| **PRODUCTION** | BUILT | DEVELOPS |
| **ACTION** | FOLLOWS OR-DERS | INITIATES ACTION |
| **BEHAVIOR** | DETERMINED | FREE WILL? |
| **MATERIAL** | INORGANIC (**SILICON**) | ORGANIC (**CARBON**) |
| **REPAIR** | OTHER | SELF (HEALING) |
| **DESIGN** | ENGINEERED | EVOLVED, GROWN |

**Table 3: How machines and brains differ. With efforts to make computers match human attributes, these distinctions blur.**

Obviously the brain is functioning constantly as a correlator and user of input from numerous simultaneous sources. A football player is waiting for a pass from the quarterback, but he also has to keep alert for players of the opposing team who threaten to tackle him if a successful completion occurs. This information compared with an internal program consistent with plans for a play accentuated by practice. Then he has to position himself to make a run for the goal post. His vestibular system and cerebellum need to keep him upright and moving, but most importantly, he has to somehow estimate an optimal position for his body, and hands in order to successfully make a catch. Just some inputs include the timing and angle of the quarterback's release, estimated velocity, his own velocity and direction, all this computed instantaneously, and well beyond the capacity of any mechanical contrivance. In plain words the brain acts as an executive, receiving input from disparate sources and putting them together in order to accomplish certain goals. It must process all of this multimodal internal visual auditory input and then to issue orders for a play and run to the goal post.

The brain is an associative instrument, correlating and processing in parallel input from disparate sources. In regard to visual operations alone: "Considering the processing that takes place with visual input, sequential processing would not be possible. If it takes 500 milliseconds for a person to respond to a visual recognition test then there must be no more than 100 synaptic steps between the input and the output. Accordingly, a hypothesis that envisions a serial processing unit for visual recognition with 300 to 1000 steps cannot be right. This observation is usually followed by the inference that the brain, unlike a standard electronic computing device, is a massively parallel machine The point is 100 steps in a serial processing program is far too few to do anything very fancy."[29]*

## Marriage of Carbon and Silicon

Computer scientists are not only using biological modeling for new parallel processors. They may incorporate carbon-based molecules in computer switching devices. Switches in the form of semiconductors are the heart of computers in semiconductors and storage devices. The state at a given moment is simply the sum of on-off states in storage devices made of semiconductors and information is stored on magnetic and optical media also as a series of 0's and 1's or otherwise put, "on" and "off" states. The Holy Grail of computer technology is to find complex switches having a few basic characteristics. 1: *speed* : A device needs to switch, in other words change from the "on" to the "off" state at incredible speeds in the modern computer somewhere in the range of billionths or trillionths of seconds$^{\tau}$ . 2: *Stability or* reliability are critical in order to guarantee the states of the switch do not change unless we purposefully change them and the switches need to be durable enough to survive ordinary environmental hazards with no breakdowns. The switches must accurately record changes we make in them and preserve those changes until purposefully altered. 3: *High storage density* is critical. Huge numbers of such switches need to be placed in a very small volume in order to make storage devices and processors useful for desktop and notebook computers. Storage devices must be small. Hopefully a light or laser will be able to alter the state of the device, in other words, to *write* a series of "on" or "off" state changes into memory also to be able to *read* written changes on memory devices. In order to accomplish these goals, researchers seek to incorporate biological molecules into silicon computer devices. Molecules such as Rotaxanes which are unusual large molecules whose structure may be altered by beams of light, and Rhodopsins molecules that are multiply altered by beams of light used in the eye and by organisms to store energy. Silicon devices may be impregnated with biological molecules and beams of light used to "write" to these molecules, in other words to alter their state. Some of these molecules change variably, depending on the color of laser light shown on their surface and seem to be very stable, holding an alteration of their state until changed by other beams of light. These molecules may then be placed in a three dimensional structure appropriately addressed for location so that a storage device is produced.[30]

In the field of *Optimization,* biological strategies of genetics and evolution apply to design of computer software. You may wish to create a model strategy to improve financial return. To do so you write a formula. This formula achieves maximal financial gain under current financial conditions. You can write a formula but you have no idea how it will do in the real world, a financial "habitat".

Why not take a hint from biology and use the methods of survival of the fittest. Hold onto the financially lucrative parts of the formula and jettison the weakest concepts in a real financial milieu instead of trying to fly by the seat of your pants and create a theory that may or may not benefit you in the real world. This formula, is the most *fit,* that is, it achieves the greatest financial return. A formula is either more or less fit than other formulas, which means that it either is a better or worse strategy for survival within a specific financial milieu. In biology we mean by *fitness* the ability to pass down the largest number of viable offspring carrying our own genes, the ability to pass down one's own genetic endowment. But this is a useful topic computationally as well and here is why. Let us suppose the financial environment changes, stocks no longer are a good investment because of inflation or something of the like, then that was once the most adaptive fit formula now no longer is and another formula, will give us a better financial return. Chances are the second fitter formula shares a lot of the characteristics with the first formula, that it is related to the first formula almost genetically. These optimization or fitness formulas have similar characteristics. They are quasi-biological entities within a financial world and may be interpreted genetically. As the financial environment changes these formulas for optimization of return need to *adapt.* In order to do so they will have to change slightly, to mutate. Or, possibly some of the inherent structure of this optimization formula will be borrowed from another formula in a form or *translocation* in much the same way as genetic material from one chromosome moves to another. In the financial as well as the real world there is the survival of the fittest. Many characteristics of these formulas may be borrowed, passed down and recombined in just the same way that genetic characters are. Financial models may borrow from biology and vice versa. The biological method thus turns into a new means to seek truth and to pass it down. More than this it gives us a design plan that adapts to changing habitats.

What we are experiencing is the intrusion of biological concepts into computation and computational models into biology. This is beside the point of attempts of computer scientists to achieve a level of parallel processing that is commonplace in the brain. Though the brain is a Carbon based organic structure and computers are Silicon based, change is inscribed in both devices in a similar manner, electrically through transfers of charges and alterations of chemical molecules. It is reasonable to expect that combination devices will be employed in computers incorporating carbon based biological molecules but also even more tantalizingly, silicon based devices may one day be implanted within the human brain. These devices may aid in functions that are deficient in most humans such as computational and analytical skills. If so human characteristics would be altered for good and there is the very real probability that the genetics of an individual, which is the Carbon-based living part of a person may not be the only characteristics that need to be preserved in future generations of progeny. Such science fiction movies as "Total Recall" and "Johnny Mnemonic" have already begun to incorporate such concepts as Silicon based structures placed inside the cranium.[31]

If such fanciful mergings of living organisms and machines never come to be, carbon and Silicon are bound together anyway. The computer revolution is unfolding right before our eyes and needs no amplification within these pages. The computer will magnify human cognitive capacities in much the same way as the invention of writing or the printing press. These inventions allowed us to record our thoughts and inventions to develop and communicate a collective consciousness, to build upon the past and work on complex concepts one painstaking part at a time. Extemporaneous thoughts and music are primitive by comparison to recorded words, plans and music. You might store in your own mind some sort of vague notion of the shape of an airplane. But write these plans down, find a way to experiment and manipulate these plans, and work on them part by part with the input of experts in various fields necessary and you will design a real flying machine. In music compare the primitive percussion of tribal music to the opera or the symphony.

Today a single human brain can be connected to information in any corner of the world. In his head is a certain picture an organization of his world, but inside his personal computer, is an alter-organization, a different world view which is also his. With a minimum of effort and skill, an ordinary human brain can be connected to the total knowledge of the rest of the globe[32]. The major impact is expansion of consciousness. We use an appliance outside the skull, not a piece of a biological organism inserted inside a computer, or of a Silicon instrument inside the skull, but some form of intimate contact between brain and machine with any of a variety of communication devices that may range from the traditional keyboard or mouse or touch screen or other pointing device to some kind of a virtual reality instrument. This allows a person to point with his eyes, for example, or other body parts using thousands of tiny points within a virtual reality suit called 'tactiles' (as opposed to visual pixels). The purpose of this information appliance is to extend the abilities of the brain. At each contact the person would focus his attention or consciousness on the contents in the device.

The concept of computers seen in science fiction novel is and on television is that of an advanced logical processor. Responses can be predicted as a function of hardwiring and software and computer logic is entirely deductive. By contrast, humans are more capable of inductive reasoning, able to recognize patterns. Humans can intuit and go from one topic to another fasciley considering a number of aspects of a problem simultaneously rather than being tied to a sequential method. Reasoning is frequently done through analogies or may even hang from a thin thread of similarities or symbolism as often happens in dreams, myths and stories. Pure logic is only mode of mental operations. Mechanists fail to see the entire spectrum of human mental operations.

The effects of such nonlogical parallel mental processes sometimes surprise me as in a delirious or schizophrenic patient moved by thoughts that appear to be irrelevant or contradictory. Yet these abberencies show how thought is driven by a different engine than machine logic. Human thought may also be saltatory or jumping after the method by which electrical impulses are most speedily conducted in nerves. Long nerve cell extensions, the axons, are covered by myelin a fatty electrical insulator. Electrical charges cannot cross this insulated barrier. But at certain intervals over the axon, the myelin insulation is interrupted by discontinuities, the nodes of Ranvier. At these nodes electrical charges cross and collect. Changes in electrical currents are conveyed down the long axon by a process of jumping termed saltatory conduction. This method for the spread of currents is extremely efficient and fast. It is similar to the most creative human thoughts which don't rely on a continuous logical process but instead occur through discontinuities and the buildup of disparate motivational and informational factors just as charge builds up on an axon membrane, which can make a seemingly revolutionary thought almost inevitable. This phenomenon has been noted repeatedly and goes by many different names depending on the field of endeavor. In psychology and religion much is made of thought processes that are essentially foreign to a computer, designated as "aha" experiences or revelations, and are considered to be bursts of understanding.

The brain is an initiator of thought and feeling processes while a computer, at best, can bring an already initiated logical process to a successful conclusion. Even where it seems to be creating, for example, in providing the first move in a computer game, it is merely choosing from predetermined alternatives. If we know everything about computer hardware and software and we can define the stimulus, then we can always predict a response. Some neurobiologists believe the same about the brain. They look at the brain rather rigidly as a hard-wired complex of conduction pathways and circuits. Upon this is superimposed human experience, perhaps learned patterns of response analogous to computer software. Know everything there is to know about how the brain is built and functions and also its experiences and you will always be able to predict accurately, its response. There is a certain smugness about this mechanistic all-knowing approach, also a certain amount of backward reasoning, an assumption that computer circuits simulate the brain's function. It's much easier than admitting that human thought and action is not determined or at least that we have inadequate data to have an opinion about whether or not it's determined. To the biological mechanists human thought, feeling, and action would be one hundred per cent predictable, if only we knew more. Our inadequacy in prediction comes purely from a lack of knowledge. The observed randomness of behavior is only an illusion or false perception following from our ignorance.

Early in the twentieth century when physiologists had finally described the simplest of all nervous system responses, the deep tendon reflex, they naturally became infatuated with the idea, and an artificially inflated notion about the level of their own understanding. The brain and nervous systems responses, they reasoned, must function only as a complex of simple reflexes. If we knew everything about all reflexes then all of the brain's responses could be understood. As it turns out, higher nervous centers serve mostly to dampen the stretch reflex. The higher order neurons of the pyramidal tract synapse directly with spinal cord motoneurons, not only to convey commands from the brain, but reduce muscle tightness or tone. Others are connected to control of muscles changing the stretch receptors themselves (termed the gamma efferent system). This brings up a basic principle of nervous system function. Higher order neurons that control simpler lower order circuits do so mostly through modulation of the hardwired lower order reflex response, in other word through inhibition. These physiologic understandings are expressed clinically in various neurological conditions. When the upper part of the spinal cord is interrupted, the stretch reflex in the lower part of the cord controlling the legs, is liberated from inhibition of the brain. The stretch reflex in the legs functions with impunity. The afflicted individual has a very active deep tendon reflexes and very increased muscle tone, termed spasticity. Many people think they are healthy if their reflexes are active. Actually, the opposite is true. Muscles become very tight even when moved passively by an examiner suddenly giving way in a clasp-knife manner through their excursion around a joint. The spastic will experience at best a very tight scissors gait and at worst, even at rest his legs will tighten or bounce continuously as his uninhibited stretch reflexes express themselves.

Drugs may be used to help such patients decrease muscle tone. These may inhibit motoneurons or affect gamma efferents or sometimes may impair the mechanism of muscle contraction itself. Neural circuits are deceptively simple and predictable only when studied in isolation.

Hard-wiring neural circuits are in higher animals and even man means a predictable series of electrical responses. We see this in a record of tiny electrical potentials in the auditory pathway of the brainstem. In a test called the Brainstem Auditory Evoked Response an auditory stimulus of short duration (a click) will reliably produce a series of 5 electrical bumps or waves as the nervous system responds by conveying the message from the lower brainstem to higher centers. This series of waves will always occur, each bump representing a way station or synapse that corresponds to an anatomical point in the auditory pathway (Figure). This is most useful as a test in medicine. If a bump is absent or delayed this is a sign of a problem in the specific area corresponding to the wave. After the impulses travel through this well-established pathway, electrical responses become much less predictable. This corresponds to the role of higher nervous centers (the cortex) where sounds registers in consciousness. This is a pivotal point that should not be lost. Another thing about nervous function at higher and higher levels is that responses are less stereotyped and predictable. The higher one gets within the nervous system, the less predictable the response. As a sensory impulse travels through the nervous system over the first few synapses the pathway is determined. However when you get to the cortex the electrical response is widely distributed and cannot be predicted at all. More primitive organisms have only this stereotypical nervous response. It is a sign of more advanced nervous function that you stop being able to predict a response one it achieves a certain level. It seems to me this is a general comment on higher and lower function in general.

Computer scientists are becoming less naive about nervous function as they discover that machines just cannot duplicate nervous function or reason even at a child's level. There has been some appreciation of the function of networks of neurons and function of such units. Each individual nerve cell is literally connected to thousands of others. A neuron's output may directly connect with hundreds of others through axonal branching and dendritic spines provide a much more complex array of inputs. The decision whether or not to fire may be influenced by many thousands of nearly simultaneous excitatory and inhibitory inputs from other neurons. In many instances a single neuron's output may return to act as a feedback mechanism inhibiting further firing. Scientists are working on ways to monitor the output of arrays of neurons rather than single cells. Cultured single layers of neurons have been monitored using arrays of tiny electrodes. With this apparatus it is possible to "listen" to the integrated output of groups of neurons. Some information has already been obtained showing patterned firing in groups of cells in these primitive cultured networks. Though this does not reproduce by any stretch of the imagination, true brain function, it is an attempt to get at the sociology of neuronal response.

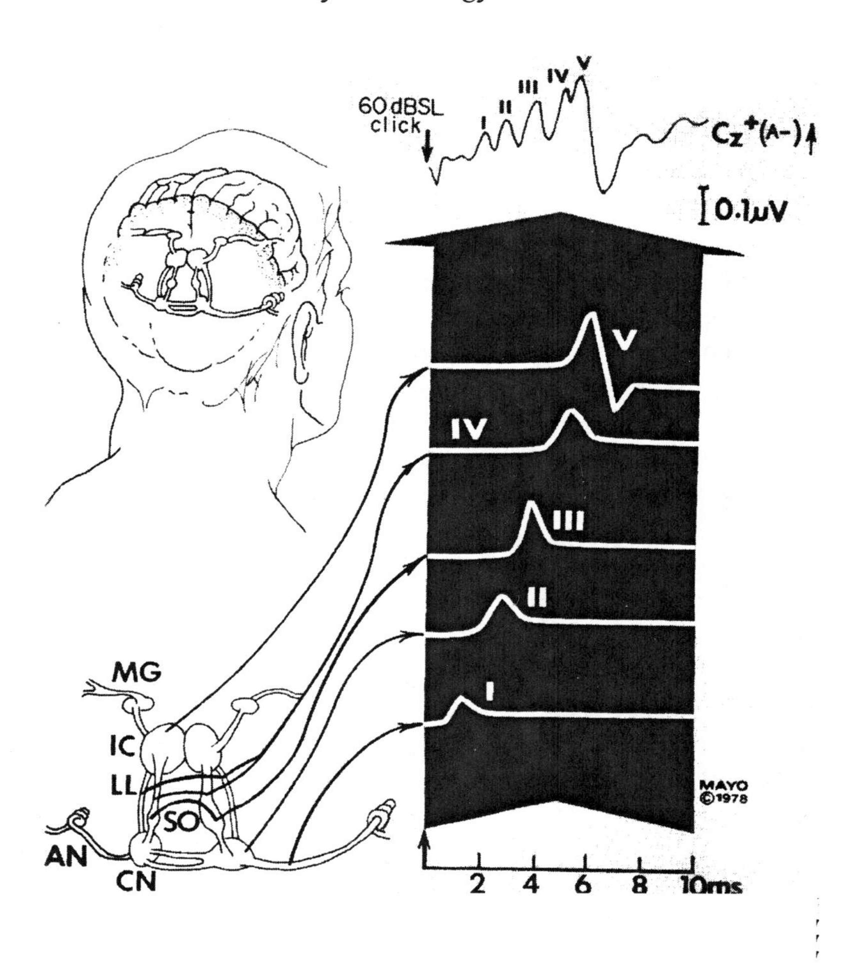


**Figure 10: The auditory evoked response. Each electrical potential corresponds to an anatomic waystation in the auditory pathway. Each wave is a reliable stereotyped electrical reflex beneath the level of conscious awareness[33].**

As we have seen the idealization of brain function in terms of sequences of on-off responses is suboptimal. There have been attempts to set up arrays of circuits which may simultaneously affect each other's output as neurons do, which is more like how neurons in the wild state interact. Higher thought sometimes involves the combined effects of numerous stimuli with which we have to make due even with incomplete information. Any child can recognize a person from his voice, facial characteristics even given such constraints as inadequate light, camouflage, etc. A manager has to decide on an optimal strategy for completion of a task, simultaneously considering different elements of that task, for example, differing abilities of various staff members. This is fundamentally different than strategy in a chess game, which occurs one move at a time each move affecting only subsequent advantage. Such arrays of circuits, each of which may simultaneously affect the output of other circuits, more precisely mimics actual brain function. As of yet such arrays function only on a primitive level, but they more faithfully reproduce brain processes that are usually nonsequential.

The nervous system is not static. We know that synapses constantly break down and form anew. Each time a nerve cell dies, thousands of synaptic connections are destroyed. Learning also reflects in anatomy as new synapses, connections, form. You lose mental capacity with age, Alzheimer disease through the misuse of drugs and degeneration you are really destroying more and more synaptic connections between neurons. Thus although you generally lose neurons with time and with them abundant synapses, you form others through learning and mental exercise. If you continue to learn as you age by doing problems, reading, and expanding your vocabulary, you will form new synapses. Adult life is a race between neuronal loss and synapse formation. It may be the total number of synapses formed that determines the net change in mental capacity. This is exactly the same as preservation and increase in muscle mass and bone density with physical exercise which can also retards aging and preserves or even increases capacities. Hence it may be that with mental exercise Alzheimer and related CNS degenerative diseases are staved off!!

After a nervous system injury even one producing large-scale damage and nerve cell death, functions that are lost mysteriously reappear. It helps to foster healing through physical and cognitive therapies. This is the macroscopic picture. The real change takes place on the level of the individual nerve cell. If the nervous system functions as an executive for other organ systems of the body, the neuron performs this function in miniature. On the grand scale sensory data must be organized, perhaps mulled over somewhat, but then acted upon. The neuron must also organize its response to complex inputs. Thus it may be viewed as a single molecular unit of nervous system function.

There are anywhere from ten billion to one trillion neurons in man. This estimate depends upon an accurate count of nerve cells (which is hard to come by) then the assumption that the density of neurons is everywhere about the same[*] (which it is not) and that this number can be integrated over the entire volume of the nervous system (which varies). The number of neurons in the brain is similar to the number of stars in a galaxy. Neurons and stars are energetic systems. Stars exert their affect upon their fellows gravitationally mostly over long distances and occasionally (in supernovas) with the explosive transfer of matter from one star to another. Neurons make more intimate contact with each other and also have their own entourage of glia and other supporting cells. The central nervous system, containing the great majority of neurons, is somewhat insulated from other organ systems by an advanced and highly organized blood brain barrier. Yet if there is a reason for the biological existence of man and all his physiology, it lies within the nervous system.

The human nervous system controls or influences all bodily function. Quite a lot has been made in popular media, by authors such as Bernie Siegel and Deepak Chopra of nervous influence on immunity. The thought has been that by influencing psychology (hence, indirectly brain function) a person's immune system may be made to play a greater role in diseases that depend on immunity, for example to help destroy certain malignant tumor cells. The logic here may be far-fetched, but the general idea that the brain is the major locus of organic control of all such functions is not far fetched at all. Nervous function arose in evolution alongside other organs systems performing their own specialized functions. The nervous system seems to us to be physiologically all important, but it very likely arose as an afterthought not the primary goal of biological process at all, an epiphenomenon, in other words.

Nervous tissue arose from the need in primitive animals composed of small groups of cells for these cells to somehow communicate. As soon as you have more than a few cells in an organism and specialization among them, you create the need for information transfer. The development and use of nervous tissue as an arbiter of information transfer between cells is but one of many strategies for survival. Brains and neurons have achieved prominence only in a small branch of the whole biological tree, namely among certain vertebrates and mammals particularly and among these especially in primates that have evolved only recently from the timeframe of biological evolution. Animals employ different strategies for adaptation. The vast majority adapt to their environment by changing their own biology over many generations, in other words genetically. Some few animals, mostly advanced animals, are able to accelerate their adaptation and change along with their environment by learning. In so doing, they may alter their response within a single or within few generations - a useful trick for organisms having a longer lifespan or long generation time. Mammals rely to a much larger extent on advanced nervous structures and have a knack for adapting over the lifespan of an individual animal. Even in the biological sense then, an individual assumes much more importance. By contrast, the most recently evolved invertebrates, highly successful ones from the standpoint of numbers, competition and speciation, especially insects, de-emphasize plasticity and learning. Thus we have two entirely different formulas for competition. Both obviously work. Insects' nervous system responses are reflexive, stereotyped and dependable. Though complex behaviors can certainly occur, these are hard-wired responses, genetically endowed. It is no big deal that adaptation can occur only over generations. The insect's generation time is short, the number if individuals extremely high, the rate of adaptation and differentiation fast. Among insects the individual is de-emphasized. In fact, for many insects individuals are very nearly genetically identical, true in particular for social insects. The individual nerve cells of insects look and function very much like ours, but the system allows for little plasticity or learning. Mammals and insects are in hot competition for global hegemony. The farmer's constant struggle against insect pests illustrates this. As insecticides and other techniques are used insects speedily adapt, changing genetic features facilely over many generations. It's our brains and planning against insects phenomenal ability to adapt, that is, survive all of our attempts to reduce their numbers.

Bacteria and other microorganisms also show how living things can compete by genetic design. Biology pits this genetics against human brainpower and no one can guess which strategy will prevail. In the race for survival, the human brain doesn't always win out. Take a look at the mess we've made with antibiotics. There must be well over one hundred of them in common use just in the United States. Bacteria have an uncanny ability to mutate into resistance. It's getting so some strains have to be treated with two or three antibiotics at once, especially hospital-acquired infections, because these bacteria living in hospitals descend from strains that were exposed to and survived antibiotics. The most dangerous infection to get is a so-called nosocomial or hospital acquired infection because these bacteria have already seen and are resistant to commonly used antibiotics. For bacteria living in hospitals, their ecosystem contains our best and most potent bactericides and they have adapted to survive our most potent drugs. Bacteria commonly pass down enzymes that deactivate antibiotics or the bacteria themselves may be infected with rings of genetic material (plasmids) that code for these enzymes and allow for survival in an antibiotic laded ecosystem. Even taking into consideration the community outside the hospital, antibiotic exposure is rampant as patients demand to be treated for infections they don't have, minimal sinus complaints treated as sinus infections, discomforts in the urinary tract erroneously treated as urinary tract infections and so forth. Organisms that survive antibiotics especially fungi such as

Candida then have the advantage as competing bacteria are killed off and we then see minor complaints such as vaginal discharges and itching being over treated with fungicides and so the problem propagates. Nursery schools are filled with toddlers all with middle ear infections all on chronic an recurrent courses of antibiotics that serve merely to foster reinfection and antibiotic resistance. Our profligate use of antibiotics mirrors overused of pesticides and defoliants so that it can be said without exaggeration, our very worst enemy is ourselves with overzealous over used of drugs and misuse of chemicals.

Sometimes the bacteria win out. We're having a terrible time with multiple drug resistant tuberculosis even though we've developed six or seven antibiotics (some quite toxic) in common use. These have arisen from incomplete treatment of a relatively few cases but the multiply resistant TB Bacillus once acquired, is almost impossible to eradicate. TB is one of those organisms that lives within host cells for years and so is hard to get at with antibiotics or with our own immune surveillance mechanisms. Other diseases of this type include Herpes viruses, leprosy and most importantly Malaria which still kills well over one million people a year and infects 300 million persons. Bacteria and viruses that have an intracellular existence are often dormant and assymptomatic for years until they declare themselves with a chronic recrudescent infection or a fulminent acute clinical attack as does malaria.

The neuron responds to information converging upon it from one thousand or more other cells through up to ten thousand or even more synaptic connections. Neurons have their own functions, but have needs and properties at once similar and yet different form other cells. The most basic thing you can say about the neuron is that it is excitable. Thousands of converging inputs add up to a resultant firing or non-firing state and the ultimate decision rests with the neuron. Whether the cell will or will not fire is the question determined not only by input, also characteristics of the cell. Admittedly, as we shall see, some neurons particularly those involved in interpretations of sensations in the periphery, have a graded response, but most central nervous system neurons either respond with an action potential or they don't, giving rise to the expression, an "all or nothing", response. What is involved in this decision?

## Inside the Neuron: How the Neuron Works

A neuron will fire under certain limited conditions. That is properly what distinguishes it from other cells. Like other cells, it is surrounded by a cell membrane that delimits it from its environment. This bilipid (fatty) layer separates two watery environments, the inside (protoplasm) and the outside of the cell. The membrane must discriminate between various substances it will let pass. Consequently, there are differences in concentrations of various substances inside vs. outside the cell. Sodium is roughly fifteen times more concentrated outside the cell, but a similar but larger singly positive charged ion, potassium, is nearly thirty times more concentrated inside. A chemical pump moves these ions maintaining these critical concentrations against electrical potentials and a concentration gradient. Because there is a tendency for substances to diffuse from areas of high to lower concentration, force (and hence, energy) is required to maintain any concentration gradient on the two sides of the membrane.

The sodium-potassium pump gets this energy from the nearly universal final common pathway for cell's immediately accessible energy, the molecule, adenosine triphosphate (ATP). ATP is produced out of the energy supplied by other chemical compounds, especially for the case of the brain, out of glucose. This chemical energy conversion is the job of the intracellular mitochondria. These independent organelles are fascinating because they contain their own genes unlike no other animal cell organelle. Plant chloroplasts actually have a very similar structure and do the opposite of mitochondria. In plants chloroplasts fix energy absorbed from sunlight into chemical energy storage in the form of sugars. The mitochondria's job is to extract the latent energy from compounds especially sugars.

Mitochondria and chloroplasts are membrane bound and have an advanced lamellated structure, almost as if they are organisms unto themselves. They are the only organelles that contain their own DNA and they reproduce independently. Mitochondria have almost a symbiotic relationship with the cell. A lot of mitochondrial protein is coded for in the nucleus of the cell, produced in the cell cytoplasm then imported into through the mitochondrial membrane. These proteins are attached to a special series of amino acids, in other words a short peptide, that acts as a code which tells the mito membrane that the protein is meant especially for the mitochondrion and needs to be imported. After importing these critical proteins, the mitochondrion uses other special proteins that cleave it, leaving its active portion. But certain other critical proteins are encoded by special mitochondrial DNA. There are a few interesting diseases that are inherited problems relating especially to mitochondria.

These disorders tend to affect muscle including the heart, preferentially. This stands to reason when you consider that mitos deal with energy utilization and muscle's tissue is the most energy intensive tissue in the body. Second these disorders are passed down only through the mother. Your mitochondria are passed down through the ova alone and are not inherited through the sperm which carries only DNA from the nucleus. The mitos reside in the cytoplasm of the ova. This shows that the sperm merely is a vehicle that carries information from the nucleus of the father. On the other hand the mother's ovum does carry many genetically divergent mitochondria. In some of these disorders mitochondria are extremely plentiful. The cells especially muscle cells can be full of them. Maybe there is a feedback mechanism that tells mitos to reproduce if their job isn't getting done. Alternately, they may appear to be wildly abnormal under the microscope. As you might expect, they don't work well either. But there are found to be specific problems when we look for them, ordinarily with coding for a single protein as in other genetic diseases. Persons with these diseases may have something wrong with their heart rhythms, may have muscle weakness or fatigue or may not be able to process some energy containing foods especially fats and so become weak and fatigued on that basis. Just one typical example is a disease that affects only the transport of certain fats into the mitochondrion so energy can be derived. This involves a substance that helps to transport these fats called Carnitine. The brain usually functions abnormally as well. Mental retardation is a problem and the tiny muscles that move the eye will often be affected so that eye movements can be grossly abnormal and a person may have double vision. The diseases are often progressive and severe. Under the microscope a typical abnormality is what is called a ragged red fiber with grossly abnormal and proliferated mitochondria. (Figure). But the mother's ovum passes down some of these diseases. Others are not passed down probably because they are so devastating that a person with the disease does not typically reproduce. We think that there is an alteration in the Mitochondrial DNA early in life (a mutation) and that the disease occurs sporadically (non- genetically) on that basis.

Mitochondria and chloroplasts may be remnants of primordial bacteria or other small organisms that continue to live in a symbiotic relationship in every animal and plant cell. Many of the cell's organelles seem almost to be organisms unto themselves. Another example of a structure in the cell recently discovered is the peroxisome. These structures went unnoticed for many years, but are plentiful in kidney cells so that after a certain period they could not be ignored. They get their name from certain enzymes they were found to contain that help process peroxides and help to detoxify and protect the cell against free radicals that can harm it. Later these structures were also found to be important in the metabolism of fats especially long chain fatty acids and disorders in these structures among other things cause diseases of fatty acid metabolism the adrenoleukodystrophies. These diseases are often present in males predominantly (are x-linked) and cause adrenal gland insufficiency and affect the deep white matter of the brain where these fats reside. Peroxisomes may one day prove important in detoxifying free radicals which are implicated as the cause of neuron destruction In many chronic diseases especially brain diseases. But the point is the Peroxisomes are similar to mitos in some ways being delimited by membranes and almost separate from the cells containing their own enzymes made in the cell but targeted for these structures specifically, most probably by a mechanism that is similar to the mitochondria described above. Mitos though are so basic to biological organisms. All of them have to provide energy for the business of life and so mitos are present in every cell and even in the most primitive cells. Cells so primitive that they do not have a separate nucleus, still have mitochondria without a membrane delimited nucleus ie bacteria, proving that they must have become intrinsically important to life, very early in evolution. If they come from a certain symbiosis, it took place extremely early in geological history and is extraordinarily basic. In fact both mitochondria and Peroxisomes and chloroplasts too are thought to have evolved from an early bacterial or blue green algae invader of ancestral animals that eventually became a symbiote, its DNA encoded now in the nucleus of the host cell. Peroxisomes may have been in the cell earlier[#] and were useful once because oxygen was lethal to the primitive anaerobic cells that were at the beginning of evolution as oxygen is even today to anaerobic bacteria such as Clostridia. As time passed the peroxisome took on other chemical roles such as handling long chain fatty acids. Indeed peroxisomes seem to have lost some of their reason for being with the addition of the mitochondria which utilized oxygen for cellular energy production. The earth's life was initially anaerobic before plants with chloroplasts began to produce oxygen in large quantities that can Kill an ordinary cell.

Mitochondria handle energy for the cell and are electrically active in a sense carrying information for the cell much as the neuron does for the body. Much of what mitochondria do is to transport electrons so that in a sense mitochondria are in multiple electrical and or energy states depending on the energy cycle as are neurons. Since they handle energy needs they may well determine at least partially the particular state of excitation and the ease or difficulty of stimulation for the neuron. In their role as energy and electron handlers for neurons one may speculate that they serve some function in information handling since, as far as we know all, neuronal information is translated into an energy and electronic code.

Mitochondrial derangements cause a wide array of ills. Not surprisingly though all cells depend on mitochondria for energy handling, mitochondrial diseases are manifested as abnormalities in nerves and muscles the most energy intensive cells of the body. The first is a disorder of young children called Reye's syndrome, which almost never occurs anymore. No one knows what was the actual cause of this deadly disease which would cause brain and liver failure, brain edema, and coma and severe neurologic impairment of young children who would survive the disease. The best theory is that certain benign viruses such as influenza and chickenpox particularly would affect mitochondria and that aspirin added to the virus' effect in children only who somehow had different or immature mitochondria. Treatment of these kids was a medical emergency and soon they would be fighting for their lives with low blood glucose, liver failure and coma and increased intracranial pressure. In the 1970's particularly we had a run on such cases which would take normal children and ruin them. Then it was realized that aspirin was part of the cause, The specific mechanism was never fully elucidated and the aspirin manufacturers at first balked at the prospect of losing an important market for their fever-lowering medicine. When the relationship was made between aspirin and Reye's after some fits and starts aspirin ceased to be used in children and the Reye's epidemic became no more.

The cell works very hard to maintain different concentration gradients for different ions inside and outside the cell membrane. In the neuron this produces a voltage gradient critical to the cell's excitatory function. The inside of the cell is kept some 70 microvolts negative with respect to the outside. Whatever energy the mitochondrion produces is expressed primarily in molecules of the immediate energy carrying chemical ATP. The ATP-consuming potassium sodium pump removes some three internal Sodium ions for every two Potassium ions that enter, maintaining cellular homeostasis. Nerve cells communicate and change their status by altering these electrical charges determined by membrane ion voltage gradients, so these membrane potentials are extremely critical.

This ion exchange is more than a theoretical voltage gradient maintainer. Water molecules diffuse into any cell along with Sodium. When this ion pump breaks down as it occasionally does when the cell is sick or does not get enough of a chemical energy supply for example, with the interruption of brain circulation, the cell becomes swollen with water. Multiple swollen brain cells accumulating water lay the basis for a dangerous form of brain edema. In the cranium an increase in volume within a closed space always causes an increase in pressure that disturbs brain function and ultimately leads to brain injury and death. We treat this situation by trying to stave off swelling with dehydrating agents. These chemicals create a concentration gradient between the brain and blood to allow water to be sucked out osmotically from the pulp of the brain. By using simple molecules such as Glycerol and Mannitol, carbohydrate derivatives, that build up a concentration gradient of solutes on the outside of the brain cells in the bloodstream as an emergency measure the cell's swelling are slowed until the basic pathologic process that caused the problem in the first place can be definitively treated. These chemicals eventually seep into the brain and compound difficulties in creating brain swelling, so that all we're really doing when we use them is to buy some time. In the meanwhile the swelling and increased intracranial pressure can be monitored closely in the intensive care unit. We use sensitive pressure devices that can be screwed in under the surface of the skull and other devices to look at the pressure inside the cerebral ventricles that contain spinal fluid.

What are some of the processes causing the individual brain cells to swell? These are serious conditions that disturb the neuron's ability to utilize energy and their Sodium-Potassium pumps. Some disorders directly affect mitochondria such as Reye's syndrome, discussed above. Reye's syndrome damages mitochondrial function and affects the brain primarily via severe swelling and increased intracranial pressure. Although the precise mode of damage is not completely understood, energy utilization especially of glucose and fats likely suffers and the individual cells cannot maintain concentration gradients. Much more commonly any process that cuts off the blood supply to the brain causes the same problem. Here I am referring to stroke where a blood vessel occludes and chemical energy thus ceases to be delivered to starving neurons and glia. Trauma and inflammatory swelling or inflammatory swelling alone as in encephalitis may induce pressure changes that left to their own devices will eventually block the circulation to the brain. All of these situations cause a malignant form of cell swelling called cytotoxic edema. Supporting cells such as glia also participate.

The neuron cell membrane separates the inside and outside of the cell. Lipid (fat), insoluble in watery environments outside and inside the cell, makes the backbone of the cell membrane. The two layers of lipid and this membrane have individual long molecules each with a water-soluble and a water-insoluble component. Like dissolves in like and the water molecule is polar, slightly charged as is one end of each lipid molecule. Each molecule thus lines up so that the polar portion is closest to water. In the meantime the nonpolar portion is hydrophobic or water-avoiding and this lines up with other lipid molecules. The membrane thus is made of two layers with polar portions lining up along watery environments on the outside and inside of the cell and nonpolar portions closest to each other (Figure1) As we have seen, maintaining the separation between two watery worlds inside and outside the cell takes considerable energy. But the membrane's function is not simply separation. It has to selectively allow certain substances, both small and large molecules, to pass into and out of the fortress which is the cell in both directions and under special, well defined circumstances a considerable problem in chemical molecular design, when you come to think about it. It seems all parts of living systems, even cells and components of cells, play an active role in a larger plan. Proteins that traverse its bilipid layer help the cell membrane. Each of these proteins serves a specific function. Some are structural supports. Others are receptors and channels or tunnels to allow passage of certain molecules into and out of the cell.

Thousands of these channels stud the cell membrane to allow passage of the Sodium ion alone. This ion whose major importance is its positive charge, flows across the membrane in different ways depending on the exact status of excitation of the neuron.

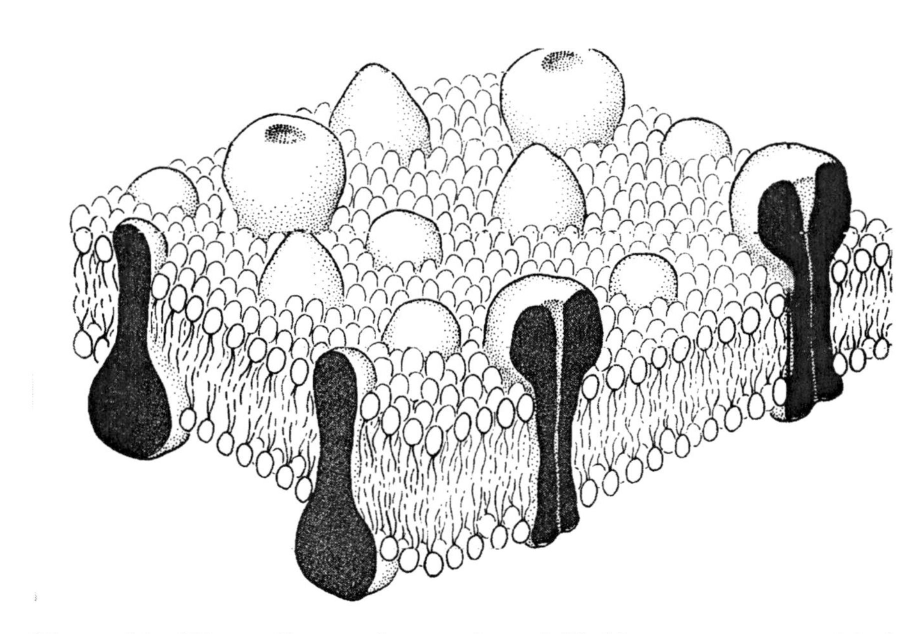

**Figure11: The cell membrane is a bilipid structure studded with ion channel proteins that respond to receptors with Membrane pore.**

One such receptor protein has been studied in detail resides at the neuromuscular junction. This is a synapse that is relatively easy to study because it's so accessible. The neuron secretes Acetylcholine in order to signal a muscle to contract. The muscle cell is similar to a neuron in many ways. It is electrically active. When excited it becomes depolarized the way a neuron does. While an electrically excited neuron ordinarily secretes a chemical, a muscle is made to contract when it gets excited.

The muscle membrane has certain receptor molecules for ACh. Only recently have we really been aware of the complex structure of this protein. It is composed of four peptide (amino acid chain) subunits. The ACh receptor protein structure twists and turns, as do other proteins, assuming a "tertiary" structure. Even though proteins are composed of peptides, most often multiple peptide chains, they do not stay straight. Instead the electromagnetic attraction or repulsion between portions of the chain, causes it to twist on itself. Electrical charges in the surrounding medium also affect the protein's conformational properties. For example, the ACh protein twists so much that it ends up traversing the neuronal cell membrane many times. By looking at specific amino acid sequences within the protein structure you can actually predict where the protein will cross into the cell membrane. Amino acid sequences that are polar (which have large separation between positive and negative charges) will naturally be attracted to a polar medium, namely the mediums inside and outside the cell which are relatively watery because they are like water and like dissolves like. By contrast, nonpolar protein regions will want to stay within the membrane composed of lipid tails. This is much the same thing as oil not mixing with water. Polar and nonpolar substances like to stay with their own kind. Water is polar, oil nonpolar.

So the ACh receptor molecule ends up criss-crossing the cell membrane four or more times. But it also has a large portion outside and a large part inside the cell. What happens is that Acetylcholine fits perfectly into a part of the receptor molecule outside the cell that is specifically designed to cradle Acetylcholine and nothing else. But this portion of the receptor molecule does more than cradle ACh. ACh also changes some electrical properties of the receptor protein and changes the protein's shape. As it does so the receptor which is going through the cell membrane literally opens up a pathway so that Sodium ions can begin to pour into the muscle cell. As discussed, Sodium is in much higher concentration outside the cell and only the cell membrane ordinarily keeps the Sodium from coming in. If Sodium is allowed to come in its positive charge will make the inside of the cell become relatively more positive and thus it will be electrically excited.

Lest someone who hasn't been exposed to physiology may think I'm going a little overboard describing obscure things like receptors for acetylcholine and its protein chemical receptors, they should be warned how extremely powerful these principles are. One thing we haven't gotten into is acetylcholine esterase (AChE) that break acetylcholine down after it has been released. Saran and other 'nerve gases' block AChE irreversibly allowing ACh to accumulate a small thing you may say. But these are extremely potent chemicals when absorbed in incredibly minute quantity. Such nerve gases were released in a Tokyo subway and luckily there were few deaths. What happens to a person exposed to nerve gas was portrayed graphically and without exaggeration in a recent movie, THE ROCK. Subjects would simultaneously salivate, have incredible cramps, diarrhea, sweat, and convulse hard enough to fracture bones and die within minutes. The antidotes may lethal themselves. They have to be injected directly into the blood system immediately. Some antidotes are Atropine that blocks the effect on internal organs and Curare[ψ] to block the effect on muscle. Using these chemical weapons which were mostly perfected by the Nazis, fortunately is a problem on any battlefield. The reason is that it is hard to account for the spread of chemical with prevailing winds and other factors and despite your best efforts, you could easily end up destroying as many or more of your own forces when you use them. However they are easy to make and not that hard to deploy, so that the worry in a situation such as the Gulf War with Saddam Hussein, was that he was just crazy or desperate enough to deploy them, even all the while being fully aware that his own forces could be annihilated.

## The Single Bit is Really the Individual Channel Protein, not the Neuron. The Neuron Makes a Decision Based on Complex Input

The ACh receptor is actually like a pore or opening which always is in one of two states: Either it is closed and doesn't allow Sodium to come in or, it is open and does. Furthermore it is in its open state only when Acetylcholine is bound to it and closed otherwise. The ACh molecule very neatly lays the basis for the binary state (either on or off) that is described above. The receptor protein, which is stimulated or not stimulated, porous to its ion or not porous to its ion, is in a binary state which may be described as a '0' or a '1' in the same way as can the state of the whole cell, excited or not excited. Thus the receptor is in a binary state that partly contributes to the state of the entire neuron which has thousands of such receptor molecules studding its membrane. Synapses too, each utilize many such protein receptor molecules. We have binaries determining binary states so the actual mathematical description of the state of a single neuron isn't so simple anymore. Each neuron does emphatically not provide a byte of information but is itself influenced by binary protein receptor and synaptic devices which makes the neuron an executive, an administrator, a decision maker such that this is a situation we may call "double binaries" in honor of binary stars, or the famous double bind which puts you in a position where it can be difficult to make a decision. These are binaries determining other binary states, but he basic determining binaries run in themselves, from the standpoint of the individual cell, into the thousands.

Therefore computer models that seek to describe the individual state of a neuron in an all or none way as either firing or not, mathematically as a '0' or a '1', vastly oversimplify. Really on some level we may view the individual neuron not as a simple binary element as computer persons have tended to look at it, but actually as a microprocessor in an of itself, a little living executive. In fact I am not sure the role of the individual nerve cell can be precisely translated into a computer binary model at all. The complexity of the neuron with double or triple binaries and all permutations of the status of receptor molecules is near impossible for the mind to fathom. Now, when they take the next step and use arrays of neurons to model brain function computationally it is with a little bit of hubris, the thought that even these simple arrays of neurons can possibly model processes served by groups of neurons, say ganglia or nuclei within the brain. Furthermore, on the other side of this issue, the notion that arrays of binary elements such as switches and microprocessors mirror cerebral function is also presumptuous. These electronic arrays perhaps model the membrane of a single cell, studded, as it is with binary elements, the ion channels. Even here it is debatable that an electronic analogy can ever be made. The point is the level of generality of abstraction in computer science is such that persons in the field flatter themselves into thinking their methods reproduce in some way function of sets of neurons whereas what hey are really mimicking is the working of a single membrane of the cell, if that!

The individual neurons is always in a more or less excited state, that is pretty much of a continuum. How do you measure the degree of excitation of the individual neuron? In at least two ways. The neuron is more or less "polarized" that is the negativity of the neuron differs by more or less of an amount with the outside of the cell. If it is more polarized or *hyper*polarized, the inside of the cell may be 70 microvolts negative with respect to the outside, less polarized and it may be 60 microvolts different than the outside and so on. Also the neuron may be more or less *excitable.* It may be harder or easier, again on a continuum, roughly proportionate to the polarization of the cell, to stimulate it, in other words get it to have an action potential. In very rough terms brain *excitability*, sleep and wake states etc corresponds to some average excitability of a large group of neurons and depending on a task, certain regions of the brain may be more or less excitable. More excitable regions generally draw more energetic compounds such as glucose for their function which can be seen on such functional tests as the PET scan which looks at regional utilization of glucose in the brain. The EEG too, described above shows slow waves when large areas of the brain are less excitable, with neurons hyperpolarized and fast waves (high frequency) when de-polarized or more excitable.

It happens that the muscle cell has an excitable membrane just like the neuron with its own action potential. Like most neurons it is at any one point in time either excited or not. The muscle cell is also a little different because it is not responsible for integrating a lot of signals from disparate parts of the membrane and sending a message along to the next muscle cell down the line as is the neuron. in the case or muscle a single synapse will determine whether the small fiber that is directly connected to the nerve cell axon will contract or not contract.

Ultrastructural studies show the receptor molecule is shaped like a pore or hole in the cell membrane. This interesting finding links proteins to actual structural characteristics. Proteins are not mere amino acid chains but in terms of three-dimensional "microscopic" structure that reflects function. This single protein has a remarkably specific design. Its non-polar portions snake in and out and through the cell membrane; its outside portion cradles ACh, and ACh specifically which will end up changing the conformational properties of the rest of the receptor protein causing it to act as a gate for the flow of Sodium ions responding to ACh binding, laying the basis for the on or off status of the muscle cell. One receptor molecule is far from sufficient as many have to respond, each letting in some Sodium ion before the muscle is depolarized enough to cause the muscle to have an action potential. The muscle action potential signals it to contract, in a process known as excitation-contraction coupling. Significantly, certain neurons also have ACh receptors similar to those in muscles. Other chemical receptors share many characteristics with this particularly well known receptor molecule so that it serves as a model for receptor function.

Research on the ACh receptor was made possible by scientist's ability to extract large amounts of the stuff from organs that naturally contained loads of it. These were the electrical organs of the stinging ray, Torpedo. This animal has adapted muscle like membranes to be strung along in series rather like very low voltage batteries so that a significant charge can be generated to stun prey.

Among other things, the cell membrane is a capacitor that arrests the jump of electrically charged ions in the watery environments on both sides of the membrane. If the membrane's two molecule thickness is taken into account, which is extremely thin for any charge separation, this capacitor stores a charge of some 100 thousand volts per centimeter. This electric field changes the conformational structure of the proteins that pass through it. The channel proteins are designed to allow different fluxes of Sodium depending on changes in the surrounding electric field.

These particular proteins are made to fit a number of constraints. First of all they allow passage of a single positive ion that carries a certain electrical charge. They do, for example, exclude the larger Potassium ion, which has its own membrane channel protein. Most of the time the Sodium ion cannot penetrate its channel protein. Its permeability changes dramatically within small fractions of a second, first increasing at the right times, then decreasing again. Recall that Sodium is much more concentrated outside the cell. When the channel opens Sodium will rapidly rush in. In addition there is an electrical force attracting any positive ion to the inside of the cell because the inside is negatively charged. When a Sodium ion with its positive charge rushes in the inside of the cell is less negatively charged in relation to the outside, thus depolarized, as the cell becomes excited enough once it reaches a threshold of depolarization with enough Sodium ions to finally have an action potential. Within hundreds of thousandths of a second (milliseconds) the Sodium floodgate opens, allowing a gigantic influx of positive charge to momentarily make the inside of the cell some twenty millivolts positive compared with the outside. Then the channel will close also within a very small timeframe. The electrical milieu is responsible for making the necessary conformational changes in the structure of membrane channel protein because it contains portions with differing electrical charges. Suddenly the permeability to the Sodium ion (Sodium conductance) is severely decreased. Almost simultaneously, the Potassium conductance in its own channels increases allowing the single positive charge carried by each potassium ion to flow outside of the cell.

The concentration gradients for the various ionic species, mainly Sodium, Potassium, and Chloride determine the voltage gradients inside and outside the cell membrane when ion channels for the individual ions are in either the open or closed position. Suppose the Sodium Channel protein alone allows Sodium to free flow inside the cell, then the membrane potential, the electrical voltage gradient on the inside vs the outside of the cell membrane will move closest to the voltage gradient for Sodium alone. Given that Sodium is almost 15 time more concentrated outside the cell, the ion wants to flow into the cell and the voltage gradient for Sodium alone is closest to +70 mV. For Potassium, which is about 30 times more concentrated within as opposed to outside the cell, this membrane potential is closest to -98 mV. In recent years a whole host of diseases that are caused by abnormalities in these channel proteins has been described. These are genetic disorders that most commonly cause defects in brain or muscle, though Cystic fibrosis is caused by a defect in the Chloride channel and is not a muscle or brain disease.

This is the simple model of receptor protein function. In some other cases, which include most molecular receptors, the receptor protein initiates action of a second messenger, either a simple molecule or protein itself. This second messenger will have some modulation effects on the post-synaptic neuron causing a change in the flow of ions such as Calcium inside the cell or by some other mechanism. A couple of examples of 'second messengers' are the nucleotide GDP turning to GMP (for Guanine di and Guanine mono phosphate) which in turn affects other proteins changing conductances of channel proteins or other proteins and thereby affecting the excitability of the cell. When acetylcholine from a motoneuron reaches a muscle membrane ACh receptor it excites the muscle through the influx of Sodium ions and initiates *excitation-contraction* coupling that ends in the muscle cell contraction. The Calcium ion is made to traverse a part of the muscle cell, the sarcoplasmic reticulum, and this ends by exciting other molecules, Actin and Myosin within the muscle to slide over each other (a process consuming energy in the form of ATP) and the muscle cell will contract. This Rube Goldberg like harebrained scheme or chain of events, or whatever you may choose to call it, lies behind all muscle contraction in the entire animal kingdom. Such cascades of events, even far more complex ones, are nearly universal in biology, by the way, underlying nearly all important biological processes from clotting to inflammation. Unlike what we experience in real life, in our macroscopic world, where complex schemes tend to fail, in biology complex cascades seem to extremely reliable for some reason, but it means that something could conceivably go wrong at any phase or stage of the cycle. However such complex schemes have the advantage of working, and working well, where you need to have an extremely reliable response occur under very narrow limited circumstances. One of the best examples is clotting. Blood needs to clot to stop a hemorrhage in the event of injury obviously and to keep out microbes where you have an open wound. But imagine what would happen if blood clotted freely as it flows in our arteries, veins an capillaries. We would die instantly. So it has to be made to clot under extremely narrow, well-defined, situations but not the vast majority of the time. Consequently we have a cascade involving at least 12 steps and conversion of different enzymes or factors at each step along the way from an inactive to an active moiety. The very same holds for muscle contraction. Right now, as you are reading this, what percentage of your muscles do you think are contracting and under how much force? You need to control them precisely, maybe not at the level of the Olympic athlete but muscle contraction and neural excitation for that matter need to occur according to narrowly defined parameters.

An interesting pastime popular in Japan is eating Fugu or raw Japanese puffer fish that contains one of the most potent toxins known to man, tetrodotoxin. Tetrodotoxin poisons the neuron's, voltage sensitive Sodium channel. This channel is most important to the neuron because a change in membrane voltage first lets a rush of Sodium ions in creating an action potential and little useful can occur if one can't generate action potentials. Skilled chefs prepare it so that only a trace of the poison remains, enough so that the thrilled gourmet will come from an inch of losing his life. He may feel a certain tingling in his tongue and mouth. In rare cases a paralysis results which does not cause an alteration of consciousness or memory for the event (perhaps not enough of the stuff actually seeps into the brain but instead mostly impairs peripheral nerve and muscle cells).[34] We have now defined disorders of channels, so called channelopathies as a class of diseases[35]. I've already alluded to cystic fibrosis, a disorder of the Chloride channel of certain cells. There are other diseases, some caused by animal toxins but more important and common, genetic abnormalities affecting one or another channel, for example the Potassium channel in muscle cells may be affected causing paralysis

Proteins are very long combinations of twenty-one amino acids molecules produced en masse in different areas of the body far from their areas of use in the brain or in other organs. Ten “essential” amino acids are not actually made by humans and have to be consumed from animals and plants. You can take the reductionist’s view of humankind that we are nothing but a complex of chemical reactions. It is a point of view that makes sense when thinking about proteins and their functions. As enzymes, proteins control the type of chemical reactions and their rate. Proteins also play a role in making up the basic building materials of cells, organs and extra-cellular substances in the body that determines structure. It is a basic tenet of biological thought that the determinants of protein structure create the entire form and function of the organism, responsible for the entire phenomenology of life. This hypothesis is worth closer scrutiny. Everyone knows that living creatures are composed of various kinds of chemicals that bear little resemblance to proteins determining living forms including fats, chains of various sugar moieties, water soluble substances and ions in specific concentrations arranged in a certain way to define the organism. Proteins can act as structural elements to compose for example the skeleton of cells (as in microtubules) but more importantly they increase the rate of certain chemical reactions. Protein enzyme systems for example catalyze the joining of two sugar molecules and help in the formation of simple sugars before they are joined, making possible the chains of sugar building blocks that ultimately will become familiar starch and cellulose. Other proteins will aid the organism under the right circumstances in digestion of these formed products. Still others aid in the formation and breakdown (anabolism and catabolism) of other proteins. They are ubiquitous helpers of biochemical function.

## Proteins Make the Organism

Thus protein is the start of a chain of events entirely determining an organism's form and function. The information system that dictates the formation of protein molecules is thus all-powerful in the biological sense. Strands or multiple strands of amino acids in a precise conformation, control everything in an organism. Miraculously, proteins formed of these 21 simple amino acid building blocks have proven to be extremely diverse in their function. Th only information necessary to completely determine the form and function of an organism, spells out the structure and quantity of its various proteins. Just determine all the proteins and the whole animal or plant will come together of its own accord. Fruit fly is different from human only by virtue of a difference in their protein vocabulary. A protein can change as it functions as the ACh receptor and second messengers. Other enzymes can help add side chemicals to the amino acid backbone of a protein. A protein may be composed of hundreds of amino acids, in multiple chains, but if you exchange or leave out just one amino acid you may alter its function and even render it useless.

For example sugar derivatives (hence glycoproteins) may be added to increase protein diversity. Hence proteins which determine structure and form in a diverse biosphere are built out of simple components that are relatively few in number. It is the basic work of science to divine a set of simple unifying principles or elements that explain all diverse phenomena. For physicists the eternal dance of the heavenly spheres if one wants to describe this so whimsically, is brought about through the relatively simple workings of just four simple basic forces, Strong, Weak, Electromagnetism and Gravity tentatively welded into equations called unified field theories. Then there is the bewildering array of elementary particles comprised of quarks, leptons, gluons etc. It seems that striving for grand simplicity explanations have become more and more complex. But the striving to find just a few simple elements and processes that determine the diversity of phenomena is the grand goal of science. In chemistry interactions between some 103 elements of ordinary matter composed of just a few different types of particles, electrons, neutrons, protons on a more macroscopic scale, is the topic of study.

On the surface biologists are less overt about a quest for ultimate beauty and simplicity that would explain biological observations. Yet they seem a lot closer to their goal than chemists and physicists. Only in the last thirty or forty years has this goal begun to be realized. Permutations of just a few simple elements produce all biological diversity.

One of the great debates when you start off in school taking your first biology courses is what should be the proper subject for study, what is living. What properties imply that the subject Is a living thing? Everyone agrees that the one-celled animal or plant is a living thing. What about something more basic like a virus, that can't reproduce without commandeering the machinery of another host cell, an obligate parasite in other words, Is a virus close to a mere combination of chemicals? A virus is primitive living creature in that it competes in evolution, indeed is altered by changes in its environment. Viruses are composed of just a few types of chemicals, protein and nucleic acids. About the only useful thing a virus does is make more of its own kind. If this is true then reproduction defines life. Perhaps in addition an organism as opposed to a thing, should struggle for existence, whatever that means, strive in some way as it is subject to evolution and natural selection. But we could devise a machine that makes copies of itself. In that case would the machine be alive?

Nature has played a little trick here I think. Living organisms don't just reproduce. They almost never make exact copies of themselves but rather close but inexact reproductions. If you look on the molecular level, DNA seems to make second copies of itself with a remarkable fidelity but we know sometimes, rarely, there are errors and when there are, a single nucleotide may be deleted or replaced by another, altering the genetic code for that cell just slightly and we have a mutation. This happens rarely and even when it does there are enzymes in most cells that correct the error and make the DNA normal again. Even so, some mutations do occur and survive the corrective process. Mutations may be made to occur with greater frequency under certain conditions, for example under chemical or radiation exposure. You can expose bacteria to low levels of ultraviolet light and over many generations they will change form.

When a cell replicates it seems to make perfect copies of itself; this is how a cell line, a species is preserved and lays the basis for how organisms survive, after all the fittest organisms are the ones best abel to survive and reproduce, but the replication is inexact. Close but inexact. The slight difference, which seems to be mere detail, makes all the difference. As we learn more about genetics, we have discovered that the inexactness of replication is far more extensive than we had ever imagined.

In addition to mutations, here are a few examples of inexact replications: Viruses infect host cells, and in doing so incorporate their own genetic instructions into the genome of the host cell. As these cells reproduce, it is conceivable that in many cases, at least some of this genetic material will be included in the host cell's DNA for good, particularly if viral DNA carries with it, instruction that will help the host adapt to its new environment. Even in a complex animal, there is every reason to believe that some of this material will eventually be incorporated into the germ cell line, may end up, in other words, in ovum and sperm cells. Living in the antibiotic era, we have seen how quickly bacteria of all kinds, acquire resistance to antibiotics. I've described this tendency above in terms of a war between our minds, which continuously invent new antibiotics to add to our armamentarium, and the "mindless" but rapidly reproducing bacteria (and insects too which resist insecticides) organisms which have to advantage in evolution of a short generation time that allows them rapidly to adapt. Humans may well be losing this war. One major reason for all of this is the frequent *exchange* or movement of genetic material between bacteria and viruses. For example, penicillin resistance may be passed from one bacterium to another. They actually pass on genetic material which can travel from one bacterium to another even between species. The gene for penicillinase which breaks down penicillin and thus causes penicillin resistance passed with a circular piece of DNA called a plasmid. A person or animal given penicillin will be selecting for bacteria carrying the plasmid for penicillinase - those bacteria will survive in preference to their compatriots, the billions of other bacteria inhabiting the gut and elsewhere. Resistance genes for virtually all antibiotics in almost all species of pathogenic disease causing bacteria have surfaced and natural selection has helped humans to "breed" in essence, ever more resistant and virulent strains of bacteria and viruses. The genetic complement of the organism can no longer be seen as a static entity, the organism reproducing itself with ultimate fidelity. Instead we have come to appreciate the genome even in the natural non-genetic engineered state as a continuous exchange of information with occurs even between species. This information is constantly reexchanged and reshuffled. It moves and helps determine which organisms will survive[36].

In our own bodies, we have in the past, viewed each cell, as a part of a single person partly because it contains the exact same genetic information. This is generally true. But here again there is abundant reshuffling and change in material. Our immune system is capable of producing perhaps a million different antibodies that help us to fight off invaders. It does this by reshuffling, recombining short lengths of DNA a relatively modest amount of genetic material that achieves great diversity by recombining elements that end up in plasma cells, which make all of our antibodies. B lymphocyte precursors to plasma cells make antibodies that they then "display" on the surface of the cell. That way they are showing off just who they are, displaying their true colors. They circulate in the bloodstream. When a virus or other threatening entity invades, and antibody interacts with the invader, the body has a mechanism that selects the lymphocyte displaying the proper antibody and this will differentiate producing plasma cells to produce this one antibody in abundance. In essence we have natural selection of some of our own body cells whose genetic complement will prove useful to the survival of a whole organism. This is just the same natural selection observed in bacteria with antibiotic resistance, the same natural selection seen in all of biology[37].

## Cells Reproduce, but Do Not Make Exact Copies – The Secret is in the Imperceptible Difference in Cells

When you get to the level of a single cell, it can divide to make two daughter cells. This is fission or mitosis. Each daughter cell will have exact copies of genes of its parent and seems to be identical in every way. But while the difference is so slight as to be imperceptible in a single generation, if you observe the cells after many doublings, you begin to detect differences. Sometimes you see a simple degradation in reproductive product as cells down the line accumulate genetic defects. This is a mechanism for senescence of a genetic line. Down the road, the cells may lose their ability to survive in the same environment or to reproduce. Perhaps a certain number of doublings is predetermined, written in, the genetic code of each and every cell line$^{\Phi}$ . Precise genetic recombinations can then be added to viral and bacterial genes, a genetic informational exchange, make exact copies of human antibodies in abundant quantity that can be used to treat diseases such as cancer and multiple sclerosis.

In the embryo an even more interesting thing happens. The fertilized egg divides and divides again. Now there are four apparently identical cells. But they are not identical. Each will give rise to a different part of the embryo as it continues to divide giving origin to different tissues, organs and body parts. You may say that as the embryo gets to the 4 or 8 or 16 cell stage, the blastula or morula stage or whatever, that each cell, though seemingly identical with its sisters, has an intrinsic potential to have certain restricted progeny. Even though it has exactly the same genes as its sisters do, and looks in every way like its sisters, it will produce a whole line of totally different cells. That is because many of its genes begin to shut off through more and more generations and no longer function, while its sister cells have a different combination of genes that are shut off. There are other mechanisms as well. At an early embryo stage, certainly when there are just 4 cells, you can kill off three of the daughter cells and what will happen? Apparently nothing. At the end of gestation, a normal whole animal will emerge. Nothing will have happened to the pregnancy. You can even take and 'devote' three of the four cells to three other new identical embryos and you will have identical quadruplets at the end of the pregnancy. So whatever determines which way a cell will ultimately go, is decided later than the four cell stage. At that point, all four cells, though undoubtedly very slightly different, are still multipotential. As more generations of cells come into being, these cells specialize more and more and soon will be unable to give rise to a whole animal on their own.

| EXAMPLE | ORGANISM | MECHANISM | RESULT |
|---|---|---|---|
| MUTATION | All | random change in gene. | diversity |
| MITOSIS, FISSION | CELLS, BACTERIA | systematic alteration over divisions | Embryology senescence tumors |
| TRIPLET REPEAT | HUMANS ?ANIMALS, PLANTS | lengthening of repeat | various diseases where severity varies |
| PROLONGED SYMBIOSIS | SINGLE CELLS | incorporation of symbiote | Mitochondria chloroplasts ?other organelles, co-evolution |
| SEXUAL REPRO-DUCTION | MOST ORGANISMS | Reshuffling and mixing of genes | increases diversity increases adaptation |
| ANTIBODY PRODUCTION | LYMPHO-CYTES | Economy in reshuffling small numbers of genes | Diverse antibody production |
| GENETIC EX-CHANGE | BACTERIA, VIRUSES | infection, conjugation incorporation | antibiotic resistance, increased virulence |
| GENETIC MA-NIPULATION | HUMANS | genetic engineering | curing disease "perfecting" genome? |

**Table 4: Examples of "inexact" reproduction. Organisms reproduce imprecisely. This little appreciated fact may lay the basis for all biological adaptation and success.**

The point is that living creatures reproduce, yes, and this is perhaps what defines them as being living, but they don't reproduce *exactly*. It is this non-fidelity of reproduction, that in fact makes living things unique!! Even viruses mutate, fail to make exact copies of themselves. And when you get to more advanced creatures they even stop trying to make exact replicas of themselves. They reproduce sexually. Suddenly their offspring share only half their genes on average. Even so, they are usually somewhat devoted to their children's survival and that of the rest of their clan, to which they are related and whose genes they share to a greater or lesser extent.

Now you might ask what for? Why does a living thing not produce exact copies and be done with it? The answer is that variation, diversity, is the raw material for adaptation. In order to be more or less fit in a certain environment, you should be different from your contemporaries in some way. This difference will yield either a higher or a lower probability of survival. Now suppose there is an animal or plant somewhere that produces only exact replicas of itself. It's an hermaphrodite and its children are just like the single parent. It is somewhat adapted to its environment through previous evolutionary design and seems to be doing fairly well, making its living and surviving. Then the environment changes. It does not have to change that much, perhaps by a hair. Perhaps this organism is an insect, and farmers have applied a new insecticide, or the winter is a little longer one year, or any change of this magnitude that under ordinary circumstances would kill off some but not all of offspring but might kill off all of offspring that are identical, in which there is no variation. Then that animal would die out and be seen no more. Even more important this particular exactly reproducing line would have to compete with a close cousin which does not reproduce exactly, one that has built in variation. These cousins would give rise to some creatures that are poorly adapted and would not be fit, while having others that are even more fit than the average and these fitter fellows would have an advantage and eventually take over. Therefore you can easily see that those organisms that have built in some mechanism for variation will survive, while others that have not, will die out. The system of survival, evolution, built into our biosphere demands variation. You could almost say that organisms strive to be different. That is why even one-celled animals sexually reproduce, exchange genetic material, why some insects and other animals reproduce sexually in inhospitable environments and assexually in halcion times. You could argue that the greatest mass of reproduction gives identical copies. Ironic that the residuum, the tiny remainder of non-exact copy-making, is what seems to define living creatures. So, what defines life, more than anything else? What is the most basic factor that makes an item a living thing? An organism needs to be subject and product of the laws of evolution, has to be adapted by virtue of built-in inexact replication.

## Elemental Biology

As chemistry is thought of on a certain level as a branch of physics that defines certain molecular interactions, biology deals with a certain set of chemical interactions that relate to life. What determines all form structure and function in nature is alteration of chains of amino acids that are parts of animal and plant proteins. But an even greater simplicity underlies protein production. Amino acid chains are specified by other chains consisting of just four different nucleic acids. In 1945 the one gene-one protein concept was first enunciated by Beadle and Tatum. Much of the subsequent work in biology has been aimed at elucidating the basic processes underlying this seminal idea. The human is estimated to produce only fifty to one hundred thousand different proteins as dictated by an equal number of genes. This is not to belittle the amount of information that this represents. The genetic information contained within every cell, just a listing of the sequence of nucleotides is enough to fill 500 volumes of 1000 pages each with 1000 words. Every cell is capable of producing only certain proteins. This functional repertoire is much less than is encoded in its genome. The production of most proteins is suppressed according to the specialization of that particular cell. Cellular differentiation occurs only because every cell surrenders its ability to synthesize certain proteins at the same time producing others in great quantity. Some of the proteins alter the form or appearance of that particular cell.

Other genes get involved with specifying the appearances of the individual tissue of which the cell is a part. Still others change the form of the organ and even help specify the appearance of the whole person. A switching mechanism turns on and off, usually permanently, a cell's ability to make proteins and thus determines all the different kinds of cells and cell products residing within the organism. What happens when their is a glitch in this process? One problem is dedifferentiation. This is part of what causes some cancers when certain cells suddenly undergo malignant change. Certain cells take on more primitive characteristics and often synthesize chemical that they are not supposed to. Under the microscope it's easy to see that malignant cells change form and grow in an amorphous haphazard manner. The most malignant cells have grotesque nuclei. Nuclei carry genetic information on Chromosomes that can, at certain stages of cell division be made out under a light microscope. Chromosomes may be unduly clumped and there can be more than the usual two copies of each that is found in the normal nonmalignant cell. There may be multiple copies genes that specify certain proteins responsible for cell growth.

Other proteins whose production is suppressed by fully differentiated cells, can suddenly start to be produced in great quantity. The Syndrome of Inappropriate secretion of Antidiuretic Hormone (SIADH) happens with some small cell carcinomas of the lung. This hormone is normally secreted only by the hypothalamus, a small mass of cells hanging beneath the frontal lobes of the brain whose purpose it is to control the pituitary gland and vital basic functions such as hunger and thirst. Antidiuretic Hormone (ADH) decreases the amount of water put out in the urine by the kidneys (diuresis) and thus helps conserve water.

ADH also increases thirst. An inappropriate overload in this hormone will cause water overload. If malignant cells secrete ADH the concentration of important elements such as Sodium drops and cells swell. The brain is affected clinically before other organs. Confusion a decreased level of awareness and even seizures can result. This is but one example of cells secreting chemicals in an uncontrolled fashion that ultimately exert profound effects.

The range of protein production by a cell determines form and function. Cells are chemical factories which may produce substances altering their own function and that of other cells. Neuronal chemicals most typically influence and control other cells. Moreover, proteins must be produced at exactly the right time, as with certain chemicals and enzymes whose secretion is carefully regulated via feedback loops. What is the switching mechanism that controls protein production and its timing? Anything along the complex pathway of protein manufacture may affect the rate and timing of protein production.

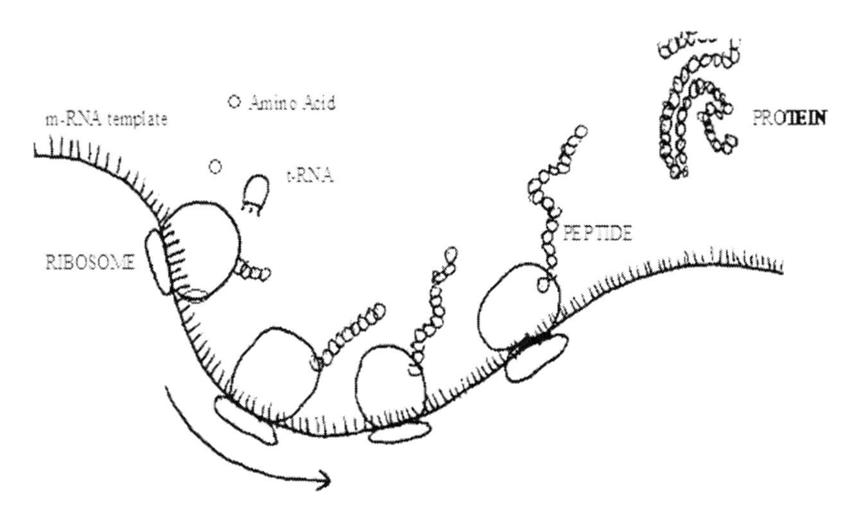


**Figure 12: Protein production by Ribosomes uses a Messenger RNA template.**

A single organism will house the same genes within all of its cells. The DNA is the same. If you find some cells in an animal or plant in which the DNA is different, that is a different being, perhaps a commensal, parasite or symbiote. The DNA inside the neuron is the same as that in any other cell. Yet neurons obviously differ from other types of cells and from each other. The DNA is a library with instructions for the production of all the proteins an organism is capable of making. The first stage in protein production is the manufacture of RNA made on the basis of the DNA code, a process known as transcription. The RNA is an exact complementary copy of DNA formed over its length in the nucleus. It differs from DNA only in the use of one of the four constituent bases with another i.e. Thymine of DNA with Uracil for RNA. Corresponding bases that attract each other line up, the RNA following the DNA template so that at the end of the process RNA, specifically messenger RNA, (mRNA) ends up being made in the linear base (nucleotide) sequence exactly specified by the DNA code. The RNA produced will also be a great deal more portable and there will be many more copies of it. While it is produced within the nucleus it later traverses the nuclear membrane to the cytoplasm. The cell will then produce proteins going according to the instructions of this messenger RNA (mRNA) in the protein producing part of the cell the Rough Endoplasmic Reticulum. The Rough Endoplasmic Reticulum is nothing more than a complex of membranes inside the cell studded with Ribosomes which are like round balls pictured in the illustrations that walk along the mRNA chains an keep adding single amino acids to peptides. Another good way to picture ribosomes is like tape player heads, playing the tape the mRNA that goes through them in order to make the final end product, the peptide.

The most critical stage controlling protein production is in the nucleus at the step governing m-RNA production. A single strip of DNA forms a chromosome contains instructions for up to tens of thousands of individual proteins. There are about six billion nucleotide bases in each human cell's complement of 46 chromosomes. The RNA formed from a tiny part of this nucleotide sequence gives instruction for but one of the proteins in the DNA code. It is told where to begin formation on the DNA template by a certain code on the DNA called a LOCATOR (frequently the sequence TATA). Just after this nucleic acid sequence m-RNA will begin to be transcribed. Another short nucleic acid strand on DNA serves to bind the enzyme that enables transcription (transcriptase). Still another short strand controls whether the DNA sequence will be transcribed and perhaps at what rate. This is the enhancer region of the DNA. Presumably a signal, perhaps an altered enzyme, changes the enhancer region of the DNA in some way to either encourage or prevent the formation of more m-RNA at that particular site. The LOCATOR, transcriptase binding site, and ENHANCER region together constitute the CONTROL region of a particular gene on DNA. This seems to be a uniform system for all organisms including animals, plants, bacteria and viruses supporting the notion of a common ancestry for all life.

The long chain of nucleotides that is RNA corresponds to the exact instructions aboard the long chain of Nucleotides that constitute DNA and a protein enzyme responsible for this process is transcriptase, which builds strips of RNA on the basis of the DNA template. This is the first step of a long line of steps in protein production. Some viruses carry their basic genetic information in the form of RNA. Why not? Strips of RNA nucleotides carry the very same data asDNA. These viruses do no make their proteins directly as we do from their own RNA. You would think that they could theoretically use their RNA as messenger RNA , go directly into a cell and make proteins but they do not. They have to depend on the protein-making machinery of the cell that they invade and become an integral part of that cell's genetic process. Therefore for RNA viruses the first step in making proteins is to make DNA from the RNA template. To this end they utilize a *reverse transcriptase* which makes DNA from RNA nucleotide chains. If these viruses were to invade a cell only to utilize the cell's resources to make proteins from their own viral RNA, they would merely be “guests” of the host cell. Viruses could probably function this way and reproduce successfully. Instead they use their own RNA to make

DNA with reverse transcriptase. This is very profound. Now they are inserting viral RNA produced DNA into the host cell, and in so doing are not merely guests, but part of the government of the cell. Their instructions are right in the cell's nucleus and they are part of it. They have control of the host cell's reproductive capacity. The advantage to the virus is great, much to the detriment of the host. The virus no longer needs to reproduce as a whole, making whole viral bodies or its protein. Instructions for viral production, new viral DNA, will be amplified every time the host cell, replicates DNA, or divides, as the viral infection remains dormant and possibly assymptomatic. Yet at some future time through a signaling process which is as yet unknown, the viral DNA, now replicated many times, will express itself, make m-RNA according to the virus' instruction, and unleash millions of complete viruses into an unwitting host. The HIV virus responsible for AIDS acts this way, and is just one of many RNA viruses using reverse transcriptase. The virus will then take DNA made from its RNA which is now part of the host's (in this case human host's) genome and take the first step to make messenger RNA using everything available from the human cell.

Getting back to the cell's normal reproductive processes, messenger RNA goes into the cell cytoplasm traveling as it does from the nucleus and create a corresponding chain of amino acids which is a peptide, part of a protein. Of the twenty-odd amino acids, each one is given a certain code which is three nucleotides long. For example, the triplet codon GCU for Guanine, Cytosine, Uracil would specify the amino acid Serine. The architecture of the protein peptide is a chain of amino acids and the form of the RNA molecule is also a chain, of nucleotides. These triplet codes evolved at the start of evolution. Douglas Hofstadter has drawn the analogy between computer information processing and what occurs in miniature in every cell and every nucleus. The permutations of nucleic acid sequences that in the nucleus represent sequences of amino acids as they will appear in proteins are in DNA, ATG and C for Adenine, Thymine, Guanine, and Cytosine. The Purine (Adenine and Guanine), and Pyrimidine (Thymine and Cytosine nucleotides form triplets specifying single amino acids. Each long sequence of Nucleic acids thus spells out a protein amino acid sequence. Nucleic acids are bonded together in incredibly long chains and two such complementary chains are loosely connected lengthwise to form a spiral backbone. The translation from the nucleic acid language (each word is a triplet that specifies a particular amino acid) to amino acid language is referred to as "Typogenetics". Given a triplet sequence of the four different nucleic acids, there are 64 different triplet combinations, more than enough than the number of different amino acids. Some amino acids are specified by more than one code (i.e. some sequences are synonyms). Also there are instructions such a those telling the beginning or ending of a sequence. This code is universal for all animals and plants, a lucky circumstance that makes interspecies research a great deal less complex and that makes possible the mixing of genetic material between species in feats of biogenetic engineering. IT also implies, that all life has a common origin.

## A Single Mistake

A single mistake in the process of specifying amino acid chains in proteins can spell disaster. There are so many examples in which alteration of a single amino acid within a single protein causes a severe disease. On first glance this whole system seems to be a rather absurd scheme. Just a single Nucleotide replacement, a simple mistake, will code for a single different amino acid in a protein and will most likely destroy the protein's function, endangering life. Similarly one nucleotide addition yields a missense reading which throws off the entire sequence. But you have two copies of most every gene so that even if you have one bad copy of an enzyme you should easily be able to get by with your one good copy. One of your copies is inherited from your father and the other from your mother of course. For recessive traits you need to inherit two bad copies of a gene, perhaps a single nucleotide replacement from each of your two parents in order to have that disease. The best example is cystic fibrosis, a disorder of the Chloride channel protein in cells. If you have one bad copy of the gene for this protein, you will be OK, two bad copies and you have this terrible disease. Still another example is that of blue eyes. You need two copies of the blue eyes gene in order to inherit the trait which is also thus recessive. Fortunately the blue eye gene yeilds simply a variation and should probably not be considered to be a genetic disease, though you might consider it as such given the proneness of blue-eyed persons to have sunburns, melanomas and other pigment related diseases.

Translation is the process by which m-RNA specifies an amino acid sequence. We have transcription, the copying of an m-RNA chain from a DNA template, and translation, the change in language from nucleotide sequences to amino acid sequences, peptides. Enzymes help turn peptide chains into full proteins. The peptide chains bond with each other to form proteins, some made of 4 or so different chains that bond together, then side chains of other substances, especially sugars and fats can be added. These compounds determine the final shape and other chemical properties of the protein, in other words its personality. The various parts of the chain or chains of amino acids will fold multiple times depending on electrostatic forces, that will determine its the final protein structure.

*Introns* separate protein encoding portions of nucleic acid chains, from exons. Introns aren't merely separators or spaces between pieces of DNA that actually code for proteins. Introns carry information that controls quantities of protein produced and the method of stringing peptide sequences together. The information specifying all biological divergence, the demarcations that separate individuals within a species, even information telling the function of each cell within an organism, is contained within the nucleus of each and every cell. Added to this are specific instructions for the embryonic development of the animal, the migrations and differentiation of cells that will determine the final anatomy of the adult organism. All of this information is in contained in a volume about one micron cubed. A micron is a millionth of a meter; Hair is about

100 microns thick. The information contained in such a small space is reproduced with extreme fidelity. There is no information storage facility that stores so much data in so small a space. No miracle can match this feat of scientific reality. Science fact is more remarkable than science fiction. Biological information processing thus outstrips any system of information storage thusfar conceived by man.

## Ribosome As Metaphor for Consciousness

Protein production is a metaphor for consciousness, especially that part productive of complex structures. At any given moment, it is possible to focus on just a single element, just exactly as a ribosome does in picking out a single m-RNA nucleotide triplet and assigning to it a corresponding amino acid. Then that amino acid will be added to the peptide chain. It is so easy to come up with myriad analogies in human action. The mason adds but one brick at a time, the musician, performer or composer one note or chord at a certain instant. When we discuss vision we will learn that the fovea focuses on only a tiny portion of an entire visual field, and that the brain has somehow to be responsible for forming an entire image. Gradually a peptide structure is sequenced over time. Peptides become polypeptides are bound together, modified, much as melodies and harmonies begin to come together sequentially in time. We are left with the dichotomy of an entire structure, a protein, a building, a symphony, which function as a whole, yet are pieced together sequentially utilizing an extremely narrow focus. Just as a whole human organism is making myriad proteins at one time, he is perhaps seeing, hearing, tasting, focusing on various sensory parameters all at once, still at bottom, the focus has to be narrow and particular, and considered at the level of the ribosome at the level even for a fraction of an instant, at a single nucleic acid triplet. At any instant, consciousness is seen to have a narrow focus resulting in formation of structures, buildings, symphonies, scientific theories that may be more and more complex and beautiful, even as they reach higher levels of abstraction.

| Template: DNA, RNA | Template: Musical idea, emotion | Template: Idea |
|---|---|---|
| Amino acid | note | letter |
| peptide | measure | word |
| polypeptide | melody | sentence |
| peptide with side chains | melody + harmony | paragraph |
| multiple peptide chains bound together | connection of themes | work (story, essay) |
| “tertiary” folded structure | symphony | book |

**Table 5: A linear rendering of levels of Abstraction in creating complex structures, proteins, and music literary works. Ultimately we piece something together starting from a narrow focus:**

This is relevant to the stream of consciousness, what Paul Churchland calls the Joycian stream of consciousness[38] referring to the literary technique used in Finnigan's Wake. Why does time for us seem to flow one event at a time even though we function taking many stimuli in parallel. Our consciousness seems to work like a tape-player head or a Ribosome. We have a single uninterrupted series of events like m-RNA being played like a tape over a ribosome[Φ] . If, on the subcellular level many of proteins are made at once, on a macroscopic level, many mental processes are also taking place simultaneously in parallel and the best computer model involves, as previously mentioned, arrays of microprocessors, each taking in a single instruction at one time.

The creation of worthwhile complex forms requires laying down of a structural idea, a recording process, for which the DNA or RNA molecules are also a model. You can create appealing music without resorting to writing, but the most complex and interesting forms of music are written down. That's why a symphony or opera are a lot more rewarding to listen to than a simple song or improvisation. There's something to sink you teeth into. (I realize in this I am expressing my own musical bias.) Likewise chances are stories and literary works, the Iliad and the Five Books of Moses come to mind, were passed down orally for hundreds or perhaps more than a thousand years. Even in those times there was recording in memory of the original stories. Despite many person's best efforts to recall these stories faithfully, they were undoubtedly subject to embellishment in being passed through myriad generations and human channels. The invention of writing allowed permanent recording, but more importantly, a great building upon a basic structure into a complex literary form.

What I have just described is protein production as a natural process. Today it has become possible to make proteins by design, and for proteins to be produced inside cells that, in their wild natural state never produce them. Technicians take Eschericia coli bacteria, usually harmless inhabitants of the colon, and teach them to produce certain substances such as Insulin and interferons for medical uses. They do this by introducing genes into the bacterium. Then it will double as bacteria do many times to produce quantities of the desired protein that is then purified.

Some disease that represent enzyme deficiencies are beginning to be treated in this way. Adenoviruses, RNA viruses that infect respiratory epithelium can be enlisted to carry the gene making the cystic fibrosis Chloride channel and that may be introduced to some of the most affected cell, namely the respiratory epithelium which is the most affected tissue in this disease of children[ℵ] . The rare disease ADA or Adenosine Deaminase Deficiency. Because of a loss of this one enzyme the immune system is non- functional.

## Elemental Biology Gives Man New Powers of Design –The End of Randomness

Bioethical concerns arise out of the fact that this kind of genetic manipulation, which ameliorates disease is useful but certainly doesn't stop here. It is extremely powerful. Once the meaning and effect of each gene is deciphered and scientists discover reliable methods to introduce genetic instructions into cells and for those cells to be able to go on expressing this information, biology as we know it is transformed. We've been given the keys to the kingdom, in essence. It will start with the introduction of enzyme production that will ameliorate symptoms or cure disease as in ADA deficiency, cystic fibrosis, and disorders such as Gaucher's disease where certain kind of lipid molecules build up on account of the body's inability to digest and excrete them. These diseases and hosts of others, occur because of defective proteins and all we need do to help those afflicted is introduce instructions to produce normal proteins in the proper cells at the right times. Other disorders like Alzheimer disease develop because of a certain proneness or tendency to accumulate substances, in the case of Alzheimer's Beta-Amyloid in neurons. These persons may be helped by providing a pathway to decrease production or accelerate degradation of the offending chemical. This could possibly be done by genetic alteration that reverses the inborn tendency for this substance to accumulate. The next step is the elucidation of strands of genetic material that regulate protein production, introns and other exons that code for proteins that regulate cell division and protein production. This is the control and command machinery of the cell. In these DNA strands is information that bears on all kinds of uncontrolled cell production as in cancers and may even be involved in the buildup of material in atherosclerotic plaques inside arteries.

Knowledge of these processes again, will merely cure diseases. Then locked inside every cell is instruction on senescence, information bearing on why certain cell lines will stop reproducing only after a finite number of divisions. Sooner or later we will learn the facts and be able to manipulate clones of cells, from neurons that stop dividing, to other cells particularly blood elements. Technology has already developed to the stage that regulator proteins have been discovered, produced in quantity, and used to control and increase red and white cell production. For example, Erythropoetin, made by the kidney is given in kidney disease and after chemotherapy and with various kinds of anemias to stimulate red blood cell production.

## Is this Knowledge Good or Evil?

The processes controlling normal aging, as the body falls irrevocably into a state of disrepair are also slowly being uncovered. At this point you may shout, "Wait!!, all of these things, genetic disease and variation, the cycle of life and death have been around throughout the history of biology and are not for us to control. We already know more than is good for us. These processes are for God or nature to control. Perhaps, but there is no way to stop people from using this information. It is debatable that we should even try. Regulate this powerful knowledge here and somewhere else someone will take advantage of this information use it. But you may protest, certain "developed" societies such as in the U.S. and Europe perhaps have such a lead on this research and development that we will hopefully be able to legislate some control in these strategic regions of the world on some rational basis. But the pendulum has swung away from government and toward private entrepreneurial control. Considering the rapid growth of knowledge and new means for dissemination of information (the computer and Internet) we will not have anywhere near the time that we had in the era of atomic energy our last great dangerous technological expansion. One hope: persons who can use this information, and discover how powerful it is, how much they would have to gain financially, will begin to keep it as a secret; there would be limits on dissemination of information by design and copyright.

It doesn't stop here. Beginning with our ability to change superficial characteristics before or after birth, such as skin pigmentation and eye color, which should be a simple matter of adding certain genes into the mix, researchers will start to manipulate certain human characteristics such as muscle mass, endurance, intelligence, emotional tone and level. At some point, perhaps in the very distant future along with elimination (and thereby also creation) of certain diseases, lengthening, perhaps indefinitely the human lifespan, we ought to be able to alter individuals, to create subjects of great beauty and monsters[∅] too if we desire, beasts of burden and nobles, plebeians and patricians, perhaps various breeds of patricians designed by competing entrepreneurial entities. Men may very well not have the same appearance as they do now and may not look like their brothers. And yet we will be the ancestors the progenitors of these various different races.

When I saw the movie Star Trek -Generations, I had to consider how false a view it presents of the very distant future: Humans that are just as we know them today inhabit the scenes of the movie. Perhaps this is to help us identify with human passions and needs. But as mankind becomes adept at manipulating physical entities, so we will be able to change our distant progeny. Advances in this kind of knowledge are for the most part ignored in our future projections. We want to imagine a future world inhabited by us essentially so it will be possible to identify with the main characters, the same beings we know today. Chances are that will not be the case. Various models of humanity will be designed and tried out.

Perhaps the only hope for the future if we are to recognize it in science fantasy, is that design changes will come about, but will not be found to work as well as the model that we have today, the one that has come down to us by natural selection. This is very likely to be the case in fact, for the very distant future. But for me, at least, Kirk and Picard are mutual aliens. If ultimately the general biological form of a human is not a constant, if our form can be altered at will, then we have no base. Humanity harder to define. In that case, where the most basic picture of a person can be altered and may not resemble what we in our time view as a human at all, perhaps mankind should be defined by a historical mode , that is whatever we consider to be a human has human forebears or origins, but in current form, may resemble what we think of as human, very little if at all.

## What is Man? Surely Not His Current Form or Incarnation.

I'm fully aware how much I have gotten ahead of myself. Understand, I am not talking about what is theoretically possible in the future, but getting into the inevitable, that is, if we end up surviving future man-made and natural upheavals. According to this model, the very idea that man is a part of nature is passe. To the greatest approximation up until this time, man is in nature and the great majority of natural processes are certainly not under our control. And of course without the continued working of natural processes, without human life, in other words, nothing else can take place, at least as far as we know - leaving aside religious speculations about a life apart from the real biological life that we know in this world. Also all conscious thought and speculation as we know it, occurs because of intact workings of biological machinery. And without life we have nothing, at least as far as what we can see objectively in the real world. So you may say that life, biology exerts a *permissive* effect, it permits all conscious action and speculation. Apart from that one can now perceive an element of design and control and even alteration of biological processes that are fruits of the brain, a part of brain work, but also still something else that reflects back, even alters brain function, some undefinable will or design or mental process that may even control or change biology.

Without biology, a living structure no speculation, no manipulation of nature is possible.

## The Bioplatform :

One way to picture this idea is that biology provides some kind of platform, or underlayment without which no further action is possible. Without a basic supporting structure nothing else is possible. Once that platform is provided however, the possibilities are limitless. The platform is our biology. In the future we will appreciate more and more how this platform is mere launching pad of sorts. Up until now we haven't been able to do all that much, being unable to escape its confines. If a person had a disease, it would have its way with him. If his habitat did not provide enough food, he would die of starvation because his level of knowledge was limited. Slowly we are liberating ourselves from the limits of our biology. At present we are learning to reach down and change some of the substructure of this platform. In the not too distant future, it will transform into a launching pad, and we'll begin to escape from it, in very much the same way that we've taken our first steps above and outside the confines of mother earth. It happens though few of us think of it this way every time a person determines he will not just let a disease happen to him, and he takes an active role and does something about it. There is, in other words some residuum or part of us, a will, a part of conscious thought or action that is **beyond biology.** At a certain phase in mental processing biological considerations cease to define the individual and we have to look at something else. This proves the case that there is more to mankind than biology alone. It seems to me reducing mankind to biological considerations will usually work, given enough biological knowledge and is a good approximation in the world as we know it, much as is Newtonian mechanics in physics. But we are demolishing the biological model as a complete system. There is a will which supersedes biological processes and seeks to control them, seen most clearly in the practice of medicine but in other endeavors as well. And in recent years, technology has developed to the extent that we finally have the wherewithal to express this quest to control biological processes. Perhaps this is most apparent in new knowledge of genetics. What is important here is not how far we've come in this process, but that we are involved in biological control at all, something no other animal has been able to do.

It is not always possible to do a real experiment, to know something empirically. Physicists are used to the concept or a thought or Gedanken experiment. It should be possible to devise an experiment utilizing reason, practical knowledge without actually having to do the experiment itself. The best example is Einstein's famous twin experiment that helps prove relativity. One twin is traveling at a high speed close to the speed of light and he comes back younger than his stationary brother. It's hard to do such an experiment but one may figure out the results that have in fact been proven at the level of subnuclear particles. In biology, since we don't yet have all of the advance warning of what is to happen in our own time perceptions of our future constitute a thought experiment. Such considerations bring ideas to life about us manipulating our own basic structure and our ability to separate ourselves from our biological underpinnings, helping us to realize how dependent on our biological machinery we are, and yet that as persons we have stepped quite beyond our biological endowment. We have already widened our horizons, stepped away from our biological heritage, but much of this has been imperceptible or at least it has gone unnoticed. Certainly we've come a long way from our origins as the erect ape of the African savanna. Humans have radiated into all of the world's inhospitable climates, learnt about shelter, and agriculture and medicine. Paradoxically the exponential increase in biological knowledge, promises to teach us even more about ourselves, especially our non-biological essence.

The cell is a protein factory, proteins being used for structural repairs and to form enzymes and hormones signaling other cells to action. Neurons, like other cells, need to produce only certain chemicals and manufacture them within certain precise time frames. Luteinizing Hormone, and Follicle Stimulating hormone are made and released precisely to bring about a normal woman's menstrual cycle. One mechanism occurs at the level of transcription or RNA. These are regulatory schemes that can intervene at different stages of the protein production sequence. Other regulatory processes come into play in translation, the process by which RNA sequences are translate into actual amino acid strands, peptides, and also in the release of these proteins inside or outside the cell.

The protein thus produced is often packaged in specialized structures made of lipid membranes, part of and similar in structure to the Endoplasmic Reticulum. This Golgi Apparatus named for Emilio Golgi, a turn-of-the-century anatomist, was first discovered in neurons where this organelle is very prominent. These laminated piles of membranes separate certain substances, particularly active newly manufactured proteins, from the rest of the cell. Some proteins and chemicals, made by the cell, would be dangerous to it if unleashed within the cytoplasm, and also for various other reasons need to be made and stored then released under controlled circumstances. Digestive enzymes like Trypsin, would reek havoc within the cytoplasm digesting everything in sight unless kept from the rest of the cell. Golgi bodies are systems of membranes that form free vesicles composed of lipid material. These lipid bubbles hold chemical that may later be useful to the individual cell or package substances later exported from the cell. Neurons are specialists in packaging chemicals for export because they are used as signals both to other neurons and also to non-neuronal cells. At a certain point when these substances are released the Golgi produced membrane of the vesicle merely fuses with the cell membrane and molecules are dumped out into the environment in a process termed exocytosis.

Numerous cells specialize in chemical production for export, guns ready to fire in a sense, except as we have seen, deciding when to fire based on myriad inputs. Some cells are secreting cells or glands, those with ducts in the digestive tract for example, and ductless cells that release hormone directly into the blood, such as the pituitary gland, the controller of other ductless glands. Plasma cells make protein antibodies in prodigious quantity. They may be considered quintessential protein factories and show it with their rough, ribosome laden endoplasmic reticulum that dominates the cell structure (Figure). Each plasma cell or clone of cells produces one specific antibody for export. The antibody recognizes and attaches to only a very specific individual chemical, foreign to the animal and needs to be neutralized for later destruction.

Neurons, like plasma cells, produce a variety of vesicle-stored chemicals for later release. These substances, a few examples include Serotonin, Acetylcholine, and Norepinephrine, carry messages. They are typically produced in the cell body, then transported down the long axon to a bulbous swelling at the end of the axon, the axon bouton. At that location, the vesicles bind with the cell membrane and release a small packet of material, remarkably an almost constant number of molecules of the material, into the extra-cellular space.

As this happens at a synapse, the junction between two nerve cells the chemical messenger will either excite or inhibit the nerve cell membrane exposed to it. At the neuromuscular junction or synapse between nerve and muscle, Acetylcholine, excites the muscle into a state of contraction. Small numbers of vesicles are always being released at the axon bouton, each containing a specific quantum of neurotransmitter. Some quantity of chemical is always spilling into the synapse and it seems just as if the neuron is always poised to release its substance. But when the neuron is excited that is invaded by an electrical action potential, then there is an enormous orgasmic release of its substance after which comes a short refractory period at which time the neuron is not excitable at all. The chemical released into the synapse in the meantime affects the electrical potentials in the adjacent cell. The released neurotransmitter will make the adjacent cell membrane (the post- synaptic membrane) either slightly more or less excitable depending whether the particular chemical excites or inhibits.

Other neurons release their own chemical messengers directly into the blood, as in the hypothalamus, which is a neural structure controlling endocrine glands. The hypothalamic neurons release short sequences of amino acids, commanding the pituitary gland which hangs from the bottom of the brain. A special circulation, a portal system, carries these commands in the blood between hypothalamus and pituitary, the blood dumping regulatory hypothalamic peptides right into the pituitary, itself a controlling endocrine organ. The pituitary, in turn, controls all of the ductless endocrine glands of the body. That is how the brain exerts control over the entire endocrine system, via carefully modulated chemical production and release of chemical products, starting in the brain.

The neuron is like other cells. It has organelles that function just like those of other cells. The neuron's structure reflects specialized function. No other cells are elongated as some neurons which can be over a yard long, yet their cell bodies are microscopic. This is part of their unique form. In fact there is such a great diversity of the structure of neurons, some of which are tiny like the granule cells of the cerebellum, others long, like sensory ganglion cells. The classic model of a neuron includes an *axon* which is a great length of cytoplasm and easily the most distinguishing feature of a neuron. All neurons do not necessarily contain axons, as such, perhaps not even the majority of them; some have two, some many. But the standard neuron model is that of a cell body and an axon, or cell extension. That's because the major job of a neuron is to communicate, and the axon will carry a signal from the body of the cell sometimes over a very long distance, to communicate with another cell. As an example you can take the motoneuron, a cell inside the spinal cord, which when excited, will send an action potential, an electrical signal all the way down its long axon. This action potential will reach the end of the axon, the axon bouton, causing Acetylcholine to be released into the neuromuscular junction. Acetylcholine excites the muscle membrane causing the muscle cell to contract. What you are doing is sending information, command and control, over long distances, just like a message over telephone wire.

Having a distant outpost like the end of an axon is a great responsibility. Not only electrical messages but nutrients and supplies, especially the neurotransmitter such as acetylcholine, are made, in the cell body and need to reach the distant end of the axon. As in larger animals, cells maintain their shape with a cell skeleton, a *cyto*skeleton. The major component is the microtubule. This is a rather complex structure comprised of peptides of tubulin subunits bonded together into protofilaments which, in turn, form the entire microtubule. The microtubule architecture is highly conserved and used by motile cells in cilia and flagella, as in sperm flagella and in basic processes such as cell division. The mitotic spindle that pulls the chromosomes apart, each daughter cell taking for itself a portion of chromosomal material, is made of microtubules. Substances made by the central cell body travel, often inside vesicles, back and forth much like railroad cars over tracks, on the microtubules that at once, preserve the shape of the cell and axon, and serve to supply the distant axonal outposts of the cell. This is a process known as axoplasmic transport or axonal transport. Diseases of neurons may derange this process. It is these structures and related ones that get involved with neurofibrillary tangles seen under the microscope in Alzheimer disease and other associated proteins, such as "microtubule associated protein" MAP which is tested in cerebrospinal fluid in Alzheimer disease. In Alzheimer's disease something disturbs the use and deposition of these proteins which alters the function of the cell, causing cell death.

**Figure 13: Neurons and trees look similar. Branches are like dentrites that coalesce at the trunk, analogous to the cell body and axon. The cell body may not be prominent.**

If axons are great cytoplasmic elongations, *dendrites* , taken from the Greek dendron for tree, are just like branchings and re-branchings of trees that are closer to the cell body and nucleus. Dendrites communicate electrically with the cell body. The dendritic input is from other cells, often from thousands of axons that transmit through them to the cell that at any instant correlates these thousands of inputs and decides whether or not to fire. Or, failing to fire the neuron reaches a certain level of excitation and functions as a communication device. The pictures or some types of neurons illustrate the critical link of form and function.. The best example is the large neuron within the cerebellum the Purkinje cell. This cell that lines the folia or cerebellar convolutions, receives input from thousands of synapses over its elaborate dendritic structure but also along over its axon. All of that input, some of it excitatory, some inhibitory, is integrated. The final output of the Purkinje cell axon is inhibitory or controlling. What does the cell control? Fluid and precise movements, fine hand movements and balance as of a ballet dancer, illustrating how wrong it is, how far off one must be to consider this complex executive merely as an "on' or "off" state at a given moment in mathematical representations. Each Purkinje cell is an executive.

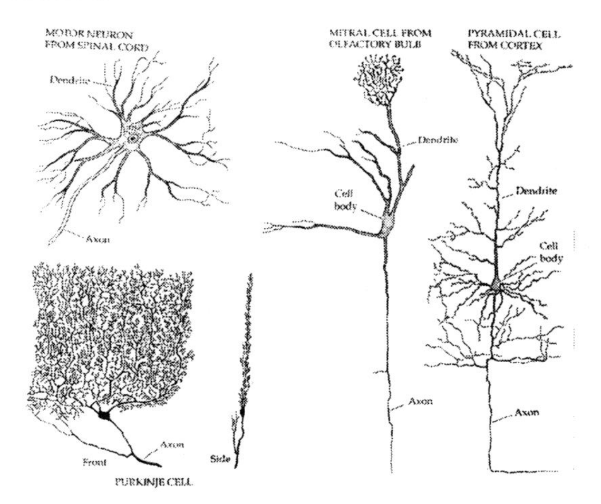


**Figure 14: Shapes of neurons. Drawn by Deiters and Cajal. To give some idea of the variety of shapes and function of cells[39].**

A large portion of a neuron's function is determined by its shape. Consider the tens of thousands of branchings that make up its dendritic system. Also the typically long axon that may be over a yard in length. The architecture or the cell determines just how information will flow through it and to other cells in this vast machine. While there are exceptions, the information flow is generally from dendrite to cell body to axon.

The overall structure of the neuron can't be left to chance, for the neuron is in the business of establishing contacts. It is a social type of cell that never works in isolation. The significance of the cell's specific shape is that some synapses lie closer, some farther from the center of the cell. Given thousands of inputs, some will have more, some, less influence on the decision for the neuron, whether or not to fire. The single synapse has ordinarily, very little influence on this decision. At any given instant, the sum total of influences determine what the neuron will do. These are summated spacially, that is some synapses are closer, some further from the central sphere of influences, the inner circle, and temporally, there is the sum of inhibitory synapses and excitatory ones affecting the decision of the neuron over time. With transmission across a synapse the effect on the cell will be felt only for a finite amount of time, then if it is without influence the effect will cease. The overall sum of synaptic influences determines the degree of excitation of the cell, especially its state of polarization, whether depolarized or hyperpolarized.

Other synapses are on the axon of the cell and their influence is different. Some, that are close to the release of neurotransmitter laden vesicles, no longer affect the action potential but modulate the number of vesicles should the cell fire. Some of these synapses modulate the amount of transmitter released in cells which release transmitter without firing. Synapses structurally then are of many types, axo-dendritic, perhaps the most common type, axon to cell body and axo-axonal, dendo-dendritic.

Cytoskeletal proteins in part help to form the nervous system and maintain the structure of nerve cells so important to the neuron's gregarious nature. The neuron is the only cell whose cell body is stretched so as to provide connections with fellow neurons. It does not function as an isolated executive, but lives, much as a whole person does, as part of a complex network with its neighbors.

Electrical impulses need to travel considerable distances in the central and peripheral nervous systems. The CNS has its own myelin on axons formed by oligodendroglia whereas peripheral nerves have their own chemically distinct myelin made by Schwann cells. Myelin is a fatty insulator blocking the egress and ingress of ions over the long axon. The larger the diameter of the axon, with its thick myelin, the faster is the conduction. Touch receptors depend on thick axons with myelin, whereas pain fibers are unmyelinated or lightly myelinated, hence conducting information slowly.

In the myelin covered zones Sodium, Potassium and other ions are unable to pass through and an electric charge is felt over these zones, instantaneously creating an electric field spreading over the volume of the axon over a short distance. When an action potential invades the axon covered with myelin there is a fairly pronounced alteration of charge that goes instantaneously over a distance in the axon. However over the myelin covered area that electrical difference cannot result in any flow of ions across the cell membrane. This must occur only in the denuded uncovered node of Ranvier. As soon as the field is felt in the nodal area Sodium rushes in and there are other ionic changes similar to those described above with neuron excitation. Again, there is the spread passively (electrotonically) of a field of electrical charge influencing the next node which will in its turn respond with its own flux of ions. The passive electrotonic spread of charge in the internodal (covered) area is almost instantaneous and to the extent that it is possible for electric currents to travel in this way, transmission will be very fast. This can only be achieved over a very short length of axon before the spread of electrical charge degrades and will no longer be a factor. The nodal denuded zone serves to recharge the battery to create another area of electrotonic spread.

Axons also function without myelin, but in that case electrical conduction is comparatively slow. Take away myelin from an axon and you may even find its conduction completely blocked. If the diameter of the axon is larger one can make up for some of this lack of speed because a larger diameter will conduct impulses faster. This is the strategy employed by invertebrates such as the squid which has some "giant" axons studied intensively in the past because of their accessibility and sheer size. Larger diameter axons in humans too tend to conduct impulses the fastest but this is also because these axons are invested with the thickest layers of myelin. You can feel the difference in speed of conduction of various different axons which depends of the axon thickness. If you smash your toe you are almost immediately aware of it because of the initial non painful sensation felt that travels through to the brain over the thicker touch sensation axons. You may anticipate a wave or pain coming only later, because the axons carrying painful sensation are so much smaller and slower conducting axons. These groups of axons will be traveling in the same pathway. A group of axons runs in parallel making up a cable of axons which is a nerve (called a tract in the CNS).

If myelin is affected serious disease is the result. In peripheral nerves Guillian-Barre disease affects myelin and slows or blocks nerve conduction. Here one's immune system is made to attack the myelin covering of nerves. Inflammatory white blood cells invade the region surrounding axons and help to destroy myelin. Signals can't get through. A severe paralysis can result as well as numbness and other changes in sensation. Some similar problems occur in the central nervous system with a similar process resulting in inflammation and destruction of myelin in multiple sclerosis. Both Guillian Barre syndrome and multiple sclerosis are diseases of myelin.

When you consider the central importance of the shape of the neuron in its function, you begin to wonder what determines the architecture of each cell. In part, this is built into the genetic code, but think of the enormous amount of information would have to be stored genetically were genes to specify exactly the hookup of each and every cell. This would be impossible. Connectivity and shape needs to unfold and develop within the lifetime of the organism and is part of a cascade of events that unfolds in the embryo. Later, this modifies further as the effects of learning and experience are reflected in the formation of new synapses and alter the cell shape, perhaps by creating new axon and dendrite sprouts. The exact positioning and structure of supporting cells like blood vessels and glia have their own effect in determining the final structure or design of the neuron. Undoubtedly there is a constant feedback loop of interaction between neural transmission and structure that determines it. The structure determines function which determines structure in a sort of continuous dialog.

Let us take the case where the axon synapses with a dendrite (illustration) the axon of the first cell will carry an action potential that will invade the terminal. As described we would then expect a release of a chemical neurotransmitter from the axon bouton. Vesicles containing quanta of this chemical would fuse with the cell membrane and many molecules of this powerful communicating substance would be liberated into the immediate space between cells, i.e. the synapse. the transmitter will reach the membrane of one (or more) of possibly thousands of dendrites of the second or post-synaptic neuron. The transmitter is not absorbed by this membrane but instead exerts and effect on a receptor protein. this post-synaptic membrane is studded with such specially designed receptor proteins (specially designed because they fit stereoscopically pretty closely to the physical and electrochemical shape of the transmitter.

What happens next is that after fitting closely the receptor protein is altered in some way, usually so that there is a conformational change in the membrane and certain ions, for example Sodium are now allowed to traverse the membrane in small numbers. This will change ordinarily minutely, the electrical properties of the post-synaptic neuron. Alternatively the receptor protein will affect other intermediary proteins ("second messengers") within the cell to exert an effect. A direct electrical effect will be felt just very close to the actual area in which it occurs. This is why the exact structure of the postsynaptic neuron with all its dendrites is so critical. A single event at a single axodenritic synapse means next to nothing to a cell with thousands of such contacts. This signal will ordinarily be lost and have no effect. What becomes of it strongly depends on what is happening in other dendrites in the area who will also be involved with their relatively tiny bursts of transmission and ion fluxes. If enough of them get together within a short enough space of time what will happen is that this individual neuron will be made to fire with its all or none action potential. Anatomy and cell structure help determine just how these interactions occur within the matrix of space and time. Each cell comes to a decision based on precise spatial and timing characteristics, determining whether or not to fire. There are inhibitory synapses and excitatory ones so that negative and positive influences are made to summate. Add to this the effect of internal and external chemical and electrical modulators determining the state of excitation of an individual neuron.

The concentration of Calcium at an axon bouton helps determine the amount of transmitter released. If there is less Calcium, less transmitter will be released with each excitation. Certain changes can take place, with repetitive stimulation that ultimately can alter behavioral responses. An example is given in experiments of habituation in the sea hare[40] (Aplisia) that occurs on a cellular level. Habituation refers to behavioral responses that decrease with repeated stimulation. You're reading and you suddenly hear a loud noise. You have to get up to check it out. You go back to what you're doing and hear some other sounds. Perhaps then you look up. Pretty soon you are going to continue working and ignore the extraneous noise. You've habituated. This is an obviously adaptive neurological response which it has been shown can take place at the cellular level. Aplisia will contract its gills when its mantle shelf is stimulated. An axon coming from a sensory nerve excites a motoneuron controlling gill contraction. But with repeated stimulation, pretty soon the sensory axon will start to secrete less transmitter. This is attributed to an altered Calcium concentration at the sensory axon bouton. The point here is that a rather advanced behavioral phenomenon is explained on the basis of a subcellular event.

Habituation is one kind of learning, a change in neural response that happens as a direct result of exposure to a stimulus. Habituation in Aplisia thus may model how learning takes place even in more complex situations. Ultimately a change has to happen at the level of the individual cell. But even habituation which is only one kind of learning has wide variation. It happens with a disorder of the middle ear causing vertigo due to a signal mismatch between balance organs of the two ears. Eventually the brain adjusts and dizziness and vertigo diminish or habituate. Exercises that cause vertigo are routinely prescribed to decrease the response of the brain to the ears signals and thus relieve vertigo. Habituation plays a pivotal role in drug addiction. The body begins to respond less to a given dose of an addictive drug and requires more to get the same effect , increasing craving for the drug. The neural mechanisms are very different in each of these instances to habituation, which is a type of learning, and no one is saying the specific process found in the sea snail relates to these kinds of habituation.

The very same configuration of synaptic influences exciting the cell at one point may not produce an action potential at another time. There are chemicals secreted within the individual neuron that modify the cell's response, changes related to such factors as fatigue and other modulating influences coming from higher and lower levels within the nervous system. The axon secreting the transmitter is subject to influences as well. Some of them have presynaptic synapses that modify the amounts of transmitter released. There are humoral factors i.e. hormones that affect cell's level of responsiveness.

It is apparent that individual neurons can respond at much higher levels than they do in an intact healthy person. Higher nervous control occurs mostly through inhibition. As we have observed, the main effect of higher level input of the brain on the spinal cord is to decrease the deep tendon reflex. Cut off the influence of the brain or higher centers (for example by cutting the spinal cord) and you will have wildly active deep tendon responses. In patients with severe spinal cord injuries there is a host of severely active uninhibited responses. The urinary bladder may contract violently causing urgency while the sphincter muscle, also uninhibited, contracts, blocking the outflow of urine. Slight stimulation of the foot which in the usual case would cause no withdrawal at all, can cause a violent triple flexion response and even a pronounced sympathetic output, including profuse sweating, blood pressure fluctuations and even erection. Lower spinal cord mediated responses, now liberated from higher influences.

Inhibition of the same type occurs within higher centers, even the cerebral cortex. Certain degenerative diseases that cause a loss of cortical neurons produce an overreaction to stimuli. For example a sudden stimulus may cause your whole body to jump. In some diseases sudden total body jerks called myoclonus happen even with minimal stimuli. These are disinhibited cortical responses and have to be considered as part of an orientation response to new stimuli. For example, if you suddenly hear a loud sound, your whole body will jerk. This is bound to occur particularly in a quiet contemplative environment. Repeat the same stimulus and very soon, no sudden jerk will occur. Similar total body myoclonic jerks occur in the early stages of normal sleep. Under certain pathological conditions sudden jerking is seen in relation to any slight stimulus, stimulus sensitive myoclonus. The entire body will jerk with any new sound that interrupts the background.

These sudden starts have an electrical counterpart in the EEG which can show high amplitude sudden generalized electrical spikes. Evoked responses, which track the brain's electrical response to sensory stimuli often show giant electrical potentials over the cerebral cortex. We see this in conjunction with severe disease of the gray matter. An example is dialysis encephalopathy associated with the buildup of Aluminum in neurons of kidney dialysis patients. Another is with serious degenerative conditions such as Creutzfeldt-Jakob disease caused thought to be caused by a primitive virus like organism, a Prion. Certain childhood diseases cause similar but not completely identical high amplitude uncontrolled waves to occur over the cortex including subacute sclerosing panencephalitis caused by the measles virus.

These disorders have many features in common. They all involve degeneration and loss of cortical neurons and electrically we can see giant potentials over the cortex which clinically loosely correlate with myoclonus. Most importantly they indicate that so called lateral inhibition occurs right in the cerebral cortex, that higher cortical neurons simply inhibit other higher cortical neurons. The brain simply isn't revved up to respond as strongly as it possibly might. Rather, there are intrinsic mechanisms that exert control and specifically by the mechanism of inhibition.

The cerebellum is the major organ of motor control of postural and fine movements in the extremities. The output of the cerebellum seems to be entirely through large neurons, the Purkinje cells. Their influence on other motor centers seems to be entirely inhibitory. It's easy to see that control is accomplished largely via inhibition, after all, this is the major mechanism of political and military influence and why should the brain be any different? Perhaps control is accomplished universally through inhibition. Lest we become convinced that this is the case we should consider examples in which higher control occurs through the initiation of action as in the cerebral cortex which is the theoretic author an initiator of motor movements.

The cerebral cortex, as the major initiator of movements, has intrinsic mechanisms to inhibit them. This is partly, no doubt, in keeping with the cortex's role in control of movements. Also we know when electrical discharges go out of control as in an epileptic seizure, the cortical neurons have mechanisms that suppress such discharges. This mechanism may occur simply as a matter of fatigue. For example, we know that a neuron that fires, is unable to fire for a variable period which for each individual cell is its refractory period. The electrical properties built into each cell simply dictate that neurons are unable to fire at more than a given frequency. There are other mechanisms that inhibit neuronal firing. These include simple fatigue and the inability to function once certain nutrients such as oxygen and energy sources supplied by the blood are used up. The neuron's intrinsic mechanism for energy utilization that resides in its mitochondria, also plays a role.

What happens if neurons or groups of neurons continue firing as occurs in an epileptic seizure? Due to the above mechanisms neurons are unable to continue to fire and the most actively firing neurons are the ones most actively suppressed. certain patients may have a seizure that come from an uncontrolled electrical discharge in the right frontal lobe. This will cause uncontrolled repetitive movements on the left side of the body, say, depending in the group of neurons involved in the left arm. For a while the left arm will jerk repetitively. But later, the left arm will stop jerking. If we examine this person just after the jerking stops we may find that his left arm is paralyzed, that it just can't move. This is because the neurons responsible for the discharge are temporarily unable to function and it is called a Todd's paralysis. If a doctor examines a patient after a seizure and finds signs of paralysis in a specific area, he can safely assume that the seizure was initiated in the area of paralysis. The area of the brain corresponding to the paralyzed limb (in our case the right frontal lobe that initiates left arm movement) may be permanently damaged due to some other cause. It is just as possible that neurons may either be suppressed or exhausted from the previous vigorous uncontrolled firing from a seizure. It is clear that cerebral mechanisms of inhibition are at least as important as initiating mechanisms.

Just as in a Todd paralysis the most actively firing neurons are most actively suppressed there are all kinds of regulatory mechanisms, likely feedback loops that suppress a single region out of control. New PET data has shown how short lived normal transient emotions that we all experience such as sadness, differ from chronic pathologic emotional states such as clinical depression. Although you wouldn't think so somewhat different areas of the brain are involved. And it may be theorized, mechanisms that suppress active emotive areas of the brain, after an appropriate period of activity may be disordered. An active brain areas perhaps in the amygdala, a part of the limbic system, is not sufficiently suppressed and we have what amounts to an uncontrolled unmodulated “emotional seizure”. Emotional disorders seem then to implicate controlling, modulating, suppressing brain mechanisms. At this point much of this is speculation but it may explain how antiepileptic agents such as Tegretol and Depakote, general inhibitors and suppressors of electrical activity, have recently been found to work quite well in affective disorders such as manic-depressive psychosis.

The general trend in neurology in the Nineteenth and earlier in the Twentieth century, emphasized pathology. Most clinicians were also versed in pathology. Pathology was the linchpin of neurology training even in my own training just twenty years ago. Then what distinguished various diseases, what always defined these disorders, was a distinct pathology in the brain or the nervous system, a lesion you could define. At that point you would know if there was anything you could do to help (usually not) and there was a satisfying feeling of recognition of the disease and your ability to predict its course. I was being trained in the era of the lesion, an abnormality to be identified either grossly or under the light microscope. The lesion defined the disease and the greatest bulk of information on how the brain works, could be inferred by seeing what happens when a part of it broke down or went out of commission.

We learned a lot by approaching our subject matter in this way. Just some of myriad examples are that the anterior periSylvian region of the left brain termed Broca's area controlled the motor aspect of making language. The posterior peri-Sylvian area in the

Temporal lobe connected with language reception and understanding. Wernicke's area, the basal ganglia deep in the brain, the frontal cortex, and the cerebellum each contribute to smooth motor function in their own way, that we know about from seeing what occurs when the function of these areas is destroyed. The aim of the clinical exam was to localize the problem in the nervous system, to decide whether what the patient had, had to do with the spinal cord, or cortex or brainstem or cerebellum or some other structure. The great clinician, only just a few years ago, was the best at localizing and only after discovering the locus of the problem, would take the next step and hazard a diagnosis. We talked incessantly about localization in those days. It was with the increasing popularity of the of computer, that talk of localization got translated into the jargon of the module. The brain, when all is said and done has a distinctly modular structure. The whole nervous system, indeed the whole person, is nothing more than a series of modules, little substructures that perform their specialized routine. Ruined your knee or hip joint? Replace it. Done your heart in? Transplant it. It is not quite that simple for the nervous system or brain because it tis heavily interconnected and performs even simple task with the aid of extensive communication. This is the function of the white matter, and all of our peripheral nerves, which carry information. Today we can even watch, with the aid of technology, electrophysiolical tests such as computerized EEG or on PET scans that look at metabolic function in areas of the brain, how even the simplest sensation, ramifies or travels throughout what is really a whole integrated neural structure. As such we have started to appreciate the function of individual parts of the brain and the integrity of the whole structure simultaneously.

But to the greatest extent, analysis, the separation of the nervous system into its anatomical parts, the assignation of specialized function, is extremely useful. For the person with Parkinson disease in which you can see degeneration in the Substantial nigra, a loss of its normal blackness of pigmentation and severe dropout of neurons, you may be able to transplant cells, hopefully from a fetus whose cells are immature and may be able to grow. Or you can create yet another lesion in the group of cells that ordinarily counteract the substantia nigra's effect, with a pallidotomy which burns a hole in the globus pallidus. I mention all of this merely for purposes of illustration. Just as in a computer or a car, you strive to find out what is wrong and what needs to be replaced, or beefed up, depending on your particular problem, the hard drive, memory chip modules, and so forth, so the brain is conceived as merely a conglomerate of interchangeable modules in a "plug and play" structure.

I started deliberately with the argument of the ultimate lesion experiment, brain death. We saw that even in this extreme case there were unanswered questions. What is the organizing structure of the brain? How do we account for the sensation of consciousness, that is the feeling that we all have that we are alive? Should the brain be thought of as a conglomerate of different modules? The argument then shifted frequently from macro to microscopic. This is because each view has a great deal to contribute.

## The Brain is Not Different From Other Organs; It is Peripheral to the Purpose of the Organism

From this vantage I began discussing the concept of the brain lesion. Along the way we saw for the first time how various brain regions function. When a lesion develops, the organism will continue to function as a whole, unless that lesion is lethal of course. He will find some way to survive, often by finding some inventive way to get around the impairment and, if possible to preserve the very function impaired. If a stroke destroys a large part of the left frontal cortex controlling verbal expression, the frustrated subject will still try to express himself. As we have seen this struggle to get around the effects of a brain lesion is not at all different from a person hobbling on a broken leg. In examining individual brain functions we see a common pattern and we are forced to draw an inescapable conclusion. The brain is not different from any other physical entity in the body, any of which can be diseased enough to hypofunction and cause a deficit. The organism is left in the same boat in any case, still having to find a way around that deficit. And we have come to see the brain, any of whose parts we have so far described and mapped out function as a sort of tool or instrument of one's will, somehow peripheral to a basic purpose or desire. We have not yet been privileged to find the anatomical locus of this basic driving will, nor can we be optimistic that it will ever be found anatomically within the nervous system. Moreover we have come to see by repeated example just how much neurons and the organic structure of the brain, resembles all other organs, also subject to disease, aid function while intact, but produce their own deficits in a diseased state[Φ].

In many patients it is possible to witness the accumulation of disease in various brain regions or anatomical modules. Many persons suffer multiple brain strokes that pick off brain volumes one by one. Some degenerative disorders like Alzheimer disease accumulate damage to multiple brain regions. And so it is possible to see, "the bright lights of the brain extinguished one by one like lamps[Ψ] ", in the words of James Joyce the destruction of brain modules. But all the while, until the very end you witness simultaneously a struggle to function despite the existence to a deficit. The Alzheimer sufferer will write things down when suffering from a memory disturbance, striving to function, to limp in that broken limb. And what you discover in all of this, and this is the most profound lesson of the lesion experiment, is that that the brain like other organs, is still peripheral to a central will, still an instrument that helps us to accomplish the tasks of life, exactly like a leg, a liver or kidney, no different from any of these entities. So that therefore even after the study of the brain, there still remains some central principle, some kernel of self whose physical and anatomic boundaries are uncertain at this point. This does not diminish the fact that it is possible to build up so many modular destructions, that you will eventually see the dissolution of the whole personality, as eventually occurs in such degenerations as Alzheimer's, but again this very same phenomenon can be seen in disorders of any other organ system, any of which can spell death.

Dealing with Alzheimer's disease is difficult not just because it is, from the practical sense, for patients and families so difficult, but also philosophically. You witness the total dissolution of the personality over a relatively short timescale. In other disorders this short deterioration is even accelerated. I'm speaking about so called slow virus disease such as Creutzfeldt-Jakob disease in which we lose the sufferer in a matter of months rather than years. In Alzheimer's what happens as the months progress is that generally there is, of course, a diminishment of abilities. Whatever was once recalled automatically requires attention and needs to be written down, otherwise it will get lost. Appointments will be missed; items will not be picked up at the supermarket. Then there will be other problems in functions that we take for granted, getting the proper change, balancing a checkbook. The speech may take on a little more halting quality and may be hesitant and unsure. At an early stage, and surprisingly for a very long time in some cases even in more skilled occupations, not requiring acquisition of a lot of new information, a doctor or lawyer even may function, without any protest from clients or co-workers for a surprisingly long time. And as a general rule, most premorbid reactions and adaptations continue to hold; an easygoing person, continues to be easygoing, a depressed or irascible individual stays that way, but in some cases, spouses particularly notice that there is a personality change. In some cases a man can feel the decline in his abilities and denies this or reacts to it, so that a once likeable character becomes a bear, always angry and defensive. At some point, he retreats from former interests and hobbies, pleasures that used to make life worthwhile and leads a constricted limited existence. Performance, one's own occupation, whose skills for the most part, had been acquired many years earlier and are relatively resistant to alterations in learning ability and recent memory deficits, now become noticeably impaired. Colleagues, co-workers, and clients now find the person's performance unacceptable and our subject is forced to leave his occupation. Later on even simpler motor and sensory, linguistic, logical functions disappear. Movements and gait and sight may even be disturbed on an inexorably descendant path of deterioration that leads to death. To some degree this may describe a sort of accelerated aging, as many capacities fail in the very old. In Alzheimer's the decline is more certain to be global, in the ultimate sense, no ability will be left untouched.

What I am trying to get at though is that in seeing this time and again, one begins to wonder whether there is an enduring permanent personality structure, after all this is what is meant by the personality, enduring attributes in a human person, or whether whatever we see that makes a person, is entirely determined by function as we see it at a given moment, of nervous structures. This question comes to the fore in seeing impaired individuals and how they cope. To some degree, a person's coping strategy remains the same as he continues to experience the throes of his illness. An easygoing optimistic person may sometimes remain optimistic and rarely becomes abusive, a bear remains a bear, but this is by no means always the case. In some cases the person changes the basic way they react to things. He seems to be a different being, a different person. We have a situation where a person makes a dramatic change in self. In the general sense, not only with Alzheimer's disease, though this is a dramatic example, persons change, are transmuted, during the process of living out their lives.

We know that persons do, change dramatically. Suppose we believe that something of the person endures after life, that there is a heaven or a hell. Then there must be some notion of an enduring object some unchanging entity that is sent to a heaven or a hell. But as we have seen the personality, at least the manifest personality, changes throughout life, sometimes slowly by development and life experience, sometimes dramatically with a serious brain disease, but the personality structure is a moving target. Which person goes to the afterlife, the child, the adult, which adult the 25 or the 50 year old defines the individual. This non-fixedness of the person makes notions of heaven and hell less tenable. After considering Alzheimer disease and the powerful effects such a disorder has on the basic personality structure it is reasonable to ask if there is, in fact, anything to a person except what may be defined by organic machinery.

As we have seen the way out of this conundrum is to concede that the brain and the body, and all that is organic or biological, is a base for what makes us human, permits us to act and to feel. To a certain extent we can live with some malfunction or hypo-function of these structures. If we have a deficit, we somehow get around the problem as long as we are able and there is no insurmountable accumulation of deficit caused by hypo-function. Biology, mechanics and bodily function serve as a platform. Take this platform away and the person disintegrates. An intact platform makes possible the performance of life work, like a diving board, or even better, a sort of launchpad. We are mistaken when we see this biological function as being all there is rather than observing that it all is necessary but not sufficient to explanation.

In the latter part of our century we saw an explosion of techniques that have allowed us to dissect the microscopic functions of the cell and its various organelles. We have witnessed breakthroughs, in biotechnology, electronics, neurochemistry , genetic mapping and cloning in particular that emphasize a complementary microscopic and sub-cellular view of neurophysiology, so that in subsequent illustration we found it beneficial to alternate from macro to micro processes, have achieved a fusion of these vantage points. We have come to see the tiny neuron as a mini-executive, a decision-maker dependent on sometimes tens of thousands of different inputs, yet through all of this examination, we have not been able to come up with a specific locus of an initiator of thought or action. Indeed the best anyone has been able to come up with is a picture of the brain as some type of grid of interactive modules that vaguely sets performs unitary actions. Yet our understanding of how biological processes affect consciousness, making that critical step from the mechanical to human experience, still eludes us.

We have seen, the attribution of all thought, perception and logic to biological processes is epistomologically speaking, unsettling, for at base, if all perceptions occur automatically, result from chemical and physiological events, then how can we do we view truth? What is real if everything we sense and believe results from automatic determined biological events? On the other hand seeing biology as a mere intermediary, an instrument of awareness, is much more palatable. An idea, even the simplest idea, the thought that we might move a single finger for example, originates in some, as yet unknown place. The act , or thought, is carried out using biological instruments and may be scientifically studied with instruments sensitive to biological events, EEG's , PET scans and the like. We ought not to be deluded into thinking, in watching these events occur in the brain, that the brain initiates thought and feeling, for that there is no evidence at all. The brain is used, just like an arm or a leg or one's eyes to accomplish life, experience and action. Study neurological events as we might, we are still at a loss to explain the origin of most action and experience. What we have done, is examine intermediary events that are expressed biologically.

In the field of neuropsychology our fund of knowledge is large and expanding. We have defined illnesses mostly, that affect human behavior, and have graduated from an erroneous position common not very long ago, that all behavior results from reflexes, action and reaction upon a nervous system which reacts automatically. The deep tendon reflex (knee jerk) is less a model for behavior, than a symbol of automatic action. Neural circuits yielding automatic behaviors are modulated by other circuits upstream, closer to cortical control centers; hard-wired neurological events are only the beginning neural function. Biological and chemical models are limited giving us at best a partial view of a whole behavior. More important, they help us to find a way to intervene should something go awry in this basic circuitry, but never provide a comprehensive explanation for behavior.

Computer models of brain function are helpful, but incomplete. The computer is like the brain in the sense that it functions as a tool or intermediary and never initiates action with a plan though with some sleight of hand it can be made to appear to act on its own accord. Computer scientists have long ago past the time when expert systems were to take the place of experts. These systems again are used somewhat successfully as tools, repositories of rigid logic and information, by some experts but have failed achieve human levels of function. Similarly navigation, locomotion, visual recognition programs have inherent flaws that are unlikely to be overcome by increasing computer power, the so-called, "brute force" approach. Rather a paradigm shift will probably be necessary, perhaps more accurate knowledge of biological function which appears to be taking on a more and more pedagogical role for computer scientists, will be useful.

What is the neurology of thought? I have talked about the permissive effect brain function, a platform or diving board mode. It the brain is working, it allows a person to function at his highest level. Biology that makes awareness possible, has mortality built into it. All living things die. Thought on the other hand, especially certain more abstract or deeper thoughts, do not die. Hannah Arendt ascribes this association to Greek Philosophy.

"Part of the Greek answer lies in the conviction of all Greek thinkers that philosophy enables mortal men to dwell in the neighborhood of immortal things and thus acquire or nourish in themselves "immortality in the fullest measure that human nature admits." For the short time that men can bear to engage in it, philosophizing transforms men into godlike creatures, "mortal gods," as Cicero says. (It is in this vein that ancient etymology repeatedly derived the key word "*theorein*" and even "*theatron*" from "*theos.*")"[41]

My only criticism of this concept is that Greeks didn't think of Gods the same way we do. Greek gods, it has been pointed out, were immortal but not eternal. The beginnings for Greek gods were the same as for men. They're all born at a certain instant and were not thought as being beyond time or timeless. In fact Greek gods aren't like God at all but are rather like noblemen and women, elite personages really, who by virtue of noble birth are able to live separate from the common rabble, and being idle, not having to have to earn a living as commoners do, are able to consider higher, non-practical or worldly things. Hence man, through contemplation becomes god-like, is ennobled only in the basest sense really. He can be idle like a nobleman, but is not therefore elevated to any godlike status as moderns would understand this according to the teachings of Judaism and Christianity[Φ] . However the very notion that man is through the machinations of his intellect, escaping to some degree the mortality dictated by his biology is very attractive. It's interesting that the brain, obviously a biological organ, secretes thought, even abstractly, nobly, so that we have come full circle in our thoughts in their fullest sense. Man's peculiar biology that is the superior function of his brain is what gives him the wherewithal to escape from his biology. Mortal attributes yield immortality.

The common man is prisoner of his own biology, rarely venturing beyond the confines of a prison cell, wanting nothing but the satisfaction of his basic needs. Put another way, he ventures rarely from his bioplatform, and thus is little more than an animal. If he is hungry he will have food, which he consumes without limit. If he is aroused, sexual partners are readily available. He doesn't have to be, yet seems obsessed with the prompt satisfaction of his craving. As his trophy for his efforts, he's subject to the pathological effects of obesity, and his offspring of whom he takes no account, and women he has had, are spread over the landscape. He will never want or delay any wish or need and so remains constrained with base depravity and disease. Others are severely impaired by conditions that never allow them to consider anything except their own physical being. Some developmentally disabled or demented persons might fall into this category, but also a lot of people who are physically ill. Since curing these conditions happens rarely if at all, medical practice is mostly about helping the afflicted get beyond their limitation and condition, by lessening their symptoms, but more frequently, utilizing compassion and advice and education, providing a path to a fuller life. We ought never to complain that a condition is incurable but instead use our best formulations to find a way around it, a crutch and cast for a broken leg, a training regimen and medicine for an asthmatic Olympic swimmer. This includes cognitive techniques as well as physical ones. Persons who have stunted development due to disease are imprisoned by their physical conditions. The most rewarding part of medical practice is helping them break out. Rarely you will find a soul who though severely ill will not give up, whose inner fire cannot be extinguished and then there is shared exhilaration, symbiosis almost, between doctor and patient.

Most of us are not physically ill. We limit our own cognitive capacities to practical and physical matters anyway, turn away from a reason for being, live our lives as automatons in a purposeless existence that is sure to be extinguished when our bodies die. We know of nothing else, want nothing, strive for nothing, except acquisitions and physical comforts. Those persons who consider nothing beyond themselves are as imprisoned by their biology, as the sick patient. They are locked in the mundaneness of the real world and have never really considered anything beyond. Neuroses literally engulf them and ruin their lives. They remain oblivious to the beauty of the night sky, immune to the power in music, unvexed by any profound idea, insulated, invulnerable to anything except the mundane and insignificant. They may be happy, though their joy is superficial, their responses wooden and mechanistic. What does it matter for such persons when they come to the end of their lives? How do they differ from any other animals, a crab or an insect which has no care for anything except self-preservation? Those persons are in a prison too, so much so that your may try, without success to awaken them from their coma, rattle them, motivate them to escape.

Twentieth Century neo-modernist dictators have reveled in biology as well, at the same time being unable to escape from biology's limitations, in their interpretations and misinterpretations of it. Why should they not? They have far more technology now at their disposal. Absolute rulers from Stalin to Hitler have exploited physicality for their own ends in order to convince the world about the superiority of their regimes. They stop at nothing in competition with freer governments. Only since Glasnost have the excesses of Soviet Olympic competition been revealed. We have read stories about selection of toddlers for a servile life of athletic training, of special feedings and mistreatment used to prevent growth, the illicit use of hormones for smallness and short stature and even forced pregnancies among young gymnasts, in a misguided effort to help them absorb nutrients with forced abortion too, done before pending competitions. The pseudoscience of Nazi regime, was extensive too, ranging from transformation of Darwinism to an Ubermensche (man and superman mentality), alteration of physical anthropology to create racial superior and inferior castes, to the infamous experiments of Dr. Mengele and others. For a while there was weak debate about the ethics of using Dr. Mengele's “medical data”, on the effect of heat an freezing on human subjects until someone actually reviewed it and found it, not surprisingly, to be quite useless, not up to scientific standards[42]. Of course the Nazis were not beyond the using of the Olympics as a bully pulpit for their own propaganda machine, most notably in the Munich games. We have seen how the point of view that a human is no more than a beast is a self-fulfilling prophecy. We are inured by this time to mass graves and giant bulldozers unearthing mounds of human flesh. Religion for some is the cult of the body. The physical, the savage instinct is all.

Despite the remonstrations of ne'er do well icons of scientism like Carl Sagan and others who while hoisting scientific information to new heights, and moralism and ethics that supposedly can come from an enlightened understanding of biology and of all scientific principles, and they do make a good argument for altruism, all of their thought is anemic when you consider the workings of nature, and especially as you hold them up against more demonic believers in pure biological physicality, Hitlers, Karadzichs, Noriegas and their ilk. Deepest respect for biological principles, alone while ignoring what in humans goes far beyond biology, will have cruel results. It dehumanizes man, makes the individual far less important, denies the basic non-biological self, imprisons the person and profoundly limits ethical considerations.

Our feet are firmly planted in the real physical world. We have to go about our work make our living, bring up our families, perform our functions. But a part, sometimes a relatively smaller, sometimes a larger part of us is in a different world, less physical more expansive, non-tangible, more friable, yet eternal. Our hands and our eyes point to the sky. We are some to a greater, some a lesser extent, part of some vast web of being. Without the practical and biological, we cannot enter this second higher plane, biology is necessary, it has a permissive effect. But more and more we will see humankind breaking free of biological limitations empirical evidence of an alternate existence beyond a practical and physical biology which is by all appearances, the very reason for even our physical being.

The alternative to this vision of humankind's reaching for the sky is that we are embedded in amber biology, caught in time as some petrified insect. The consequences of this line of thought are destructive, an excuse for the bestiality we have witnessed in our Twentieth Century. I'm presenting here not a rejection of the scientific method, but a humanistic version of the pathway to the truth, a system which revels in our biological origin, if only because we can now use this information to better the lot of humankind (if not for the pure love of studying who we are and how we came to be), all the time making us aware of the greatest potential for us which is the future.

The brain is the bully pulpit of consciousness, a bioplatform not duplicated in any computer mainframe. Bioconsciousness ends with death, greatly feared by all animals, a fear reaching eloquent expression in avoidance responses and in man with desperate imaginings of eternal life. Kill an animal and you throw away a computational device more wonderful and powerful than any human contrivance. Whether it may one day be possible to reproduce by design what is accomplished in nature in the bioplatform of the brain, is a matter of conjecture. Thinking will be relegated to our own brains for the foreseeable future. The brain, aided by various assistive devices expanding its limitations, will be center of cognitive activity and executive planning for a long time to come. If we are to transfer our consciousness into a machine or reproduce it, limbic emotional, cingulate, frontal, temporal lobe, afferent sensory and motor modules among others would be necessary and even then, we have no way of knowing whether the whole may still be much more than the sum of interconnected parts.

As of this writing, no one has proved that the brain is anything other than an instrument or tool of awareness. No one has found is or is likely to find that part of our anatomy, that piece of matter authoring, initiating the simplest behavior. It's not that we haven't tried to find complete biological descriptors. We've just found biology inadequate to the task. Different methods have been employed. We have dissected conscious, subconscious, dream and even hallucinogenic states. We looked at whole and partial brain lesions, described them functionally physiologically, looked at cells and disease processes at the sub-cellular level There are still other strategies for attacking these problems, as we shall see in the next chapters. After all is said, biology is inadequate to explain all that is human. This implies that there must be a principle as yet undiscovered that is beyond biology.

Not only are biological and mechanical explanations for humanity insufficient, they are also extremely limiting. Freed of these shackles, breaking out of our biological shell or cocoon, the possibilities for our future are truly limitless. If we can simply admit that this is so, leaving aside our pride in scientific discovery and belief in mechanistic explanations for human experience for one moment, then we will find with the instrument of our human brain, such as it is, as a mere beginning.

# Chapter 2

## BEGINNINGS

**Beginnings**

Our birth is but a sleep and a forgetting:
The Soul that rises in us, our life's Star,
Hath had elsewhere in its setting,
And commeth from afar:
Not in entire forgetfulness,
And not in utter nakedness,
But trailing clouds of glory do we come
From God, who is our home:[43]
-William Wordsworth

"Where have I read that at the end, when life, surface upon surface, has become completely encrusted with experience, you know everything, the secret, the power, and the glory, why you were born, why you are dying, and how it all could have been different? You are wise. But the greatest wisdom, at that moment, is knowing your wisdom is too late. You understand everything when there is no longer anything to understand." [44]

-Umberto Eco

## Intro:

Memory is simple, deceptively simple. It is like a single beguiling facet of crystal seen in an uncut stone. Should you cut carefully through the rock that hides the crystal, you will marvel at its complexity. So it is with memory. On the surface there is nothing to it. Human memory is easily assessed. Storage in devices from notepads to computer disks, is second nature. But explore memory completely, as an element of cognition, and you will find it to be more complex than is appreciated by the average neuro or computer scientist.

Ask a person to recall three simple objects after a couple of minutes say, "hammer, three and yellow", as doctors do in a mental status exam. People look askance when I ask this type of question. What does it prove? It's an entrée into memory function. Memory is the easiest cognitive function to assess. I used to wonder how memory batteries became part of I.Q. tests. What does a person's memory tell you about their intellect? We all know people who do well in school because they memorize easily and regurgitate verbatim what is taught with little mental processing. People who don't think seem to get the best grades in school.

Nowadays we have less regard for simple memory. Schools claim to teach students how to think, eschewing rote memorization. Students are given open book or take-home tests trying to to deemphasize memory tasks. Why should a student depend on his memory when we have so many recording devices? Educators tend to lose sight of the fact that creativity is drawn from a storehouse of internal memory, images, words, combinations of words, melodies that are part of us. Creative persons recombine accumulated memory elements in novel ways.

I always admired the way my father recited poetry and literature that he was made to memorize in school. He was educated in the old days and kept these words with him his whole life. It gave him great pleasure every time he recited a relevant piece of verse, something which he loved to do. We don't give our kids that opportunity anymore. They may read a poem, but their teachers reason that this verse will always be available to them, should they ever need it, on some recording device. The teachers may be right about the availability of information, yet they lose sight of the fact that the lines never are revealed to their young minds as well, never mean quite as much, as when they along with nuances and connections become a part of a child's being, repeated and internalized.

Clinically speaking, it is quite easy to tell when someone is having trouble with their memory. They complain of losing objects, forgetting names and appointments more than in the past, going to another room to take something and forgetting what they came for. In more advanced cases they may no longer be able to learn anything new, and forget more well-established things like bridge or chess, or their way home from a close friend or their job. In the worst instances they can't be trusted alone at home for fear they will leave their door unlocked, the oven on, or loose themselves outside. Memory dysfunction is clearly visible to friends and family who notice problems early on well before deficits become obvious in other spheres of cognitive function. It all seems extraordinarily simple. Complexity arises as you consider how memory is woven into a rich fabric of cognitive function. Memory forms the basis of learning and adaptation to one's environment. How does this simplest of mental functions play a role in cognition and the development of the total personality?

Schematically we break memory into three components, immediate recall -- the ability to repeat what has been said, recent memory -- what can be reproduced a few minutes after a stimulus, and remote memory -- recollection of one's distant past. You may be surprised to learn that remote memories are harder to erase than recent memories. Old memories are difficult to conjure up, it is true. They tend to pop up when not interfered with by the laying down of new memory traces (engrams), something familiar to us from conversing with older persons who do not register new memories well. Old memories are stored differently than new ones. Well established engrams arewidely available throughout the brain. Memories that have circuited through the brain often enough become overdetermined, and reside in multiple locations to interact richly with other memories. Few of us will ever forget how to tie a shoelace. On the other hand we have only newly been exposed to recent events which are just now assimilating into memory pathways.

It's still useful to think of new memories as beating a path through a set of neurons and synapses as the process of learning takes place. This is the point of departure of the old theory of Donald Hebb who in 1949 suggested that the physical substrate of memory was a strengthening of connection between neurons, more efficient transmission across the synapse, where nerve cells communicate with each other. The average human brain contains about 100 billion neurons and perhaps one to ten thousand as many connections between neurons, synapses. A memory is laid down when the connectedness between neurons is increased, when a message passes more easily from one neuron to the next across a synapse.

We usually associate the hippocampus in the brain with new memory traces. Hippocampus means "sea horse" since it looks like one under low power of the microscope stuck in the middle part of the temporal lobe. If the hippocampus is removed on both sides or is affected by a disease, a person will no longer be able to learn. New memories cannot be formed. There is some evidence that some simple learning takes place by strengthening connections between hippocampal neurons, that learning can be thought of mechanically as the strengthening of connectedness or facilitation of transmission between nerve cells rather than, or supplemental to, a change within a nerve cell. This involves a <u>pre</u> and <u>post-</u>synaptic neuron. One process that strengthens connectedness between neurons is Long Term Potentiation. This process utilizes excitatory transmitter Glutamate. This creates a situation in the post-synaptic neuron where it can be more easily stimulated by chemical signals coming from the terminals of a neuron synapsing with it. It happens because of the influx of Calcium into the cell. The marine snail Alplysia serves as an experimental model for this process. Calcium affects an intracellular enzyme, Calmodulin. Eric Kandel and colleagues worked extensively with this instructive animal model years ago. There is a complex interaction here in that the post synaptic neuron secretes a chemical messenger that makes the presynaptic neuron more likely to secrete its transmitter. There is some new evidence that what is made by the postsynaptic neuron to affect its partner is the simple molecule nitric oxide (NO). So what is involved is communication that goes in both directions between the pre and postsynaptic neurons, communication that ends up strengthening their interdependence and connectedness. The sum total of increased connectedness between neurons is reflected in a change in the behavior of the animal that is learning. The hippocampus is involved with more advanced and rapid kinds of learning that enters consciousness.

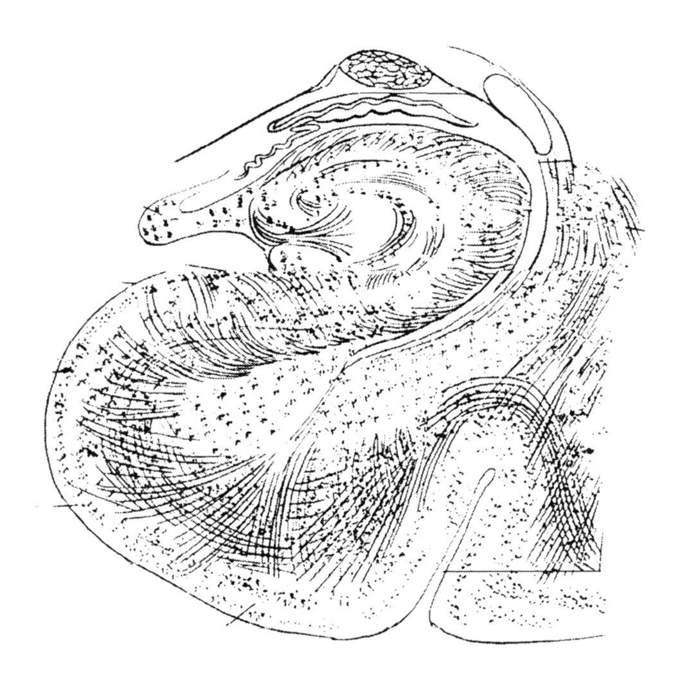

**Figure 15: The hippocampus[45] of the temporal lobe. This seahorse shapedstructure controls the initial stages of memory formation.**

Hippocampus

You detect learning takes place by observing a change in behavior. This is true for humans and experimental animals. A snail can learn to withdraw a body part, or a rat can learn a maze after a small number of exposures in order to efficiently find a food reward, probably by the very same subcellular mechanisms as a child learns to spell. His teacher seeks to observe a change in behavior on a spelling or other test. Whereas the child had not been able to spell a word before, she is able to write it perfectly now. She has learned and the test looks at an alteration of behavior.

Scientists have looked at how these learning processes take place within the individual cell on the microscopic level. In order to do this they generalize from a simplest case. After a first exposure, memory needs to *consolidate* if it is to result in a reasonably permanent effect on the brain. Memory consolidation is expressed in molecular and then in structural alterations in neurons and other brain cells. Recent work has shown unequivocally that protein synthesis is necessary for memory consolidation. Other proteins modulate production form the RNA templates. The chemistry inside the neuron utilizes chemical messengers[46] , the very same messengers involved in a host of other cellular functions, such as cyclic AMP. Why does the individual cell start to make proteins? The goal is to create new connections between cells, new synapses and parts of synapses, perhaps divided synapses, and the like. Rats learning to navigate mazes require production of specific proteins in the hippocampus. If this process is interfered with, the animals will not be able to navigate the mazes. For the first time we are able to correlate protein production and alteration of the untrastructural microscopic change, with the laying down or recording of memory traces. Once the chemical and structural changes are known this increases the promise of somehow intervening in memory disorders, or even improving capacities of normal. Learning can be made to occur on the microscopic and chemical level[47]. The new protein production is thus expressed as an untrastructural change observable with an electron microscope. The molecular, chemical and structural changes are then expressed by a final common pathway, behaviorally. The child gets and A on his spelling test.

The laying down of explicit memory is accompanied by ultrastructural changes in the hippocampus and other areas. The brain is not a static structure. It is changing, rebuilding, forming proteins and synapses constantly as it is used. We cannot speak of the brain as having only a static anatomical structure. The anatomy changes as it is used much like a muscle or any other organ of the body. In the brain this change in structure is easiest to see in memory circuits, especially the hippocampus and the mechanism for structural change involves gene products and proteins but is ultimately expressed in formation of synapses or connections between neurons. Disuse and stress may lead to atrophy. Hormones particularly cortisone like hormones and estrogens have been shown to influence this process.

It is easy to see that as soon as we learn which specific molecules aid learning then we may some day be able to influence them. We will know how they are affected by disease, and, even more importantly, we may be able to enhance memory function by altering these molecules.

The memory we think of most of the time, recalling words or methods, is *explicit* that is, mostly verbal memory. Humans want to be able to recall most of the time in task utilizing their language: names, dates, places, methods of operation and the like. *Implicit* or nonverbal memory is at least as important as explicit memory however.

Implicit memory is the second type of learning that happens on a nonconscious level. It uses a different mechanism and anatomic substrate. For example you learn motor skills like playing the piano or basketball subliminally and many aspects of this learning which involves practice do not enter consciousness. The anatomical pathways for implicit memory are quite different, especially memory which enhances motor skill. There are a number of different types of implicit memory.

For example, in addition to memory that enhances motor performance at a piano or in athletic competition, there is undoubtedly a similar kind of sensory implicit memory. What allows emotions to surge with recognition, resonate in a fashion, well after a rhythmic figure or theme has been introduced in a symphony, is undoubtedly a subconscious implicit memory mechanism. You hear a brief rhythmic figure in the Berlioz Symphonie Fantastique as he is being led to the scaffold which is "drummed" into the head over and over again. Much later, a similar figure picked up as the Finale closes, has emotion reaching a fever pitch. Most listeners don't even notice it. They just feel the high emotion. Perhaps the composer himself is not aware that he's used this technique because the music being on his mind, he will tend to use the same themes again in any case, much as the writer will, if he doesn't watch out, use the same words or even phrases again in very different contexts. Much of this occurs without a conscious thought, on the part of the composer, or the listener. It happens on a much more primitive level, of course, in popular music, as themes are repeated, often interminably, in a short song that lasts perhaps one or two minutes. What happens a good deal of the time in popular music especially, is that there is such an abundance or repetition, of rhythmic figures, simple melodies instrumentation, harmony etc. as to breed ennui. And thus brings up another type of learning, habituation. A stimulus repeated over and over again, ceases to have much of an effect. After a while your nervous system becomes so used to a familiar stimulus that you barely take notice.

In classical music there is the sonata form, consisting of exposition, development and recapitulation. One or more themes present themselves in the exposition. The development sees new expressive territory being claimed and finally in the recapitulation, the listener is transformed in a certain way, that is, he hears the same or similar themes differently. Jazz pieces use the same basic form, incidentally, with a statement, embellishment ( improvisation) and restatement of a melody. Since this process seems to be so universal, this may hint at somet basic physiological mechanism. We have a mutual maturation process involving composer and listener alike. Otherwise stated, repetition alters the brain's response. Memory has to be there for past to alter the future. The memory could just as well be subconscious, unnoticed, yet it heightens subsequent response.

The hippocampus has vigorous connection to areas responsible for consciousness and emotion[#] . Emotion areas of the brain the Papez Circuit, or limbic system, physically connect to memory pathways. James Papez published his observations in 1937 which makes him ancient as far as biomedical literature is concerned. But many of his basic observations are still extremely useful. He noted that the brain could be broken into medial and lateral sections. Medial parts connect most strongly to the hypothalamus which is involved in basic bodily (visceral functions) and is a structure that also helps organize emotion. The lateral parts connect to the dorsal thalamus that is a way station for sensory inputs. The structure of the medial of inner part of the hemispheres is intimately involved with emotion and also memory. Early anatomists connected many of these structures with the sense of smell as this was considered to be one of the most primitive senses in animal evolution (phylogenetics). In particular olfactory (smell) pathways connect very intimately with the hippocampus. Thus memory, olfaction, and emotion are closely allied anatomically. Since many other animals, especially reptiles and mammals, are capable of expressing rage, but only man shows evidence for advanced thought, with these medial brain structures being relatively older in evolution, and with simpler cellular architecture, they are considered to be more primitive, hence emotion is more primitive than cognition or thought. We now have a much better understanding that all of these functions interact, so as to be all part of a larger conscious whole that includes both thought and emotion[48]. The individual structures and details about their connections may be interesting to some readers but are not at all necessary for purposes of discussion here.

At about the same time, new work was published dealing with another brain structure, the amygdala (for almond, an almond shaped structure in the temporal lobe). Animals that lacked an amygdala were docile, hypersexual, and hypervigilant and restless having a syndrome named for the two scientists who performed this lesion experiment, Kluver and Bucy. This is also a rather ancient concept which has definite clinical correlates in humans. The Amygdala is tied to many of the structures in the limbic system and is undoubtedly involved in emotional expression[49]. It is not uncommon to see a patient whose personality has been drastically altered due to temporal lobe disease. Often, patients have some or all of the characteristics of the Kluver-Bucy Syndrome and are thus rendered refractory to any kind of medical or psychological intervention.

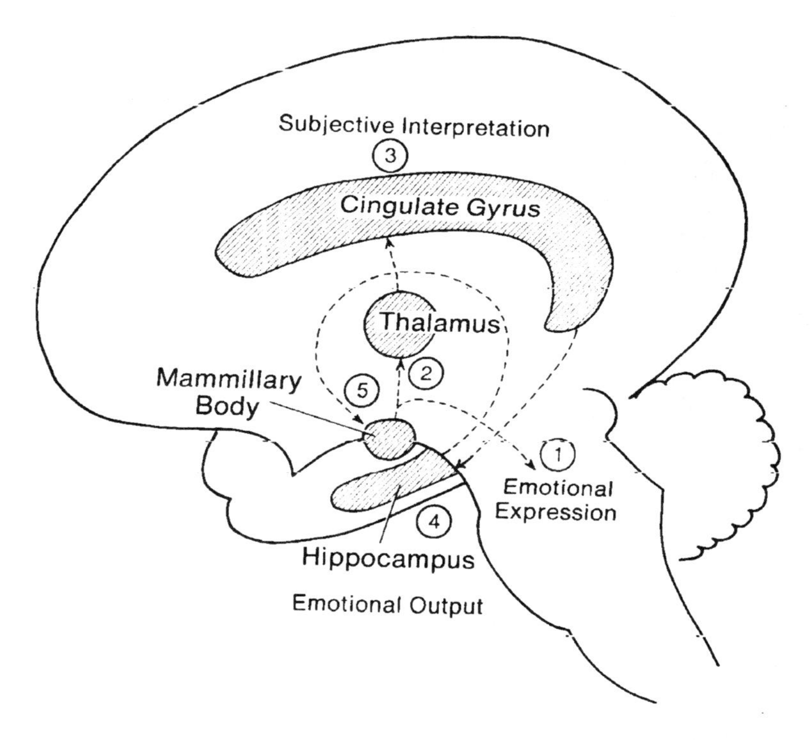


**Figure 16: The Papez circuit. Mamillary body to midbrain (1) and anterio-ventral nucleus of thalamus (mamillothalamic tract) (2), thalamus to cingulate gyrus (3), cingulate to hippocampus (4), hippocampus to mammillary bodies via fornix (5)[50].**

Papez Circuit (see above)

The connectedness of the memory circuit to the limbic system or emotion centers of the brain is neither coincidental nor insignificant. The post-traumatic stress syndrome links memory to high emotion. Vietnam veterans and accident victims have flashbacks to highly emotional experiences. What is recalled, is the extreme emotion as much as the actual event. Laboratory animals can be conditioned to associate strong emotions such as fear and rage to neutral stimuli such as an animal cage where shock had previously been administered. This kind of association occurs in all of us every day. Conditioning can also be explained on an anatomic basis strengthening transmission between neurons that connect to each other. This is used to explain part or the pathogenesis, the medical cause, of the Post-traumatic stress disorder where sudden intrusions of memories evoke high emotional states.[51][52] The emotions associated with certain memories can be used to improve memory function, Memory specialists often use this uncanny ability of the brain, that connects emotions and sensory experiences to increase mnemonic powers.

PET scans show that the amygdala is extremely active in forming memories. One may ask why this is so, since the amygdala is viewed as an emotional center oft the brain. We have seen that it is intimately connected with memory and emotional circuits. Perhaps this is why in sleep when we are working to consolidate memories which closely tied to emotion and emotional expression . Memory and emotion are closely linked.

Emotion that goes along with memory is an integral part of it. I rarely recall dreams but was able to remember and record most of what turned out to be a very complex dream because of a musical theme that seemed to be playing through the dream sequence, much as music plays in a movie. It was a theme played by the high strings and flutes of Prokofiev's Romeo and Juliet that I'd been listening to the day before. The music was supposed to be portentous telling you that something fateful would happen in scenes to follow. In the dream it evoked a totally different emotion, one of ghostlike eerie fear that I hadn't appreciated while listening to the piece. The day after the dream, recalling the music and its attendant emotion, I was able to reconstruct the complex dream surprisingly well, all through the process using the melody and emotion as a sort of mnemonic device. Moreover there since sensory experiences from the day prior are frequently used for dream material it may be that one function of sleep and dreaming is a dry run rehearsal of the previous days events that leads ultimately to the permanent laying down of memory traces. The hunter dreams of catching his prey, re-rehearsing experiences of the day before, reshuffling images, recreating scenes that will enhance his performance the day after. In fitful sleep you rehash yesterday's argument with your boss, etc. Dreams place memories in an emotional context, are intimately tied to the emotional valence of memories. Perhaps this is why the Vietnam veteran typically awakens with his frightening recollection, and why we frequently hear or persons with panic disorder awakening with in panic. It's a lot like "pavor nocturnus", night terrors in small children, who wake up suddenly drenched with the output of their adrenal glands which have been enlisted in support of the nocturnal event, screaming in dread, palms sweaty, heart pounding. Only these little folks awaken without the specific memory of a dream, only the fear and dread that usually comes with a nightmare without the nightmare. Night terrors do not typically occur in REM or dream sleep.

Forgetting is at least as important as remembering. There must be active processes that aid in forgetting, some of which will someday be described on the cellular and biochemical level, just as active memory is described. This would serve obvious housekeeping functions. For one thing you could picture that it would be impossible to function if all of our memories old and new were constantly competing for our limited attention. Imagine if all of your old memory stores kept creeping into consciousness. You wouldn't be able to handle current tasks. You are taking an exam in biology which asks specific questions. Think what would happen if you were unable to keep out of your mind's eye for at least a while what you had learnt in physics the day before. Some other memories, might be more difficult to exclude, from your current attentions, but the point is, some active adaptive process pushes even recent memories out of consciousness, mostly unclutters awareness. Older memories are pushed even further away. Somewhere in our brain (this turns out to be all over our brain and is difficult to localize) is a memory attic which contains relics of our past.

Dynamically trained psychiatrists still talk about repression as a defense mechanism, an active process, pushing old memories out of awareness. This is still a useful concept, but contrary to psychodynamic renderings of the process, it is almost always a healthy, not a pathological process. It is easiest to conceptualize repression as an uncluttering mechanism. As we have seen, dreams from the previous night are actively pushed back into unconsciousness where they belong. For most of us dream content can almost never be retrieved or is brought to the surface only with great difficulty (and very questionable authenticity). This is universal for most healthy functional individuals.

I recall a conversations with a very troubled man in his mid 50's who was plagued with numerous memories of his childhood and unresolved conflicts mostly revolving about his relationship with a now deceased father. Time and again he'd mention to me with great emotion, how his father could never tell him he loved or approved of him, something which obviously hurt him deeply. But it was apparent that he was trapped in a web of old memories. Trapped or fixated in his past, he was incapable of handling the challenges of the present. Some therapists might talk about how important it is to revisit these childhood conflicts with the rational retrospectiscope of a grown adult. The only problem is that in many cases there seems to be a total entrapment or fixation on the thought processes of childhood. Plain repression of these memories would seem to be much more adaptive. Perhaps old painful memories which we push away have subtle and unsubtle effects on our present lives as they influence attitudes and behavior. And so the debate goes on.

Through years of scientific research there have been many theories of how learning takes place. Some of these theories stress what happens inside instead of between cells. They implicate synthesis of chemicals such as nucleotides and proteins. Neurons are analyzed for these chemical constituents after an animal is exposed to certain experiences that change behavior. These older theories are in not incompatible with synaptic theories and there is every reason to suspect that our explanations regarding how memories are formed are very incomplete. Memory function is certainly a composite of many biochemical processes within and between neurons. Even the simplest discussion of memory, the most rudimentary of cognitive functions, reveals that we are trying to deal with many separate processes. We have already mentioned immediate, recent, and remote memory, conscious and subliminal motor memory. The chemical and anatomic substrates for these different processes are not the same. For example, verbal and conscious memory is the type we talk about classically being mediated initially by the hippocampus. For words and symbols that usually require conscious awareness, the hippocampus is the portal of entry into the brain until the memory engram becomes firmly established and destroying the hippocampus will affect short term memory, blocking the initial establishment of a foothold in the brain. But nonverbal performance memory, is mediated by motor systems such as the cerebellum and it is more than likely that performance learning, basketball and piano playing among other things takes place directly within these systems. It is likely that there are other forms of learning as well, but the basic breakdown is into conscious vs unconscious processing.

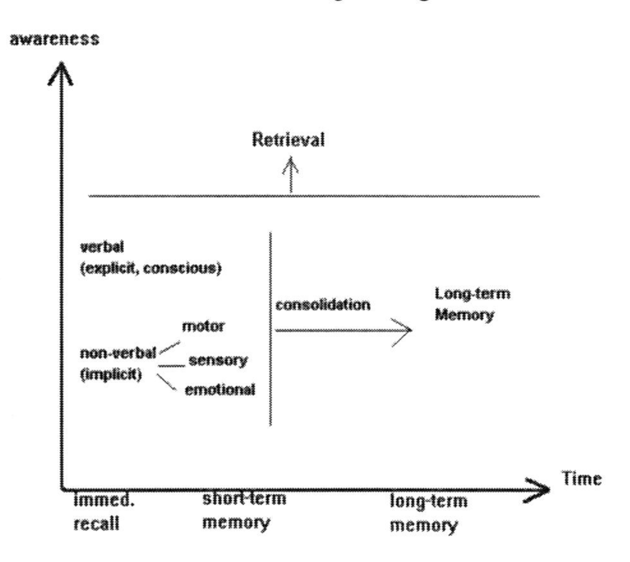


**Figure 17 Types of Memory. Retrieval brings memories into awareness. Emotions, motor and sensory engrams may not be separably retrievable but still affect behavior and performance.**

Clinically a decline in mental function seems manifest first in a loss of ability in recent memory. This corresponds to neuropathological changes that occur first in the hippocampal and parahippocampal areas of the brain, especially in Alzheimer's disease. We all know elderly persons who lose their ability to remember. In some of these persons other spheres of mental effort remain relatively unaffected and they are said to have a benign forgetfulness of aging. In others, loss of recent memory function signals a global decline in cognition termed dementia. This is what happens in Alzheimer disease which is frequently presents with a loss of memory function.

The transmitter Acetylcholine is most closely associated with this disease, and intimately tied with recent memory. A diffusely projecting nucleus in the brain, the nucleus basalis of Meynart shows the most change. The output of this group of neurons is primarily cholinergic and largely to the hippocampus.[53] A lot of people have tried to help failing memory by increasing this neurotransmitter much as one does in Parkinson's disease with dopamine that is deficient because of loss of Dopamine producing neurons in the brain. All kinds of substances have been used to retard the degradation of Acetylcholine, or to replace this transmitter, some of these dripped directly into the ventricles of the brain to improve memory, all with a very minor effect. The first of these to be FDA approved is called Tacrine or Cognex but it has little lasting benefit and significant toxicity, affecting the liver mostly. The minimal benefit can best be appreciated by sensitive measures such as psychological tests.

It's just because memory is so easy to get at that it seems to have so much philosophical relevance. Computers and libraries have vast stores of information as bundles of bits appropriately labeled, which can be retrieved when asked for. Named bundles of engrams are stored in isolation, retrieved, and operated on in ways defined by a user. It is reasonable to ask whether the brain functions as a repository of information in the same way as a computer or library. Surprisingly, The answer is that it does not.

For humans the laying down of memories is equivalent to maintaining the continuum of consciousness. Time is continuous and we reckon and follow it on the basis of stored past and newer experiences. Our past serves as a basis for comparison but it also colors perception and action. This is what we mean when we talk about learning or memory. We expect the laying down of a memory engram will make some enduring change in thought, emotion or behavior that we will be able to measure. If there is no way to determine that a difference has somehow been wrought, then the effect of memory is imperceptible. On the basis of what has been laid down before we are made to see things in different ways. Each old memory trace may be more or less critical. For example, a traumatic childhood sexual experience not only changes adult sexual encounters but colors all subsequent experience. Much of the work of classical psychotherapy is breaking a traumatic or maladaptive link between emotions and current events, and re-establishing connections with previous repressed memories. What is truly fascinating is that this seems to occur whether or not specific memories can be specifically retrieved[ϕ] . This continuity of experience is perhaps the most important determinant of personality structure. A human life is a chain of experiences encoded in memory that is a continuous flux of maintained consciousness.

It is undoubtedly true that experience alters our personality, our response to environment, even though the precise memory of that experience may not be specifically retrieved. Whether or not you can recall an event, it alters your subsequent person, it changes you. This implies that an experience may will alter the information content of your brain, but that specific alteration cannot always be tested for. Perhaps this is analogous to explicit or verbal memory versus implicit memory. Philosophically it is apparent that every experience alters us. A sinner who truly repents and changes his ways, is somehow better than a pure person who has never made a mistake; a person who has tested his mettle through suffering and has later been redeemed, is far better for this experience than the person who has gone through life without any struggle at all. Redemption alters us in subtle and not so subtle ways which may not be so easy to document on tests of explicit memory. Undoubtedly such experience changes us in profound ways. Thus it seems that sin and suffering serve a specific purpose, which is a galvanizing of the personality. This would be a classical explanation for the biblical concept of redemption. But is the ability to retrieve specific memories important or not? A tough question.

This has implications concerning the meaning of life and death. Consider the possibility that a soul or personality is reincarnated after death, not just discarded by posterity, but simply transmigrates into a different vessel, a different body. Typically persons who believe in such a doctrine reason that we are not aware of this because memories of our past lives are inaccessible. At first blush this seems to be an absurd notion. After all, a soul with no access to its past is a new creation. But the past may not be entirely cut off from the present. It may be difficult to get at, especially for those of us unskilled at the proper technique. More importantly, it may come out in subtle ways and may still color present experience in much the same way that our own living memory does. Even so reincarnation of the soul is hard to swallow at least for Westerners because memory is for us the contiguity of existence. When continuous formation of memory traces stops a new line of experience, in essence, a new being, emerges. In one sense, life and death, creation and destruction may be defined this way, a break in sequential memory. But here lies an even more critical point. There is more to human memory than simple retrieval of information bits in storage. Memories, meaning our past, even if not specifically retrieved, change present and future experience. Memory, arguably the simplest of all cognitive functions, may take on an almost mystical significance. Some persons claim that they can bring back memories of a previous life through extreme hypnotic regression, that memories inaccessible to our waking existence can be brought back via hypnosis. There is as yet no scientific evidence that hypnosis improves retrieval under any circumstances and none to support the notion that hypnosis causes the true elicitation of childhood and other forgotten memories. Observing hypnotic sessions, one comes away with the impression that the hypnotized subject is manipulating the therapist but under the best of circumstances no one has ever proved the veracity of a hypnotic regression into childhood or into past life.

## False Memory

Most supposed hypnotic recollections are really works of the imagination. This is especially apparent to scientifically-minded folks who look with a jaundiced eye at recollections of infancy and past lives. Where does memory end and imagination begin? This is a most difficult question to answer for the subjects of hypnosis, assuming that they are not out-and -out malingerers or fakes.

False memories have been subject of a lot of attention in recent years. There are celebrated cases of child abuse and sexual misconduct such as one involving the Catholic Archbishop of Chicago and The McMartin preschool case in California that saw the conviction of six preschool teachers, ruining their lives, on the basis of recovered memories induced by "therapists". The question has been how much if any of the data provided by these small and highly suggestible children were true memories and how much was fabrication. Recent work with PET functional brain imagery of Daniel Schacter and Larry R. Squires[54] attempts to differentiate true from suggested memories on the basis of the PET image. The memory of an actual memory of a word read to and heard by the subject would be retrieved utilizing the auditory portion of the brain, an actual word seen would be retrieved using the visual area, etc. indicating that memories partly reside in the sensory cortical areas involved when they are laid down and later need to be retrieved using these same areas. Numerous psychological studies have shown how easy it is to create memories, especially in suggestible children and certain adults.

Memory and imagination[55] are both abstract non-material entities comprised of pure thought. Some of this manipulation of ideas may be non-verbal and subconscious. How do we know, when we are creating something, that we're not simply reaching down into our memory stores and placing ideas in a new juxtaposition? A good deal of creativity and synthesis involves bringing up memories in a new light. An inspiration, or solution to a difficult question hits us when we least expect it, most often when we are not even working on a problem that is consuming our attention at the time, but more often during random activity. This sudden inspiration, has been noted by a number of creative persons. For example in the words of mathematician Henri Poincare[56]:

"Most striking at first is this appearance of sudden illumination, a manifest sign of long, unconscious prior work. The role of this unconscious work in mathematical invention appears to me incontestable, and traces of it would be found in other cases where it is less evident. Often when one works hard at a hard question nothing good is accomplished at the first attack. Then one takes a rest, longer or shorter, and sits down anew to the work. During the first half-hour, as before, nothing is found, and then all of a sudden the decisive idea presents itself to the mind. It might be said that the conscious work is more fruitful because it has been interrupted and the rest has given back to the mind its force and freshness. But it is more probable that this rest has been filled out with unconscious work and that the result of this work has afterward revealed itself to the geometer just as in the cases I have cited; only the revelation, instead of coming during a walk or a journey, has happened during a period of conscious work, but independently of this work which plays at most a role of excitant, as if it were the goad stimulating the results already reached during rest, but remaining unconscious, to assume the conscious form."

All of us have experienced the same process so well described above. It happens when we lose something and suddenly, never when we are looking for it, remember where it was. This recollection comes to us in a flash. And we just know that what we have recollected is accurate. Isn't it exactly the same when we have been working on some puzzle or problem, that we experience a sudden flash and the solution is ours. These points are so inspiring because in an instant we just know we have come upon the right solution; all of it just works out so perfectly we almost need no verification. You can observe this for yourself in hearing people solve puzzles on TV or radio. Many of them make the same observation. But the more interesting question is whether any of this is pure memory or imagination, or even if there is a difference in these two aspects of pure thought.

Some people lose the ability temporarily to store new memories. A condition known as Transient Global Amnesia, is most probably caused by blocking the blood supply to specific regions of brain. We know that if only one side of the brain is affected memory registration will not be affected. The problem has to affect both sides of the brain at once. The subject with TGA loses the ability to form memories for a few hours. During that time, he is exceedingly uncomfortable and feels disoriented as if in a cloud, asking questions pertaining to time and place time and again, forgetting answers given almost immediately and is concerned about this. After the episode there is no recollection at all for the period of time for which memory was affected. Sometimes this seems to occur after some kind of exertion or trauma just as if a blow or concussion has occurred. In one case TGA was occurred repeatedly in relation to sexual intercourse[57].

Another condition that breaks the stream of awareness of life is syncope, or simple faint. Of course any epileptic seizure that disturbs consciousness will do the same thing. But syncope is interesting for one reason. Interrupt the blood supply to the brain and a faint occurs. It happens with a drop of blood pressure. If the heart, for any reason stops pumping blood to the brain efficiently, you will faint. How long does it take to lose consciousness once brain stops getting blood? Almost immediately!! This is quite remarkable and one wonders why this is so.

The brain metabolizes glucose and needs oxygen continuously in order to make energy. It is one of the most voracious consumers of energy in the body. Other tissues can switch into alternative energy sources when necessary. When you exercise for a while muscle, sensing a diminishing supply of energy, with the liver overwhelmed in its ability to supply constant amounts of glucose, switches to the consumption of fats and can also depend (briefly) on anaerobic glycolysis so that energy can be utilized without a constant high concentration of oxygen. Not so the brain.

The brain requires a constant blood supply. Otherwise it will stop functioning right away and a person will faint. If left for just a few more minutes there will be irreparable damage or even brain death. This underlines a certain fact about the brain. It is an information storage and manipulation device needing to be constantly connected to its power source. The brain resembles a computer that is on and storing and manipulating data. Cut power for even an instant and the thing shuts down. Whatever program you were working on is lost. One wonders why the brain couldn't have been designed in a different way, perhaps with the ability to switch immediately to its own "battery" or energy supply for just a few moments when necessary. It seems when awake, the organ is revved up and utilizing to the hilt immediate sources of energy, glucose and oxygen. After one or two minutes or so without any blood, the old informational content can still be gotten back. When consciousness returns after this short length of time, you are still the same person.

Also in the brain, as everyone knows, neurons are, as a general rule not replaceable. This is in contradistinction with other organ systems, the liver, the gut, skin, blood and so forth where cells are replenished constantly and rapidly. Injure the intestine, kill cells by the millions and they will grow back very fast. Why not in the brain? Of course that is because every brain carries information built into its anatomy and transmitter and electrical patterns. Nerve cells are not replaceable because new nerve cells will not have the same informational content. Information and experience, life in other words has a certain *immediacy.* The same holds for instruments which store changes in time or even keep time. A constant source of energy is required. That is the reason for that little battery in your clock radio why everything is lost when your computer's power supply gets interrupted.

The chain of recollections that defines continuity of consciousness is broken, but the situation is still different from that of reincarnation in that it is still possible to reach back and recall experiences of the just slightly more remote past and also to continue to record future experiences in a stream of mental awareness. Even so, the person with TGA is very troubled by the void in memory long after this temporary process is over. This is so even when as usually happens, family and friends inform him of events relating to the specific time frame that these events occurred. Other's recollections of events in one's own life aren't satisfactory. A vacuum or discontinuity, however short, must be filled.

## Diversion: why time travel is not possible:

Memory determines a stream of awareness. Breaking this stream of awareness is very uncomfortable. An hiatus is difficult to deal with. It is only memory that changes us as we journey through our lives from our past into our future. History gets recorded in our mind. I think that's why it's so difficult to fathom, why for nuclear physics, time can stretch and contract and just as easily flow backward as forwards in some instances. The best example is the Feynman diagram in which nuclear particles collide and create other nuclear particles, for example a neutron will split into a proton and electron liberating some energy. This process can just as easily flow backward as forward as long as conservation laws are respected. This is intuitively impossible for us in our real macroworld. According to physicists the main reason time does not flow backward is the tendency of a system to increase disorder or entropy, the second law of thermodynamics in other words. Drop a porcelain cup on the kitchen floor and try as you may, you'll never be able to put the thing back together exactly like It was; things tend toward disorder, and you can't correct your mistake, turn back the flow of time. This is the humpty dumpty effect, “All the king's horses, all the king's men..” can't make things as they were. But this is only part of the reason why time flows only forward for us. The major reason is memory and the workings of consciousness and the brain, our own peculiar physiology, in other words, not physics.

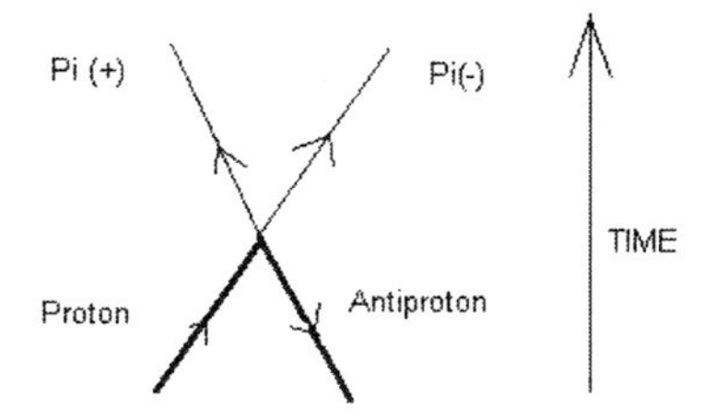


**Figure 18: Typical Feynman diagram[58] in which two pion particles are formed from the collision of proton and antiproton particles. This could just as well go the opposite direction in time with the pions forming the proton and antiproton.**

For humans there is an immense difference between the past and the future. In the past something occurred, it alone and of itself. But the future is unknown and filled with alternative possibilities, only one of which will actually happen. This is particularly true if we introduce free will into the equation. The past and future are asymmetric and not interchangeable, for the past is one hundred percent determined but the future is at best probabilistic often chaotic. As the future is subjected to time's arrow and becomes part of the past, it is recorded in conscious and unconscious memory (it becomes history) and the uncertain turns into certainty. The flow of time is thus asymmetric, and the past is recorded in our mind, altering forever future experience. The Second Law of thermodynamics seems so theoretical and flimsy, not good enough an explanation for what we all know to be true for all of us. Time travels in one direction, forward. The most important explanation, has to do with how our mind's work, consciousness and human intuition about time's irreversibility. Even if the backward flow, the undoing and redoing of events is theoretically possible, it is not possible as a mental event, not unless we can fundamentally alter mental processing, which some day we may be able to do. Time travel may thus depend as much on changing mental processes as it does in manipulating physical laws.

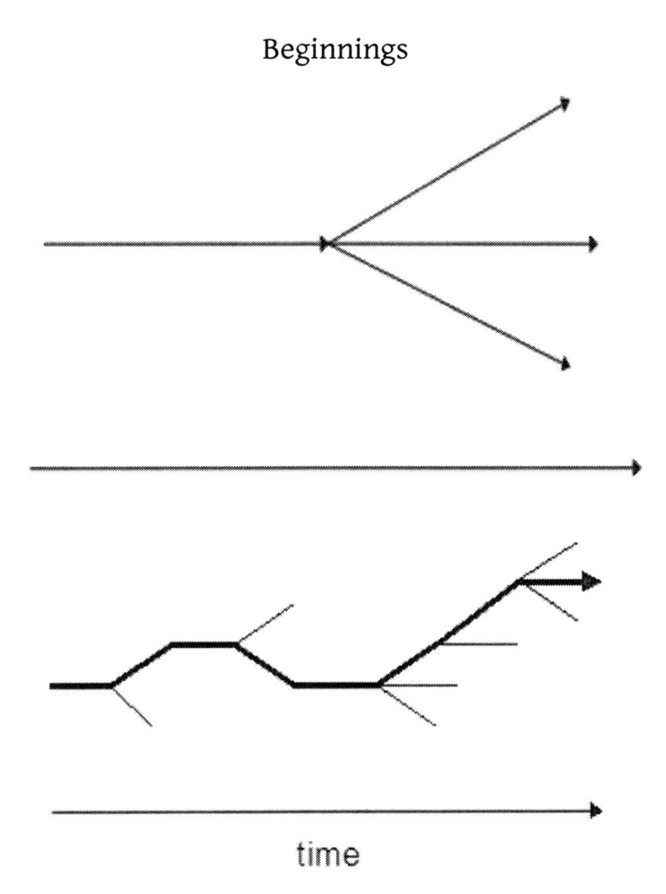


**Figure 19: Time's arrow. As the past blends with the future in our present, one becomes many possibilities, so that one can't turn back. Free will, exacerbates this process, expanding alternatives unpredictably. Below: History Diagram. At each branch point only one of a number of possibilities occurs.**

Let us suppose for a moment that one could go back in time. One of the major problems that a lot of people pick up on is that the act of going back and reporting on the past has to alter the future. The reporter who does go back then returns to the future is sure to alter the course of events. This is so even if he is a fly on the wall, invisible to those experiencing their present. When the time traveler returns to his world in the future, his present, his knowledge of the past is sure to alter his subsequent behavior as well. And the time traveler is even more apt to make changes if he is able to travel into his own future. He is sure to wish to change the course of events by his foreknowledge of what is to come. This problem can be obviated simply by destroying the time traveler's memory of what he has experienced in his travels but then things be just as if he hadn't made the journey through time at all!

But that is not the biggest worry with time travel. Every moment in time, as the figures show, is a branch point of alternative possibilities. Only one of these possibilities actually occurs. Then events continue until almost immediately another branch point with alternatives again occurs. At any point it is just about as likely that things could have turned out a different way.

Biologists have noted that life on earth as we know it is due to specific developments as branch points. Scientists have to admit that even on the earth, life and sapient life probably would never have existed but for a very improbable chain of events. Preplanetary gases had to have coalesced into what we know as the earth just the right distance from the sun making our planet like Goldilocks' porridge, neither too hot nor too cold. Floes of molten rock inside the earth had to have created a magnetic field which blocked the lethal effects of radiation to make life possible. Specific conditions perhaps heat and lightening had to have existed to juxtapose and give sufficient energy to simple organic compounds in order to make life which at any event had to eventually become self-replicating. The cell had to have evolved the way we find it today, with the aid of prokaryotic parasitic invaders which carried with them mitochondria and chloroplasts, remnants of simpler forms that make the cell and cellular energy handling possible. Photosynthesis had to have served as a chemical process that freed oxygen to make an atmosphere for aerobic life to survive. Cells had to have gotten together and specialized to make more advanced complex animals and plants come into being. Explosions and extinctions of groups of organisms had to have taken place in highly specific ways. Dinosaurs and not mammals could well have ruled the earth but for the accident of an enormous meteorite impact upon the planet etc etc. In other words, biologists sensing that life as we know it on earth is the result of highly improbable alternatives are less sanguine than physicists and astronomers about life as we know it especially sapient life, existing on other worlds. Even if you look upon life as some kind of self-organizing complexity, still one has admit that life as we know it on the earth, is the resultant of a specific chain of happenstances. Things might well have ended up a whole lot different even on our beloved earth, in other words, without us!!

We begin to appreciate an indeterminacy. If you can look back and you see that things could just as well have turned out different or you begin to see the future in the same way, all of a sudden you realize why time travel is impossible. You will not be able to travel along the precise path of possibilities you were on; you will be sure to lose your way among different alternatives. You may get stuck on a different time line or, put differently, alternate universes. Maybe you will end up in a different world that only has prokaryotes or lacks a magnetic field and has no life at all. If you ever travel in time you can be virtually certain, you cannot go home again.

This implies that the inability to travel in time is tied to the uncertainty, the indeterminacy of the universe. For time travel to be possible the universe must be deterministic, alternatives at branch points need to be fixed, established. Our lack of conception of time travel is thus as strong an argument as there could be, against determinism.

Dream memories are stored, affect us subliminally, but are difficult to bring to any form of surface awareness. Most of us can't recall them in any detail and we are hard-pressed to attach any specific meaning or context to them. Even so, they affect at the very least, the feeling tone of subsequent experiences and color perceptions. I've often had the experience of acting and feeling as if a dream event were true while in reality, it was a mental fabrication. Only after noticing that what I'd assumed couldn't possibly be true did I realize that I must have in fact dreamt it. Certainly thought involves a kind of simultaneous layering. A lot of times we find ourselves trying to manage even juggle feelings. Feelings come to us from contradicting forces-- present reality, manifest thought content, and dreams that lurk in the background of thought. We are handling simultaneous realities, our present conscious stream of reality and a subliminal illogical dream-driven thought stream. The incompleteness of recollection of active thought processes of sleep shroud them in mystery. Ancient dream interpreters were accorded supernatural powers and even in modern society are held in high esteem from Joseph to Daniel and Freud to Jung. Still, there is every reason to believe that at least two continuous streams of consciousness interact and operate upon each other even though the memories are not made manifest. We know that details of immediate past experience form the previous day, often apparently inconsequential details, are incorporated into the content of dreams. Conversely, dreams imperfectly remembered while awake influence mood and beliefs. Your wife or your child having appeared in a dream experience at night may color your interaction with your family the next morning. Mental engrams may be operated upon and may interact without even being specifically recalled.

Memories of events from actual experience are difficult to reach. Details inscribed in certain portions of our brain may still be inaccessible. Perhaps you've experienced this while at a gathering introducing a person whom you know and anxiety momentarily blocks a recollection. Tests in school engender the same kind of anxiety that blocks recall. The answer may enter your mind only after you've left the room. Details of earlier life become inaccessible merely because events are too distant to be pertinent to our current situation. This is a service that are brain provides for us, the sorting or packing away of memories, an uncluttering of immediate experience. Old useless memories are almost irretrievable.

The difference between people with super memories as opposed to the rest of us seems to be a matter of memory retrieval, not storage. To utilize a memory you have to have done two things: It has to have been written down somewhere which in the brain means there has to have been some structural alteration in the brain that results in the memory's having been stored. Next you have to use some mechanism to retrieve the stored memory. The reason why you can't recall something is frequently a matter of getting at, retrieving, the stored memory trace. Storage involves lower brain structures such as the hippocampus and some other areas adjacent areas too, the thalamus, entorhinal area that lie beneath the highest levels of the brain, the cerebral cortex. But retrieval, recruits cortex especially. Consequently it is retrieval that is most vulnerable to changes in level of consciousness.

Thus we have an anatomic model for a memory engram or any thought or sensation reaching consciousness. In order for us to be aware of anything, a memory, a sensation, pleasure or pain, it has to stimulate the highest areas of the brain, the cerebral cortex, responsible for consciousness. The stimulus has to have caught our attention, that is aroused the gray colored cell bodies at the surface of the brain. Example: an unconscious person is unconscious because the lower brainstem levels have failed to arouse the cerebral cortex which is not awake or aroused. You squeeze a finger on the hand of the subject. He withdraws the hands and winces in discomfort. But he is not awake and aware. Does he "feel" the pain? The answer is that he does not feel it. In order to experience or feel, one must be conscious to the stimulus, awake and aroused. If the stimulus does not reach consciousness and despite intact automatic behavioral responses, withdrawal and wincing, there is no pain[Ψ] . The anatomical substrate of conscious awareness is the cerebral cortex. In order for us to be aware of something, it must take hold of cortical neurons, something needs to direct our attention to that particular thing. This is what we mean by retrieval. Here widespread cortical excitation, attention, is directed to a specific memory engram dredged up from the past. It is taken out of storage, comes to our attention and is thus retrieved.

It's been demonstrated time and again that memories are very well recorded. They can be elicited with certain techniques such as hypnosis, the use of certain inhibitory drugs such as short acting barbiturates and by electrical stimulation. Some of these methods relax a person, help him get past any anxiety or distraction impairing retrieval# How are our memories ordinarily retrieved? One model assigns to each memory a specific anatomical location. Your recollection of your third birthday party may be stored somewhere in the temporal lobe perhaps adjacent to your memory of your second birthday. In thinking about your second birthday or birthdays in general you might elicit memories of your third birthday party. This has some relationship with experience certainly. Electrical stimulation of the brain would seem to support this anatomical model. Memories are retrieved by somehow specifying anatomical coordinates in a three dimensional brain, much as a song is found on a record as the needle alights on a certain track.

Another far more useful model for memory retrieval is described by the great Russian psychologist A.R. Luria in his classic book about a subject with a superlative memory, *The Mind of a Mnemonist*. This person was described to retrieve memories through synesthesia, the merging and connecting of various sensory modalities such as sight, and sound. Certain specific recollections might be connected with an explosion of vivid colors, for example. One might ask what does this accomplish? Doesn't the mnemonist now have to remember even more than he would otherwise in order to extract a specific memory? The answer is that this is not a factor because the most difficult aspect of remembering has mostly to do with retrieval, not storage.

Synesthesia has a definite psychedelic quality and as one might imagine is a common event in drug users. It is probable that the editorial, inhibitory or excluding properties of the brain are impaired and especially by opiates and sedatives that may impair the more advanced or controlling areas of the brain preferentially compared to lower areas. Dr Ober in his very interesting study *Boswell's Clap and Other Essays* goes into marvelous detail in his study of literary figures and medical impairments and their influence on literary production. He points up the synesthesia in Francis Thompson's "Ode to the Setting Sun":

From: ODE TO THE SETTING SUN By Francis Thompson

ODE

Alpha and omega, sadness and mirth
The springing music, and its wasting breath--
The fairest things in life are Death and Birth,
And of these two the fairer thing is Death,
Mystical twins of Time inseparable,
The younger hath the holier array,
And hath the awfuller sway:
It is the falling star that trails the light,
It is the breaking wave that hath the might,
The passing shower that rainbows maniple,
Is it not so, O thou down-stricken Day,
That draw'st thy splendours round thee in thy fall?
High was thine Eastern pomp inaugural;
But thou dost set in statelier pageantry,
Lauded with tumults of a firmament:
**Thy visible music-blasts make deaf the sky,**
**Thy cymbals clang to fire the Occident,**
**Thou dost thy dying so triumphally:**
**I *see* the crimson blaring of thy shawms!**
**Why do those lucent palms**
**Strew thy feet's failing thicklier than thy might,**
**Who dost but hood thy glorious eyes with night,**
**And vex the heels of all the yesterdays?**
**Lo! This loud, lackeying praise**
Will stay behind to greet the usurping moon,
When they have cloud-barred over the West
Oh, shake the bright dust from the parting shoon!
The earth not paeans thee, no serves thy hest
Be godded not by Heaven! Avert they face,
And leave a blank disgrace
The oblivioius world! Unsceptre thee of state and place![59]

There are many other techniques employed to increase retrieval so that subjects appear hypermnestic. Ancient Roman orators connected individual memories with rooms and parts of rooms in a type of topological memory system, of a mansion or structure already familiar to them[Φ] . A related technique is to assign each specific memory to a wall, corner or ceiling of a familiar room. Number these in order and you never forget about any topic in your talk. What's more, you will always deliver them in order. This method deployed visual associations and organized thoughts in the service of memory usually to help prepare them for a major oration. This helps in recalling whole lists of things is organizing individual elements and bringing them to the surface one at a time, being certain not to exclude any, and to get to every one, in order. It was used by Cicero and others for thousands of years. Widely known, these and related memory techniques were helped recount great epics such as the Hebrew Bible and the Iliad and Odyssey which otherwise wouldn't have come down to us. These great works were presented orally through generations long before the wide use of the alphabet. Even after efficient means of writing was in use an oral tradition survived in matters deemed too esoteric for the general public's perusal, secrets that it was felt, needed to be reserved for a select persons of great understanding and wisdom. The Talmud and Kabbalah are counted among such oral traditions as well as more contemporary and banal Masonic Codes. It is likely that sages of the past had much better memories than we do today, simply because they had not all the means of recording that we have become accustomed to. I couldn't imagine giving a presentation without my slides or computer or at least without some written notes. The news media reported admiringly how President Clinton once continued his speech, "without skipping a beat", when his teleprompter momentarily failed. Great orators of classical antiquity gave lengthy speeches without having to rely on any external prompting device. Only later were their orations recorded for posterity by an amanuensis hired for the purpose of writing the history of the great man.

Most other memory techniques work to strengthen associations between two or more objects. One may be considered the known or "given", to be associated with an unknown or "data". For example, a lot of us have trouble associating a name with a face. We may see a person to whom we've recently been introduced and forget the name. The face is the given, the name the data. Or, we may have trouble connecting a state with its capital. There are many common sense techniques employed to strengthen these associations. These ordinarily appeal to other senses, especially vision, in the making of the memory. To reinforce Harrisburg as Pennsylvania's capital you could see in your mind's eye a hairy pencil, a sight that you are not likely to forget![60]

Many people find the simultaneous recall of memories by two individuals mysterious. Two people may suddenly blurt out a recollection at the same time. Perhaps this means they are simpatico or really in love or that they know each other so well there is an undercurrent of deep subsurface communication. They fail to realize that both of them have been subject to the same stimuli. A more parsimonious explanation for their simultaneous verbalization is that the same memory is tied in their mind, with the same stimulus. This is much the same as our ability to recognize and extract the memory of a certain song after the opening bar, sometimes just a single chord, is played. All this means that there are myriad ways of getting to memories once they are firmly imprinted on the brain.

The purely anatomical model which proposes that each separate memory resides in a specific anatomical location is appealing. Electronic devices such as records, laser disks and floppy disks store information this way. You retrieve a memory by finding its address. Penfield's electronic probe may easily be likened to a phonograph needle. (Recall that Penfield was the neurosurgeon who tried stimulating brain gyri during brain surgery and found that electrical stimulation would elicit specific memories some of the time.) Stimulate the proper gyrus and you will elicit (play back) a specific memory.

But it has not been possible to erase specific memories in the brain merely by cutting certain anatomical regions out. In humans who have specific lesions caused by stroke, penetrating injuries, or the like, that destroy specific regions of the brain, it has never been possible to show specific memories are lesioned as well. Rather it is most likely that specific memories are not lost. The brain seems to be unlike electronic data recording systems. You can't pin down specific engrams to certain tracks, locations or addresses. In fact, as you cut areas of the brain out, even temporal lobes (as is occasionally done for epilepsy) you get the impression in seeing these patients, that their basic personality and memory structure is intact. How do we explain the fact that memories can be brought out by an electrical probe, by topiographic stimulation of the brain and yet cannot be lesioned?

An engram is a single unit of memory that makes an impact on the brain. One popular theory of memory is that engrams are stored widely rather than at specific addresses. The entire brain or large portions of the brain may store memory traces conspiratorially. Perhaps engrams are stored widely over a large volume of brain. This would fit with the experience of clinicians that it is not possible to connect specific focal brain lesions with the loss of specific memories, also with the fact that older memories are easy to retrieve and more resistant to loss. Perhaps repeated exposure (as in learning, studying and then experiencing a certain fact, produces widespread anatomical (synaptic) or biochemical cerebral changes.

When holograms (three dimensional pictures) first came into existence, cognitive and physical scientists were intrigued with the technique. Bits of information were widely distributed over the entire picture so that far flung regions of the hologram contained information about other distant regions. The hologram lived up to its name and recorded data not in an analytical fashion as most electronic devices and even regular photographs seem to do, but saw information as a complete unanalyzed package. Data about a single point in three dimensional space was widely distributed on the photograph. Another feature of a hologram is that data from the whole picture is contained in a small part and thus a good portion of a hologram would have to be damaged before the picture would be lost. These are features that holograms have that seem to fit the model of memory driven by lesion experiments[61].*

As it turns out, computer scientists are actively studying holographic storage. There are all kinds of possibilities here. Two beams of laser light will work to retrieve information stored by comparing their transmission in what is called an interference pattern. Information is stored in three, rather than in two dimensions and mechanical contrivances such as disk drives would be replaced by non-moving electronic devices. Retrieval of specific pieces of information would be much quicker. Rather than retrieving a single bit of information at time, a whole image of thousands of bits could be retrieved at once making video image storage more of a possibility and also parallel rather than serial or sequential information processing[62].

Another slightly contrasting theory, is that engrams are overdetermined or, in other words specific memories are multiply stored in various regions about the brain rather like having a record with a song stored over and over again or a floppy disk with the same file stored in multiple locations. Apart from the sheer inefficiencies involved, this explanation doesn't make sense. Why should memories be stored multiple times merely for the purpose of not losing them in the unlikely instance of a brain lesion. Perhaps just to fool the neuroscientist? However this picture turns out to be closest to the truth.

At one time I had visions of neurosurgeons extracting specific memories in their operations. When you went under the surgeon's knife you never knew what you were going to lose. In one patient he'd extract memories of married life, in another, records of childhood, perhaps a high school diploma here, a Ph.D. there. We now know (from experience) how ridiculous this notion is. By whatever means (and it is most probably quite different from a hologram model) specific memories are widely distributed in the brain.

Memories seem to be are stored, then attached by certain *handles* tugged upon during the process of retrieval. Some of these handles aren't adequate in and of themselves to extract memory engrams. Perhaps these may even break off under the weight holding a memory down. Certain forces that keep a memory from coming to the surface impair retrieval. Foremost among these is anxiety attached to a memory, but also the age of the memory, the amount of time it's been stored without any attempt at retrieval adds to its weight. When a lot of energy has been devoted to helping a memory to come out, for example, in spending a lot of time going over and over a fact that will be needed for a test, or in the case of Luria's mnemonist attaching a memory to the handle of vivid sensory experience, then this handle is firmly attached and the memory may be easily recalled. In other situations the brain may need more than one handle to extract a memory because each of these alone proves to be inadequate to extract the engram. You can see this easily when you finally remember something when quizzed and given more and more hints. The hints are individual inadequate handles.

To Review (Please see figure below): As memory gets established it is embedded, dug into, the brain, (signified by the spade in the picture). A single memory or *engram* is multiply represented over the cortex, and stored in different cortical regions through use of such techniques as synesthesia (multiple sensory involvement) and visualization. A primarily auditory engram or single memory can be visualized and is thus represented an additional time in a visual association area. That way the memory is, in a sense, reproduced over the cortex and is over-represented as visual reproductions and auditory associations accrue. The engram is thus acquiring various handles by which it can be brought to the attention of a person or *retrieved.* Some of these handles implicate other engrams and thus remembering one thing that is associated helps us to recall another. Indeed we all have embedded in our minds strong associations between different memories and find that as soon one thing is recalled, it brings to mind many others. If the handle, usually an association, is not strong, it may break as you try to get at the memory using that handle. Various forces inhibit retrieval, especially anxiety which can weaken associations in attempting recall (see the hammer and chisel destroying the handle) as when you "choke" on an exam, and anxiety can also increase the forces which keep an engram from being extracted or retrieved, holding the engram down, signified by the springs as happens through repression. Anxiety also blocks attention interfering with the laying down of new engrams. Attention, awareness of the stimulus, is ordinarily necessary in order to lay down new explicit memory engrams. You are sitting at a lecture. If you are not paying attention or fall asleep, you won't "get it", you will not even be able to take the first steps in laying down the memory in your brain. This is in contradistinction to implicit memory, for which attention is not necessary. As long as you practice, whether you are paying that much attention or not, you will improve your basketball or golf, game. Finally we have Penfield's electric probe. He's the neurosurgeon who demonstrated that a patient could recall specific memories when he stimulated the cortex electrically. These were memories that were already consolidated, dug in, well-established. This reminds us that ordinary recall very likely does happen when you stimulate the appropriate area of the brain. You're not doing it with an electric probe of course, but exciting a specific brain region through the use of those associative handles. Thus we have a complex multi-stage processing of the engram for explicit or verbal memory:

Thus a single memory engram is subject to: FOCUSING OF ATTENTION, RECORDING, CONSOLIDATION, CORTICAL SPREAD (Reproduction, overdetermination), RETRIEVAL

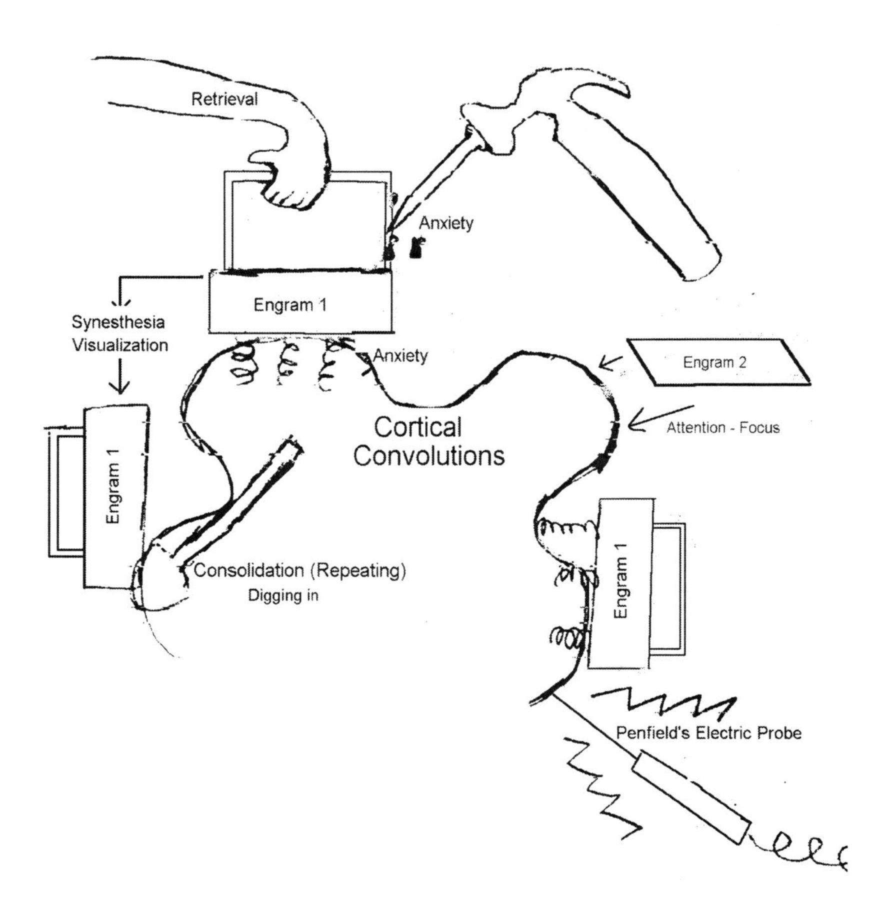


**Figure 20: Model for Storage and Retrieval of Explicit Memory**

When you cut out a specific area of the cortex, you do not destroy a memory. What may happen is that you will extirpate a specific way of getting the memory out. Let's say you stored recollections of a hammer among many other tools. If some process comes along, typically a stroke, that destroys the small language region to which you've assigned tools, then you will not be able to extract a recollection of a hammer by appealing to it as a tool. That particular handle, for that specific engram, the hammer, will be affected. Since a hammer is otherwise quite a familiar object, you will still be able to appeal to a recollection via other modalities, say a tactile or a visual association. Recent research, especially by the neurologist Antonio Damasio, mentioned previously, shows that exactly this type of categorization and anatomic compartmentalization of concepts takes place in the brain. The handle determines the anatomic locale in the brain. In other words, visual associations are likely to be stored in visual brain areas, language categories in the language portions of the brain and so forth. Individual memories, long established, are stored in multiple brain regions.

Computers store memory in files and register them once typically, unless there are multiple copies in the hard disk, CD ROM or other storage device. Computers can bring up ideas fast because they can search text for any occurrence of a word, looking through their various files. They perform these whole file searches at enormous speed, that brains aren't capable of. The age of the filing typically does not matter to a computer either. Ancient or recent filings have the same valence. The brain, relies on surface associations and ancient memories are superseded by recent ones. As time goes on, the handles are buried deeper and deeper and if never called upon and pulled to the surface, ancient memories become harder and harder to retrieve. This may be an adaptive. The brain's retrieval process brings about not a simple appearance on a computer screen or a printout, but rather requires us to focus attention or consciousness on any engram retrieved. Since one is capable of focusing attention only on a very finite number of objects, all of these compete for the individual's attention. Newer established memories are the ones bound to be the most relevant to a person's life.

Nine times out of ten when a young person complains of a poor memory one can assume that the memory is grounded too strongly and cannot be removed, or that this person is distracted from the business of storage by some process that distracts his attention, i.e.

anxiety or preoccupation. Anxiety affects memory in two ways. It blocks attention so that the original memory cannot be laid down. The original stimulus isn't noticed because the subject was paying attention to other things. Then he wonders why he didn't remember it. Or, anxiety can weaken the association that helps us to extract memory. Often anxiety is a force that keeps a memory buried so that it just cannot be extracted. This is one mechanism for the phenomenon of repression.

On the other hand older individuals are much more likely to suffer from a disease that impairs the original storage and registration of memory engrams. Older individuals can just as well have problems that younger ones do, but in addition are more likely to have degenerations that preferentially affect the hippocampus like Alzheimer disease. Strokes and seizures and many drugs which older folks are much more likely to be taking can also block attentional processes instrumental in laying down and consolidating engrams. In other words older individuals are more likely to have an organic neurological problem that impairs memory formation and retrieval, yet old and young may present with the same complaint.

Certain unpalatable recollections of childhood are actively suppressed, particularly the ones attached to high emotions. The inability to recall details of our past or conversely aberrant recollections may be the cause of serious psychiatric derangements. On the other hand, when a person is helped to bring such recollections to the surface an explosion may well occur. A war veteran may have suppressed horrible recollections of the battlefield or his participation in human slaughter. While specific memories cannot be recalled guilt or other emotion attached to these memories may impair this person's ability to function in his current situation. He may feel unworthy to participate in normal life, suppressing his normal aggressivity that he needs in order to compete in real life situations; he may drop out of life entirely until he has worked these problems out perhaps becoming addicted to alcohol or another substance, compounding his sense of worthlessness.

Memories that are essentially irretrievable may thus affect our present life in profound ways.

Unacceptable recollections of childhood such as participation in a sexual act may be actively suppressed may cause symptoms that can ruin lives. One woman afflicted with a variety of apparently psychosomatic ills including among others chronic neck pain with losses of consciousness also apparently prone to a variety of drug addictions recalled that she'd been sexually abused as a child. Other life circumstances include in psychiatric hospitalizations, her son's tragic suicide by hanging pointed to pervasive maladaptation in this highly intelligent and articulate woman, a remarkable illustration of the biblical observation that the sins of the fathers are visited upon the children.

But I noted, the abuse that this woman had suffered as a child was relegated to the very end of her history. This often underlines its importance pointing to its basic role in her life circumstances and symptoms. Indeed there seems to be a subliminal realization in many people of the fundamental importance of certain life events that are not at first brought to the surface in a conversation or in a medical history. Very often, these come to light at the very end of a history with appropriate probing or these details may finally make their way through at the end of an interview as if they somehow are of critical importance and just have to come out. Sometimes such critical details literally come out just as a person is about to leave. It is as if this person can't leave the room without first telling you what is really on their mind.

When dark volatile memories finally do reach the surface, there is often an emotional explosion that profoundly alters a person's world view. In immunology we know of a mechanism whereby an organism having once been exposed to an object, which has excited his immune system is exposed to it again. On the second exposure, immunity which may have brought but a slight reaction on the first exposure is poised to react with great force, almost as if frustrated at inaction on the first encounter. A person given a Tuberculin skin test, may have a mild swelling in his skin. That is how to determine that he has had some exposure to the Tuberculosis bacillus. But give him the skin test a second time and his reaction may be extreme and can literally (on rare occasions) cause him to slough his entire arm. This quite appropriately is called an anamnestic response. It is analogous to one's mental reaction to certain memories.

Immunity is highly analogous to learning on many levels. The immune response is altered by repeated exposure to an antigenic stimulus just as our behavior is altered by repeat exposure. As we saw in chapter one, antibody production is combinatorial. Plasma cells are capable of producing an enormous variety of antibody molecules that fit molecules on microbes capable of causing infection. On first exposure to a microbe a particular clone of plasma cells producing antibodies against the invader is called into action. The antigen is displayed on the surface of activated t-cells and macrophages that attack and incorporate and digest the invader. The initial immune response is limited, but continued or repeat exposure will cause antibody production and an active immune response to redouble. The invader is not likely to gain a foothold a second time. That is the principle employed in vaccination and the reason why we ordinarily do not succumb to an a viral infection a second time. As we have seen antibody molecules are pieced together from a combination or repertoire of component parts. Clones of antibody producing plasma cells are selected for, the ones that make antibodies to the invader du jour. These clones will put out large numbers of descendants.

What we see in the immune system is an altered response based on experience, learning in other words. Learning takes place in the immune system, learning - a change in behavior based upon experience. It is possible that similar mechanisms control an alteration of responses in the central nervous system, although this remains to be elucidated. For example the CNS may well be capable of generating a large array of molecules via combinatorial techniques analogous to antibody production in the immune system. Learning may then be induced on the molecular level after repeated exposure to a stimulus. There are other analogies too. Our immune system can be desensitized to an immune stimulus is presented repeatedly. The nervous system tends also to dampen its response in the same way when a stimulus is presented time and again. This is termed tolerance or habituation.

Human awareness is a continuous thread of memory wending its way through time. More accurately, it is a whole multidimensional fabric with complex interweaving of recollections, patterns and trends. Our brain is capable of looking at this fabric in a wide variety of ways gazing at individual threads delineating simple chains of events in normal time, or stepping slightly farther away to look upon how events are interwoven or even further still, to examine patterns or mosaics of patterns that comprise a life. Details and the big picture are hard to appreciate simultaneously but describe the recording of experience that is human life.

Korsakov's syndrome happens with destruction of areas of brain responsible for registering recent memories, the mesial temporal lobes on both sides, the mammillary bodies, and/or the anterior thalamus (illustration). This can happen after oxygen deprivation which tends to affect some of these areas preferentially, after a Herpes virus encephalitis that also likes this area of the brain or in thiamin deficiency that causes Wernicke's disease often related to alcohol consumption. Since memory is affected permanently in these conditions the problem is more severe than just the inability to recall names, the time, learning new skills or finding one's way around. The patient is lost in a mental limbo disoriented in time and space, a mental cripple, because his whole stream of mental awareness is broken by this process. We tend to underestimate the effect of losing even permanently a single mental function. After all the personality is left relatively intact because such patients can and do reach back into their store of remote experiences and utilize previously learned methods of dealing with the world. The fact is though that such persons are devastated by this simple inability to store new experiences and form memories.

Remote memories are more resistant to processes that affect mental function. By the time remote memory is affected we observe a more profound and global deterioration in personality structure. By referring to our memory model it is easy to see why this is so. Most old memories are multiply determined and stored according to large numbers of handles in widespread brain regions. For a basic remote memory to be lost, a lot of the brain has to be affected. A person who unable to recall even remote events, or who is unable even to recreate his own methods and habits and ways of performing simple acts has nothing lost his entire personality. Yet many of us are oblivious to our own personal past and also to the history which allows us to place our own lives and daily events in true perspective. We may be blinded to the context of events in our own lives and a feeling of orientation of lives in human history. Without some appreciation of history we are lost in an amnesiac limbo.

We have seen how memory in living organisms differs from information storage and retrieval in machines. Living things have a history and memory that affects them. It is part of a complex fabric interacting with current experience and continuing registry of experience. But from the human perspective, memories are tied to an emotional valence, memories mean something, they aren't just stored.

Humans aren't manufactured like machines. They develop. Humans contain within themselves a record of their history and as such history is indelibly embedded in the present. Each stage of development bears upon all later stages as the fully formed adult human, still an evolving creature, emerges. This happens not through any noncontiguous quantum process wherein all past forms are rejected in turn in favor of a final finished perfect product, but rather as the obverse of this process, through continuity, modification and reworking of past forms. It is true that we see many examples in human development where children and adults appear to make quantum leaps. Suddenly, or so it seems a baby who has been crawling begins to walk, sometimes quite well. An adult may rarely make certain sudden changes in his life, perhaps giving up alcohol or cigarettes. Of course there may be signs of a more continuous process that are difficult to appreciate at first blush. This continuity or inability to forget is the basis of all biology. Evolution is inscribed in the development of every embryo and our mental development over the course of our lives contains within it a history of past ideas that are reworked and reshaped over time. Ontogeny recapitulates phylogeny, as was first enunciated by the German evolutionist Ernst Haekel. The evolution of an animal is replayed in its embryology. The same holds for mental development. Hence, the understanding of a person's present is imperfect without appreciation of his past. As we observe a person's behavior, we can only see very little of what he is about. If only we could live his life as he has lived it, developed as he has and still function as an objective outside observer of his behavior and experience. Our actual powers of human observation are so limited.

All of this information is recorded in our genetic endowment which unfolds to make a final organism. This unfolding programme is given by the genetic machinery of the cell. This is how the living differs from the inanimate.

"Organisms are unique at the molecular level because they have a mechanism for the storage of historically acquired information, while inanimate matter does not. Perhaps the was an intermediate condition at the time of the origin of life, but for the last three billion years or more this distinction between living an nonliving matter has been complete. All organisms possess a historically evolved genetic program, coded in the DNA of the nucleus (or the RNA in some viruses). Nothing comparable exists in the inanimate world, except in man-made machines. The presence of this program gives organisms a peculiar duality, consisting of the genotype and a phenotype. The genotype (unchanged in its components except for occasional mutations) is handed on from generation to generation, but, owing to recombination in ever new variations. In interaction with the environment, the genotype controls the production of the phenotype, that is, the visible organism which we encounter and study.

The genotype (genetic program) is the product of a history that goes back to the origin of life, and thus it incorporates the "experiences" of all ancestors, and Delbrück said so rightly. It is this which makes organisms historical phenomena. The genotype also endows them with the capacity for goal-directed (teleonomic) processes and activities, a capacity totally absent in the inanimate world"

-Ernst Mayr[63]

This is well-taken but seriously underestimates the ability of humans to store information and evolve. Our own information storage capacity, indeed our very evolution is increased and accelerated by our ability to store data in the brain and in recent human history to store engrams extra-cerebrally, through the invention of writing and more recently with the aid of ever- more efficient and information-dense storage devices such as hard disks and CD-Roms. All this makes possible change that occurs faster than by biological methods alone which occurred first through acculturation and dependence on ordinary mechanisms for memory storage utilizing language, and later through use of writing and ever more sophisticated storage devices.

Physicist Frank J. Tipler gives a cogent argument in his book, PHYSICS AND IMMORTALITY, that the brain is nothing more than a "finite state machine[64]", that is, if you had extraordinary knowledge about the state of all of the neurons and synapses in the brain, the connections between neurons, the current status of all the huge number of elements, say 10 to the 10th neurons each with up to 10 to the 5th synapses in various states, all of these states, both of neurons and of synapses can be assigned a numeric value, and although as we can see there is a very large number of neurons and synapses, even so, this number is finite. Given the numeric description of the brain's state at any time, and a description of input or influence on the brain, the outcome will always be known which is an alteration of the brain's state at time (t+1). This model ignores the effect of history, experience and memory on the brain's response, or at the very least, it de-emphasizes it and gives history diminished importance. Tipler would probably say that the brain's or a neuron's history and development can also be expressed as some numerical descriptor of the current status of neurons and synapses. The past is expressed as a numerical alteration of present status that admittedly may alter the response to a stimulus. Perhaps.

In the embryo, our nervous system starts out inauspiciously as thin tubular structure. The brain, spinal cord and peripheral nerves are modifications of this basically tubular design with all the major components being well formed by the fifth or sixth week after conception.

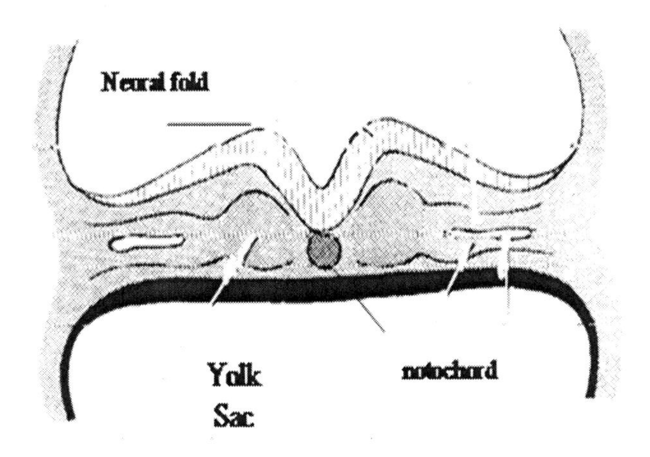


**Figure 21: Early Embryonic Disk.**

How is this accomplished? In the embryo the a flat structure the embryonic disc, forms a central neural groove. Cells well up and proliferate next to this groove to form 2 neural crests that rise beside the groove and then finally join at their center (see figure) eventually forming the primordium of the nervous system the neural tube. So the nervous system ultimately comes from a basic tubular formation. From there all else is a modification, a reworking of forms.

Only the spinal cord retains its basic tubular form by the time at birth. It is a rather flattened ovoid structure, especially in the neck when the originally hollow center has been filled in by the growth of neurons composing its gray matter. The brain's structure which also is ultimately derived form a simple tube, is a much more elaborate. But he brain's development involves the proliferation and growth of neurons and other supporting elements. The circular hollow structure eventually becomes the four fluid filled chambers (ventricles) of the brain.

Nervous system formation from a a tiny streak that begins to form in an embryo only three millimeters long depends on the presence of notochord, a semirigid structure that is present in some of our evolutionary forebears. The notochord keeps an animal rigid, but the notochord is given by vertebrates including humans in favor of a bony vertebral column. This acts as a structural support for animals with an elongated architecture that is symmetric about a single axis. The notochord degenerates by the time our development is complete. Cartilaginous discs are remnants of notochord between bony vertebrae. Yes, these the same disks that herniate to cause back pains, nerve root disease, so that people lose work and wages and otherwise cause other mischief. Notochord remnants may even be present in a mature animal as tumors called chordomas that can compress the spinal cord.

The notochord is the first use of an endoskeleton to maintain rigidity of body form, specially useful for forward swimming. Cartilage often matures into bone. Bone is more complex, more rigid and requires calcium. Evolutionary development mirrors maturation. Cartilage precedes bone in more immature animal forms.

Moreover the presence of notochord cells prefigures and causes first segmentation, the breaking of an elongated body form into segments and then for the laying down of bone that forms a repetitive structure that is composed of vertebrae. These similar repetitive structures are enumerated in anatomy so that we can easily locate nerves, ribs, arteries and other structures that arise within each segment of the animal. The skeleton with muscles attached can now accommodate controlled bending. In the embryo it marks our kinship with other animals in the phylum Chordata. These animals contain a cartilaginous notochord during some stage of their development. this animal group includes a primitive fish-like animal, the Amphioxus that is mentioned in basic biology courses.

But the notochord plays an even more important role. It induces or creates the proper signals to bring about the earliest formation of the nervous system. Were it not for the development of the notochord the primitive nervous system could not commence formation. In more basic animals such as Amphioxis a nervous system forms above or dorsal to its notochord just as it does in the human embryo. This illustrates two basic embryologic principles. Firstly a structure once formed gives signals for the formation of other structures. Embryologic experiments in which such structures are cut away show that development depends on preexisting parts. It is likely that preexisting embryologic structures send out chemical signals to adjacent cells that cause these cells to develop in an exact way. This is a process of induction.

We now know what some of these chemicals are, proteins, specified by certain genes called homeobox genes. These genes are remarkable for a number of reasons. First, much of their D.N.A. pattern appears to be preserved virtually throughout the entire animal kingdom even in such disparate species as worms, insects, mammals, and in humans, virtually all animals that have an elongated back to front structure. Second, the genes lie in an order on chromosomes, that is exactly the same as the front-to-back order on the animal! One can only speculate that the original zygote, the one fertilized cell that is present at conception, is already oriented in a certain way to bring about cell divisions that will dictate the front to back development of the final animal, depending on the orientation of its homeobox genes. The proteins specified by these genes appear in specific foci on the developing embryo. Around these foci, there is a gradually decreasing concentration gradient that also signals specific development. One chemical might signal that a forelimb should develop at a specific place, while still another homeobox protein might center about the hand so that there will be two opposing concentration gradients of proteins, from the shoulder down and another from the hand up that each serve to signal specific anatomical developments. Homeobox proteins determine differentiation of embryonic cells from front to back along the longitudinal axis, from head to tail or in anatomic lingo, in the rostro-caudal direction, the first most basic factor that distinguishes developing cells in different locales.

Early in life the embryo is a totally noncommitted ball of like-appearing cells. Later a certain group of these cells differentiates into a flat sheet that will eventually form the animal itself and another group of cells responsible for the environment and nourishment of the developing animal including the yolk sac, placenta, and amnion. The flat disc of cells that develops into the actual living animal cannot be distinguished on any basis from the cells that form the rest of the embryo, yet the process of differentiation happens with perfect fidelity as if by magic. These cells at this stage have not yet committed themselves on the question of the exact part of the animal that they will form. Experiments in which these first primitive cells are moved to different locations within the embryo show that these first cells are multipotential, that is are capable of forming any of various parts of the nascent animal. If a few of these cells are cut away, chances are the developing animal will not miss them at all at an early stage because the others will just fill in, or, otherwise stated the same cells that are cut off can at this stage form a whole new organism.

Slightly later in development three germ layers form designated mesoderm, ectoderm and endoderm. The mesoderm forms structural elements, bone cartilage and muscle including the primitive notochord. Mesodermal cells seem to signal adjacent cells to differentiate into ectoderm which in turn, forms the nervous system and the skin, while the endoderm eventuates in the viscera or internal organs especially the gut.

At this stage cells are multipotential still. The cells that are descendants of a single cell may form any of a wide variety of organs, only their options are somewhat more limited than at the previous stage of differentiation. Some ectodermal cells in a given location may be capable of forming any of various parts of the nervous system. For example at this still relatively early stage a whole wide area in the front of the animal may contain cells capable of forming part of the eye, cells in this particular area designated as "the frontal eye field". The frontal eye field becomes smaller in relation to the size of the animal as many of its cells make commitments to form other organs. This commitment of cells is determined at every stage by location. It is as if each cell determines what it will become by asking where it is. Although the exact mechanism responsible for this commitment is imperfectly understood, it is only reasonable to assume that the process has something to do with concentration gradients of proteins, some of these homeobox proteins.

Even after birth parts of the brain become committed to serve certain functions as localization of abilities develops. In most people, the left hemisphere performs the bulk of language functions; this is what we mean by left hemisphere cerebral dominance. Certainly the left hemisphere is responsible for the bulk of language function in most right handers. This hemispheric preference is preordained and mostly genetically determined. Left handedness does tend to run in certain immediate families. Very early in life, though, it is possible for a person who was originally destined to be right handed and left brain dominant to instead become left handed and have language function predominate in his right hemisphere. This happens in the event of an early cerebral injury. In these persons who may have had a devastating event in utero or a brain injury during or shortly after birth there may be weakness, sometimes profound, on the right side of the body. Scans may show atrophy or damage to the left side of the brain and they will also presumably be right hemisphere dominant for language function (though this is difficult to prove.) Presumably this also occurs in persons originally intended to be left handers but this is harder to diagnose because of the relative infrequency of left handers. If a person is left handed and has no family history for left handedness, an early cerebral injury may be the mechanism.

This altered pattern of cerebral dominance will not occur unless a brain injury occurs very early, especially before the person's first year of life. Later, different portions of the brain are much more committed to perform their specialized function - committed in the same way as embryologic anatomy described above. In a younger person it is easier for non-damaged parts of the brain to assume function for damaged parts, a property known as *plasticity* within the human brain. It is clear plasticity partly depends on the degree of commitment of function and that this depends on the age of the organism, one good explanation for why younger brain-injured persons may have better prognosis that older persons.

Development is in part about commitment. Every cell which gives rise to other cells must at some time decide upon its destiny, commit itself to play a certain role that is determined by its relative position in the embryo. Some will hold the exalted function of a neuron, others will have a more modest existence as liver or gastrointestinal cell. Once totally committed, there is no turning back. Once a cell commits itself, its descendants, other cells, products of cell divisions, differentiate on the basis of this decision. Early in embryo's life specific cells are committed to be endoderm cells, those that comprise the viscera. Some of these endoderm cells separate off and give rise to cells of the liver, others the stomach, still others, the gut and so forth. Some of the liver cells will break away and form the hepatic ducts that transport fluids inside the liver, others hepatocytes that mediate chemical reactions, others supporting cells and so forth. Sometimes other specialized cells actually migrate into an organ from afar as is the case with certain glandular tissues. The process is commitment of the ancestral cell or anlage in German, then differentiation or specialization of its descendants. Once committed, and especially once differentiated, the cell has embarked upon a process that cannot easily be reversed. The cell can not turn back and perform another function. It is no longer plastic.

Notice here that in making this commitment the predicament of the individual cell is not too different from the individual neuron described in the last chapter, which at a certain point, given multiple inputs, mostly from adjacent cells has to make a decision to fire or not to fire. Here in development an individual cell is made to commit itself to play a specific role, to give rise perhaps to a line of cells that will eventually form a certain organ or perhaps even degenerate.

If you begin with the observation that the single zygote cell gives rise to cells committed to perform every function in the body then you have to conclude that the zygote cell and hence every cell in an animal's body which has exactly the same genes, is multiply capable. Every somatic cell contains all the genetic machinery necessary to become any type of cell in the animal's body. What happens, of course, is that many genes are turned off, not expressed. The liver cell does not produce the neuron's specific proteins, assume the shape of the neuron; the neuron does not make proteins specific to the hepatocyte and so forth. Therefore commitment, differentiation, specialization mostly are mostly about inhibition, masking of certain genetic elements. Only some of a cell's full genetic potential is expressed and that is how cells become what they are going to be. If scientists could learn all about the processes that control gene expression, that could be every bit as useful as mapping out all of our genes, the current focus of research.

Well after all of our chromosomes and genes are mapped, researchers will be trying to unravel the mysteries of gene expression. Just as one example, as we have seen with the plasma cells and likely many other cells create proteins in a mix and match process combining information, pieces of information from different genes in novel ways. New plasma cell clones compete for survival and a license to reproduce in a Darwinian fashion right in our own bodies. Those whose antibody products match the antigens displayed on the foreign invader, divide and make more of their own kind, thus increasing their fitness. This is but one unusual mechanism of gene expression and there are likely to be many others.

It is clear also that this is only one example of natural selection that occurs inside an animal's own body among its very own cells. In the embryo certain cells are selected to divide for a few generations just to form a supporting matrix, then targeted for mass destruction. These cell lines are necessary for development only for a while, then sculpting takes place as organs take their final form. This happens to many cells of neural origin in the brain, whose precise form is so critical to final function. Precise neuronal connectivities will determine exactly how the brain will eventually perform.

Certain cells in development are destined to be destroyed. A perfect example is tail cells in animals that are not destined to have a tail, as in tadpoles that mature into frogs. Lately scientists have discovered that cells go through a specific cycle before they die. These cells literally sacrifice themselves for the good of the adult intact organism in a process termed *apoptosis* [65]. This is a preprogrammed cell death cascade. The interesting thing is that it appears that embryonic processes that sculpt the final form of the organism in which cells are programmed to die, resemble cell death in disease as well. An entire cascade of events is set up both in the embryo and in certain diseases such as degenerative neurological conditions or in the immune response to an attack from an outside organism. If you can define the cascade of cell death, then it should be possible to intervene in conditions where programmed cell death is pathological. For example in stroke we know that neurons die thorugh a course of events that involves the chemical Glutamate, a so-called excitotoxin. Stoke therapies are aimed at inhibiting Glutamate and excitotoxins or in otherwise intervening in the cascaded of events that occurs when cells die during a stroke to save brain cells a process termed neuroprotection.

It is characteristic of a mature differentiated cell line, that cells reproduce slowly in controlled fashion. During development in the embryo, on the other hand, cells may reproduce like crazy. In total considering the final number of cells that make up a person, there is only the equivalent of perhaps 50 doublings of cell numbers from zygote to mature person. The differentiated cell has a limits its repertoire of what it will become, what descendants will be. The uncommitted cell is pluripotent. Its descendants may do anything. Occasionally a committed cell will regress and act like a more primitive progenitor cell. That is how cancer happens in most instances. A mature cell line will give rise to a renegade immature pluripotent cell line. These cells may even produce chemicals that are characteristic of other very different cells. Internal mechanisms that control cell reproduction and growth, no longer work and the cells start to reproduce wildly. In very young children some solid cancerous tumors called neuroblastomas form from neuronal cell elements when some cells fail to mature and differentiate. Neuroblastomas are swiftly growing lethal tumors. Neuroblastomas are quite different from ordinary tumors that develop typically in older adults. Most tumors in adults come from immature cell lines that have dedifferentiated from mature cells. In the case of neuroblastomas, cells that give rise to the tumor have failed to differentiate in the first place. In some cases their chromosomes are defective. They have a deletion in part of the first chromosome. However some Neuroblastomas having a better prognosis have more normal chromosomes In some of these advanced cases of neuroblastoma, the tumors will miraculously regress. How? The involved cells will suddenly differentiate giving rise to a mature cell line.

The neuroblastoma cells, immature forebears of sympathetic ganglia, send a signal that causes immature Schwann cells to migrate into the area. Schwann cells are the elements that add the myelin covering to a nerve. Neuroblastomas with a good prognosis, those that will eventually regress and mature, have abundant Schwann cells. Evidently there is an exchange of chemicals, between the neuroblastoma tumor cells and Schwann cells in the area. The neuroblastoma sends out signals that call in the Schwann cells, and then the Schwann cells send signals that hasten the maturation the embryonic normalization of development in the tumor and it regresses and matures rather than kills the tiny patient. Here we have an excellent example of the constant crosstalk that is the exchange of chemical signals that causes maturation and differentiation of cells in the embryo[66].

The best example of cells that have partly determined the destiny of their descendants is the *stem cell*. Blood is produced in the bone marrow by various mesoderm cells destined to give rise to different blood cell lines, the red corpuscles that carry oxygen, the white cells, polymorphs and lymphocytes that fight infection, and platelets that aid in clotting. Stem cells are the partially committed immature, not yet differentiated cells that will give rise to all of these cell lines. Harvesting, growing, finding and using the chemical signals that control the differentiation along these individual cell lines, has become a major industry. Stem cells may turn out to have all kinds of uses. One would be when it becomes necessary to destroy the bone marrow that makes blood cells as a treatment for cancer, reimplanting stem cells will provide new blood cells and save the life of the patient. Knowing about chemicals that control blood cell production will help in a host of diseases where blood cells become deficient. For example, erthropoetin a protein that encourages the production of red blood cells, has been synthesized and is in wide use in sick patients who are not making their own blood cells. Some of these persons have had chemotherapy for cancer, some are otherwise chronically ill and anemic, some have severe kidney failure and the kidneys are the main source of erythropoetin. Other proteins signal the production of white cells, others platelets, making cell transfusions a thing of the past.

The cell line that is differentiated, the specialist, has a hard time turning back and into multipotentiality even if called upon to do so; it is less plastic. Plasticity, the ability to assume a different function, to fall into the breach in the case of a lesion or injury, is the opposite of differentiation in the sense that a cell which is more plastic is less specialized. In most healthy animals in which no pathological process occurs, differentiation is the key to success. It determines how well specialized cells will work l together. For the whole organism, maturity is largely a matter of increased specialization among elements of its structure which develops from a single pluripotent generic undifferentiated cell.

The organism develops from this finite inauspicious beginning, a singularity, that is, from a single perfectly pluripotent cell, one that can give rise to everything within the organism. Over time the organism is in a constant state of developmental flux determined by previous less differentiated states. The anatomy reflects this continuous embryologic development. A final structural integrity is exactly analogous to the contiguity of memory. The intact integrated person is a composite of continuous memory as the organism is determined by its complex development. The present is a product of the past.

Systems are built on the basis of preexisting parts. the organism forms on the principle of adjacency. If not for the notochord the central nervous system would fail to materialize. The notochord, which came from the need for an elongated body to swim forward in an aqueous medium, now induces adjacent cells to become a part of a segmented structure. Some differentiate into bony vertebae, others into nerves, and vessels. These components would not “know” their role or how to form but for the structures that happen to lie next to them. You can experiment by cutting adjacent structures away, and you will see that areas will misform or not form at all. A part such as the notochord that affects the formation of another part is said to have induced the formation. *Induction* is a basic embryological principle, induction by location, by adjacency. Sometimes the inducer puts out a chemical that causes the adjacent cells to differentiate or grow in a certain pattern. Developing muscle secretes a substance that causes a nerve to innervate it. That nerve that will ultimately control muscle contraction. Vessels form in the same way rivulets do, subject to forces from the flow of liquid blood.

On the sides of the primitive notochord along the axis of the developing animal, segmentation occurs as individual nerve roots innervate their own specific segment. Vertebrae and ribs are also mark animal segments. I recall as a student being overwhelmed by the sheer wealth of anatomic detail and especially the precise choreography of cellular reproduction and migration in the embryo. It is difficult for the student to even get a handle on the basics of anatomy and of embryological organ formation simply because the process requires the transfer of vast quantities of information as reproduction and migration of each cell is precisely spelled out. How could all this information, not only the final detailed anatomy which I had some trouble learning, and the texts had so much difficulty describing, but complex choreography of development, be housed in the genetic material of the nucleus of a single cell, the just fertilized diploid zygote? It all seemed incomprehensible to me. Part of the answer is that the embryology and anatomy are not two different things, much as it seemed that they were to me, a student who had to learn all about each as if they were two separate subjects. They aren't. Anatomy is determined by development. Development and structure are one and the same. The past is what determines the shape of the present. The design is in the development and the final form is determined by it. The process of development and thus form is what is codified in the gene. Also a single piece of information is amplified or repeated so that it only has to be specified once. Segmentation is an example of this. Examine the leaves of a mature plant. Plant cells must specify the leaf form only once. Once recorded in the nucleus, the pattern is used again and again as new similar leaves are formed. Individual leaves on the same plant are quite different, but this difference owes not to the basic specifications for the leaf's formation, but mostly to the position of the leaf on the plant and the leaf's relative age. At the same time you only need have a slight error in design specification of a repetitive structure to spell complete disaster, since a basic code is repeated so many times.

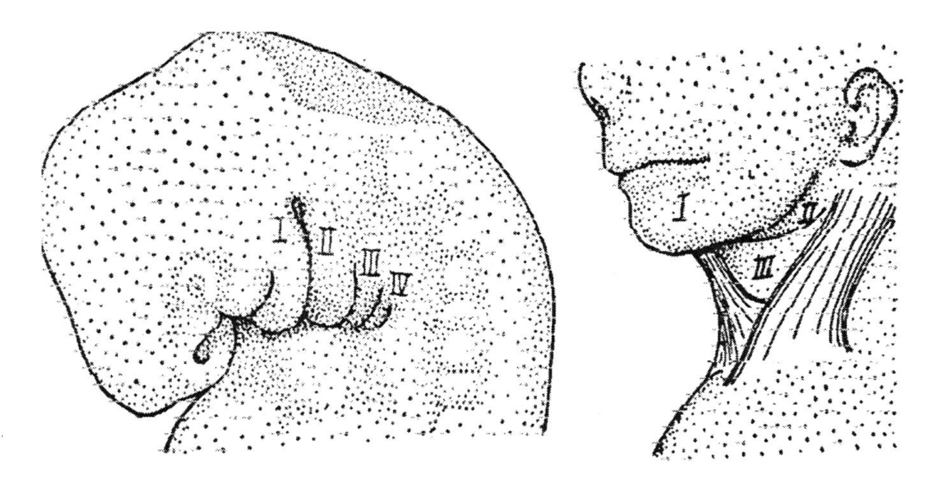


**Figure 22: The embryologic branchial arches evolve from gills of lower animals. Note segmentation of the spine as well.[67] And to the right with a surface view showing what becomes of the branchial arches which govern makeup of both deep and superficial structures. The first branchial arch forms the chin andd jaw area[68]**

Still it is remarkable that biological patterns unfold repeatedly with so few mistakes. What we witness in the embryo is repetition and amplification of basic processes. There are modifications and remodifications of primordial patterns. Vertebrae and discs are such repetitive structures so are the branchial (=gill) arches which owe their origin to fish like ancestors but in the human develop into parts of the face and neck. With the repetition of primordia evolutionary history is recollected in human development. Ancient adaptive patterns aren't just discarded. They are used, modified, remembered, are today written into our genetic code, only they specify our own form. The same is true of our nervous systems which retain simpler structures, and less evolved circuits such as peripheral nerves and spinal cord, but pile on them, layers of neurological control.

In fact as has been recently found that most animals, even distantly related species like vertebrates and insects share most genes. The difference between animals is mostly accounted for by regulation of gene proteins, that is turning on and off genetic machinery and controlling the number of repetitions as in vertebrae, segments, and bone elongations. Organisms develop by reaching into a toolbox of genes, most of these elements shared by practically all animals and even plants, picking out the genes determining their exact form. Genes are expressed or not expressed controlled by switches and then expressed to a greater or lesser degree. That determines the major differences between animals. That humans and chimps share over 99% or their genome only some genes, for example those determining vocal expression, and brain size are expressed to greater or lesser degree. Not only do organisms share common language or code in their nucleic acids for protein production, they largely have nearly identical genes. Thus evolution becomes a process of reaching down into a toolbox or repertoire as needed and choosing out varying expressions of genetic material. [69]

The human embryo has much in common with first an Amphioxis, then a fish and a tadpole, then a chick embryo but develops modifications that unmistakably define it as a human form as embryology replays evolutionary history[ϕ] . It is as if we could watch a new car being built by witnessing modifications in its design, from a Model T, through Edsel perhaps then to an old winged Thunderbird finally to the aerodynamic jellybean that is a the modern Ford Taurus. But machines don't contain within themselves recollections of their history. Lifeless forms have no perception of their past but for animals the past is inscribed in the present.

Talk to most people who have never appreciated embryology and it is not at all obvious to them that there is a plan or pattern that relates all life forms. Embryology proves the point. We can see over time what has been grafted onto previous structures. It is as if we have a video replaying an unfolding of form. Most people have little to no perception of this. People who advocate for animals haven't heeded many lessons from biology, mainly that life fits into a larger plan or pattern. They tend to treat all living things as equivalent and to lose sight of their differences. Well-educated persons either haven't thought about this subject or aren't convinced either that biology culminates in humankind. I learned a long time ago that some things that I accept as almost axiomatic, aren't so obvious to others.

I was talking to one of my physician colleagues who is interested in this topic. He'd just read some work by Stephen Gould, the evolutionary biologist. It was late at night and both of us were rather exhausted from having made rounds all that day. We stood in the intensive care unit, under the eyes of all, and debated the topic. Even lowly bacteria whose cells lack a true nucleus may be by his account, considered to be as biologically advanced as us. I could see some of his arguments but couldn't believe my ears, as we stood under the light created by human invention, without that light we'd be in the dark, perhaps relegated to sleep by that time, in a warm building on a winter's night, also the product of an human invention--bacteria are as advanced as us!! By what system of reckoning? In the bugs' favor they have modified themselves in a milieu of natural selection well after humans had. In other words some bacterial strains are much “newer” than the human species. Many of them were in that very intensive care unit, having adapted to the newest antibiotics we could use on them. As a group, hominids, apes similar to humans had been on the scene only very recently, for perhaps 4 million years, the genus Homo having arisen some two million years ago in Homo habilis. Primitive bacterial species had already evolved over 3 billion years ago.

Very soon both of us realized that the discussion hinges on how you define what is more advanced. That's where the whole argument gets hung up. I admit from the start that for me this idea of advancement is not a difficult topic. I “know” which kinds of animals I consider to be more advanced from the beginning (admitting that such foreknowledge is anti-scientific--in science you are supposed to "find" the truth, which can be very surprising, using objective evidence.) Below would count as some of my own criteria. This is important n the sense because in the method that I view evolution, there is a progression from less advanced life-forms, all the way up to those who reflect on the nature of such things. There is a great bulk of creatures who make their own living and survive some well, some not so well, and others who function on a level of making a living also but go a step beyond, transcend the mundane every day business of staying alive to consider something beyond themselves. In this category I place humans. We've reached a point that we are not only concerned about immediate survival, but can plan for the future, not just by making up a nest and a home under hormonal pressure according to a pre-wired plan but to influence what is to come according to an inner plan. And then to contemplate about what all of this means.

From a scientific viewpoint how reasonable it is to dismiss the concept of evolutionary advancement. Advancement is difficult to define. Depending on what criteria you use you may come to different conclusions as to which creature is more advanced. And why should anyone care about which life-form is more advanced anyway? It's just that we like to think that things progress, improve, evolve. Biologically we have a very hard time proving that humans are more advanced than other creatures. After all early and simple life forms such as bacteria exist in much greater numbers than more recently evolved and complex creatures and in a much wider array of habitats. They are, even today, more successful, judging by their numbers. And we have many examples of more complex creatures further adapting into a relatively simple state, arthropods and bacteria that evolve into obligate parasitism, sighted creatures that lose their lose the ability to see as they adapt to dark environments. Yet it is possible to find a few criteria for advancement which come out in our (human's) favor so that at the end of the day we may continue to see a conventional progression in evolutionary history from "lower" to "higher" life forms.

What determines whether an animal will be considered more or less advanced, what are the criteria by which we may state objectively that one organism is more advanced than another? An organism could be considered more advanced if it competes more effectively in its ecological niche. But this would be confusing what is advancement with success. Lots of species win out by sticking to a simpler strategy. Still another criterion is recency of appearance of a life form. Certain bacteria in our hospitals have even more recently evolved antibiotic resistance, insects, insecticide resistance. Indeed some might argue that lately man has almost ceased to evolve biologically because the weak survive and the physically strong or most aggressive don't necessarily make out better in our societies. But humans and human like creatures have appeared very recently in earth's history and out-compete every other species of plant and animal in a largest array ecological niches. Some large groups of organisms, most importantly bacteria and insects, are giving us a good run for our money, but humans have been able to employ certain members of these groups for their own gain, to make antibiotics, proteins, or using certain carnivorous insect species to eliminate crop eaters, even infecting certain pests with bacteria; our ingenuity knows no bounds. No other single species is as ecologically sophisticated or sociologically prolific as humans are. Many animals have evolved advanced social organizations but no species has the diversity of social interactions that we have. We aren't so happy that our efforts have apparently resulted in the extinction of many species of animal and plant by this time.

Still another criterion, perhaps the main one, is complexity. In order to decide which organism is more complex we need estimate its informational content, how much information one would need to reproduce a single individual. This point of view is enhanced by the computer, which has given insight into recording and expressing information. If as we have indicated, the human genome contains roughly $10^{10}$ bits of information or four thousand 500 page books, we may call man the crowning achievement of biology on the basis of the amount of information in each cell, far more than other organisms. It's also reasonable to ask the question for a human, how many bits of information does it take to encode the totality of being, that is including biological information and brain content, and there are a number of ways making an estimate. Consider that that brain has 10 to 100 billion neurons which have up to many thousands of synapses. Each synapse may be considered to carry one bit of information. As revealed in chapter one, I maintain that this will also give us a profound underestimate of the information stored within the brain, mainly because many bits of information are stored on the subcellular even the subsynaptic level, but for our arguments here, in which we are only trying to find the life form that most complex this estimate is good enough. Frank Tipler, in his book THE PHYSICS OF IMMORTALITY uses an estimate of $10^{15}$ possible states of the entire brain and this is close enough for our purposes, though it may in fact be far off.[70] Another disadvantage of this particular informational state concept, is that as I hope we have seen in the present argument, the current state of the human organism, does not tell all about all future behavior which is partly a function also of its past and past memories as argued in this chapter. One can counter that even in that case, the one in which we need take the past into our consideration of the present, the current status of the organism as a function of past events is still expressible in terms of bits of present information. Fair enough. But it should be clear at all times that the past does at all times make a difference perhaps seen in alteration of current expressions of informational status in the present.

The vast majority of living things are simple (in comparison with ourselves.). There are far far more things alive on earth that lack a true cellular form, with nuclei and specific organelles such as mitochondria, than there are that are composed of true cells. The technical terms are prokaryotes vs. Eukaryotes, precursors to cells vs. True cellular organisms. Prokaryotes are blue-green algae and bacteria mostly and of course there are even more primitive living forms such as viruses that contain nucleic acids and you might even include Prions a type of mostly protein parasite that appears to contain no nucleic acid among the living if you like. There are many many more of "them", organisms without true cells or nuclei, than "us" who are made of cells. Then we have the one celled animals and plants, and the protozoa. Bacteria in particular, may according to some of our criteria as given above, be counted as more advanced. They are, and have been for billions of years, extremely successful, adapting to all kinds of environments form the hottest sulfur springs, and places deep inside the earth's crust to ecosystems deep inside other more complex animal's intestines, to the polar caps, bacteria may be considered *as a group* to be by far better adapted, and to have radiated into more biological situations than any more complex creature.

**Table 6: Criteria for deciding a species is more advanced.**

| SPECIES ADVANCEMENT | |
|---|---|
| CRITERION | VICTOR |
| Numbers | Many examples: bacteria, insects, variety of other organisms |
| Success in single niche | Many examples, as above |
| Success in Multiple Niches (diversity) | Man |
| Recency in Evolution | Bacteria and insects acquire new forms most rapidly |
| Complexity | Man |
| Embryology | Man |
| Consciousness | Man |
| Artifacts | Man |
| Transcending Earth boundary | Man |

Whole groups of cells got together only recently in biological history, by current reckoning only about 600 million years ago. The earliest life that has left a trace is estimated to be 3.5 billion years old. Still, relatives of the earliest life forms far outnumber multi-celled organisms. It is not at all clear that more complex life forms dominate the earth. Simpler life forms, prokaryotes, have pushed the life's envelope to the limit, having been around in time as soon as the earth became earth, that is just as soon as it had cooled off enough to remain solid, and perhaps well before, and in all habitats on the planet including deep within the earth's crust. But multi-celled organisms are more advanced according to a number of criteria. When cells get together, it gives them a chance to specialize while maintaining a common identity primarily by virtue of an identical genotype. Multi-celled creatures are more advanced by virtue of complexity and specialization. Not only information theory, but embryology argue in favor of these creatures being more advanced forms, as to some extent you can follow their development through previous evolutionary forms. This is not to say that the complexity of embryology of an animal is the sole determinant of level of advancement. Otherwise it might be possible to state that certain insects that metamorphose into quite different forms might be considered to be more advanced. Here I'm not just talking about butterflies from caterpillars, but also various insects that go through complex forms, larva to pupa to adult and various molts in which they take on a number of different shapes.

The major feature that humans have of course, is the advance of the nervous system and we are uniquely freed to some extent, the degree may be debated for many of us, from pure biology, adaptation of structure and form and dependence on one niche. As can be seen from the table the criteria for advancement go from the more to less purely biological as you go from top to bottom on the table. The brain has allowed us to be successful in many environments and at some time in the future we may expect to live away from our mother planet. Of course we need not evolve gills to travel underwater, wings in order to fly, hair and subcutaneous fat to travel to the poles. Humans can now be found in all these diverse habitats thanks to our own ingenuity, though few of us want to stay for any length of time in most of these situations. And again, we are the only biological organism that we know of capable of considering of our own predicament, that is who are self-aware.

Having made criteria for biological advancement helps us to see a trend, a purpose or teleology. It allows you to state that the "purpose" of nature is the development of more advanced organisms or involved neurological structures. Simpler creatures such as bacteria and insects are in some ways much more adapted, numerous, fecund, successful than are we. Here Stephen Jay Gould contends that the advent of humans is by no means inevitable in biological evolution, that we are, in fact, more accidental.

"It is tempting to say that the victors won by virtue of greater anatomical complexity, better ecological fit or some other predictable feature of conventional Darwinian struggle. But no recognized traits unite the victors, and the radical alternative must be entertained that each early experiment received little more than the equivalent of a ticket in the largest lottery ever played out on our planet--and that each surviving lineage, including our own phylum of vertebrates, inhabits the earth today more by the luck of the draw than by any predictable struggle for existence."[71]

Gould makes a good point. The best example is the great cataclysms that happened a number of times in the earth's history, that of the Cretaceous period that saw the sudden extinction of dinosaurs, perhaps due to a meteor landing on the earth and also the end of the Ordovician and other mass extinctions, advents of partial clean slates if you will few of these extinctions being well explained. At these times it wasn't necessarily the strongest and best adapted creatures that survived. A massive explosion of biological forms suddenly came to be in the Cambrian period in which all of the known phyla evolved over a short space of time In an "explosion" of diversity. This Cambrian explosion is analogous to the inflationary model of the universe in which, shortly after the Big Bang the universe expanded suddenly in size. There is no good scientific explanation for the sudden Cambrian expansion in life forms. And the four or five or so mass extinctions of biota remain unexplained as well and don't fit well into any current evolutionary paradigm, making the earth in its current form, and our own existence, far from inevitable. There are a good number of scenarios of worlds of fundamentally different form that may just as probably exist, and quite possibly do without we humans gracing them, considering them, writing about them and these parallel worlds probably do exist right within our own galaxy (there is no need to invoke parallel universes in this context), only no one is describing them because they do not included a mechanism of self-awareness. For the same reason, we can say little about countless human societies that left no written record. There is no witness.

A number of scientists have written about this lack of inevitability of evolution from the less complex or advanced to more complex organized biological systems and that Darwinian principles would fail to predict a world such as ours with people and complex biological systems on small and large scale. They have allowed themselves to explore if furtively, into teleology or purpose in their writings, though such thought is forbidden in the strictest scientific sense. Scientists don't argue teleologically . They are supposed to consider only the outcome on the basis of antecedents.

Stuart Kauffman and Michael Behe have noted that the level of complexity couldn't ever be predicted using our current understanding of biological and physical laws. At best, evolutionary theory, is descriptive of diversity, but doesn't help much in yielding up the complexity of our world. Levels of organization we see cannot be predicted or even fully explained with our current scientific tools. Whether you're observing microscopically at enzyme systems, or macroscopically at social organization and human inventiveness, scientific understandings are inadequate to the task of predicting and explaining and you have to conclude that either there is a self-organizing principle, or that there is some one who is the author and controller of this complexity, teleology.[72] It is important to note that neither author is a religionist or suggesting a belief in God. Both would are searching for scientific explanations for their observations and find them lacking.

Animals develop which means grafting a present state onto their past. The past is not discarded but rather recorded in embryology and signs of it can be seen in a current structure. In order to achieve an adult form, animals have first to pass through an embryologic process. All of this information too, is recorded in the nucleus of their cells. Deciding which organism is most advanced, as unpleasant as it may be for certain political arenas, is a rather easy task. All of our methods, basically agree. It is man. What is more, like other animals, man's indelibly inscribed in the present. Indeed this is the most important argument for the concept of advancement.

Anyone who maintains that humans are the most advanced creatures on earth, endowed with reason and evolutionary and developmental memory, imposes an awesome challenge. Our intellectual endowment, our competitive advancements do not entitle us to rape the environment, don't license men as super-competitors. Quite the opposite. With intellect, free will comes a burden a responsibility to care for the earth and everything living.

Each person is made to replay his evolutionary history in his own development, and his cells, more specifically the nuclei of these cells, carry the specific information that makes this journey possible. Evolution is written into our present form, our anatomy. But if there is an embryology of form, there is equally one of ideas. An unfolding plan in both spheres must be specified somewhere. The more advanced the animal to human form is, the greater the need to have this information. The greater the amount of information needed in the developmental process, the more advanced is the organism.

As adults we carry not only primitive atavistic childhood thought processes but also patterns derived from mythic historical thought processes that are a part of our cultural heritage, what Jung referred to as archetypes. Succeeding thoughts graft onto older thought processes. Though ideas are overthrown, very little is thrown away. We function with reference and reverence for the past. It is as if we have purchased a garment that no longer fits but yet we do not discard it. We modify it to follow current cognitive topography. Changes in the embryo appear to be complex and even the final anatomy is difficult to master, yet the blueprint for development is specified within the very small volume of a cell's nucleus. Instructions for a ballet are contained within a small number of pages.

As we have seen, the homeobox protein products specify front to back, rostro-caudal, development within the embryo. Very soon after it is first formed, the fertilized egg, or zygote determines its future axis, where will be its head, and where its rump or tail. At the time that all of this is determined, perhaps after the very first cell divisions into 2, 4, or 8 cells perhaps, the front to back development of the embryo will be specified by varying concentrations of homeobox proteins.

The other major direction of differentiation is dorsal to ventral, back to soft under belly over the thickness of the body. The embryo starts out essentially as a sheet of cells, a flat disk of cells hanging over a yolk sac. Future dorso-ventral differentiation will be determined mostly by induction. Notice that development is specified along two axes rostro-caudal (head to rump) and dorso-ventral (backbone to underbelly). A third orthogonal axis is less well known about from the embryologic standpoint since it is generally symmetric[Φ] , namely side to side or right to left. In a sense development is specified on three axes, length (cranio-caudal), depth (dorso-ventral), and width (left to right).

The neural tube comes from ectoderm, the outermost layer of embryo that forms folds. These pile up largely through some reproduction of cells, and mostly via cell migration. Cells also elongate and internal skeletal structures within the cell the microtubules and ensure that they maintain this shape partly determining the ultimate folded structure within the embryo. Ectodermal folds later meet over the top or dorsal aspect of the embryo to fuse into a primordial tube from which the central nervous system will develop. Meanwhile cells lying just above this tube bud off to form a crest that will be the peripheral nervous system. They form long nerves that carry signals to and from distal reaches of the body.

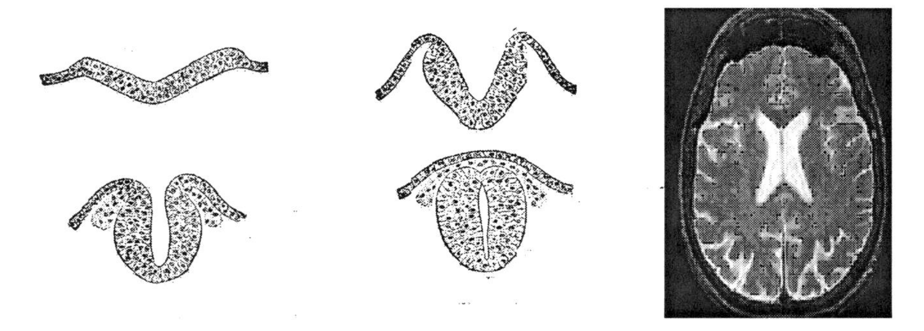

**Figure 23: Infolding embryonic development of the nervous system from the neural tube from folds and crests and resulting fluid filled ventricles of the brain on MRI scanning. .**[73]

The spinal cord remains in the form of a thin tube, but the brain's development at the head end of the animal is much more complex. The brain is very much only a modification of the basic rostrocaudal tubular structure. Though the spinal cord remains basically straight the brain will form by taking certain bends, near the midbrain and again at the pons and at the junction of brain and spinal cord. Brain formation rostrally occurs as neurons grow and then migrate around what will become the four spinal fluid filled ventricles.

Primitive nerve cells find a resting place a short distance away from the surface lining of the ventricle. When they reproduce they have a neat way of migrating to the ventricular surface. Primordial neurons have a tendency to form extensions adumbrating later axon and dendrite formation. While preparing to divide the nuclei of these cells begin to travel in down the cell body which changes shape like an amoeba. Nuclei travel toward the ventricular surface. This movement is aided by glia or supporting cells and the withdrawal of previously formed neuronal cell processes. The neuron assumes a smaller form with the nucleus near the ventricular surface and here divides to form daughter cells. The nucleus again remigrates back to its resting place remote from the ventricular surface as the neuron again grows new long processes during maturation.

Basic principles of form and function established during this embryonic stage presage basic principles of neural function that being now established, will persist throughout the life of the individual. There is a specific division of labor.

The first general principle of nervous function is that it can be broken into three divisions, an afferent or receptive limb, an efferent or motor limb which effectuates adaptive responses, and an associative component most important in the brains of advanced animals.

Early in the development of the embryo separate afferent vs. efferent functions are mirrored in the anatomy of the embryo. In the spinal cord, afferent or sensory components take on a more dorsal (toward the hard backbone) position while motor or efferent components assume a position ventrally (more toward the soft underbelly). In the adult the dorsal horn in the spinal cord will receive all sensory impulses from nerves over the body while the ventral horn will send out motor impulses. The nerve cells that effectuate movements termed the *Alpha Motoneurons*, reside in the ventral gray matter of the cord. The cord's gray matter is shaped like a butterfly (figure) with the bottom (ventral) portions of the right and left wings housing motoneurons. In the thoracic spinal cord there is a third horn (the intermediolateral column) that is just above (dorsal) to motor neurons, comprised of effectuating cells of the sympathetic nervous system. Well above (or dorsal) to these cells, the axons carrying sensory impulses enter the cord. There are sensory neurons that reside in this general area, the upper portion of the right and left wings of our butterfly. These modulate or act on sensory impulses beginning the first stage in sensory processing. Some of these neurons are in the tract of Lissauer. When a sensation is extreme it is felt as pain. More moderate sensations are just felt without being interpreted as painful stimuli. In general the more energetic is a stimulus, the more extreme the sensation, until it reaches a certain threshold that is interpreted as being painful. The classic example is heat. Your feet might feel comfortable on a pavement that is 80 degrees, but a 150 degree pavement would be a different story. There are cells in Lissauer's tract that modulate mild stimuli so that they do not reach the brain in the same way that painful stimuli which should initiate avoidance do. Some cells have axons that have collaterals (branches) that synapse or reflect upon themselves via an intermediate neuron. Where the intermediate neuron synapses with the sensory neurons an inhibitory transmitter is secreted. so that a neuron inhibits its own transmission! The intermediate neuron is called a Renshaw cell and its discovery has started many debates about how painful Stimuli are processed. If a sensory impulse is extreme enough (as in our example of a hot pavement) the Renshaw cell's inhibition becomes ineffective and the painful signal gets through. Stimuli of moderate intensity, however, end up inhibiting pain perception. In patients with chronic pain we use stimulators called TENS units (for Transcutaneous Nerve Stimulators) or implant dorsal column stimulators that deliver little shocks constantly serving to inhibit pain perception at the spinal cord level.

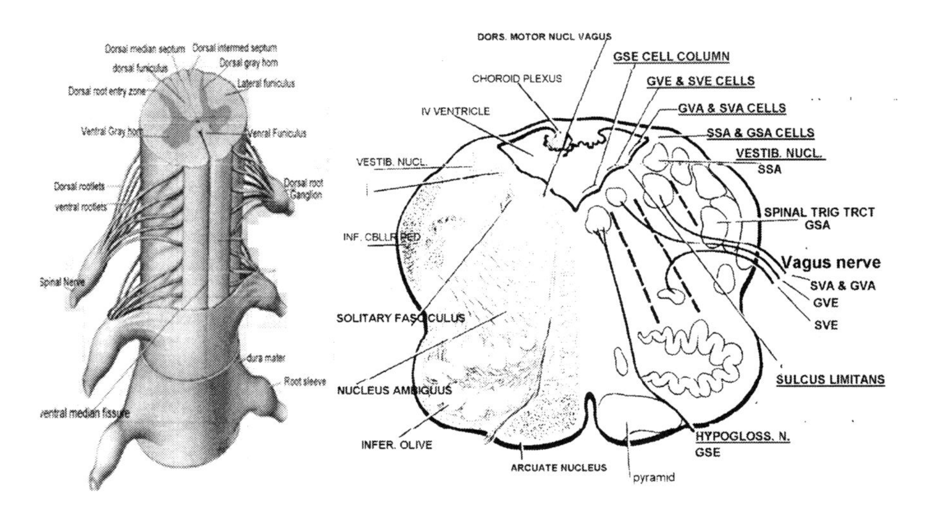


**Figure24:Structure of the Spinal Cord. Sensory neurons are dorsal; motor neruons are ventral[74]. And of Medulla showing the importance of the Sulcus limitans separating sensory from motor nuclei in the brainstem. [75],[76]**

Most sensory nerve cells reside outside the spinal cord in clusters of neurons, the dorsal root ganglia. These neurons are bipolar, that is, they have two long axonal extensions. One process goes out into the periphery to sensory end organs collecting sensory data. The second process goes into the spinal cord and helps convey sensory information toward the brain. The Dorsal root ganglia are all derived from neurons that eventually separate from the neural tube, part of the neural crest. They lie above or dorsal to the neural tube. The axons that go into the cord conveying pain, synapse or connect with other neurons within the cord, practically at the same level and then another neuron takes over to carry the impulse toward the brain. As this neuron does so, it crosses to the other side of the spinal cord. The axons that convey non painful (and non-temperature) impulses are thicker and have more myelin, hence conduct impulses faster entering the spinal cord continuing to travel toward the head on the same side, in white matter denoted as the dorsal columns. The dorsal columns carry information about position, touch, vibration, etc. Not surprisingly, such perceptions arrive at the brain somewhat faster than does information conveying pain. This explains why when you burn your toe, for a split second you know what you've done and that your about to have an extremely painful perception. Then the actual pain is felt. Your foot's withdrawal however, has already taken place, before you feel the pain, because much of the mechanism for withdrawal resides within the spinal cord itself and is carried out on a very basic primitive level.

The general structure of the nervous system as determined early in evolution and also in development is carried forward in the brainstem where the principle dorsal = sensory, ventral = motor is carried forward. The brainstem controls many automatic and repetitive activities and also serves as a conveyance for sensations toward the upper regions of the brain and motor impulses to the spinal cord. The same tubular structure applies except, as described, there is a bending or flexure at the midbrain and at the junction between the brainstem and spinal cord also still later in the pons. The tubular structure remains, only the top of the tube becomes flattened and spreads out rather like a garden hose with a bend. A special demarcation line, the *sulcus limitans* separates the dorsal sensory component from the ventral motor portion. The dorsal or sensory part is designated as the alar plate while the ventral motor portion is the basal plate. Considerable variation in this basic form is produced by all this flexing, yet the principle that separates sensory and motor function persists.

In the central nervous system clusters of nerve cells form nuclei analogous to nerve cell clusters of the neural crest in the peripheral nervous system, the ganglia. Another basic principle of nervous function is that there are both sensations and actions that are carried out automatically without the intervention of consciousness. You sense, for example, that you have to breathe more when the carbon dioxide concentration in the blood increases and when oxygen decreases and both the sensing and the breathing are done automatically, freeing your mind for higher mental operations. These totally automatic functions are termed general visceral, referring to basic bodily functions that have to continue on an automatic basis. There are other sensations and movements that are semi automatic and under more conscious control. The best example is movement of facial muscles of expression. You laugh, smile, and cry semi automatically, effortlessly, though you could exert more precise control if you wanted to. There is a sort of gradation of volition and perception involving more or less, as a matter of degree, awareness in action and perception. This gradient has its foundation in the structure and development of the nervous system. Swallowing muscles and muscles of facial expression though voluntary, are homologous with and derived from the gill arch muscles in water-dwelling animals. A better example is the stapedius muscle in the ear that automatically tightens the eardrum in response to sudden loud noises in order to protect the delicate hair cells in the ear. Movements of these muscles are not very much controlled specifically, but rather work in patterns of activation so that they may be considered only partly voluntary, and are partly controlled by reflexes. not on a par with larger hand and leg muscles for example, or muscles of eye or tongue movement. Branchial muscles are controlled by cranial nerves the trigeminal, the facial, glossopharyngeal, and the vagus. A third category of nuclei in the brainstem both sensory and motor are those which have the closest relation to conscious perception and action. These eventually connect with higher brain centers i.e. the neocortex. These are termed somatic afferents and efferents.

All the afferent (sensory) nuclei are separated from the efferent (motor) nuclei by the sulcus (fold) limitans. The major separation to be made is between perception and action. The afferent and efferent automatic nuclei, sensory and motor i. e. the visceral afferents and visceral efferents, lie next to each other on either side of the sulcus limitans. These are flanked by the somatic afferent and efferent motor and sensory nuclei. Thus a structural separation is complete, a separation that reflects functional differences. First, afferent nuclei are separate from efferents. Second, there is a gradation from inside out on both sides flanking the sulcus limitans (diagram). Progressing farther out from this major dividing point, the nuclei become more voluntary. A person is more "aware" of sensory perceptions transmitted; they are more conscious, and motor activity progressively becomes more voluntary and deliberate.

The best example of a nerve serving automatic function is the Vagus which helps regulate such functions as heart rate, blood pressure, and gut motility. We can't be bothered with conscious control of such functions. Signals arrive into certain nuclei (for example the Solitary nucleus which receives data on blood pressure, carbon dioxide concentration in the blood and taste, a visceral afferent nucleus) and actions are taken by the dorsal motor nucleus of the Vagus, a visceral efferent nucleus. Another visceral efferent nucleus controls salivation. Nerve fibers from each of these nuclei contribute to the Vagus nerve. This means, as in a lot of cases, that a number of different nuclei, way stations of gray matter within the brainstem, contribute to a single nerve, which is nothing more than a cable conveying messages to and from the periphery.

An example of a nucleus that has connections with consciousness is the Cochlear nucleus that receives data about sounds in the environment. Hearing is a special sense that impacts upon consciousness, hence the cochlear nucleus is a special somatic afferent nucleus. If this weren't an obvious fact considering auditory functioning alone, we could still know this by looking at the position of this nucleus in the brainstem. It is one of the most lateral of brainstem nuclei, establishing its role as a somatic afferent nucleus. The nuclei controlling movements of the eyes and of the tongue are under conscious control and are considered by virtue of their function as well as by their position far away (medial) to the sulcus limitans, somatic efferents. Thus we see by example the profound and beautiful organization of even the lower areas of the brain.

Finally some of the muscles that control voluntary movements of the pharynx and little muscles in our ears that protect us from damaging loud sounds come down to us through evolution as remnants of gill arches and are called branchial muscles. These are partially under voluntary conscious control and fall into the category of special visceral efferents.

The picture of the nervous system that emerges through all of this anatomic and developmental description is one of tight organization. It is built upon the basic structure of a simple hollow tube. At maturity, only the spinal cord preserves this general structure. In the cord we have two divisions of nervous function, afferent and efferent limbs, which are only minimally embellished with any but the simplest reflex associative function. Afferent (sensory-incoming impulses) are generally dorsal, to our back, and efferent (motor) impulses are sent out ventrally toward the underbelly. Automatic behaviors reside right in the cord. We have discussed at few, the stretch reflex, withdrawal with contralateral extension of the unaffected limb to maintain an erect posture. The major function of the cord is, as a complex cable, to convey information from the brain down to the body and limbs and from the limbs and body to the brain. The outer cord (outside the gray butterfly --see illustration) contains all white matter with fibers that run, for the most part, from head to tail and tail to head.

Higher in the brainstem the nervous system maintains its basic tubular structure but it is somewhat flattened out so that motor fibers are more medial or more toward the center and sensory fibers more laterally. The brainstem is further organized about a central sulcus limitans that separates afferent (lateral) from efferent (medial) cell groups. The further away are these afferent and efferent nuclei from this sulcus limitans, the more they are involved with conscious action and perception. The closer fibers have to do with less conscious, more automatic, action and perception. The types of nuclei (grey matter cell groups) are designated as somatic (voluntary, conscious) and visceral (having to do with organs, unconscious and further by designations special and general, as in special somatic afferent, having to do with the special sense of hearing, for example. The brainstem is thus a tubular more complex, embellished spinal cord retaining a lot of design features, but modified.

Traveling higher up to the thalamus, basal ganglia and the brain, we still retain some design characteristics of even the by now primitive spinal cord structure. The brain too is afferent, efferent, associative. Generally the design is that the efferent limb is still more ventral, the afferent, more dorsal, but there are all kinds of more complex embellishments to all three components. The archetypal afferent or sensory component in the brain is vision, which sends its fibers as dorsal as any fibers can go, all the way to the back of the brain in the occipital lobes. Vision is the most precise and highly organized of the senses, the most logical and developed sense. Seeing is believing and all that, all to be discussed in its own chapter. Also the extreme front (ventral) part of the brain, the frontal lobes, get their own chapter later on in this text. Here the associative limb of nervous function reaches its peak, with motor plans and schemes, philosophical thought and wonder and interpretation of high emotion. The associative limb of nervous function is especially overgrown in the brain. Embellishment of the same tube? Yes only things have gotten a little out of hand which is why I suppose, you and I are who we are.

We mentioned the function of the Notochord in organizing the nervous system. It first separates the cells destined to perform nervous system function then the cell above the Notochord organize into a tube. Interesting that the notochord at once defines our phylum, the major taxonomic division that is right below the level of a kingdom, which is where we are on the evolutionary scale, and is, embryologically, the major organizer of the spinal cord and nervous system. This is not an accident, but a sign of a major strategic departure, in evolution into a rigid vertical head to toe design plan. Unfortunately, the notochord eventually mostly degenerates becoming part of the cartilaginous disks between vertebrae. The notochord also does not go through the entire length of the embryo. Just in front of (rostral) to this embryo organizing structure neuroectodermal cells organize to form the brain. The brain also begins very much like a tube but then its development is a flowering or embellishment upon this structure. Diverticulae develop, for example for the special senses of sight and vision. There are bends or folds that occur, and finally the tiny area analogous to the central canal of the spinal cord becomes the fluid-filled ventricles.

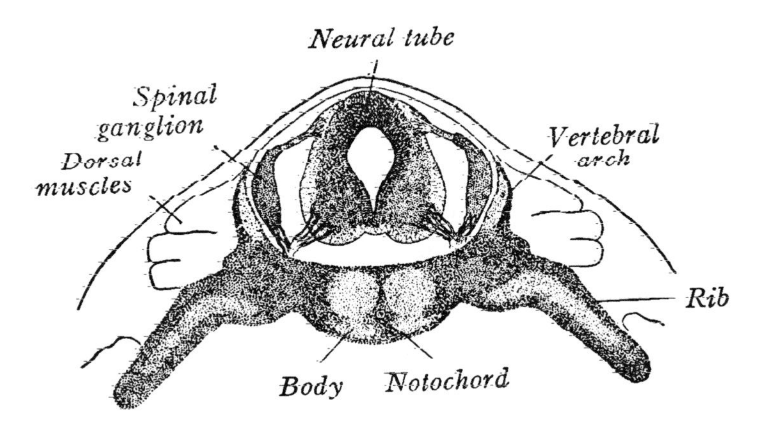


**Figure 25: Relation of spine and notochord.**[77]

The hindbrain which includes the medulla and pons, is more or less similar to the spinal cord in structure. In fact its a little hard to see exactly where the spinal cord ends and the medulla begins, at first, as you travel slightly toward the head of a mammal. Later on as you travel closer to the top of the head, it's a different story as abundant complex structures are added. But the notochord may still be an organizing principle.

At 28 days of gestation or so for the human the previously open neural tube is essentially closed. The middle closes first forming a rostral and caudal residual neuropore, the last open part of the neural tube. Bone encases neural tissue with cells descended from mesoderm. Some muscle, also of Mesodermal origin, surrounds the bone, and all of this is covered with fascia and then skin. Eventually the neuropores too close in turn. But all may not go right. The embryo is vulnerable at this point.

Spina bifida is one of the most commonly encountered anomalies. This refers to a failure of fusion or closure, of elements around the spinal cord. It can be of various degrees. In some cases only the overlying bone for perhaps one segment doesn't close completely but the skin, meninges that cover the spinal cord etc. do make a complete seal. This common finding on lumbar X-rays and usually causes no symptoms and is termed spina bifida occulta. In other cases elements of bone and skin fail to close over the spinal cord over a variable length usually leaving the rear most portion of the cord, and some variable portion in front of that uncovered and most often not functioning. It appears closure first of the neural tube is necessary for subsequent closing and sealing of overlying tissues. In some very severe instances the neural tube doesn't close and then the overlying tissues also are malformed. Abnormally formed and nonfunctioning nervous tissue is surrounded by coverings which do not protect it adequately. The main danger is that the clear spinal fluid that is contiguous within the nervous system and bathes the brain and spinal cord, is exposed the outside and bacteria enter it to cause life-threatening infection. In some infants all that is present is an oozing bag near the rear end just waiting to become infected. These defects, meningomyeloceles (Named for the coverings of the spinal cord, the meninges that are often part of the tissue in the defect + 'myelo', meaning spinal cord. A 'cele', is a bag or an open space.) should be closed as quickly as possible.

The other problem is that the nervous tissue within the defect does not function. Almost always these caudal (near the back) nerves affect the function of the bladder and there is incontinence and repeated urine infection later in life. These nerves also go to muscles in the leg. If the defect is long, it will affect correspondingly more nervous tissue. To make matters worse there is many times some defect closer to the head as well termed an encephalocele or sometimes a more minor misdevelopment in that area. This echoes the embryologic closing of the rostral neuropore and caudal neuropore at about the same time. Something must happen at a certain time to affect this particular stage of development but we have no real clue as to what this may be. Is there a toxin that affects the embryo from the outside? We now know that giving an expectant mother extra Folic Acid, a B vitamin, prevents, though not completely, these defects, called dysraphisms, in newborns and that some drugs increase their incidence.

There have been many theories. We do know that these defects have been for some reason most common in persons of British descent and there is in some cases an increased incidence in certain families. Embryological development is a house of cards the normal formation of one structure leads to normality in others. Just one little error though will have dire consequences that cause other malformations because of the mechanism of induction in which the proper formation induces the development of other structures.

When the bones lengthen as the person grows the cord which originally extended past the bottom of the lumbar spine ends considerably higher up at about L1 or L2 or so. (There are 5 lumbar vertebrae.) What's left near the tail is a space containing a bag of nerve roots having the appearance of a horse's tail or cauda equina. In a lot of cases though with malformations at the bottom of the spine the nervous structures are bound too tightly (tethered) to the spine and with growth the spinal cord ends up being stretched and the bottom of the brain can even be pulled down past the bottom of the bony skull. When the spinal cord is tethered in this way to the bottom of the spine, function can be compromised, so that surgery may be necessary to untie it and decrease stretch on the cord[Ψ] .

At a certain stage the neuron will stop migrating and dividing. Apparently it receives a chemical signal that tells it that its time to stop division., or perhaps only a certain fixed number of cell divisions are programmed into the developmental process. In the process of repeated division and maturation the cells become committed determining the location they will assume and ultimately the role to play in the adult animal. By the time of a human's birth most neurons have lost their potential for cell division and their anatomic destiny is irreversibly determined even though the final appearance of each individual neuron is not yet finalized.

Some neurons are destined to die. Those that survive often dramatically change from growing more (or sometimes less) dendritic processes and changing their exact pattern of intercellular connection. While connections between neurons, ultimate microscopic intercellular anatomy is critical to the understanding of how the brain functions, the process determining how cells are interconnected is not at all understood, nor are the signals that tell a cell it should finally cease to divide.

Glia called astrocytes are critical to the migration of neurons. These essentially star shaped cells share many characteristics of neurons and also embryologic origin being derived from neuroectoderm. Glia, like neurons, tend to form long cellular extensions. In the Embryo these cellular processes serve as scaffolding on which neurons migrate. In the cerebellum, a complex structure modulating motor control, small neurons, the granule cells, constantly migrate during development over the processes of a special type of astrocyte, the Bergmann glia which pathologists pick out easily in post-mortem sections of cerebellum in an adult.

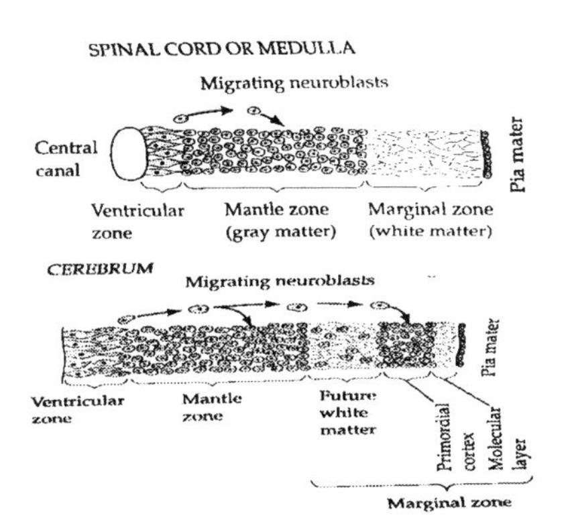


**Figure 26: Neuronal migration. Future neurons move, directed by glia from the shoreline of the Central Canal or ventricle (in the brain)to their final home[78].**

The glia, like their neuronal cousins∅ , are long cells. Many of these stretch radially from the surface of the ventricle or central canal in the case of the spinal cord, a structure filled with spinal fluid and lined with ependymal cells that delimit the fluid-filled ventricle. They stretch from the ventricle, to the outer surface of the spinal cord, or inside the head, to the covering of the brain, the pia, from inside to outside, in other words. Neuroblasts, forerunners of neurons, reproduce like crazy at the inner ventricular surface. All the newborn neurons accumulate in a mantle layer (see Figure 14) and then begin to migrate to their final place hugging onto the radial glia as a guide[79]. Their is a general flow of migrating neurons from the inner ventricular surface to the outer pial surface in both the brain and spinal cord. The glia guide the migrating neurons but other chemical substances determine final relationships and connections. This is how the organization of the brain and layering of neurons will be determined. The whole human enterprise, the root or origin of intelligence and all human endeavors, lies in this connectedness and layering. As we have seen with whales and dolphins though their brains are larger than ours, our mental capacities far outstrip theirs only by virtue of this cellular layering and architecture.

Once cells have found their final home the relationships with their fellows are firmed up. These relationships continue to evolve from a less to a more committed state. The method of growth of interneuronal connections is best worked out for the visual system. In classic experiments performed by Roger Sperry, eyes of frogs were removed then replaced, but rotated 180 degrees from their original position. When the optic nerve connecting the eyes to the brain and conveying visual signals regenerates the axons grow exactly into their previous positions. Remarkably, since the eyeballs are turned around, the frog behaves as if objects are 180 degrees in the wrong direction and the poor frog will then leap away from a tasty fly. This shows that by this stage these specific axons have already been committed to grow in a precise path. Similarly in monkeys the precise projection of each eye upon alternating stripes of visual cortex in the back of the brain (the occipital lobe) are well known. There are alternating stripes representing visual projections corresponding to each eye within the visual cortex. If one eye is covered or is not used, neurons corresponding to that eye's projection area develop to a lesser extent. When that eye is then uncovered it is still virtually blind. The same process occurs in the cortex of cats reared in the dark. Occipital cells fail to develop properly. Research has linked this cortical development to changes in a certain cellular protein, microtubule associated protein 2 (MAP- 2) whose active state has been found to be altered by the lack of synaptic neuronal input. The alteration in brain function in dark reared animals only occurs if the animals are placed in the dark during a critical period, indicating that cellular formation and commitment occur in a specific time frame. We may be observing the same phenomenon in infants and children with crossed eyes (strabismus). Because the eyes don't focus together and visual objects end up falling on different parts of the retina, these children would have double vision. Ordinarily they adapt by suppressing the image of one eye. When an eye's image is continuously suppressed neurons within the brain that handle this eye's visual input likely fail to develop and blindness or low vision in that eye becomes permanent.

Bell's palsy causes paralysis of one side of the face due to an inflammatory process that damages the facial nerve. In the healing process, the axons that are partly destroyed begin to regrow. They tend to follow a specific course reinnervating muscles that had been previously deprived of their nerve supply as functional movement of the face returns. Most likely the muscle sends out a chemical signal to the damaged nerve telling it to regrow in the muscle's direction. Actually another scenario that is even more likely is that the axons of the nerve sprout into the tubular axon cylinders. Axon cylinders are empty myelin tubes left after the dead axons of the facial nerve degenerate. These myelin cylinders very probably send out a chemical saying, in essence, "Come to me" stimulating the damaged axons of the facial nerves to regrow. As it happens, the facial nerve also controls salivary and lachrymal glands and sometimes axons that had previously innervated one set of glands grows aberrantly into the other. Thus recovering patients cry "crocodile tears" when they really should be salivating over food. What is even more common is that nerve regrows into the wrong muscle so that the muscles of facial expression are not as easy to control. Closing an eyelid may cause a corner of the mouth to elevate at the same time, as two muscles that were not initially supposed to move together do so in a *synkinesis*. This is an example of the variable tendency of neuronal extensions to grow into more or less specific paths. Similar principles may apply for the growth of nerve processes both in development and in regeneration after injury. Long after birth and development, some regeneration is possible, utilizing many of the same mechanisms that applied during embryogenesis. Nerve cell axons can still sprout even in old age, depending in many cases, on the very same chemical signals that grew them in the first place. Both nascent and damaged nerve axons respond to chemical signals arising their sites of future innervation muscles, glands, and other structures as if imploring "Come to me, come to me." Presumably these chemicals are so-called "nerve growth factors" and other substances and some perhaps are similar to cytokines as well, chemicals that call lymphocytes and other immune cells (troopers) into a battle zone created by bacterial or viral invaders. Identification and utilization of these chemical signals will be crucial for treating diseases where there is nerve damage and also in finding the cause of other conditions where nerve components grow uncontrollably in genetic disorders such as neurofibromatosis and for certain tumors of the nervous system.

As neurons migrate out to the surface of the brain and continue to proliferate the surface area of the outer areas designated as the cortex, increases greatly and there are multiple infoldings. Gyri or convolutions separated by sulci or folds, form. Some infants have been born with relatively few gyri or folds Their brains are fairly nonconvoluted and smooth and they are said to have lissencephaly. This problem produces profound mental retardation and apparently occurs after the 22nd week of gestation, the time when folds begin to develop. The problem relates to abnormal neuronal migration in that the normal layering of the cells of the cerebral cortex is very abnormal, usually resulting in severe mental retardation.

In our conception of the human cortex various portions are demarcated by sulcal folds. For example the frontal lobe is separated from the parietal lobe by the Central Sulcus. This infolding or convolution of the brain (sulcation) adds greatly to the surface area of the cortex. The whole surface, folds and all, architecturally, has the cell layers that we have been talking about and thus increases the processing power of the brain. The cerebral cortex of lower mammals is a great deal more smooth and infolding occurs late both in evolution and in the embryologic development that reflects it. Neurons proliferate near the ventricles then migrate up a scaffolding of glial extensions. With more convolutions and infolding comes more cortical surface and neurons. These also require more cabling (white matter) as they project downward to effect sensation and motor control. Whit matter tracts also connect interacting cortical areas with each other. The highest brain centers in man are very specialized. We can see this not only clinically when for some reason a patient destroys certain specific areas of his brain and loses a specific function. You can also see changes in cellular architecture that define areas of the brain.

The highest brain centers have six layers of neurons. This pattern is defined late in the embryo. This association cortex or neocortex is also a marker of our humanity. Going from the surface of the brain over increasing depths, layers II and IV with smaller granular shaped cells, are primarily receptive and are most developed in cerebral areas concerned with special senses, particularly hearing and vision. Layers III and V are to a large extent, efferent. In some areas of the brain giant Motor neurons, Betz cells, reside in layer V, and are concerned with initiating movements. In the visual cortex of the occipital lobe layer IV has so much input it is broken up into two separate sections by a Stripe of Gennari. This stripe is responsible for giving this area its anatomic name, the Striate (striped) Cortex. As we began to appreciate in chapter one. It is this multilayered cellular design, that you begin to appreciate under the microscope, which distinguishes the much less intelligent dolphin and whale brains (cetaceans) many of which are actually much larger and heavier than humans from the incredibly greater computational capacity of our own smaller organs. (For a fuller description see Chapter 1: “Inside the Neuron”.)

Below is a rendering of the famous Brodmann Area map of the human cerebral cortex. Regions of the brain, surface were demarcated by Brodmann, a German Neurologist (1868-1918) who carefully classified these areas under the microscope. The names and functions of specific areas need not concern us here. The point is that generally the functions of each separate area correlate with the cellular architectural design and layers as seen under the microscope. It should come as no great surprise that structure affects function in the brain or any anywhere else. The more advanced areas of the brain have the most complex cellular relations and the most levels of cells, generally six layers. More primitive cortical areas have less layers and are less complex.

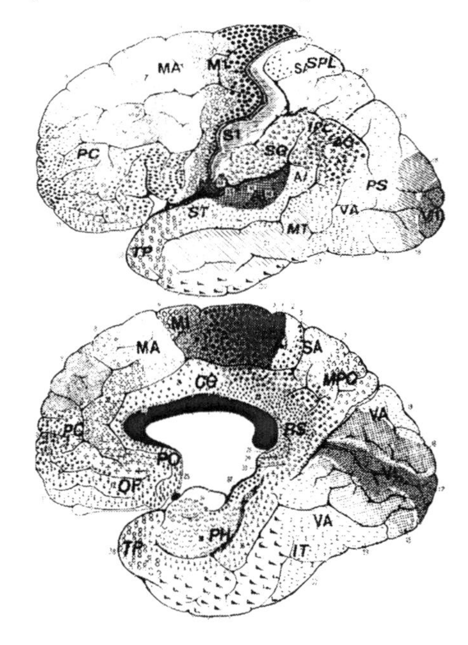


**Figure 27: The Brodmann areas. Regions of the cerebral cortex were classified using cytoarchitecture. We know these areas correlate functionally with sensory, motor, and association areas[80].**

Late in evolution and by implication late in embryologic development, we begin to see that brain anatomy becomes more complex. As we have seen this goes with advanced cerebral function which was late to develop. The brain of animals and humans have their similarities but human uniqueness is reflected in anatomy. Differences may be subtle, may be missed by a less discerning eye. In broad terms though, we humans posses certain association areas inside the brain, or at least our areas are much larger than those of lower animals. Once you begin to see changes in the architecture of cells under the microscope you can compare similar brain areas in man and other animals.

This is not to minimize the primary and secondary sensory and motor areas of the brain that are eloquent, that do affect, the neurological exam. These areas too are architecturally much more complex in humans than in other animals. The areas that has been studied the most in this regard is the visual cortex, also explored later on. Suffice it to say here that the general scheme in cortical structure is a column of cells. This vertical column may be composed of as much as six layers of cells. One vertical column, sometimes referred to as a barrel, will take on a specific sensory or motor function. In primary cortices, those receiving or sending our specific information one column or barrel may receive information from a specific body part, say a portion of the tip of the pinky finger. That is this column's receptive field, or, a primary motor column may control of portion of contraction of the little flexor of the thumb or perhaps a grasping response of the hand. Specialization and localization of function is quite specific. While in the *primary* sensory and motor areas there is specific localization of function, in the *secondary* areas localization is less exact. This is because *secondary* cortices are concerned with less specific, more abstract, higher order levels of function. Motor wise secondary areas might control more of a pattern of movement rather than a specific muscle. For example, in the striate, cortex or primary vision area, many cells have a receptive field. This means that they will respond only to a stimulus that occurs in a small area of one's vision. A single cell may respond only to an object within a very small part of one's visual field. Only in the human brain are such association areas so well developed. We have just hit upon a whole third limb of human endeavor (in this case almost specifically human) reflected in development and anatomy, association. thus we have afferent, efferent, and associative limbs that together define neural function.

In the human brain the above breakdown into afferent, efferent and associative gives an incomplete description. There are regions that are meta-associative, the areas that are phylogenetically (in evolution, and in development) newest. Not surprisingly, these are the regions of the brain that make us the most distinctly human. They are also hardest for the clinical neurologist, biologist or psychologist to get at.

For example, it has been relatively easy to define function within most areas of more primitive parts of the brain, say, the brainstem or areas of the cerebral cortex serving specific functions. We described such areas. It 's easy for a neurologist to see that a patient who has a stroke in the right motor cortex has an obvious weakness on the left side of his body where his arm is weaker than the leg. Thus we have the typical hemiparesis, a half body weakness. Moreover, since analogous areas are present in lower animals, lesioning them or stimulating them to obtain information about these areas is no problem either.

But there are other areas in the brain that appeared late in the evolutionary scheme of things. The best example is the prefrontal areas of the brain. This is a region that is in front of the motor and speech areas of the frontal lobe. For generations, this area was considered a clinically "silent" or "ineloquent" area of the brain, because ostensibly, patients with lesions in this region did not appear to have any neurological deficit. In fact, patients with schizophrenia and other severe forms of mental illness were subjected to gross, primitive neurosurgical procedures during the 1940's and 50's called prefrontal lobotomies[81] Clinicians and psychologists had trouble finding specific deficits in such patients especially using such crude measures of mental function as psychological tests

(IQ. tests and so forth). This is what is meant by the notion that such areas of the brain are hardest to get at, to get a handle on and assess function.

It's easy to appreciate that our measurement of human cognitive function must be skewed toward those areas that are easiest to assess such as memory, and even some forms of simpler arithmetic and verbal ability (for example vocabulary tests) and away from more complex abstract areas that really separate the men from the boys. Stephen Jay Gould has made a study of this fascinating topic [82]. Science and pseudoscience has naturally focused on quantities that are easiest to measure such as skull shape and volume (and by the way they haven't always measured even these quantities accurately) and away from the more difficult less tangible areas such as acculturation to draw erroneous conclusions. The best way of proving this is that there happen to be whole huge areas of frontal and temporal lobe, especially in one's non-dominant hemisphere, whose function is not established and that we have a devil of a time assessing. I can't say how many patients I've seen whose neurological function appeared to be intact, with gigantic tumors in one of the silent areas of the brain, the frontal or temporal lobe especially on the non- dominant side. These persons usually present late in their disease with intractable dull headaches or, most commonly, seizures, or their spouse may notice a vague alteration in mental function that really results from deformation of the rest of the brain by a large mass or increased pressure within the skull.

Quite obviously, the prefrontal areas are not a part of the human brain for no purpose. In fact, being that these areas are so new, they have to serve an extremely adaptive function for humans particularly and this area must be critically important for out biological success. This means that the frontal lobes and all other such newer area of the brain, the very areas that were so long felt to be clinically silent, must be the areas most important. As with a lot of things, what is subtle, not visible to the casual observer, turns out to be the most important and intriguing. The prefrontal area, and other parts of parietal and temporal lobe constitute the physiological substrate of our humanness. The brain is divided then into "silent" or ineloquent areas and eloquent ones. The eloquent areas speak the most and the loudest. When there is a lesion in these areas it's obvious. Because our measures of such things are crude, the ineloquent regions, namely the huge areas that are phylogenetically the newest, ones that appear only In humans are barely missed when diseased. Huge tumors are found in these areas without an obvious deficit, that is something gross like a hemiparesis, yet it is in these quiet regions that the personality resides, the frontal lobes, the temporal lobes. They suffer in silence.

Memory function works this way too. Whatever is less obvious turns out to be the most important. Packed into our being are all kinds of memory engrams, data from development (that we are not just made as some final product, but unfold, develop), data passed down from previous civilizations and our own society, stories, myths legends, information from early childhood and adult memories as well. All of us function at this moment $t_0$ on the basis of information incorporated into our being from all previous time, $t_{-\infty}$.

The higher areas of the brain develop and are myelinated (insulated so as to function) last, another example of ontogeny recapitulating phylogeny. For a long time it was thought that myelination ends at birth. It has been found that the prefrontal areas are still being myelinated until the age of 16 or so.

If myelination and other maturational processes still occur in children, this presents a great theoretical obstacle in testing young children to determine their intellectual abilities. Basically you can test a five, six, or seven year old and label him at that time. You may put him into a gifted, average, or learning disabled (slow) category as many of our schools do now. But at a young age, you are testing a not completely developed brain which is still plastic and has not reached any final plateau. Testing will successfully stratify five and six year olds. But the ones who come our on top may merely have matured faster than their same-age colleagues. Later on, when developmental processes end, the slower kids might be superior to the faster ones. What a pity to test kids at a very young age, then pigeonhole them, put them on an academic track *for life* as has been the practice in many European countries and Japan. What a waste of human potential! Our own schools depend far too much on early psychological testing, achievement and aptitude (IQ) tests. The reproducibility (chances of getting the same result again) is notoriously poor even when kids are tested by their own psychologist sitting in front of them as individuals and is terrible when tested en bulque in a classroom or group situation. The way we close doors for a lot of our young people is heartbreaking. Always we should strive just to let a people do what they desire to do and then just see how they actually perform. The real world is the acid test. Projective tests are severely limiting.

It does no good to classify kids on the basis of which ones are the fastest myelinators. Some tasks notably some gross motor and lower extremity tasks, measured at an early age, steer you in the wrong direction. For example its often true that toddlers who walk early and are the best on their feet, perform more poorly in the classroom[Ψ]. Lots of kids with Attention Deficit Disorder function much better than their slower colleagues on the playground, making much better use of slides and monkey bars than less adventurous tykes. Look at lower four legged animals. Many of them can walk on the first day of life. Humans take a lot longer to learn to walk, typically around twelve months. If we were test such motor abilities in babies and then jump to conclusions about their ultimate abilities we would be sadly mistaken.

Even after birth humans spend an inordinate amount of time immature and unproductive. This is the phenomenon of **neotony**. Kids of a certain age are good for nothing but learning, and playing (at which time they are learning) and developing. It goes without saying that their motor and sensory systems and minds are not in their full adult form for a long while, perhaps 18-20 years. Their faces are so cute and infantile, for a very long time, reflecting how they are on the inside. They are born undeveloped. That's what makes them so much more advanced! Almost all other animals make productive citizens of their offspring within a much shorter space of time. So do all of the societies where none of us would care to live. But of course, animals never get to experience the heady the heights of human performance. Development occurs quite slowly in humans. Some kids will run earlier, some speak or read earlier which may tell us nothing about their later abilities. In my own case, I was rather poor in running and reading at an early age, yet as I look at most of my same-aged cohorts today, most have all but ceased performing these very same functions and I enjoy them every day. C'est la vie.

This is only one of many reasons why early testing and performance, that too many schools depend on, is not predictive of a child's final development (or a measure or future academic or any other success). The very best thing that can be said about early testing is that if you have a child with a certain score, you at least know he will be able to perform at that level. Even this is deceptive, since, for a variety of reasons, performance may later deteriorate. It can be said that so many child prodigies don't just burn out. The problem is that although they may be slightly more rapid myelinators or developers, or may be pushed to perform to gain their parents love and approval, poor little fellows (sort of like trained circus animals, when you think about it), this says very little about their adult capacities. Many of them turn out not to have anything out of the ordinary. Indeed it is amazing given all of the advantages that these kids receive in the areas of attention, expectations, allocation of resources etc., that so few of them seem to make any marks or to perform outstandingly as adults. Of course there are always the Mozarts and Mendelsohns who are really exceptions, but the Einsteins and Newtons seem to be more the norm so far as genius is concerned and the point is one can't predict ultimate outcome on the basis *rapidity* of development. Only in certain very specific fields is early childhood performance an indicator of adult attainment. These are in pure areas or systems of thought, in systems that are internally consistent, mostly music and math. Prodigies surface early in these specific areas. Chess is archtypical. The brain of a Bobby Fisher seems fit for chess at a very early age and you know this be the age of 5 or 7 or so. This may just as well be on the basis of something pathological in some sense, but that is another matter.

The pituitary gland the master endocrine gland of body controlling secretion of all other ductless gland hormones, arises by the joining of ectoderm from two separate sources. Actually most of the active gland as we know it isn't even derived from brain. Rathke's pouch consists of a small group of ectodermal cells that originally lie at the roof of the embryonic mouth at about three weeks of gestation. This group of cells grows upward slowly toward the brain as a hollow group of cells contiguous with the mouth to eventually join another group of neuroectoderm derived cells that grow from the diencephalon (the part of brain laying just beneath the hemispheres). Many of the mouth- derived cells degenerate or die so that the connection between the mouth and brain is broken. The cells that are left form the so-called adenohypophysis, the actual glandular secretory pituitary gland. Since these cells secrete their chemical product directly into the blood (not into a body cavity), the pituitary is a ductless or endocrine gland. Follicle stimulating hormone and luteinizing hormone help to control the testes and ovaries, also ductless glands. Other hormones include prolactin that helps to control the breasts' secretion of milk, thyroid stimulating hormone, adrenocorticotropic hormone that stimulates the adrenal glands, and growth hormone which are all produced by cells derived from Rathke's pouch in the foregut. The pituitary is housed in a bony cavity, the sella turcica which is part of the sphenoid or wing shaped bone of the skull. The nerve cell derived neurohyphophysis cells are connected with the adenohypophysis cells from behind and hang down below the hypo (=below) thalamus. These are intimately connected to the adenohypophyseal cells by an extensive blood vessel network so they can exert their own control. In addition to two known hormones, oxytocin and vasopressin that have a direct effect inside the bloodstream this system of capillaries and veins, a portal system, carries chemical that instruct the anterior pituitary to secrete or not to secrete hormones. These chemicals are the releasing and inhibitory factors. What we have in the pituitary gland is the joining of two groups of cells. The anterior lobe of the gland, the adenohypophysis is not derived from brain and is really not a part of the brain even though it ends up inside the skull. Only the neurohypophysis or posterior lobe is truly a part of the brain.

Gland or secretory cells are little different from neurons. Both neurons and secretory cells liberate chemicals when excited and both can be excited electrically. The neuron, it is true, is a specialist in excitation and affecting other adjacent cells. It can be stimulated as a general rule over a wide portion of its surface and can affect adjacent cells over a large surface as well. For the secretory cell, things are only slightly different. When stimulated it will secrete a chemical not in a very specific region over its surface but instead in order to affect other cells, ordinarily to send a signal to other cells, not adjacent, but distant. The message from glandular cells usually has to be carried by the blood.

In the development of the eye part of the brain ends up outside the skull. Optic vesicles evaginate from the forebrain and end up outside the cranial cavity. These vesicles are connected to the brain by the optic nerves. The hollow tube invaginates to accept the lens and cornea which are derived from non brain ectoderm while brain derived cells form the retina. The retina forms in two parts separated by an intraretinal space that obliterates in later development just as the hollow center in the optic nerve does. The retina consists of many layers of cells and constitutes the initial processing station for visual signals that turns light into a language of electrical impulses. It is easy to see that receptors of sensation of all types turn various forms of energy into specific electrical input for the benefit of the brain. This turning of one form of energy into another is the job of a transducer which all receptors really are. The familiar rod and cone cells of the retina are actually derived from brain. the rods, responsible for peripheral and low light vision contain the pigment Rhodopsin that changes on exposure to light; the cones contain three similar and similarly derived chemicals that change their structure best on exposure to either red, green, or blue light. The red and green pigments are actually proteins encoded on the X chromosome accounting for the most common forms of color blindness involving red and green that are X-linked or sex linked genetic defects that occur predominantly in males.

In the central nervous system axons invaginate into a glial cell, the oligodendrocyte. This distinction hits some snags, however. In the spinal cord motor nerves the cell body is within the spinal cord yet the axons form peripheral nerve. Groups of axons going from one particular locale to another within the central nervous system are designated as tracts, not nerves. Therefore the optic nerve is a misnomer. The optic nerve is really a tract.

The lesson learned from the pituitary and the eye, is that what is in the skull, is not necessarily a part of the brain. The adenohypophysis is glandular tissue that is non-neural. On the other hand the retina, which is outside the skull, is still part of the brain. Examine the retinal with an ophthalmoscope, and you may learn what is happening in the brain. The retina's blood supply comes from the brain's and the optic nerve has myelin of the central nervous system, not peripheral nerves. The tiny pituitary gland is a fusion of brain and body. The controlling ductless gland is brought up to the brain from the periphery so that the brain has more intimate control over it. Indeed the entire hypothalamic-pituitary relationship is a fusion of brain influence and the periphery. The function of the hypothalamus is to process information about body temperature, satiety, blood pressure, even emotion, and other internal functions and wield control over these processes. A retino-hypothalamic tract that running from the eye to the brain carries information about light to the hypothalamus' supra-chiasmatic nucleus. This data entrains the brain to keep a person in sync with night and day even as this varies when we travel or the seasons change. Part of this process implicates the hormone, melatonin.

The culmination of all of this prenatal development is the birth of electrical activity in the brain. We've witnessed a lot of controversy in recent years about when human life begins, when a human fetus becomes human. There is little controversy about the fact the as soon as the sperm meets the ovum to form a zygote we have a *potential* to form a human, but when does the developing baby actually turn into a human? Arguably, this is when organized electrical activity begins to occur. More specifically it is when this activity is *organized* enough to define separate levels of awareness as defined electrophysiologically, when it can be seen on the EEG and other electrical tests. It may be said that a person exists when darkness is separated from light, when arousal is recognizable as a separate state, from sleep. It should not come as a surprise that this distinction is made roughly at the full nine month gestation. One can at a certain stage of development see differences in levels of arousal from waking to slow wave and REM sleep. This happens at roughly 28-30 weeks conceptual age, that is after conception. When waking activity is defined and sleep is separate from wakefulness, then a person can be said to be. This makes eminent sense because that is when we can see arousal in an infant. Arousal is necessary but not sufficient for awareness. This is not that far from natural birth which is at about 40 weeks of gestation. Most vocal religionists today, pro-lifers, insist that life starts at conception. This is by no means a point of view uniformly held by religious persons. Some maintain that life starts at the time of birth. From the scientific point of view the latter position seems closer to correct. When alertness is separated from sleep, satisfyingly the same as when light is separated from darkness on the first day of the biblical Genesis. Indeed if we consider the time when we can reliably correlate behavioral signs of levels of arousal with EEG changes, when we can see reliably that baby is awake, asleep, and in REM sleep and are able to correlate these electrical changes with behavior, that we are remarkably close to the normal 40 week's gestational age, the normal time of birth and also when the breath of life begins. Would it then be reasonable to absolutely forbid abortion in the third trimester of a pregnancy but allow it to occur under special circumstances before that time?

Human Cognitive development is too often presented as an unvarying succession of milestones and so it is though post -natal development often occurs in fits and starts. There is not always a smooth continuity. Suddenly an infant who has been only crawling, gets up and walks, and he may be able to take more than a few steps. Some of the reasons for this have already been alluded to. The myelin insulation has to finally cover the tracts descending and ascending in the spinal cord so that the brain can finally exert its control and primacy.

What we do know in the macroscopic sense, is that development is basically unvarying, that humans have a tendency to build on past successes, in all spheres of development. There are exceptions to this rule, but very few. Probably the best example is the infant we all know or hear about who learnt to walk, but never learned to crawl. Even here, what most likely happened is that the crawling phase was incredibly short for that particular infant.

Infants and children also regress, especially under emotional tension (the best example is with a new sibling) or when they are sick. This probably serves also as a mechanism to practice old skills and to build in other ways upon them. The interesting angle here is that there is an dialog between the new and the old which never really dies. Probably the best example is the reversion to atavism or primitivism found in much modern art. It seems to me there's an active dialog between the advanced and the primitive in modern art, that is, the basic and simple, seems somehow informed by the complex in European and American Art whereas it is no so informed or changed in natural African, for example, or native American Art. This happens in the dream world as well. The most basic thing to be said about dreaming is the lack of distinction perceptually and in relationships with what is real and how we want or desire it to be. This is child- like logic, as is seem in Piaget's stage of concrete operations when there is more reality testing age 7 or so.

Regression is an important academic tool couched in terms of a review. In September of the third grade, children who were exposed to multiplication ordinarily come back to problems in addition and subtraction, something undoubtedly less threatening to them at that stage. Also there is forgetting and the need to relearn, but again the need also to build in different ways upon old knowledge, a dialog with history.

It's been often observed that a baby is born with all the 10-100 billion or so neurons he will ever have. There is some new information that indicates this isn't quite so, that some neurons will still be formed after birth, but it is basically so. The 350 gram newborn brain is about one fourth of its adult weight. By the age of three, the brain is three quarters of its adult weight and by the age of 5, 90%. I've heard some psychologists say that 90% of our intelligence is determined by the age of five. This is a meaningless and therefore unprovable assertion when you think about it. Scientifically it is meaningless since you would be unable to design an experiment that would prove or disprove the assertion. It therefore should be considered to be untrue. Most probably though its been propagated in the literature since it sounds either profound or reasonable to some persons and it may have been based on the above brain weight estimates. The thing is that our mental activity develops upon pillars built through a continuum. Brain weight can't be translated into intelligence. Besides, as we have seen, there are variations in development that make early measurement of cognitive function before the age of five perilous at best.

As an example, it has always been thought that language acquisition for one's native tongue needs to happen at a very early age, certainly before the age of five. The left hemisphere of the brain in most of us is dominant, being responsible for the generation of verbal thoughts and language function. But some information indicates that there is not a “critical period” for language acquisition. For example, in a little boy with Sturge-Weber disease, a malformation of blood vessels that affected the function of his entire left hemisphere, treatment, the removal of the affected left side that caused epilepsy, was delayed until he was almost nine years old. This little fellow who was virtually mute until that time, finally acquired, essentially normal language at about the age of nine. We also observe patients with stroke who relearn language utilizing their right or non-dominant hemisphere.

When does mental activity truly begin? Some extreme religionists and psychologists agree on certain points - that mental activity begins early, probably well before birth. The religionist, especially anti-abortionists, not only maintain, as everyone does, that in the embryo there is the potential for the eventual emergence of human consciousness as we know it, but also at a very early stage, perhaps at or even before conception, the potential human is invested with a soul. This latter contention, I'm truly unprepared to comment on, but there is the very interesting open question about when mental activity begins. The psycho-analysts say, at least some small minority of them, that experiences occurring early, very early, even prenatally, may alter a person's development in years to come, even as an adult. You find such outlandish practices as babies being born under warm water for its gentle soothing effects, and rushing really without good reason or evidence, for the newborn baby to be bonded with its mother right at birth. What does seem to be true, is that premature newborns seem to develop and grow more rapidly in a neonatal intensive care unit, when, over the long term, they have some tactile and physical contact. Very much has been made in the media of these findings and most neonatal units have, tried to incorporate regular physical contact schedules into their treatment plans. What any of this says for the infant's long term development is uncertain at this point.

An infant is considered to be born *at term* at about 38 to 42 weeks post conception. Due to the miracle of the neonatal intensive care unit, it is now possible to record the EEG in infants after their 22nd week of gestation. From a scientific standpoint, it is reasonable to attribute the beginnings of life to the point where EEG activity begins to occur in the brain. When is the embryo a person? When some form of electrical quickening or activity occurs in the brain. You can still argue that if there is any awareness at this stage, the start of cerebral electrical activity, that awareness has to be minimal and that is a good point, but at any event this is a convenient time to date the beginnings of mental life.

Some mammals such as a rat and mouse are normally born premature. In them we can see that they evolve from having essentially a flat EEG with no discernible activity at birth, to a burst suppression pattern. Activity begins to emerge but starts to be present alternating with long periods of flatness or electrical quiescence. This paroxysmal activity is disorganized but it is the first thing to appear as the brain matures. We can see the same pattern in premature human infants (figure). Electrical silence alternates with high voltage paroxysmal activity. This is the first sign of electrical life in the developing brain.

Later on, at about 30 to 33 weeks post-conception, there is a general trend for activity to become less paroxysmal and more continuous regular and organized. There are shorter periods of electrical quiescence or silence but there are still spikes that mark initial activity in developing neurons. Faster and rhythmic activity then is seen including slower waves superimposed by faster more rhythmic activity, so called spindle delta bursts or delta brushes, possibly the first evidence for some electrical rhythmicity in the developing brain.

Some short time after birth different levels of consciousness or sleep wake cycles are noticeable and can be partitioned off, the forerunners of these same stages as in the adult. Basically, we are always in one of three basic stages. We are either awake or asleep and when we are asleep we are either in slow wave restful sleep, or paradoxical=REM or rapid eye movement sleep. REM sleep is when most organized active dreaming occurs. We all have altered states of awareness, that we all live in at least two worlds (whether we can actually recall one world when we are in the other or not), the dream or REM (for rapid eye movement) world and the waking world of "reality". For some unfortunate persons with narcolepsy, characteristics of REM sleep intrude into daily waking activity. What are some of the features of REM or paradoxical sleep? It is called paradoxical because breathing and heartrate are irregular and can be somewhat more rapid resembling a waking pattern yet this is the stage of sleep when a person is the hardest to arouse. So physiological patterns resemble wakefulness in the deepest stage of sleep. Secondly there is a loss of almost all body muscle tone. The person isn't capable of using voluntary muscles with two exceptions, their eye movement muscles which are moving actively (hence REM or Rapid Eye Movement sleep - Could it be that a person is seeing a vision and searching with his eyes?). Narcoleptics while sitting quietly, may suddenly doze off for a few seconds which they cannot control, and enter dream or REM sleep. They may get an irresistible urge to fall asleep during critical periods while listening to a lecture or when driving a car. If they are laughing or at other times they may suddenly fall to the floor.

‘[Later there is maturation into patterns of the full term newborn where waking activity is distinguishable (finally) from sleep and different stages of sleep. One pattern that marks quiet sleep in the infant is the Trace Alternant pattern in which there are again some bursts of paroxysmal activity which usually occur in all areas of the brain at once. This harks back to the very premature pattern present in the early premature. The infant at term or shortly later, is just beginning to differentiate or partition off sleep and different stages of sleep from wakefulness. His partitioning is at first somewhat pathologic. The infant shows signs of going from wakefulness right directly to REM or dream sleep instead of going through the stages of slow wave sleep first as adults do. Sometime after birth at about 44 weeks or so post conception, the infant will develop the adult sleep stage pattern and REM sleep will not occur until he has gone through the appropriate stage of slow wave sleep.

Embryology is emblematic of all the other historical processes that contribute to our makeup or current state. My review of this as far as the brain is concerned is meant to at one blow expose anatomical principles and give a thumbnail sketch of biological development, to make large patterns more visible. At the same time there are so many other developmental considerations packed into our current makeup all of which influence current behavior: thought, action, feeling.

Development continues after birth of course. At birth, an infant can barely see. Acuity is estimated to be about 20/800 on a Snellen chart at two weeks, 20/70 at five months, and 20/20 by five years old. In the famous "visual cliff" studies an infant asked to cross the gulf between two halves of a table covered with clear glass to reach its mother will manifest anxiety at about the age of 8-10 months of age.

Infants and children progress over a well-known and orchestrated series of developmental stages. They acquire a long series of reflexes and behavioral responses, often casting off old reflexes such as a Babinski, Moro, forced grasp, and tonic neck response as they develop and as axons myelinate. What is fascinating is that with destruction of certain parts of the nervous system and in certain diseases, many of these once cast-off reflexes, are documented to reappear. It is as if the nervous system develops by a layering process. Each new stage of development is merely added to a former phase, in just the same way as we have seen, the cerebral cortex modulates the automatic responses of the spinal cord and lower brain centers. This model carries forward to a myriad of other developmental processes as we will see below.

Just one is human history. This is less a discussion of history than it is of embryology, except to say that the average person seems to have historical conceptions all wrong, which is the reason why most people are thoroughly bored with it. The man on the street, if cares one wit about history at all, which usually he does not, will give George Santayana the credit for what is worthwhile about studying it, "Those who do not remember the past are doomed to relive it.." This standard reason given for the purpose of studying history is all wrong. We know that the study of the past will almost never yield up a remedy for our present predicaments, nor prevent us from making mistakes of the past. Hitler was well aware of Napoleon's failed attempt to invade Russia and the Communists redid so many mistakes of the Czars of the past. Indeed that best model for history is given by the seers Cassandra and Tiresias. Those who are foretold the future are condemned to carry prophesies out. No, the study of history in no way emblazons a path for a brighter future. If the study of our past won't do us any "good" then why bother with it at all?.

Let us do a quick survey of history in one paragraph. We descend from animals which reproduced sexually over eons of time. Ancient hominids ate both meat and vegetable matter and were hunter-gatherers of the plain. Savage Cro-Magnon men from which we more closely descend, probably existed alongside Neanderthals who we now understand, had a primitive religion as well as reverence for their dead. Neanderthals may well have been brutally annihilated by more intelligent Cro-Magnon men. The biggest advance in human history came with the domestication of animals and the cultivation of crops making it possible to make a permanent home. Next came the raising of cities and development of writing, civilization, roughly 6 or 7 thousand years ago that was so momentous, that the Bible reckons it as the beginning of time. Writing coincides with a higher human consciousness or self awareness allowing us, for the first time, to build upon a past permanent record without relying entirely upon our own imperfect memory and word of mouth. For the first time, men were able to keep a written record, outside the limits of the human brain. After this we have abundant historical accounts of war, though it seems incontrovertible that organized savage physical struggles took place well before the emergence of even of our species. When these struggles occurred and the invaders prevailed, the city, Jericho, Troy or Rome, was annihilated. Weak and strong were slaughtered together, the women raped or killed. Human history is the story of deception and plunder in which the most ruthless groups survived to tell the tale. There are also stories of the triumph of good and saintly persons though not too many of them and most tend to be exaggerations serving the purpose of the story teller. That is history. All the fun is in the specific stories and details.

What's the point of going into all of this? It is who we are. Let's say that a religious sect requires male celibacy and sexual abstinence. Then it is denying something human and is going to have to expect to run into some trouble. Apart from the obvious, that such groups are not self-propagating, sexuality is quite basic and may be expected to spill forth in some way whether through fantasy, masturbation, homo-eroticism, pederasty or other channels. Such constraints may bring about much more "sinful" behavior than the original behavior they were used to minimize, but also increase focus on sexuality which, naturally expressed, would never have been raised to such a high level of importance.

Less basic perhaps is meat consumption. Again, the rule is that vegetarians, while noble in their intentions, end up being much more hung up on food, which really is quite second nature and not much of a concern to the rest of us who live in an environment of plenty.

How to handle aggression. Aggression has been part human nature throughout history. There is no way to deny it and it does no good to repress it. Groups that survive today because of successful plunder. We are not the descendants of sacked cities but mostly offspring of the sackers. Granted the erecting of cities gave many groups the possibility of surviving by their own intellectual whiles and stealth as much as by physical strength. Numerous well-intentioned utopian experiments in which all persons work and wealth is evenly distributed have failed. Perhaps there are a few short-lived examples that do survive if you count in certain convents, kibbutzim, and Amish societies. But for the most part, aggression and self-interest have to be counted in the human equation and dealt with, not denied. Denying these tendencies will in the end cause an eruptions of aggression a riot, or a war. This is one human characteristic that needs constant management. Aggression has instead to be *channeled* in constructive ways. If a persons could be rewarded for self interest as long as it benefits his society as well, that would be ideal. The enlightened society is one in which human characteristics are acknowledged and incentivized to work for the common good and survival of the whole. It is the way, for the most part, that our capitalist societies function and the biggest part of the story of the collapse of communism and failure of socialist regimes. "Enlightened" self-interest, behaviors that benefit society, need to be rewarded. Repression and loss of freedom may work over the very short term but are not likely to be successful over long timeframes.

I don't recommend we deny who we are. We need to study ourselves more and work effectively with our discoveries. In the ideal situation we could then have an opportunity to advance past our immediate predicament to demand even more from ourselves. That is part of being born and not manufactured which means our past is part of our being. We need to know how we came to be to determine who we are.

Whether or not one looks at history as a saga of progress into more advanced phases of development of mankind, and I admit that that is the way I see it, the lesson from our embryologic origins how the infant unfolds and morphs from a past state, is applicable to an interpretation of past events and without this knowledge we know so much less or who we are, because we are the embodiment an unfolding of our past. That is the real value of a study of history. That is why we remember. As conscious wondering beings we crave a knowledge of who we are. If we have no idea where we come from we are less alive, less conscious. Our present is a product of our past and how we came to be. While rare is the occasion where knowledge of history will be of any practical use, memory places us on a tapestry of time and space, providing data about our particular space-time coordinates. It helps us put our lives in the context of the human experience. Without it we have disorientation. Historical distortion yields a kind of vertiginous disorientation.

There are myriad other contributors to our memory and our past. It is not my goal to review development here, only to point out in general terms how these considerations lead to a fuller appreciation of who we are and where we fit in the scheme of things. I provide a very brief summary mainly for purposes of illustration.

Jean Piaget (1896-1960) was one of the greatest contributors to cognitive development of children. He represented cognitive maturation as a series of four successive stages, Sensori-motor (0-2 years), Pre-operational (2-7 years), Concrete Operations (7—12 years), and Formal operations (12-15 years). The child is held to progress through a series of intellectual milestones in development. For example, in the sensori-motor phase he masters the concept of *object permanence* between 8 and 12 months of life and is able for the first time to try to find an object temporarily hidden from view. Before that time, a hidden object, apparently disappears forever from the infant's regard. Similarly, in the concrete operational phase "object conservation" is mastered. A child realizes that the quantity of clay is the same no matter what its shape, whether it be rounded into a ball or elongated in a sausage shape. In the formal operations stage, the youngster is able to handle rational and systematic meaning about hypothetical problems, is able for the first time to think mathematically and symbolically. Each stage is a prerequisite for successive thinking modes.

Whatever one may think about Sigmund Freud (1856-1939) and he is not looked at kindly in this day and age, he did look at emotional development of the young child, interpreting this to be essentially sexual. He invented the well-known and worn, oral (1st year), Anal (2nd year), Phallic (years 3-6), Latent (6-12), and Genital phases, proposing that it is possible through incomplete resolution of conflicts, for a person to be mired or fixated at one or another of these successive phases.

Freud's views were enhanced by Erik Erikson (b. 1902) who identified developmental stages according to successive life tasks or conflicts viz. Trust vs. Mistrust (infancy- 1st year), Autonomy vs.

Shame and Doubt ((2nd year), Initiative vs. Guilt (preschool), Industry vs. Inferiority (School age), Intimacy vs. Isolation (young Adult), Generativity vs. Stagnation (Prime or life), Ego integrity vs. Despair (Old Age). These descriptive observation of conflicts presented by life are at once obvious and bear more than a grain of truth, but again, reaching successive stages depends upon success in resolving the conflict presented by the preceding stage and unlike his predecessors, Erikson rightly extended development throughout an entire lifetime, instead of restriction to childhood. Humans experience an entire lifetime of maturation of thought processes.

Whether or not one agrees with the accuracy of such specific staging of development, there is little doubt that cognition continues to morph throughout a person's lifetime and stages may be represented according to an anatomic developmental model as successive layers or laminations.

I am presenting here a layered or laminate structure of consciousness based on memory. Each successive stage is built upon a bedrock or former stages depending on the underlying stage as a support. and prerequisite. Children either are or are not *ready* to learn about a new concept. Readiness depends on their biological development. Before puberty, they may not be able to understand a lot of concepts about sex. Or they may be unready for advanced concepts because they have not yet mastered more elementary ones. It's very hard to teach calculus before you have mastered algebra. Or, they may be unready for information due to a combination of biological and experiential factors. A 9 year old child was shown films of comatose patients on ventilators as part of program teaching them about death at his school then asked to vote whether they would want to be kept alive under these terrible conditions. He voted no. Soon thereafter his grandmother died. For some reason he was unable to sleep at night. As it turned out, he thought he'd killed his grandmother when he voted to terminate the comatose patient. Young children whose parents get divorced very frequently feel responsible. Perhaps there were terrible fights and the child may have wished for an instant that his parents would split apart. At a certain stage of development a wish is tantamount to making something happen. A substantial part of evolution into adulthood is expanding abilitiy to separate fantasy from reality. We know that well after adulthood, there are concepts that can only be dealt with at more advanced state of development. We call this wisdom by which we mean a state of advanced readiness.

At any given time, it is possible to reach downward into our personal treasure- trove of memories, past former stages of development which are very much still present, to regress, in other words, into former stages or to just use the information still there and incorporate it into the engram sediment that is our daily lives. Sometimes this happens after an injury, as in some multilingual stroke victims who have a tendency to revert to the original language of their childhood. Usually this is not the case, but regression may occur after a psychic trauma or other process.

Examples of regression into pre-logical thought abound. For example much (by no means all) religious thought is at once, pre-logical yet so basic and therefore all the more powerful. The more mythical stories stretch the boundaries of logical adult thought processes, the greater is the power driving their belief. The most unbelievable stories bring about the most resilient belief. This would stand to reason, since a person who believes strongly in something illogical or disproven by experience is the most unlikely to be dissuaded from his belief by logical argumentation. An inverse correlation exists between reality testing and the power of mythical beliefs, so much so that I hesitate to give specific examples for fear of violent repercussions. I do have faith that the reader, on being exposed to this principle, or perhaps having already made the same observation, will be able to come up with abundant examples on his or her own.

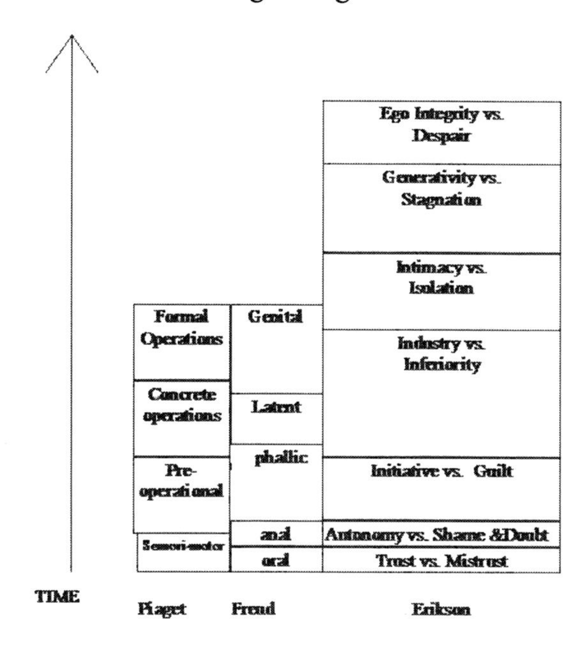


**Figure 28: The laminae of Development (examples).**

In summary, I have developed in this chapter a different, more comprehensive view of memory, which is far more than the recollection of details just presented, much more than the storage of bits on a computer disk. Rather memory in humans is a dynamic interaction of the past with a challenging present, the interface of our makeup with life experience. Memory expresses itself In learning by which we mean the alteration of behavior consequent to experience, behavior comprised of perception, thought, emotion, action.

Up until now our concept of memory has been superficial. We have been mired with overly simplistic models of memory and have failed to see it for what it is. Consequently, the importance of memory has been underestimated. When we view memory in all of its shining glory as the sum total of storage mechanisms defining what we are at any given moment, interacting with our present to produces all future experience, the complex depths of life, then we see ourselves as we really are not mere biological machines or automatons. No longer are we fooled into accepting a simple biological model for the laying down of memory engrams, or influenced by the disk drive or CD ROM model of memory, as entirely a material. Memory is in part our innards and assumes myriad dimensions and great depths. A better model than a disk drive might be Jacob's ladder with hundreds of retrieving angels continuously climbing and descending from the cortical empyrean to the layered depths of past conflicts and experience. This is what makes us who we are, unique beings whose life will be easily reproduced or resurrected.

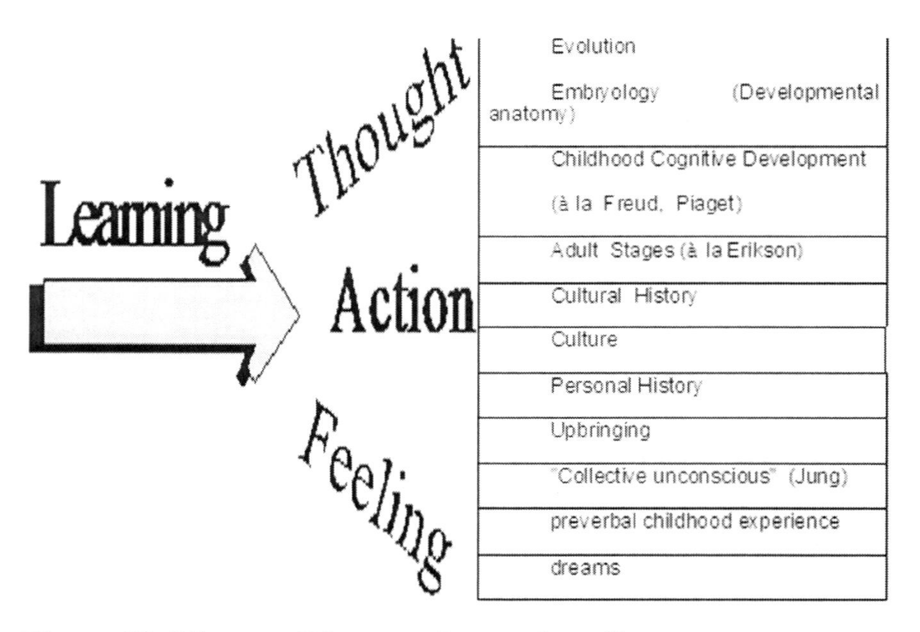


**Figure 29: Memory Diagram. Input of totality**

## Epilogue: What we leave to our children?

All of this is meant to show that memory, something superficially very simple, is a whole lot more complex. It is the sum total of our being up until the point in time we call the present. We live in the computer age and it is reasonable to see all of this in terms of information systems. After our lives are over, and even now, what are we leaving to our children?

Our life span is finite. But parts of us are passed down for a much longer period of time. Our legacy is quite obviously in part genetic. We will leave certain traits and form and physiognomy. We will leave certain weaknesses and strengths as well. Perhaps a tendency to develop heart disease of diabetes or robust of powerful muscles and physical stamina. As of this point in our history we have little control of the genetic characters passed down. Perhaps this may change sometime into the future. As we have seen our final form unfolds in a dance that is embryologic development. We will also leave perhaps some financial wealth, information as well that may be a blessing or a curse to our children.

The most important legacy we will leave is a record of our own life, well or poorly lived. Our whole life exists within certain space-time coordinates as we have seen. Certain aspects of our existence may or may not be discoverable depending on the status of memory and technology used to extract this information. Most of this data will be unavailable forever, though it will continue to exist of course. Even the pyramids and the Taj Mahal standing now as witness to the past, will eventually collapse, and at some future time there will be no record that they ever were. Certain particularly resilient ideas may well survive over much longer time frames, even though they are not palpable or tangible.

If we see this legacy as information, the next step is to construct tiers of information that become a record of our life. This information is remembered, extracted, extractable or not, as is all information registered or not registered, retrievable or irretrievable in memory. This discussion provides little more than a survey of what is possible the tiers of information recorded. Information in the form of genetics, development, personal and collective history, events and choices, cultural and personal ideas and the course of life. Our brain merely makes possible new forms of information which may be passed down to our children, culture, events, ideas as well as the genetic and structural and physiological information that lower animals pass on to their offspring. We need to be mindful that we are passing down a good deal more than biological heritage.

# Chapter 3

# VISION

## Eye and Brain

"The eye...the window of the soul, is the principle means by which the central sense can most completely and abundantly appreciate the infinite works of nature; and the ear is the second, which acquires dignity by hearing of the things the eye has seen.

-Leonardo da Vinci

What's most remarkable about vision is not the pure mechanics of seeing a visual image but explaining how we can see so much more than the basic equipment is designed for. The eye and brain would seem to have a limited specific capacity, yet we see so much, the gaps being filled in by mental processing, imagination and understanding. At the base of this process is the apparatus of visual perception, the start of a journey of the mind's eye.

The retina is something like film in a camera. The front of the eye does focus a good quality image on this screen. There ends the analogy. The retina is far more than the lifeless photographic film. It is sophisticated neural processing tissue, a complete neural way-station, first repository of vital visual data. Retina is a part of the brain sent out to gaze upon the world. The visual image transmitted from the retina has complex origins. It does not form a single whole image like the film in a camera neatly transmitted to the brain. Light images need to be deconstructed into electrical signals that the brain can use and process then somehow by a method poorly understood. We make mental images again from chunks of piecemeal data sent into the brain. The retina thus transmutes light to electrical signals that the brain can use. Neurons work with electrical data, not visual images. Vision is not different than touch, hearing, smell, and taste which all require transduction. In the brain all of these sensory modalities have a common language too and can be related to each other which means we can begin to gain a solid correlative singular conception that is part of consciousness.

For those who learn how to look at it, the retina can tell a lot about what is going on in the brain, as if the brain sends out a part of itself to display to the world. You have to look at it with an ophthalmoscope, first used by Helmholtz in 1850. The process is so routine, doctors aren't aware they are actually examining the brain, which they are. In severe hypertension you see hemorrhages from bursted swollen arteries and a bulging optic nerve head telling of increased arterial pressure. This reflects the same trouble in the brain. Looking with the ophthalmoscope you may sometimes be lucky enough to see blood clots floating their way through retinal vessels. You may conclude that similar blood obstructing clots are traveling through other brain arteries. Dangerous clot material comes from the heart, great vessels or the carotid arteries in the neck and causes strokes by plugging up arteries in the brain. These tiny clots are thrown all the time in open heart surgery if you continually examine the retina, something rarely done.

When there is increased pressure in the closed cranial cavity, you see that on the retina also, by looking for swelling of the optic nerve, visible on the retina. On a flash photograph you can see the retina in the form of a red response that is the bane of the photographer. The pupil of the eye is dilated in a dark environment where the flash is often used and the retina reflects light quite nicely.

## DEVELOPMENT:

Human embryology proves the retina is an extension of the brain. The first sign of the retina in development is at about 22 days post-conception, when two slits or sulci appear in the front part of the developing brain. Three or four days later an optic stalk already forms and an infolding to form an optic cup. The optic nerve carrying impulses from the retina to the brain is misnamed because it is really not a peripheral nerve but a *tract*, part of the central nervous system. The difference between a tract and a nerve, it will be recalled, is that a tract is a cable or bundle of axons that conduct impulses inside the central nervous system, actually part of the brain or spinal cord, while a nerve is in the peripheral nervous system. The blood supply is part of the brain's blood, the myelin of the optic nerve is brain myelin, not nerve myelin. The retina is the only part of the brain outside the cranium. [٧]

The next thing that happens in the embryo is that the front of the optic stalk folds in. This structure will eventually accept a lens whose origin is the ectoderm that forms skin. The cornea of the eye does most of the light refraction or bending focusing a beam of light on the retina, not the lens. That is why you can correct your vision with a contact lens or by radial keratotomy, an operation on the cornea. The lens in humans is there mostly for adjustment of the focus to accommodate various distances of objects.[83] The optical refracting equipment inside the eye has to be at least as good as the finest optical instruments developed by man. Our cameras, microscopes and telescopes utilized advanced materials technology and the most advanced engineering and planning available. But the eye's refracting equipment is comprised only of biological materials, water, proteins and other chemicals.

One of the basic questions about the development of the eye is what determines whether or not a person will need corrective lenses. From the biological perspective our eyes are mostly suited for distant outside use, scanning the horizon for prey or dangers and indeed in most of us a distant object will focus directly on the retina without any need for adjustment of the eye's lens or accomodation. Only close objects will require the lens to focus. But a quarter of us are near sighted which means we are more suited for near vision, distant photographically infinitely distant objects, focusing in front of the retina instead of right on it. This is because of a pathological elongation of our eyeball that seems to be a developmental defect. This difficulty, taken care of by corrective lenses, seems largely to be inherited. Near-sightedness runs in families. Certain experiments have shown that at least in experimental rhesus monkeys, near sightedness can be induced by environmental intervention, especially any action that interferes with adequate formation of an image[84]. When images are blocked out or otherwise interfered with the eyeball develops in such a way as to be too long to focus the image properly. A similar process might occur in children raised in a close environment looking mostly at televisions and books rather than the great outdoors. Control of elongation of the eyeball, near or far sightedness resides in the brain. Early in development, brain and eye interact, to determine the length of the eyeball. There is even some speculation that the secretion of a peptide called vasoactive intestinal polypeptide (VIP) may play a role in this abnormal development. Some day a chemical antagonist may be able to prevent nearsightedness.

Soon, an inner and an outer layer of cells forms on the nascent retina, the inner layer closer to the lens of the eye, coming from brain neural tissue will house many kinds of neurons and become a visual processing center. The outer layer is the pigment epithelium. Between these in the embryo is an intraretinal space. As the eye develops this space will close as pigment epithelium is apposed to the inner retinal layer and outer and inner retinas fuse. But it is as an adult that this embryonic space becomes a problem. There is always a tendency for retinal disease to cause detachment of the retina that occurs just along the border between the inner and the outer retina. (Diagrams) . A certain quirk in the retina also comes from the embryo. The light receiving cells the rods and cones, as they are called because of their shape, lie at the hindmost part of the retina. Cells that the rods and cones transmit to and which connect to the rest of the brain actually lie in front of the rods and cones. This means that light has to go through a lot of other cells and tissue on the retina in order to get to the light-sensitive cells. Because of how it is originally formed the retina is layered in such a way that light has to go through cells and processes in order just to reach the light gathering rods and cones. This presents an interesting dilemma. Either turn the entire structure around so that light reaches the sensitive cells first, or build the structure to be exquisitely thin and transparent. The retina is designed according to the second option.[85]

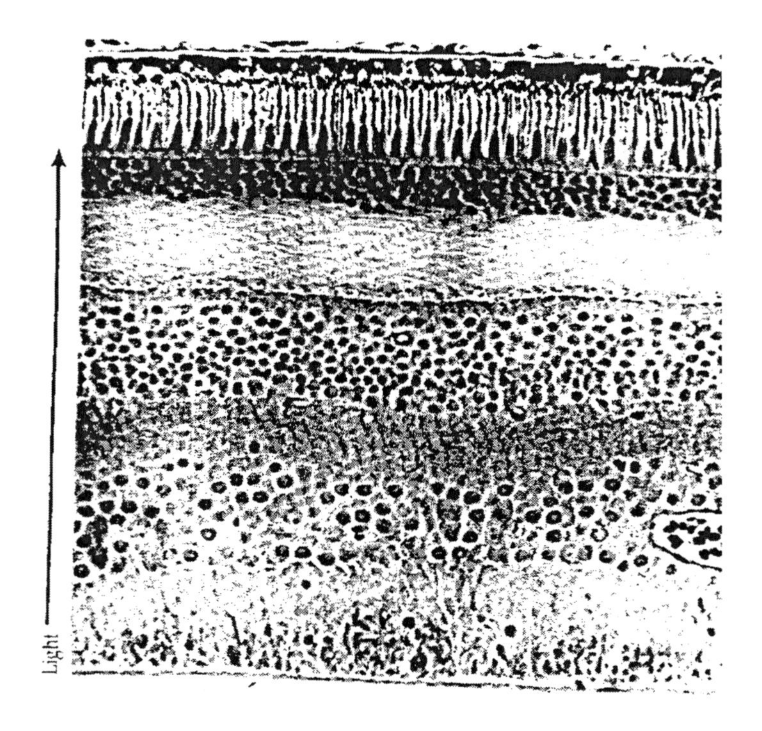


**Figure 30: Microscopic picture of a retina. The pigment layer is on top. Just below are rods and cones, then other cell bodies and processes. Light must go through the other layers to get to light receptors. Bottom is toward lens or front of eye.**

## RETINA:

As seen in the microscope, the retina has 10 layers of cells and processes. And light has to go through almost to the hindmost layer to be detected. The outermost layer is made of melanin pigment, the same thing that makes our skin dark and makes the eye a dark light processing center. In some animals who depend on night vision, this layer reflects light back into the retina instead of just absorbing it. Albinos lack pigment and see poorly. Or, the pigment can proliferate too much and be out of control, as happens in forms of retinitis pigmentosa. Tumors of melanocytes, the pigment making cells can also start in the eye. The light detecting cells, the rods and cones, are found just in front or internal i.e. towards the lens to the pigment layer. Rods detect the presence or absence of light and as such are merely black and white sensitive but they will react to trace quantities of light and as such are useful for low light and night vision. They lie primarily in the periphery of the retina. An electrode in the rod cell can show the rod responds to the tiniest quantum of light, a single photon. Rhodopsin is the key to rod light sensitivity. Rhodopsin is really a cooperative relationship between retinal that comes from vitamin A, and a long chain protein that sits on the cell membrane called opsin. Retinal + Opsin = Rhodopsin. The trick is that *retinal* deforms with a twist in a chemical bond, it becomes a *cis* isomer only when light hits it, then reforms again when the light is taken off. The close association of retinal and opsin weakens when light hits it and reforms again in the dark. Opsin is a large molecule a protein that essentially cradles the small retinal. This sets off a chain of events that involves second messenger, other substances that change the charge of the inside of the rod or cone cell. The light shines on the cell an the cell changes its electric charge. This is the course of events that begins the odyssey of light sensitivity in the nervous system. Virtually the same molecules and the same scheme holds in all of the very different light sensitive organs in other animals that are very different eyes such as in the insects or crabs with compound eyes, other mammals, even primitive animals that have simple "eyes" of only a few cells. These molecules and chemicals are highly conserved. That means that they developed early on in evolution and then were inherited as more advanced animals came onto the scene. In fact, some bacteria share a similar molecule designated bacteriorhodopsin raising the possibility that higher animals inherited their ability to make similar chemicals via bacterial infection (see similar discussion as relates to mitochondria and chloroplasts "Inside the Neuron"). This light response may have evolved in parallel, or alternately, according to more conventional thinking, the bacteria and us may have descended from the same progenitor which carried its own type of rhodopsin molecule.

By a strange twist, computer scientists are harnessing the bacteriorhodosin molecule which is much more stable than human rhodopsin and has interesting conformational properties as a wonderful memory storage device. Simply the cis and trans isomeric forms of the molecule which change on exposure to light sources are exactly analogous to computer 0's and 1's or "on" and "off" states. This means that light beams can be used as computer writing and reading devices over a biological computer chip impregnated with biorhodopsin molecules which change conformational properties in response to light. Today's computers use nonorganic molecules which are spread over a surface two dimensionally, limiting storage and data access. One may conceive of a three dimensional read and write optical memory device that can not only store much vaster amounts of data but which may access that data at unheard of speeds. Robert R. Birge of Syracuse University one of the pioneers estimates that while perhaps 100 million bits of information (0's and 1's) per square centimeter can be stored on conventional memory devices, one trillion bits per cubic centimeter can be placed on a biomolecular device[86]. When you think of it such machines are the union of the material and the biological, organic with inorganic matter which may at some stage, produce superior thinking (or at the very least data storage) machines, never dreamed of by science fiction writers.

In each eye about 125 million Rods serve as light dark detectors and they don't distinguish colors. We have only about 900,000 retinal ganglion cells that go to the brain proving that a lot of rods converge on each ganglion cell, about 140 and about 6 cones go to each ganglion cell. The 5.5 million cones are concentrated in the center of the visual field, the *fovea centralis* which is a pit in the center of the macula (see below) in which blood vessels, cells, and their processes are pushed out of the way so that light can get to the part of the eye responsible for our most acute vision. Our color sensitivity resides in the fovea. They do so because of slight changes in their pigment protein opsin the protein that holds the retinal molecule. Minimal changes in the structure of this protein make it preferentially absorb light of red, green or blue that makes it absorb light of different colors. This has only occurred recently in our evolution as primates who forage for foods mostly during the day. Other mammals with a nocturnal existence have little need for color vision. Because these pigment proteins are encoded on the X chromosome, the different forms of color blindness are x-linked and occur almost exclusively in men. The cones are in the area of highest acuity, reading and what is tested with a wall eye chart. The foveae perceive light differently than the rest of the retina. Though light has still to travel to the back of the retina the fovea is kept as clear as possible from all obstructions. Axons and other structures are kept out of the way so that light falls on the cones directly. Even so the foveae where the cones lie comprise a very small part of the retinal surface. The retina is as screen and what is projected on it is our visual world. Our whole field of view if we had receptors all the way around our head would be a sphere through 360 degrees of vertical and horizontal angle. At any one time with our eyes stationary, we can only see perhaps 120-150 degrees of visual angle. Our foveas, where we see the clearest, subtends less than three degrees of visual angle going according to its size. You can calculate that one millimeter of retina subtends a visual angle of about 4 degrees. We have an excellent internal picture of our whole 360 degree spherical visual world. We know this because we can negotiate around it, pursue run or be pursued almost without a thought despite the obvious liability of design built into our eyes and retina. All this is done automatically with the constant motion and scanning by our eyes but it involves working with visual memory to form a working picture of our environment even though the part of it that we can actually see at any one time is minimal.

The rod and cone light receptors generate a dark current. They ooze glutamate that excites the neurons they connect to. If light shines on them, there is a decrease in this dark current and they secrete less glutamate that stimulates the next cell in the chain, the bipolar neuron. Light causes a change in the flow of ions into the rod or cone cell, mostly a decrease in the amount of positively charged Sodium that is allowed to flow in. Light has no meaning in the nervous system. It has to be translated into the electrical language of the brain. The first step in this process is an alteration in the flow of charged ions in the receptor rods and cones. That change is mechanical but is brought about by a sort of Rube Goldberg system of second chemical messengers and changes in enzymes. The flow of currents is ultimately reduced in the receptor cell.

Neither the rods nor the cones are cells that fire. They do not have that all or none action potential but continuously graded response. As we've seen the "on" or "off", "yes" or "no", "1" or "0" response is much like a *digital* or computer type response. But in the eye and this is also true of various other type of receptor cells elsewhere in the periphery, for example in the ear, the initial response is *analog* or graded and is only later translated in the central nervous system to a digital signal. This allows for various modulations to occur before the data are interpreted centrally. Some of this processing takes place right in the retina. In "higher" animals, primates and humans, some of the work done right in the retina tends to be done more and more by the brain and even by the cortex in a process known as *cephalization,* Whereas in the rabbit or the frog this work is done in the retina itself. One could speculate that in the frog, most of the visual responses need to be quick and automatic, the tongue needs quickly to flick at the fly picked up visually, whereas in the human visual responses need to be more considered and slow. When we see the central hookups of the eye, we will discover that only in some animals is there really advanced cerebrocortical or higher representation of visual input.[87]

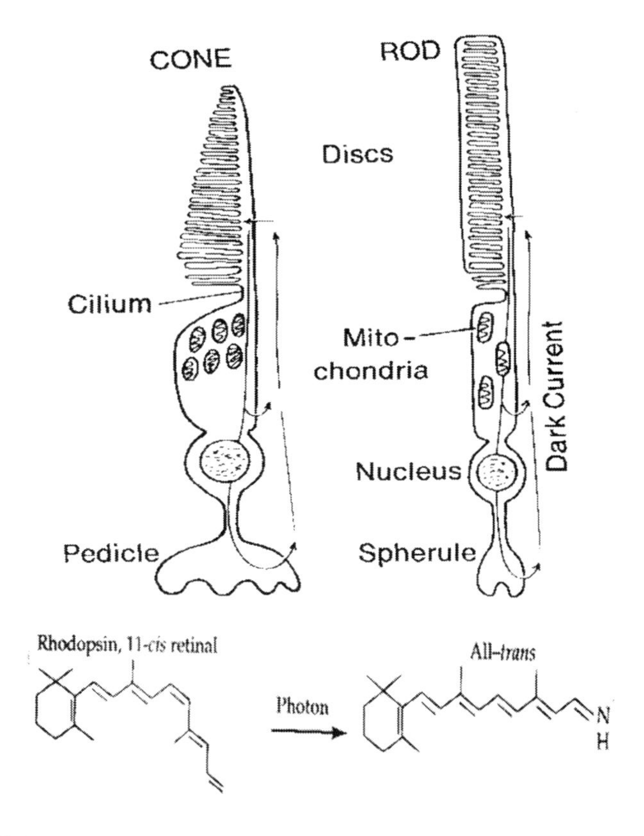


**Figure 31: Rods and cones are the initial light receptors, named for their shape. Rhodopsin pigments reside in the enormous surface area here seen at the top of the cells. Note large number of mitochondria too used to provide energy. Rods and cones are derived from ciliated cells and the tops of the cells are really modified cilia. Retinal is closely held by the protein Opsin. The key to light sensitivity is twisting of the molecule.**

# EYE:

The eye is always compared to a camera to which it bears a superficial resemblance. There is an iris that changes to accommodate more or less light and helps to respond to near and far objects as well. But the eye is living tissue so much more adaptable than any camera when you consider all of the conditions in which it works. Cameras could not survive many of the conditions to which the human eye is subject such as extremes of heat, humidity, under water work etc. The eye is adapted not only to handle images of enormously varying distances or focal lengths but also light concentrations varying over 100 thousand fold from the dark night sky to midday in the summer time or bright glistening snow in the winter. But at any one instant, we can see clearly over only about three degrees with our retina and fovea.

What is striking to me as I pursue an activity such as jogging outside, avoiding cars and dead animals along the road, is that I have a complete picture of my environment in my head. This is an illusion. At any moment, I can see only a tiny part of my 360 degree surroundings, yet I always think I have a panoramic view. How is this accomplished?

At any moment we have an excellent picture of our environment and are capable of reacting to any change. Think of a quick lunge with of your opponent's sword in fencing or an attempted tackle in football. These are primarily visual responses. Yet the eye can only see clearly about 3 degrees of visual angle. One trick is that special circuits are built right into the retina involving primarily rods that respond right away to any motion. The rods connect to bipolar cells which then go to ganglion cells and then through the optic nerve to the brain. The intervening cells perform additional processing functions in the retina. Part of these processing deals with movement. As a general rule in the nervous system intervening cells are there because further processing is required. Vision transmission is from the rod or cone, to a bipolar cell, to a ganglion cell. The ganglion cell will fire when it gets the specific input from the other cells that connect with it. One favorite laboratory technique is to put an electrode into a ganglion cell, then see what it responds to. Will it fire when light is shone on one tiny area of retina or to some other stimulus? It is easy to find out. What you discover is that different ganglion cells can be made to fire for different reasons. Some indeed will only fire if there is movement within an area on the retina. They won't respond to a constant light. Some will respond if a light is turned off or turned on. The specific stimulus type that will make a ganglion cell fire or respond is the *receptive field* . Mostly we are dealing with a specific area of retina that will make a specific ganglion cell fire, but it can also be conceived as a specific movement or color, whatever has been discovered to make that cell respond. That is its receptive field. In vision the scheme is that certain cells are wired to respond to certain stimuli. As a matter of fact, this principle is universal in all reception, not only with vision. A pressure receptor on the skin will only respond when you touch the skin within a specific area and won't respond if you stick it with a pin etc., a cell in the nose may respond only to a certain odor. This is its receptive field.

In fact most ganglion cells respond best to one of two situations. The rods and cones are wired up in such a way that the ganglion cells will most easily fire when there are concentric circles of light or dark. Some fire best with light in a central circle of a small size surrounded by darkness. These are "on" center cells. Others fire best when the center circle is dark, "off" center cells. What they respond best to is *contrast* between borders of light and dark. Ganglion cells will be more excited if there is a contrasting border and inhibited partially when light is surrounded by light or dark surrounded by dark. It is a common phenomenon in sensation that two adjacent receptive cells excited by the same stimulus, a large area of light for example inhibit one another. On the other hand two adjacent cells that have contrasting stimuli, a light area surrounded by a dark area, for example might mutually excite each other. This scheme accentuates contrast within the field of each receiving cell, ignoring sameness is *lateral inhibition.* While the retina accentuates contrasts and borders so do other receiving cells such as skin pressure receptors and even hearing functions by accentuating auditory stimuli of acute onset and offset. We notice a click or acute change in volume in preference to a constant droning sound. In our retina, ganglion cells aren't apt to fire unless there is some difference or border between light and dark that can be perceived. This is true of colors as well. Long ago it was discovered that the eye will accentuate opposites which for colors are complements, red is best next to green and so forth something noticed in the nineteenth Century by Eugene Chevreul a French chemist. Our perception of colors is colored by the color that is next to it or near to it. That is precisely the same process. This allows for better acuity for borders and contours.

We see clearly only in a small part of the retina which is structured for high acuity known as the *macula lutea* or yellow spot. Something immaculate is spotless but although the macula looks like a spot it is actually the clearest part of the retina. Here's why. In all other parts of the retina light has to go through all those other layers of cells and cell processes but not in the macula where they are cleared out of the way. And the macula has primarily cones tuned for color vision. Because there are so many fewer ganglion than light receptor cells Rod and cone inputs converge on ganglion cells but only about 6 on average for cones in the macular area but some 140 for rods that are not in the central area of the retina. Cones are tuned for color vision and work best in good light. When light hits the macular area we see the object clearly. Processes that affect the macula, such as Multiple Sclerosis and macular degeneration affect our ability to read, see colors and see clearly. But we see clearly in only a small part of the retina.

Eyes and cameras are different in other ways. With a camera we are trying to get a picture of a large field, to get everything into focus over a wide angle. As anyone who has used one of older (not automatic) SLR cameras knows, you often mess up as you try to focus on many objects at different distances from the lens and with varying light and shadows. The eye doesn't work like that. It takes in with clarity only a very small part of the visual world, or visual angle and thus can focus clearly on only a very small area. That's why you have to move your eyes around so much when you read by a complex process designated as foveation, getting the more sensitive and precise fovea centralis of the eye that contains the macula, to focus on the area of interest. The eye thereby eliminates the problem of trying to get the light and focal length right over a wide visual area and also there are fewer valuable clear vision cone cells and retinal cells used for precise vision

The retina's design of central rods and peripheral cones has an analogy in the ear.. In each ear there are inner and outer hair cells sensitive to vibration. There are only about 3500 inner hair cells and these are the most precisely tuned to specific frequencies of sound (pitch). Just the same way that the cones in the central eye are tuned to light frequencies (colors). The inner hair cells synapse with many more secondary cells than the outer hair cells. There is much more convergence of input of the outer hair cells of the ear just as there is a lot more convergence of rods than cones in the eye. Approximately 20,000 outer hair cells connect with perhaps 1000 secondary neuron cells. The reason is that outer hair cells convey much less precise information. The outer hair cells are situated in such a way as to influence the sensitivity of inner cells. Both kinds of cells may respond to vibration so as to translate vibrations into electrical polarities and the flow of positive currents (ions) into and out of the cell resonates with the actual frequency of sound stimulation. Eye and ear are quite analogous when you examine them. The outer hair cells correspond well to the rods in the retina, the inner hair cells to the cones. The cones are precise and a few of them influence more secondary cells.

If you see clearly only with the macula then what is the rest of the retinal surface for? Even when you are looking at something intently something else can grab your visual attention. That something is likely to be a moving object or a flash of light, something that signifies change. Groups of rod cells in the peripheral retina are hard wired in a conspiratorial way to respond to movement and ganglion cells that only fire when an object moves within a certain area on the retina. It's a fact that a frog will starve to death even if surrounded by tasty insects that fail to move. Steven Spielberg capitalized on this in his JURASSIC PARK where human children were prey for flesh eating reptilian dinosaurs, and though shaking with intense fear and wanting to run away were told not to move. Perhaps the immobilizing effect of fear is adaptive. As prey, we survive best by failing to contrast enough to draw attention away from nonliving nonmoving surroundings. Outlining the scheme of ganglion cells for their response to movements and other changes is much of what has been done by researchers in vision over the last 15 or twenty years.

As the inner hair cells are precisely tuned for sound frequencies, cones in the eye are tuned to respond to colors. Large groups of rod cells are grouped together to stimulate ganglion cells that rods connect to, to react or fire only when there is a certain visual stimulus. These ganglion cells are thus also tuned to specific stimuli that can be called their receptive field. What are some ganglion receptive fields? One is *motion* in the periphery. Something moves and this ends up by stimulating a certain ganglion cell. Another may be a flash of light or change in light. A light may suddenly go on or off. Another may be simply a concentration of continuous light. These are the stimuli that will excite retinal ganglion cells.

The strategy here seems to be that if a visual stimulus, say motion, can be made to excite a ganglion cell, there may be something there of enough interest to make the eyes turn to focus with the macula right on that point. The eye is constantly moving or roving pointing its macular vision into different areas of visual field as if probing constantly. Over time, a complete visual picture *accumulates* so that even though we can actually see a tiny portion of our environment, still we have the illusion that we can see almost everything. A total picture of our environment literally accumulates through constant visual probing and darting. Since this picture must accumulate and is not present in real time, it requires not only memory but the intervention of some form of advanced cognition just in order to give the illusion of complete vision. A camera takes a picture over a wide angle in real time, the eye and the brain work together in some kind of virtual reality.

Mechanisms behind visual hallucinations hint at the mechanisms for putting visual images together. A hallucination is the percept of something that isn't there. The visual sense is more exact than hearing and visual hallucinations point more strongly to a neurological or brain derangement as opposed to a psychiatric cause more the rule when something is heard. Visual hallucinations illustrate how we ordinarily need to construct a complete image from data extracted piecemeal. A lot of old folks and persons with Parkinson's disease particularly, have visual hallucinations. Some of them seem to construct an image out of an amorphous stimulus, say a spot of light in the dark. They seem to have a problem magnified by adverse visual circumstances especially dusk and low light situations. During the day there may be no problem. In a random spot of light or looking at a chair or some other object, they may see a person or some intruder, even a long lost relative. Another person with more faculties can easily convince them that the hallucination isn't there. These old folks are impaired. They may have troubles with the lens or the maculae of their eyes. The Parkinson's disease affects eye movements that are critical to bring specific objects into the realm of foveal clarity and what they do see is very unclear.

I sometimes go jogging in the dark even in the middle of the night. At those times I have noticed that rarely I will interpret an object that I don't see clearly a mailbox post or a spot of light on the pavement, as another better formed object. I especially tend to see another person for some reason. When I see it more clearly I realize there is nothing there. I have noticed that this process is worse under more adverse circumstances, in the middle of the night when I am sleepier and if I should jog without corrective lenses the problem really a hallucination, becomes particularly bad. This is why ghosts and visions and religious experiences tend to occur at night. I take it that old folks with their myriad visual and focusing impairments may have even a worse problem. This does not mean that they are necessarily demented. The complete visual image requires close cooperation between eye and brain.

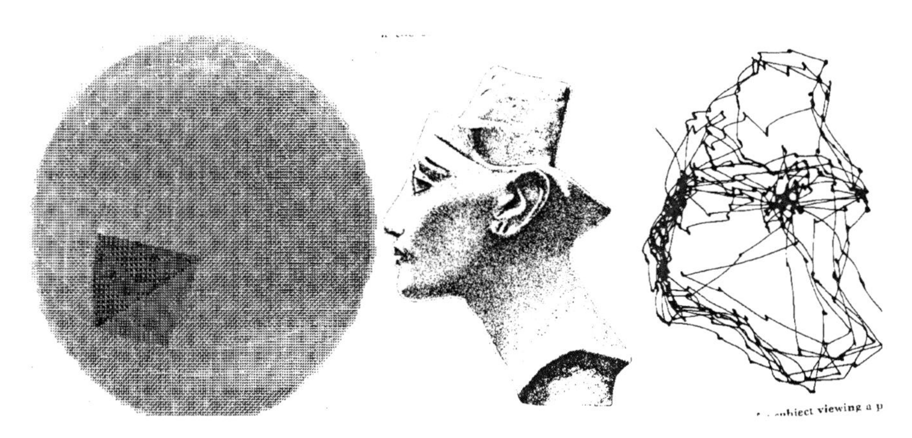

**Figure 32: In order to form a full image, the brain needs to store a composite of many images through object scanning.[88] It stores a composite image storage a small portion of a visual image over time.**

The retina cannot give us a complete picture of an object as a film image does. A complete image results from a cooperative effort between eye and brain. Our vision is very unclear in the peripheral retina. The peripheral retina is designed to help point the fovea at areas of importance often at movement or changing light. This implies that there is some kind of *smart pointing system* that tells the brain what is relevant so that eye motor systems can point the fovea at pertinent parts of our visual field. The brain will then use memory and other faculties to create an illusion of full vision. In order to get a complete picture of an object the eye will scan the whole scene since the foveal vision covers so little of it. The eye jumps or saccades at the rate of about 2 to 3 times a second to take in a whole picture spending about 10% of its viewing time in these jumps.

## Aside: The Computation of Eye Movements

This section gets rather technical and is not necessary to make the main points of this chapter. However, there is a relationship between our sense of balance provided by apparatus in the inner ear and vision. The inner ear apparatus or labyrinth is specially designed to maintain posture and to keep your gaze fixed. Eye movements have a rather complex and fascinating relationship with balance organs whose main purpose, it seems, is to maintain gaze in spite of any perturbations of posture. For example, in playing basketball, your main goal is to keep your "eyes on the prize", that basketball, even if you are fouled by another player and as you get involved with running, dribbling and other activity.

As persons who have lost their vestibular function find out the hard way, this can devastate their vision. Think about it. When you move, anything that you happen to be looking at will seem to move also. Unless you have a mechanism built in which makes adjustment for your movement and keeps the visual image still. That is what the vestibular system does. Certain antibiotics (aminoglycosides- streptomycin, Kanamycin, gentamycin) can literally wipe out the inner ear apparatus. When that happens the poor victim's visual world is a blur, every time he moves. Walking down the street, he will not be able to see people's faces and recognize them-their faces will be moving as he moves-a problem called oscillopsia. If you are reading a book and you shake your head back and forth you will still be able to read the printed page, not so a person with oscillopsia. The reason why the image remains still, despite your own movement, is that the vestibular system sends precise data to the visual brain about your head movement which helps adjust visual perception. Under ordinary circumstances your brain always assumes your environment is still and that you're moving which is true most of the time. That is why if you happen to be sitting in a rail car and the train next to you starts to move, it always feels as if your own rail car is moving. In any event, nine tenths of the time the world remains still, and we move in it, we walk or shake our head and when we do, everything has to remain visually as if it were still, which it is.

We live in the computer age. Many of our most distinguished scientists accept nowadays, the notion that the brain is nothing more than a complex computer. Nowhere is the picture of brain as computer more applicable than in the vestibular system.

Vestibular function is a fitting model for brain computation for 2 reasons:

1. The inner ear brings about eye movements and positional change *automatically,* without conscious intervention. We speak of vestibulo-ocular and vestibulo-spinal reflex mechanisms. We can predict that certain movements within the labyrinth cause specific eye movements and postural adjustments.

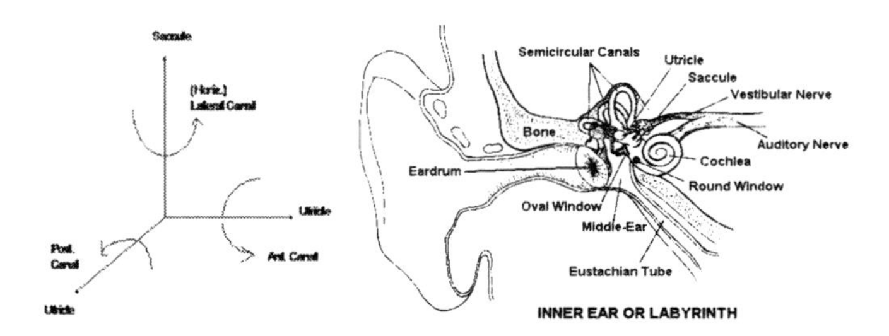


**• Figure 33: Three semicircular canals are orthogonal and deal with angular turns. The saccule with vertical linear movement, the utricle with horizontal movement.The orthogonal planes of the three semicircular canals in three planes.** [89]

2. Eye movements and postural linear and turning movements may be described mathematically. So are events within the labyrinth, and over a similar Cartesian coordinate system. Position and movement are translatable to the three familiar Cartesian orthogonal (90 degrees from each other) axes:

> x,y and z, with x being horizontal, y vertical, and z depth. As it happens the three semicircular canals, designated the superior, lateral and posterior, are also practically orthogonal, and it is thus tempting to postulate that the canals communicate with the brain using some type of Cartesian code. The two ears are mirror images of each other, of course, and the semicircles of the canals of one ear complete full circles in the same planes as the other. Between the two ears the 6 semicircular canals make up 3 full circles in three orthogonal planes.

The canals are round because they are designed to detect turning movements, angular acceleration. A complex of data integrated over the three planes will describe completely any angular acceleration.

The canals, the membranous labyrinth, named for the twisting structure from Greek mythology, sit in an outer casing, the bony labyrinth. These tiny (they are less than the size of a quarter of a dime) delicate transducers of movement, are surrounded by liquid perilymph which dampens random vibratory noise that will contaminate signal transduction. The semicircular canals are filled with a liquid of their own, the endolymph. Inertia holds the endolymph fluid in one place as the head starts to turn (accelerates). The fluid, moving in relation to the membrane, thus exerts a shear force on the tiny hair cells[μ] . These cells are sensitive to the movement of the fluid. Two kinds of hair cell cilia, stereocilia and kinocilia, are exposed to fluid shear forces.

Hair cells sensitive to pressure of endolymph are arrayed on a sensitive part of each semicircular canal, the crista or crest on the cupula. When the endolymph applies a shear force on the cilia or hairs of the cell in a specific direction (toward the kinocilium) the hair cell depolarizes, (becomes excited). The cells oriented in such a way as to be sensitive to the direction of these forces. Movement in one direction will depolarize or excite the cell, changing its membrane potential. It communicates this excitation to the next neuron whose firing rate is altered. This second neuron has a slow constant basal firing rate. Stimulate the hair cell and the second neuron will fire faster, inhibit it and it will slow down. That is how direction of the head turning in the labyrinth is converted to the language the brain can understand, firing rates. The position of each cell and the signal derived from it are codes for specific movements which are integrated in the vestibular nuclei and other brainstem structures.

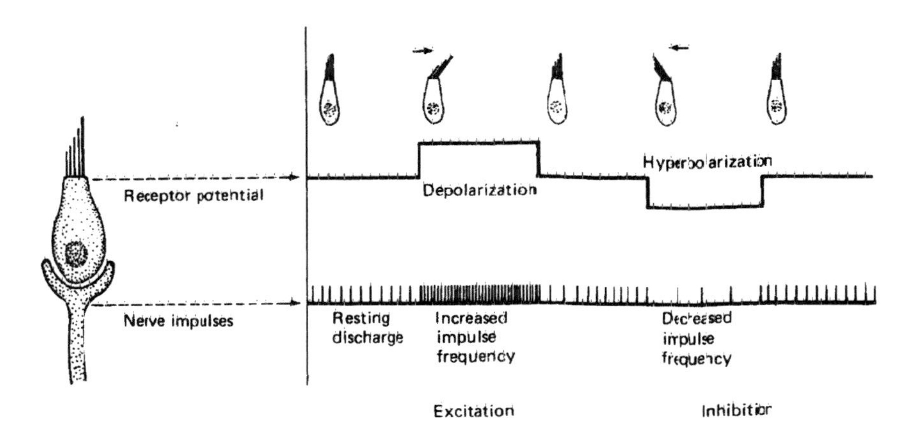


**Figure 34: This shows how the cilia on hair (ciliated) cells in the labyrinth sense motion. Note that the rods and cones of the eye are also modified ciliated cells. Sensory cells convert energy to electrical signals.**

Each sensitive region or cupula houses cells that fire all the time at a quantifiable level. The orientation of the semicircular canal of that cupula determines the direction of stimulation. Thus we have a magnitude and direction of force which in physics defines a vector. This vector can be graphically represented on Cartesian coordinate axes whose orientation at any time is determined by the head position at that instant. Under ideal circumstances each of 6 cupulas send the brain a information about a vector, direction and magnitude describing angular acceleration. The six vectors may be added to determine a single vector that influences eye movements. That there are eight vestibular nuclei, four on each side of the brainstem, attests to the complexity of computing movements on the basis of 6 angular acceleration vectors over time. By comparison, visual imaging is taken care of by just one pair of lateral geniculate bodies, though a great deal of complex visual processing is done both right in the retina and also at higher cortical levels.

Suppose the head turns suddenly to the left. Then endolymph fluid will stay still and move against the cupulas, sensitive structures in the right ear. The exact angle of torsion is written into the array of three orthogonal semicircular canals and is computed in the brain. Soon the head turn to the left will stop and the endolymph will then move in the opposite direction, that is, away from the cupulas of the right ear and toward those of left.

When the left ear is stimulated the eyes will be made to move to the right. A stimulated canal generally pushes the eyes in the opposite direction. Thus there is constant push-pull relationship between the labyrinth and the eyes. The left ear canal pushes the eyes to the right, the right canal to the left. There is a balance between these two countervailing forces in our eyes. One canal, the lateral canal, lies practically horizontal, and is most important. The lateral canal is actually horizontal as you lay supine with your head up 30 degrees. Stimulate it in this position and the eyes will move to away in a straight line. You can do this by moving the head or putting cold or warm water in the ear. You can observe the resulting eye movements and quantitate them with an electronystagmograph. Tilt the head up or down from this specific 30 degree position and the horizontal movements will become rotatory due to the combination of vector forces determining the movement of the eyes.

Observe the direction of the arrows on the Cartesian axes represented in figure 6. That axis will be serve as the same axis of nystagmoid eye movement for the eyeball. Stimulate one semicircular canal and the eyeball will rotate directly through that semicircular canal's axis line. The problem in clinical assessment is that head positions which determine the orientation of these axes, constantly change and so do the positions of the eyes. However, it is possible to assess the semicircular canals individually by keeping the head at a certain tilt, for example 30 degrees forward from the supine position to isolate the lateral semicircular canal and at the same time, control for eye positions.

A second component of eye movement comes from the upper brain and this is corrective. The cerebral cortex, noting deviation of the eyes, will make a rapid corrective movement. Looking at nystagmus there is a slow component induced by vestibular forces, and a rapid component in the opposite direction courtesy of the cortex.

Disease processes usually reduce the influence of one or another canal or labyrinth. (Some diseases cause stimulation. In Ménière's disease the diseased ear is stimulated by the accumulation of endolymph fluid.) As mentioned, the major purpose of all of this is to maintain visual fixation ("eyes on the prize"). One outcome is that absent the influence of the upper cortical brain, the eyes will move much like a doll whose head is turned, a so-called "doll's head response". The eyes will appear to be fixated on a spot despite head movement and will move in a direction opposite to head movement. This is the vestibulo-ocular reflex.

The Utricle and saccule, in the meantime, are primarily concerned with maintenance of posture. Recall that these structures have calciferous otoliths, tiny rocks, inside the ear that are very heavy (specific gravity 2.5) helping to detect the direction of gravity and fall on hair cells in the maculae (sensitive regions) of these small organs. These organs perhaps are descended from ancient statocysts that also function with the aid of heavy particles in much lower animals. Hair cells are the sensitive end organs for the detection of gravity. Hairs (cilia) are deformed and signals sent to the brain. The utricle affects the level of the eyes. In certain patients whose surgery has damaged one utricle, the eyes are on two different levels. Semicircular canal impulses then go more rostrally (higher) to control eye movements and their nerve is mostly the superior vestibular division which goes to the more rostral vestibular nuclei in the brainstem (there are four on each side.) while the utricle and saccule send signals primarily via the inferior vestibular nerve division and to the spinal cord.

One influence of computer technology is how it has influenced our conception of physiology. Brain systems including balance mechanisms have come to be viewed as individual automatous but interacting modules, each computing some function and affecting neighboring processes to perfect sensory and motor function. In the brain, modules are typically bundles of neurons, either nuclei or anatomical structures. The cerebellum is a primary example of an automatous computational device that perfects movements. Cerebellar inputs are multiple simultaneous and complex deriving from such diverse sensory sources as proprioceptors, vestibular apparatus, visual and motor systems. The cerebellum connects through cerebellar nuclei to other structures primarily over white matter tracts, cables, within the central nervous system. The cerebellum has to process diverse data from many different systems all at once, in other words, in parallel. In most modern computers information flows through a single microprocessor sequentially though with many operations per second. The cerebellum and other neuronal systems process large numbers of inputs in parallel, and so there is a fundamental difference in style of processing. The brain is considered a massively parallel device, the computer as sequentially processing device. In recent years, computer scientists understanding this basic difference in data handling and having great admiration for the accomplishments of biological systems (biocomputation of gait and balance is great example), have sought to emulate processing in the brain utilizing parallel arrays of microprocessors in their instruments. We in biology on the other hand, have benefited from computer science by acquiring a modular view of brain processes.

The cerebellum receives information on posture and movement form the vestibular apparatus, and processes this in context with other data it receives from the spinal cord and visual system. Information from the utricle and saccule affect mostly the midline cerebellum (flocculus) and the cerebellum, in turn, uses this to compute instructions on balance and stance. This data is transmitted to the motor output areas in spinal cord via cerebello-spinal connections and there is a reciprocal feedback relation with the cerebellum via large spino-cerebellar tracts that relay information from stretch receptors in muscle. The vestibular apparatus also connects widely with higher brain centers affecting gaze mechanisms through the medial longitudinal fasciculus which also connects with the cord and brainstem to affect balance directly. Vestibulospinal tracts make a direct connection to the spinal cord. Motor systems also modulate the basic function of the vestibular apparatus itself, through up and down regulation of sensory cells, making these cells more or less excitable or sensitive to stimulation. Data from the vestibular and all other sensory systems help determine general arousal levels though the reticular system responsible for general arousal in higher animals. Sensory information finally reaches the conscious cortical levels after processing in the thalamus. The cortex also helps initiate movements directly and with the help of the cerebellum and basal ganglia which perfect the rather crude motor plans of the motor cortex. Hence we have autonomous but intensely interactive modules that comprise motor functions. Underlying all of this, as we have seen are crude hard-wired reflex mechanisms built right into the spinal cord.

We may fully expect an intense cross-fertilization of neurophysiology and computer science in the coming decades. Silicon (computer) devices and Carbon based (biological) modules will certainly interact more closely. Not only will communication between brain and computer be easier, but biological and silicon modules will likely mix. There is nothing to prevent Silicon devices being implanted in the brain and biological devices from being embedded in computers.

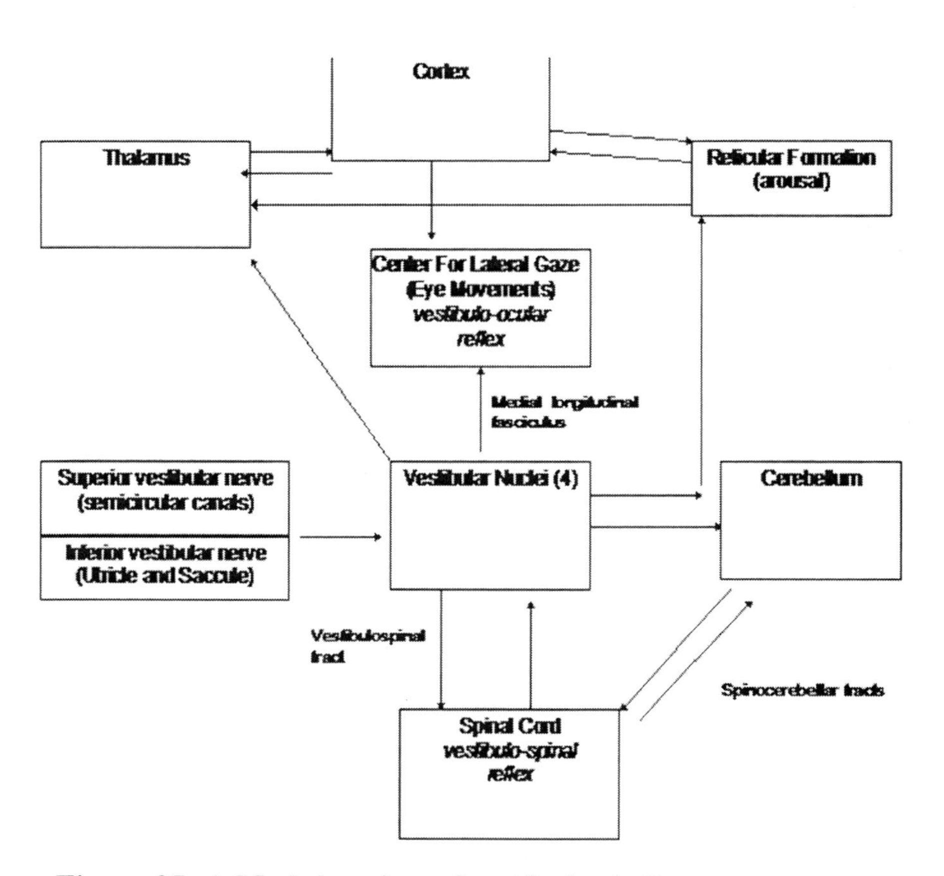


• **Figure 35: A Modular view of vestibular influence.**

Hallucinations are likely to occur if this system, either the visual system or the motor tracking system breaks down in some way. In the absence of sufficient input, the brain will still be trying to complete its image and hallucinations may occur.

When you look at how the cells in the eye and brain respond, you find specialization. Cells only respond to certain physical aspects of vision. You can always tell this by impaling a cell with an electrode and carefully analyzing what the cell responds to. For example in the retina when researchers examined ganglion cells with electrodes they found that the cells fired only under specific circumstances. I have already alluded to the receptive field of a ganglion cell. Some ganglion cells would only fire if an image appeared with a certain pattern, a light central area and a dark surround, or others would respond only in the opposite situation dark surrounded by light and so forth. The rods and cones respond to light with a reduction of the so-called dark currents. The rods and cones connect mostly with two types of cells. There are the *horizontal* cells that communicate with other rods and cones mostly. Horizontal cells partly process and influence other receptor cells. Many of them secrete an inhibitory chemical onto the membrane of rods and cones, GABA that in turn influences the currents. This is complicated but horizontal cells are excited by primary visual receptors then end up doing almost the same thing that light impulse does, reducing the dark current, perhaps ending up by heightening or tuning in the response of a stimulated rod or cone light receptor. Once a receptor is stimulated we are, more or less, further "tuned in" to it.

Rods and cones connect most directly to the next line of cells the *bipolar* cells. Bipolar cells also have no action potentials and instead have a graded or analog response. The bipolar cells connect directly with the *Ganglion* cells. The ganglion cells bring the message from the retina into the cranium and are the first digital or action potential, all or nothing type cells in visual processing. However these in turn are influenced by the *amacrine* cells.[90]

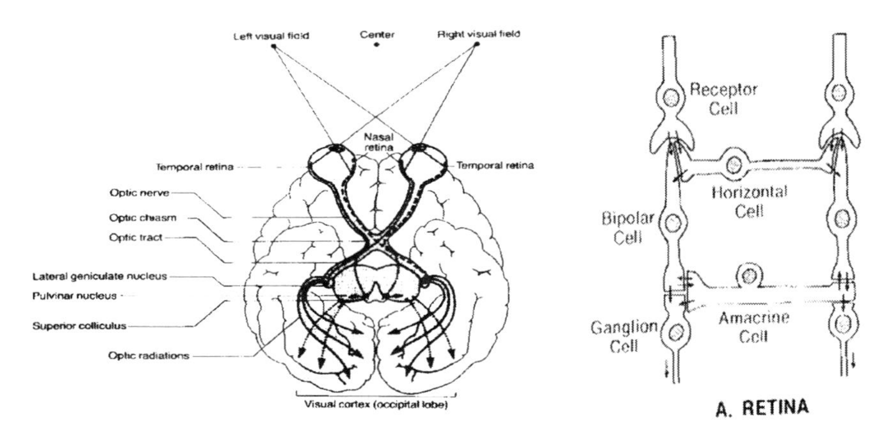


**Figure 36: The circuitry of the retina. Note that cells are more than relays. Signals are worked on and modulated before they are passed on. That is the function of the Amacrine and Horizontal cells. The more cells involved, the more a signal can be processed. Further Along in the brain, neurons don't just simply pass data on. Visual information is processed at every way station. Visual images reach the opposite side of the brain simply because the retina gets light from the opposite side of the visual field, Left is right, up, down[91].**

The ganglion cells extensions or axons gather into a cable that is the optic nerve. The optic nerve carries visual impulses to two closely related but separate parts of the brainstem for processing. In lower animals most of the fibers go to the midbrain the area of the superior colliculus, while in humans and higher forms, a separate area has developed from the same plate of nerve cells, the lateral geniculate nucleus of the thalamus*. The thalamus is a way station to the ultimate termination of visual impulses in the cerebral occipital cortex. As we have seen in higher animals there is increasing cephalization of visual input. For frogs, fish, and more primitive forms visual processing is largely retinal or at best in the dorsal midbrain but not in a cerebral cortex. In man, these lower midbrain connections are preserved and serve some purpose, but the majority of visual input goes to that lateral geniculate nucleus in the thalamus and eventually to the cerebral cortex where visual stimuli are more extensively processed, i.e. reach conscious awareness. Let's explore how this comes about.

## EYE-BRAIN CONNECTIONS:

While in humans and higher mammals, the optic nerve sends axons to the evolutionarily older midbrain tectum (roof), in lower animals this area is the ultimate terminus of visual processing. It is an area responsible for simple reflexive motor movements processed rapidly, mainly movements of the eyes themselves. As we've seen, animals respond quickly to movements perceived visually, for the frog, to flick his tongue out at a fly, or the wildebeest to withdraw his neck from a lunge lion's powerful jaw. Movements perceived by rods in the peripheral retina have to be foveated, brought into more precise inspection by the central retina. An interesting thing happens when you are out at night looking at the stars. You get a glimpse of a faint object in the periphery of your vision. Then as you reflexly focus on it, drawing the star into your central gaze for inspection, it's lost. Your more light sensitive rods in the retinal periphery perceive the dimly lit spot, but your cones in the central retina adapted for precise vision of well-lit objects, fail to respond to the dim image of a faint star.

In the midbrain as elsewhere in the central nervous system, cells are organized into groups or nuclei. These are anatomical aggregates such as the Superior Colliculus, meaning upper hill, which implies that there is a lower hill below or inferior colliculus which happens to be involved with auditory processing. The superior colliculus mediates some involuntary eye movements along with, a favorite for medical students in neuroanatomy, called the Edinger-Westphal nucleus. Edinger-Westphal helps to control pupils as when the pupil constricts to light and you may notice the pupil also constricts when you accommodate to a near or close object. In syphilis there is an Argyll-Robertson pupil that is small in size and does not constrict to light but will get smaller with accommodation. It is like a prostitute it accommodates but doesn't respond. The midbrain tectum is involved with these and other reflexive but necessary eye responses and it is where the connections are made with the nerves that bring these reflexes about, most importantly the third or oculomotor nerve.

What we are describing here is the difference between simple reflex vision mediated by the brainstem in lower animal forms and its embellishment, perhaps the full appreciation of as visual scene, true perception rather than reflex motor function, enjoyed by more advanced brains. The cortex contextualizes visual input and correlates vision with the other senses and experiences.

In some animals the visual pathway essentially ends at reflex visual responses but no conscious embellishment. Humans expect more from our vision. In lower animals the midbrain tectum is the major terminus of visual input but in men, there is more crosstalk with other sensory modalities especially touch because touch fibers also project to this area. In lower animals things are designed so their responses will be more automatic and rapid. In humans some of this projection of the optic nerve and midbrain areas is preserved performing various automatic adaptations of the eye, especially eye movements. A pinealoma an abnormal growth on the pineal gland can compress the superior colliculi of the midbrain (the pineal gland lies close to the midbrain collicui with cells of similar visual origin and light sensitive itself in lower animals). This creates difficulty looking up with your eyes over the horizontal plane, a "Parinaud's syndrome", the main sign of such a tumor. A person with Parinaud's syndrome would be unable to gaze upward enough to walk upstairs.

The pineal gland that lies close to the mounds in the midbrain, the colliculi, referred to above, secretes melatonin, a hormone much written about in recent years that helps regulate sleep wake cycles. Visual reflexes in part, also control melatonin secretion. It turns out that light impinging on the retina (as normally occurs during the day) blocks melatonin secretion and that darkness, conversely increases melatonin secretion. Melatonin may have many effects, but in part, it acts as a sedative, encouraging sleep. Visual control of the sleep wake cycle again is likely far more important in lower animals than it is in man whose life is tied to artificial light situations and other regulators or inhibitors of sleep.

The midbrain tectum connects to neurons from other brain areas that move the eye. Lower animals essentially have no visual cortex, but in man there are close connections in this area with cortical eye movement areas. In the left and right frontal lobes are separate frontal eye fields that when stimulated push both eyes to the opposite side in a rapid eye movement (saccade). In the occipital or visual back part of the brain can be found movement cells as well that move the eyes more slowly and help with visual following of a moving object. When you are sitting in a moving train watching telephone poles fly past you fixate on one pole and as you pass the next pole your rods catch sight of it and jump to that one as the first falls out of view. If someone is looking at you at the time he will see your visual fixation jump from one pole to the next. There is a slow movement of following as you fixate on the first pole while the train continues to move, followed by a jumping saccade in opposite direction as you catch up with the second one. This is repeated as you catch one and the next pole in your passing train, a slow visual following movement followed by a saccade in the opposite direction repeated over and over again. This is nystagmus and this specific kind is optokinetic nystagmus. Both the frontal and occipital areas connect in man with the midbrain tectum in two groups of cortico-tectal fibers because they connect the cortex and the tectum. Eye movements in man are under increased cortical control as is visual reception so you can appreciate that there is cephalization of eye movement as well as reception. In man and advanced animals everything is more cephalized.

As animals evolved they developed higher connections that start with a more advanced diencephalon or thalamus and end by developing the cerebral cortex where we localize conscious feeling and perception and operations between percepts, association and then action. But it starts with the thalamus. For vision, the vast majority of retinal ganglion fibers project to the lateral geniculate nucleus[#] . The retinal ganglion cells project topographically or retinotopically. The way they do so is rather complex. Cells from the left side of the visual world go to the right lateral geniculate nucleus. For the left eye retinal cells that pick up light from the person's left are in the nasal half of the retina and project to the right lateral geniculate. In other words nasal fibers cross. However, in the left eye, temporal fibers (recall that images are projected backward on the retina so temporal or outside fibers from the left eye receive vision from the right half of the visual field do not cross. They go to the left lateral geniculate nucleus. Each lateral geniculate thus receives impulses from the opposite visual field,[Ψ] the right lateral geniculate gets impulses from the left side from both the right and left eyes.

## THE BRAIN ORGANIZES VISUAL INPUT:

Inside each lateral geniculate nucleus you can see a layering of cells, a lamination, under the microscope. Fibers from each of the two eyes are kept separate and project to specific layers that are dedicated to one eye; nasal fibers from one eye and temporal fibers from the other are still separate within the nucleus. The cells then synapse with other cells inside the lateral geniculate and are still segregated retinotopically as they reach the primary visual cortex in the occipital lobe in the back of the brain. The left hemisphere receives impulses from the right visual field, the right occipital lobe, receives fibers only from the right lateral geniculate sensitive to goings on in the left visual field. Certain brain neurons are hard-wired to receive specific inputs. This point was dramatically brought home in the classic experiments done by Roger Sperry. He cut out the eyes of a frog, then reattached them through the optic nerves only upside down. Most of the axons grew back to their original neuronal connections. What happened is that the frog, now seeing a fly on its upper left would jump or flick its tongue in the wrong direction, as to its mirror image!! Most disturbing for the poor frog but illustrating that neuronal connections respond to some specific stimulus.

Here we come to another universal in brain organization. We see a small spot of light in an area of our visual field. That impinges on a tiny area of our retina that is connected to specific ganglion cells that project to specific (retinotopic) areas of the lateral geniculate nucleus. This *retinotopic* organization is preserved as still a second set of cells in the lateral geniculate project onto the visual occipital cerebral cortex. You are lightly brushed on a tiny area on the right index finger. It turns out that little impulse is faithfully projected through a nerve, up the spinal cord to specific cells then up to the cerebral cortex where the touch receives some conscious awareness, All throughout this complex projection of the touch the topographic representation (this time somatotopic not retinotopic) is preserved. If someone should stick you with a pin in the same place, an impulse would eventually arrive in almost the same location on the cerebral cortex. That is how you localize it and process it; that is how you know something happened in just this place on your finger. Impulses are faithfully projected in an organized manner to primary sensory cortices. What will happen after that, where the sensation ends up may be very different. A slight touch from someone you love is not the same getting jabbed with a pin at the same location. That will depend on further processing in secondary and tertiary cortical regions and also by some initial organization that separates sensory modalities. The same holds true for the temporal areas involved with hearing. These areas are tonotopic. Neurons project in highly specific ways. You can appreciate the homology between somatotopic organization for pain and touch over the body where the receptive field of a central neuron is a certain area over the body and tonotopic organization where a given neuron will only respond to a sound of a certain tone, and in vision, retinotopic organization where a neuron in that lateral geniculate or the occipital cortex will only fire when there is a visual stimulus on a certain region of the retina. It is interesting that with smell (olfaction), one sensation that does not project through the thalamus there is no known topographic projection. But then there is nothing specifically spacial about smell apart from graduated increase or decrease of an odor's intensity as you get farer or nearer to the odor-producing object.

In the cortex then the right hemisphere processes visual input from you left visual field. That is another way of saying that the right hemisphere processes visual information from the right half of both the left and right retinas. Central or foveal vision is projected to both cerebral hemispheres. There is extensive cross talking between hemispheres through the corpus callosum that interconnects them. But as a general rule this retinotopic organization is preserved in the whole visual pathway from the retina to the occipital lobe. Parallel groups of axons are gathered into tracts of white matter, nothing more than big cables. From the lateral geniculate to the calcarine cortex where visual inputs are initially processed, we have the geniculocalcarine tract. This is a white matter bundle.

We know that the geniculocalcarine tract carries retinotopic information and how the cortex receives it, mostly by observing the effect of lesions. If a person has a large stroke in the back of the right hemisphere of the brain, he will not be able to see to his left side, a left hemianopsia. You can also monitor electrically the response of neurons in the right occipital cortex. In doing so you will find that first of all, the cortex is arranged in columns approximately 6 neurons deep. The columns of cells continue to be retinotopic that is, it is composed of a group of cells that only respond to a visual stimulus from one area of the retina. Most of the cortical cells no longer respond only to a single eye's input, as do specific cells in the layers of the lateral geniculate. They are place coded. Not only that, many of them respond to a bar of light or a bar of darkness only in one specific orientation, that is only a certain number of degrees from the vertical. These cells of the primary visual cortex that respond to stimuli of a certain orientation are simple cells. Others respond only to various visual characteristics, for example, movement in a specific orientation or color and other features so that they are more specific with regard to movements and features and higher order of complexity therefore appropriately termed complex cells. [92].

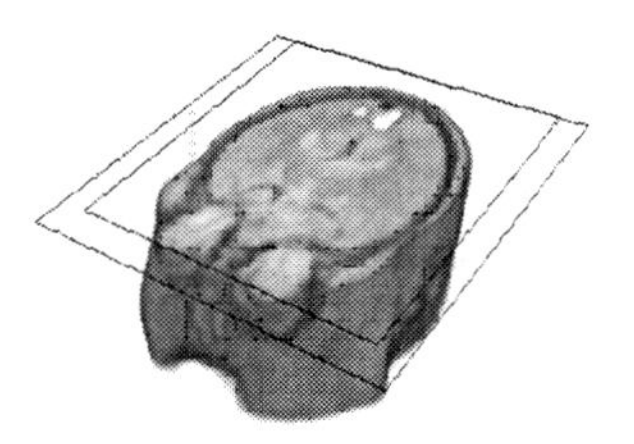

**Figure 37: Activation pattern in the calcarine cortex at the back of the brain as seen on a PET scan image.**

When you look at something there is a whole matrix of cells over the visual cortex, not only in the primary visual area that initially receives the input, but spread over a wide area of the cortex. Many of these cells respond to specific superficial characteristics of the stimulus. If you are looking at a nose of a person, the specific orientation of a vertical bar of light, color characteristics etc. And as you gaze over and explore the entire face, a huge group of specific cells is called into action, each responding to a fairly simple characteristic of the face, maybe as complex as the way the smile creases move as a person smiles. We can appreciate how the whole scheme of seeing comes together, a merging of visual input and motor output to the eye. We've talked about the limited field of view of precisely seeing cells, the cones in the fovea and the motor output to the eye allowing the eye to foveate over a whole visual scene. This information is ultimately conveyed to the primary visual area in the occipital lobe then to secondary and tertiary visual areas of the brain. The system is designed to respond to rather superficial properties of the visual stimulus. Experiments have shown that these characteristics are even more widely distributed over the cortex. Not only do they ramify to visual areas of the brain, but also all sensory inputs are distributed multimodally, mixed with auditory, tactile and even olfactory images in widely separated regions of the brain. By the time you have a picture of a friend, you have auditory associations with his voice and even emotional associations based on your arguments and good times etc. As the visual image interacts widely with an enormous storehouse of experiences and other sensory inputs the depth of the interaction is limited only by the complexity of cerebral circuitry and past experiences. No wonder that if our mind is set free to ramble we can come up with wide ranging ramblings and associations mirrored and perhaps caused by widely ramifying cortical representations of visual images.

## HIGHER PROCESSING:

I've discussed visual processing, starting from the initial effect on retinal rods and cones. From the point of view of neurons in the cerebral cortex, there are basically two broad groups of connections. The center of this activity we conveniently place at area 17 known as the primary visual cortex. From this vantagepoint the two groups of connections are corticopetal (those going toward) and *inter-cortical* those fibers connecting the primary visual area to other cortical areas. Corticopetal fibers we have already described. These come primarily from the lateral geniculate bodies to the calcarine visual cortex and comprise the geniculocalcarine tract. A lesion here, such as a stroke especially or a tumor will cause a restriction in visual fields because, as we have seen, fibers are segregated according to an area of the visual environment. A total wipeout of fibers in the left side of the brain will cause loss of vision to a subject's right side, a right hemianopsia. The lower fibers, going through the deep temporal lobe of the brain, carry visual impulses from the upper visual field so a lesion affecting just these fibers will cause a cut in vision in the right upper quadrant, a so-called "pie in the sky" defect called a quadrantanopsia. Neurologists, ophthalmologists and other doctors spend a lot of time evaluating these visual field cuts because they are very obvious and localizing for lesions inside the brain but they are not that interesting or subtle. The real challenge comes when you try to evaluate lesions affecting cortex-to-cortex connections.

Cells that do simple visual processing aptly termed simple cells, and more Complex ones that are also hard-wired to respond to certain visual characteristics, Complex cells. Then there are also hyper-complex cells. The Retina of man projects to the back of the brain or occipital lobe (Figure) where all of this processing takes place. There we find cells that are tuned (in other words respond by firing) to specific visual characteristics, for example there are cells that only respond to a light bar of a certain length surrounded by a dark area. Our visual system is set up to increase contrasts and what the retinal ganglion cells respond to ordinarily are light areas with dark surrounds and vice versa. Again all of these tiny images are somehow integrated with specific borders built in to somehow form an image. Actually we know little of how this image is finally put together and enters consciousness.

After an impulse arrives at the primary visual cortex, signals ramify to other cortical areas where they are further processed. Defects in these intercortical pathways cause more subtle problems. Many nerve fibers go from the calcarine cortex to the temporal lobe and are thus designated occipito-temporal. Part of the function of the temporal lobe is to provide cross talk between vision and other sensory modalities such as hearing and touch and it is here that visual imagery connects with language function. The language area of the brain is in the peri-Sylvian area of the left hemisphere in most individuals.

Curious disorders occur with lesions in cortex adjacent to the primary visual area or most importantly, in white matter connections between these regions. Visual scanning is integral to object recognition since the whole scene or image cannot be taken in a single instant. It can only be theorized that there is some kind of short-term memory evoked by scanning a scene to produce on the basis of small pieces of information, a total picture of the scanned object. This will further serve for recognition, say of a statue (as seen in the illustration) or a face for example.

Destruction of intercortical fibers leading from primary visual cortex to the parietal lobes on both sides of the brain will cause Balint's Syndrome. These are association fibers high in the back of the brain fairly close to the surface. What you lose Is the ability to take in a whole scene, and have trouble keeping track of or recognizing more than one or two objects in a scene. This is designated as simultanagnosia. The condition is also characterized by inability to throw one's eyes at a novel object in a scene, a sort of gaze stickiness. We talked about seeing something new especially a moving novel object out of the corner of the eye in the domain of the rods, and our ability to bring the eyes to this new object (foveation). Seeing an object, we also have to be able to respond to it, to grab at it, touch it, even lunge at it. We see only a small part of a scene at any given moment. Yet in specific situations, a sword fight, or running on an open road, we have the illusion that we have a picture of the whole scene. Image storage and integration in the cerebral cortex accomplish that. Thus that we can respond at any time with a reaching or motor movement to any part of the total scene. This is what is impaired in Balint's syndrome[Ψ] Compare this with the figure of object scanning. This may or may not be associated with a cut in one's visual field, depending on the extent or depth of a lesion, but is most often associated with an inferior quadrant field cut. The interesting thing is that a person may well not be able to recognize an object on account of an inability to form a whole picture of it not because of any language problem or trouble actually seeing it. He may well be able to recognize something in only a small part of his visual field but not an object that takes up a whole field.

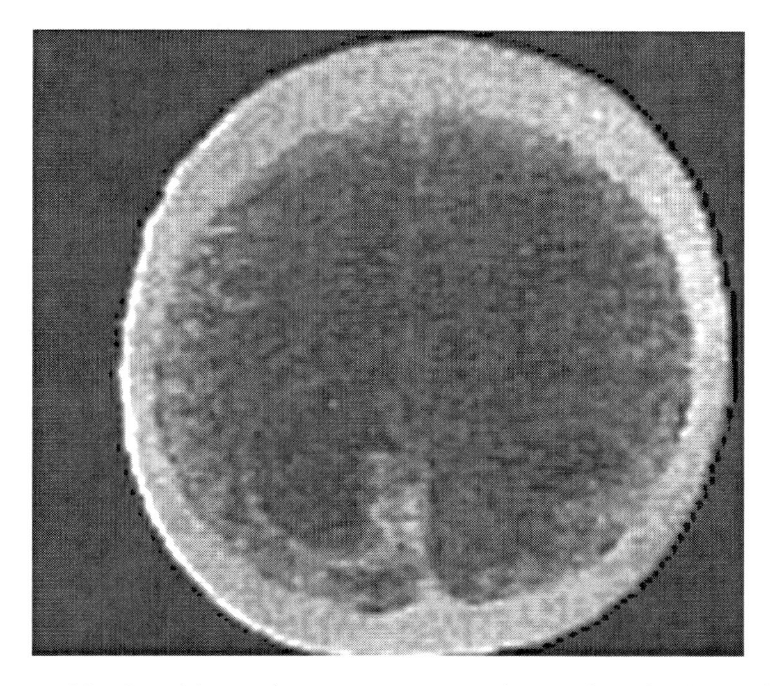

**Figure 38**: In this patient metastases destroying both parieto-occipital junctions high in the brain caused a "Balint's Syndrome"[93] which is a disorder of ocular searching.

Problems in connections between the primary calcarine receiving visual cortex and other areas of the brain that do more advanced and multimodal visual processing have been great neurological curiosities for a long while. Some patients see easy enough, but can't find their way around and unless they literally memorize their environment consciously feature by feature. They must know for example, that there are exactly six steps before coming to a corner of a wall or a chair before they are to make that right turn to find the way to their room or they will get lost. You and I don't find our way around like that. We depend on a referential inner map of our environment that is subliminal, barely noticed and automatically constructed, in other words a facility for recognition of the familiar. A person, unable to develop a total picture somehow, suffers from a sort of environmental agnosia or topor-graphagnosia., This may be called by different names but lesion-wise have something very similar to a Balint syndrome, they are unable to orient within a total picture of the environment and depend on the individual tree in a forest but have only a very poor concept of a forest.

Color recognition is simpler but is much written about. Some people can't recognize or name colors or of find named colors and so have a problem with color recognition. This problem happens with disconnection between the inferior occipito-temporal fibers that connect vision with the language area of the brain. You can lose your color reception just in one visual field, to the right typically, or on both sides. A right sided visual field cut is associated a lot of the time because the right visual field registers in the left hemisphere. Language function resides in the left brain in most of us, and so it not unusual for the connecting fibers that go from the language area to the left temporal lobe to be affected by a lesion in the back of the left hemisphere.

In prosopagnosia a person has a problem recognizing faces. One is unable to distinguish faces of friends and relatives (though he may be able to tell them apart by listening to their voice) or even to distinguish objects within a category. He may have to find his car in a parking lot by searching all license plates rather than by just recognizing its appearance. A farmer may not be able to tell on of his farm animals from another. With prosopagnosia there are usually lesions in both the left and right occipito-temporal intercortical pathways.

What are we seeing here? Disorders of visual recognition involve a higher order of abstraction than simple unprocessed vision. The visual image arrives at the primary occipital receiving station is processed here but then higher order processing and relationships are quickly handled by other areas of the brain, initially adjacent but ultimately far-removed from the calcarine cortex. We know this in many ways. You can follow a change in electrical activity evoked by simple and by complex visual stimuli over the surface of the brain. This is difficult because ordinarily such electrical potentials are small, only a small fraction of a microvolt in size and the electrodes used are far from the brain where such potentials arise, if they are ordinary electrodes placed over the surface of the skull. But such evoked responses can be observed when the skull is open during neurosurgery at which time you can place electrodes directly over the surface of the brain. This is rarely done except in epilepsy surgery where arrays of electrodes are placed in order to find the electrically active discharging focus of activity. Also during surgery, you can put a probe over the brain surface and see what sort of experience the electrical probe evokes stimulating a specific anatomical spot over the brain. This was originally done rather crudely many years ago by Wilder Penfield a Canadian neurosurgeon. The patient needs to be awake and this is rarely done anymore.

The PET (Positive Emission Tomography) scanner watches the utilization of glucose, in other words the metabolism or activity of brain regions. You can watch various brain regions metabolizing glucose as they become active using color-codes and use of various individual parts of the brain and charts the activity or glucose uptake over the brain's surface. As a person looks at or scans an object you can see what parts of the brain become activated and the sequence of anatomical activation. With that technique you see that even a simple visual object or scene, quickly activates a wide area of brain. Over the years the most important information has been culled by examining patients with certain known cerebral lesions to give a picture of localization of function and the understanding of connections within the brain. Neuroscientists have gradually evolved a picture of brain function that is *modular*. Certain brain regions perform a specific given function. There is primary visual, primary auditory, primary somateshtetic, olfactory, etc cortex and a language area of the brain. Each of these areas has to perform a certain function and also be connected to the others. Each are gray matter areas containing neurons. They communicate using white matter tracts or bundles of axons. When a problem is detected there may be an anatomical defect in neurons or connecting axons. For example, a person may be able to see and name letters, he may perform just perfectly on a Snellen eye chart with good visual acuity,

What are we seeing here? Disorders of visual recognition involve a higher order of abstraction than simple unprocessed vision. The visual image arrives at the primary occipital receiving station is processed here but then higher order processing and relationships are quickly handled by other areas of the brain, initially adjacent but ultimately far-removed from the calcarine cortex. We know this in many ways. You can follow a change in electrical activity evoked by simple and by complex visual stimuli over the surface of the brain. This is difficult because ordinarily such electrical potentials are small, only a small fraction of a microvolt in size and the electrodes used are far from the brain where such potentials arise, if they are ordinary electrodes placed over the surface of the skull. But such evoked responses can be observed when the skull is open during neurosurgery at which time you can place electrodes directly over the surface of the brain. This is rarely done except in epilepsy surgery where arrays of electrodes are placed in order to find the electrically active discharging focus of activity. Also during surgery, you can put a probe over the brain surface and see what sort of experience the electrical probe evokes stimulating a specific anatomical spot over the brain. This was originally done rather crudely many years ago by Wilder Penfield a Canadian neurosurgeon. The patient needs to be awake and this is rarely done anymore. T

The PET (Positive Emission Tomography) scanner watches the utilization of glucose, in other words the metabolism or activity of brain regions. You can watch various brain regions metabolizing glucose as they become active using color-codes and use of various individual parts of the brain and charts the activity or glucose uptake over the brain's surface. As a person looks at or scans an object you can see what parts of the brain become activated and the sequence of anatomical activation. With that technique you see that even a simple visual object or scene, quickly activates a wide area of brain. Over the years the most important information has been culled by examining patients with certain known cerebral lesions to give a picture of localization of function and the understanding of connections within the brain. Neuroscientists have gradually evolved a picture of brain function that is *modular*. Certain brain regions perform a specific given function. There is primary visual, primary auditory, primary somateshtetic, olfactory, etc cortex and a language area of the brain. Each of these areas has to perform a certain function and also be connected to the others. Each are gray matter areas containing neurons. They communicate using white matter tracts or bundles of axons. When a problem is detected there may be an anatomical defect in neurons or connecting axons. For example, a person may be able to see and name letters, he may perform just perfectly on a Snellen eye chart with good visual acuity, and he may be able to write well, but may lose the ability to read. This is alexia without agraphia inability to read with preserved ability to write. The problem is that the visual image is received and processed well enough by the internal visual apparatus, but the signals aren't sent well to the language area of the brain that lies around the Sylvian fissure in the left hemisphere. You could say that the visual module, that large area of vision processing in the back of the brain, is disconnected from the language handling module further forward in on the left and you would be right. So this modular picture of brain function has evolved.

Lot's of young children show up with specific language problems. Other cognitive functions may be intact or even superior yet they may not do well in school and have a learning disability. A lot of kids who do poorly in school have specific problems. If we can find what is wrong we may be able to get around the defect. Other kids may have more pervasive problems that are more global or markedly multifocal. These children are mentally retarded and as you may imagine trying to get past their difficulties and have them do well in school is impossible. A child with a defect in higher order visual processing, say the jelling of individual letters into words or sentences may have an interruption of certain connections in the arcuate fasciculi running between the calcarine cortex and connections to the left temporal lobe. He will present to his doctors just doing poorly in reading, math, spelling perhaps and all other subjects which is most of them that involve language function implicating visual to language connections. At first it may be difficult to diagnose such a situation. You might find that he speaks well and has a large vocabulary, yet can't comprehend sentences that he reads very reliably. In that case once you make this specific diagnosis you may find an alternate strategy for teaching him to read, perhaps involving more rote memorization of whole words and try to teach him through his ears rather then through his eyes, with audiotapes and sounds. As you can imagine, making such a diagnosis and coming up with a specific strategy is an art. You must be able to separate kids out who can be treated, namely those with more focal and specific problems, from the ones with a poorer prognosis and more global dysfunctions, also recognizing and using intact abilities where you find them. This modular view of the brain is operationally very useful.

Now the interesting thing as I point out in the chapter on memory, is that a memory of an experience, perhaps a visual image, resides in the brain. But should you try to expunge this memory by sticking specific little holes in various areas, say the image of your fifth birthday party or your mother's face, the actual memory of an event or image is cannot be found to reside in a specific place in the brain or on the cortex of the brain. Why? This is because any image in fact any percept ramifies widely over the cortex and ends up in multiple modules. Some primary aspects of angle and color and movement are in the simple visual processing areas of the brain, but then is the memory includes sounds there is some information in the auditory cortex and if there are multimodal, or linguistic or emotional evocations, those areas will become involved as well, so that any memory especially a reasonably complex memory trace calls up the efforts of widely dispersed groups of neurons. In the primate brain there are over 30 such areas involved with vision alone which communicate and interdigitate, each contributing more or less abstract properties to the visual image in an hierarchical manner. How do you recall a memory? Something stimulates, tickles maybe a single mode that brings back the memory. Maybe you taste a piece of cake that resembled that chocolate cake at your fifth birthday party and all of a sudden you will recall your friends and family singing the birthday song, or perhaps the tables and chairs the smells or even your emotions at the time.

Or, if you're a Vietnam Vet, someone will light a cigarette and blow smoke away and partly into your face, in just the same way your buddy did before his head was blown off suddenly and that will evoke the total emotion of the traumatic event that replays over and over again in your mind. This is the stuff of post-traumatic stress syndrome with multimodal emotional associations culled from widespread cortical and subcortical regions of the brain. Each of these sensory mnemonic particles that calls up impulses from widespread brain regions, is like a handle attached to a weight that is the entire mental picture. If you tug on it and it is connected strongly enough, you will be able to extract a whole scene on the basis of a single elemental memory engram. Other factors, may add additional weight blocking extraction of the entire memory such as lack of concentration and anxiety. The brain is a complex of individual interconnected (through white matter) gray matter modules.

## IMAGERY RAMIFIES IN THE BRAIN:

In visual imagery you call upon a characteristic of the image that is ordinarily more advanced rather than simple. You tickle or stimulate a visual module that is higher up on an abstract hierarchy when you recall a full image. Visual imagery or memory is thus a "top down affair" with modules in upper levels of abstraction stimulating lower level less abstract areas, the initial impetus coming from inside the brain. In some ways this is the obverse of simple perception which calls upon a "bottom up" paradigm involving first, an organ of perception and only later, abstract modules.[94]

A stimulus is thus seen to excite a wider and wider area of cerebral cortex. It is not just a matter of an image arriving intact on a surface as on a film in a camera. Various cortical areas further process this image and interpret it. We have not even gotten to cross modal representations and comparisons yet which involve even wider areas of cerebral cortex. For every image, if is complex enough, there is a whole pattern of cortical activity that will eventually involve a wide network of cells and it is this precise pattern of cell excitation that constitutes a total memory. Emotions get evoked that recruit limbic areas, the frontal lobes, and deeper evolutionarily older brain regions. Vision is not unique but is rather archtypical since it is a sensory modality that we happen to have a lot of information about. Other sensory modalities are vastly similar in their topical organization, use of primary secondary and tertiary cortices involving higher degrees of abstraction and intermodal connection and arrival at language regions of brain also contact with regions of brain responsible for feeling tone and emotion. Once a plan works, the designer uses it over and over again. Its all hooked to hooked together and then to the great descriptor which is the language module of the brain. Each of the specializing regions of the brain may be more or less isolated from the others if the connecting fibers are somehow disconnected by affecting white matter.

**Figure 39: "Alice in Wonderland Syndrome. The distortion of visual image is associated with migraine.**

A funny thing happens when some of the primary visual cortex is destroyed by a stroke one side. The higher level visual processing regions of the brain receive information from both primary areas but let's say the right calcarine cortex may have been destroyed. The person will have a visual defect in the left visual field; he will not be able to see to his left and thus have a left hemianopsia. If an object appears to his right he may see it fine. But then this secondary visual cortex receives and processes this image which may move into the blind left visual as far as the person in concerned seems to persist after the object is taken away or there may be an illusion of multiple objects even though just a single one appeared. Stimulation of secondary visual regions while the primary region is knocked out is what is felt to be the problem in palinopsia the abnormal persistence of visual images that should have disappeared from vision. An even more remarkable phenomenon is that a person can be unaware that he is blind. Let's say blood clot affects both the left and right calcarine cortices. Some secondary areas may come out unscathed as what seems to occur in migraine where blood vessel spasm seems to affect the blood supply to certain regions of the back of the brain. Then the person will hallucinate and see objects of various shapes including numerous visual distortions. Maybe there will be flashes of light, or wavy or ziz-zag wall like patterns, "fortification spectra" of bright colors are variously described that are migrainous visual phenomena.

Louis Carroll who suffered from migraine, eloquently described many of these visual phenomena which probably occur when higher order visual processing areas secondary and tertiary visual cortical regions are deprived of input from primary visual areas. Some migraneurs, typically children, experience marked distortions that they can't easily explain, close objects appearing far away or vice versa macropsia micropsia etc so that this phenomenon has been called and Alice in Wonderland Syndrome.

Some patients are stone blind but the language module that communicates with the outside world is disconnected with the visual area so they may maintain they are not blind, even though they are perfectly incapable of describing an object in their environment, the famous Anton syndrome. They are blind but they don't "know" it or at least as far as the language area of the brain is concerned, they can see. Something of an opposite situation holds in the phenomenon of "blindsight". The interesting thing is that someone may functionally see but yet be unaware that he sees. In a sense the subject is unaware of something right, as the converse of not knowing that something is wrong. Here a person will report that he is blind but when he's tested it is found he can respond to certain characteristics of a visual stimulus yet have no awareness that he's actually seen it. This is not mysterious in light of descriptions given above regarding vision. Some of the connections are intact but the language sector of brain may be not be receiving visual information that can be described. So depending on what connectors are disconnected, you can easily be fooled into believing that you can see when you can't, that you can't see at all when in fact you are still able to respond to some visual aspects of an object, and you can distort and even hallucinate[ϕ] .

In the same vein it has even been found that some subjects whose eyes are so blind that they can't distinguish darkness and light nevertheless can tell the difference between night and day[95]. Not that they can do this verbally. The retina also connects with the hypothalamus which lies just above the optic chiasm in the brain and is in charge of lot of basic functions such as appetite, thirst, temperature control, endocrine glands and sleep wake cycles. The hypothalamus through the suprachiasmatic nucleus that connects with the eye, is tied with the pineal gland which makes Melatonin, a pineal gland hormone mostly secreted in the dark. In blind persons whose melatonin secretion does not cycle properly, there may be a sleep wake disorder. Others where melatonin secretion is intact can sleep normally. Somehow these messages can reach the hypothalamus even through blind eyes. All of us have intrinsic sleep wake cycles that really are part of an internal bodily rhythm. In experimental subjects who are isolated these natural intrinsic days or circadian rhythms are slightly longer than our 24 hour day. But in a normal person exposed to ordinary cues such as light and darkness, this rhythm is entrained to be approximately 24 hours which is why some of us will arouse at exactly a certain time without an alarm clock.

Despite what common sense tells us, that a person really ought to know whether he can see or not we know of blind persons who think that they can see, and all kinds of visual distortions and partially sighted persons who report that they can not see, and that function to tell the brain about light or darkness for which the functionally blind person has no awareness. You just can't trust a person when he tells you he can or cannot see even if he thinks he is telling you the truth. Some of the very same processes must occur with all other sensory modalities, audition, smell, touch but are not as well described.

A sensory stimulus arriving at the brain will eventually spread to other cortical regions. What if it stimulates an area again and again like tying your shoelace or driving the same route to work everyday? Then these cortical connections and the whole pattern of stimulation will be strengthened. One mechanism for this might be so-called long term potentiatation (see "Beginnings") but there are likely to be many others. As you use a connection over and over it strengthens. Or there may be an extreme stimulus say one that evokes high emotions or causes pain. In that case the pattern of connections might be strong right from the beginning.

All of this becomes especially applicable to the modern understanding of pain. When a person is injured there is a whole pattern of reaction in the brain just as there is when he is scanning a visual scene. What if he continues to have pain for a while and this pain is extreme? These connections and a whole pattern of connections including his emotional reactions to pain will be frequently evoked and will strengthen. What if he has a second or third injury? These patterns will be strengthened again and his reactions to pain will become extreme. Perhaps he has strained his neck in a whiplash incident and the pain is bad for a while until the stretch injury heals. Then he is injured again in a second collision. The pain and its response is well-learned inside the brain, connections and patterns are strong and much easier to recall. This time the reaction is severe and the person is miserable and disabled with pain. This is a syndrome that I call *exuberant pain* and doctors see this frequently though it is a concept they fail to understand. Repeated headaches can beat a path of patterns of response throughout the brain, repeated injuries can make former pains and their response easier to recall and the response will be vigorous and extreme, worse than it has a right to be given the limited amount of actual injury. The pain response is anamnestic (see "Beginnings"). This is one of the most difficult situations to manage in medicine. There is little to be done in the periphery where a relatively minor injury has already healed. Even if you could cut off the offending limb (many patients literally beg you to do just that) the strength of association will be so extreme that the person will think the limb to be still there (the phantom-limb phenomenon). The only thing you might be able to do is to strengthen pleasant non-painful associations with manipulations of that limb - a difficult task and not immediately rewarding. What about techniques used by a lot of physicians, further surgeries injections and focusing on the painful area. This will make the pain worse in the long run by further strengthening these associations. Moreover in many patients pain spreads to involve other areas or even spreads out to involve the whole body. The more the doctors intervene especially if they do so invasively the worse the pain gets. If you leave them alone and don't intervene much over a long space of time, associations may weaken in the same way that you forget what you have learned last year or two years ago. Am I saying that a pain response is learned? Absolutely!

## THE BRAIN HARBORS AN IMAGE OF THE WORLD:

One of the most obvious changes in learning is an alteration of the representation of reality inside the brain. One true to life situations that brings this to the fore is the phenomenon of phantom limb pain. Many amputees continue to feel excruciating pain in a limb years after it has been amputated. This is even truer in cases where the limb was painful or painful and immobilized at the time of the amputation. Even in the situation where no pain is felt in an amputation, various sensations, touch and movement may be experienced in the phantom limb.

Every part of the body has a specific area of representation over the cortex. This is true both for motor and sensory parts of the brain. Motor functions are taken care of mostly by the frontal lobe, anterior to the central sulcus of the brain, sensory function is posterior in the parietal lobe which is behind the central sulcus. Information from the hand goes to one part of the parietal cortex, sensory data from the face goes to another part etc. if you take an electric probe during surgery as Penfield had done, sensations in specific areas will be felt by the awake subject who is undergoing the surgery, just as if a stimulus had occurred on the body part represented at the probe. In this way the cortical representation of peripheral body sensations is mapped. If the area that represents the hand is somehow destroyed, say for example by a stroke, then there will be numbness, the lack of sensation perceived in that body part. This is called the sensory homunculus (little man) of the brain. It is just as if a distorted little elf is draped over the cerebral cortex, except as you will notice from the picture, the head is inverted and certain much more strategic body parts, especially the face, hands and lips, are over-represented while others such as the trunk and buttock, warrant less cortical surface and devotion of nerve cells.

When you cut off a limb, say an arm and hand, the cortical area that will of course, stop receiving sensory data from that limb. The cortical brain region stays intact, stands ready to receive data, but receives nothing. It may go on as usual as if it the limb was still present, and confabulate, make up information, as time goes on. One of the easiest things to do probably is to continue to assume things as just as they were before the limb was amputated, before it stopped getting data from the limb. That would explain why so many patients who had been in pain, perhaps unable to move the limb before their amputation, still felt pain, sometimes years after the limb had been removed. The responsible area of cortex stood ready to get data represented somatotopically over its surface. The area of brain after amputation of the limb ceased to have afferent sensory input, was *de-afferented.* So as we see as de-afferented group of cortical cells may simply make up the missed stimuli, in other words it may confabulate.

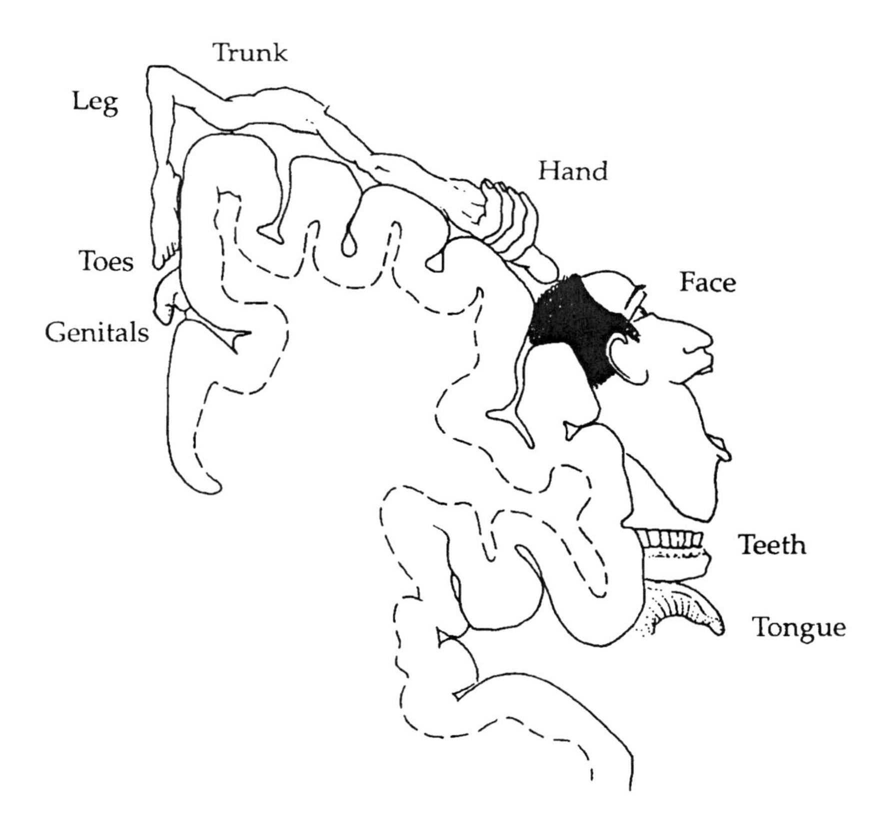


**Figure 40: Rendering of Sensory homunculus.**

After an amputation other interesting phenomena occur. A series of elegant experiments on amputees by V.S. Ramachandran, showed that this classical pattern as is taught in all basic neuroscience courses could be changed. The area of cortex devoted to the arm and hand amputated is no longer used. Adjacent areas of representation, the face, lips and shoulder start to invade the "unused" portion of cortex! It is as if actively feeling areas of the body compete, according to the level of activity for central cortical representation. Over a variable time period, the area devoted to the amputated arm functions both as the sensory area for the arm and the facial areas that invaded it. In Ramachandran's experiments, a touch over specific parts of an amputee's face will also be felt on a specific portion of the amputee's phantom limb. Touch one part of the cheek and he will feel something in his forefinger. That is because the same cortical region is now taking care of both body regions. Learning has taken place in the nervous system.

Now, as mentioned, many amputees continue to feel severe phantom pain after their amputation. Some who were paralyzed an had a limb locked in a specific position, may continue to experience this long after the limb is gone. Using a simple box and mirror, Ramachandran trained a subject to alter uncomfortable sensations in his phantom limb, alter his percept of this body, simply by using the intact arm and having the subject "see" movement in the perceived amputated limb. For some persons this could conceivably bring great relief from their painful sensation[96].

The cortical representation of sensory perception is subject to change even in adults. This has been shown elegantly by Ramachandran and others and also undoubtedly occurs in the setting of brain injury and recovery. Central processing of peripheral painful stimuli must also be altered over time by patterned experience. Painful events such as peripheral injury alter painful responses. Clinically it is common to see patients with injuries whose pain not only magnifies over time long after an injury or after suffering severe migraine headaches, but begins to affect adjacent body areas as pain persists. Chronic pain tends to spread over the body and to become less specifically characterized. The descriptions of pain given by a chronic pain sufferer tend to be less specific or if descriptors are specific they tend to describe multiple types of pain, and to be more generalized.

The classic example is reflex sympathetic dystrophy (RSD). In this mysterious disorder, there is an injury or injuries to a body part. Over time the hapless subject starts to feel what is described as a burning or aching pain. Untreated this problem persists. He immobilizes the painful limb, possibly in an attempt to splint a painful part. The pain is persistent but slowly spreads to affect adjacent areas of the body. As time goes on the pain becomes more severe and less localized. Some physicians feel that this occurs in the periphery and postulate abnormal activity in the sympathetic nervous system because it is very often the case that pain diminishes when sympathetic nerves are blocked. Also there are abnormalities in the appearance of the limb, sweating and hair growth, so-called "trophic" changes that may have to do with function of the sympathetic nervous system, hence the name reflex sympathetic dystrophy (dystrophy a type of atrophy or shrinkage, change of appearance of the affected limb which often occurs). However, we do not know what actually happens in RSD, nor do we have a good explanation for why pain tends to spread to involve a larger part of the limb and to become poorly localized. It is just as likely, perhaps more likely that such change occurs in the brain, perhaps induced by changes in the homunculus map that occur as a result of learning. The central cortical representation of peripheral stimuli over the cortex may well be competitive. Those areas of the body surface that speak the loudest, i.e. painful limbs may well take over adjacent areas of cerebral cortex, invade adjacent regions on the homunculus causing a spread or metastasis and a worsening of pain, exuberant pain.

This may bear on how such persons should be treated. One needs to mobilize the affected limb, to break the association between pain, to intervene early as the pain response is being set up and the central response is possibly altered. Invasive techniques ought to be eschewed in favor of simple mobilization and early aggressive therapies aimed at breaking the pain cycle. More importantly it points to the critical role of learning in the pain response. Something learned can ordinarily be unlearned. Conditioning is more than likely a part of the pathogenesis of exuberant pain as well as part of its cure or solution.

What is the converse of repetitive or strong sensory input is lack of input. I've mentioned what takes place in two visual circuits deprived of input. In one, language part of the brain gets no visual input. The patient, instead of being aware that he is blind when visual cortex is destroyed may instead report verbally that he can see. He does not appear to know he cannot see. An analogous situation occurs with a lesion in the right parietal lobe that receives sensory input from the left side of the body. A lesion there will cause a disconnection of the primary sensory cortex from the language module on the left. The person will report that there is no problem with sensation on the left part of the body. This is anosognosia[Ψ] , lack of awareness of a disease or deficit. The language area, disconnected from the sensory cortex, does not report out a deficit, at least it doesn't learn of a problem through a direct brain connection which is interrupted in this instance. This is a transient problem and it is likely the person learns about his problem through other means perhaps observing with his eyes that he does not feel when something touches his left body. In the meantime he will not complain that he is blind when the occipital lobe is destroyed or that he cannot feel on the left side of his body. Not only that, in order to make sense of his predicament, thinking that he can see or feel, he will confabulate. He will make up an answer to a visual or tactile question. Confabulation, neurological faking, seems to require anosognosia, unwareness of a deficit. Patients with Korsakov's syndrome have severe amnesia but they don't know there is a problem, that they are unable to register memories. They also confabulate, make up a response if questioned. Sometimes their made up responses can get very detailed. One lie can lead to another. The end organ, deprived of its nerve supply is denervated. But the end-organ here the language module or language area of the brain on the borders of left Sylvian fissure, may respond, sometimes quite robustly, sometimes overly robustly as if to "make up" for its lack of input.

Weird things happen, as mentioned, when secondary higher order visual cortex is deprived of input from primary visual cortex in other words if something temporarily or permanently interrupts input from the calcarine or primary visual cortex. Again, the secondary areas seem to react robustly and a person may see flashes of stars or bright lights or fortification spectra or rarely, a persistence of imagery, as mentioned above. This is what happens in migraine, which causes arteries to spasm in the occipital or visual area of the brain, and then we have fireworks that are, in effect a visual confabulation, also often an accompaniment of blindness in some visual area or black or neutral spots called scotomata.

## CUTTING INPUT:

In tardive dyskinesia a drug such as Thorazine used to treat psychosis may be needed for many years. This effectively blocks the effect of Dopamine an important neurotransmitter. A lack of Dopamine or dopamine blockade causes a person to look as if they have Parkinson's disease. But if Dopamine is blocked for a long time, neurons that usually receive Dopamine start to respond to abnormally small amounts of the stuff; they crave it and a very little bit will have a dramatic effect. Abnormal twisting and dancing movements result which is the what happens if neurons secrete an excess of Dopamine. The neurons that respond to it are hypersensitive; too much Dopamine and oversensitivity to it have the same effect, perpetual uncontrolled movement about the mouth and in the arms and legs, choreiform dancing movement or tardive dyskinesia. A basic principal in the nervous system is that a neuron craves and overresponds to input that it formally lacked. This is denervation hypersensitivity. If a motor nerve is interrupted the nerve gets little of the transmitter acetylcholine normally secreted by the nerve that tells the muscle to contract, A muscle with a good nerve supply is happy and silent when not asked to move. But cut its nerve supply and after days to weeks of being deprived, it will crave acetylcholine and start to respond to even the tiny amounts that it is exposed to in the blood. It will start to wiggle around under the skin or fasciculate which can be felt and seen under the skin or fibrillate which has to be looked for electrically. Here is denervation hypersensitivity in a different form. An end organ that lacks nerve input, craves it and begins to respond abnormally to the chemical secreted by the interrupted nerve.

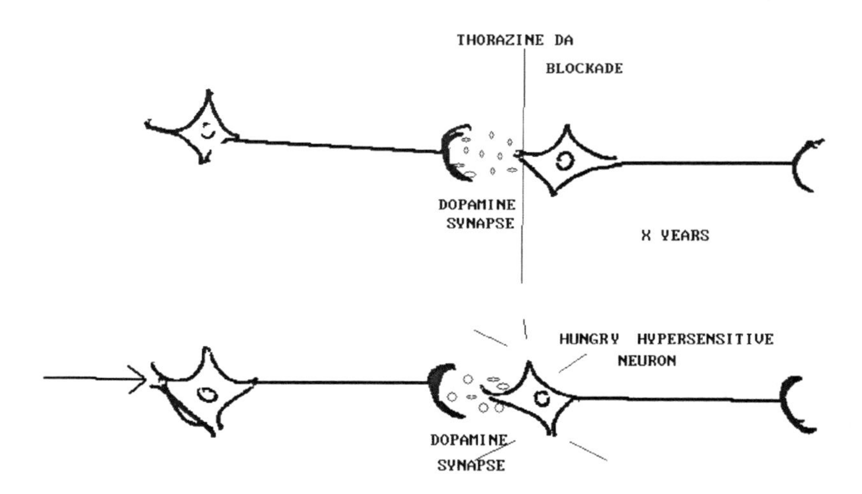


**Figure 41: Denervation hypersensitivity: Prolonged Dopamine blockade makes the post-synaptic neuron hypersensitive to any Dopamine in the synapse. The same holds for a neuromuscular junction. If the neuron is damaged the muscle cell becomes oversensitive.**

A neuron lacking input craves it. This is true on the macroscopically on the level of the intact organism. The basic neural mechanisms are reflected grossly. A person who gets no sensory stimulation craves it, so much that he starts to invent his own. That is what happens to your imagination in the dark or when you're sleeping. Lacking external stimuli, it simply becomes hyperactive. It's the stuff dreams and hallucinations are made of. Also as we have seen it occurs when there is impaired sensory input. It's why we see witches, alien flying machines and have religious experiences at night. We go to see a movie in the dark, not only because we can see it better, that there is nothing else to distract us, but our imaginations are far more active in the dark and our disbelief is suspended when we lack the light of day. A universal in sensory deprivation experiments is that imagination becomes much more active, in isolation and with prisoners of war. A lot of people testify to the fact that either they keep their minds active in a discipline that connects with physical reality, or they literally go crazy, by which is meant that they sink into the world of fantasy and hallucination. Consequently the most successful prisoners of war are the ones who incorporate reality-testing regimes into a sort of daily routine. If deprived of sensation many persons will begin to hallucinate, which is just like the confabulation described above.

How does imagination relate to sensory deprivation? The celebrated case of Helen Keller is instructive. Books on Helen Keller's life are mostly written for children because she bested extreme adversity. She was born hearing and seeing, a highly intelligent child who acquired some language function by the age of about 19 months when she suffered from a meningoencephalitis which left her blind and deaf. In one fell swoop she lost her two most important sensory modalities and effectively the ability to communicate since she could no longer get feedback from her hearing. Think for a moment what it is like to exist in a silent black world, the extreme frustration involved in being imprisoned in such a body. As a child Helen was unruly and violent at times. Yet, while young she'd also learned to sign and had enough memory from when she could see, to imitate certain things, her father reading his paper in his favorite chair with spectacles on, her aunt's bonnet strings tied under her chin, and even locked her mother in a pantry and folded clothes. She was described as a loving and lovable child, clearly mentally alive. One wonders if her native intelligence would ever have had a chance to develop under other circumstances, had she been born deaf and blind or into poorer surroundings. Her parents knew little of how to reach her but sought out and found at when Helen was 7 years old, Anne Sullivan, herself nearly blind but head of her class at The Perkins School for the Blind in Boston. There had been at least on predecessor to Helen at the School one Laura Bridgeman who was stricken with scarlet fever at an early age and was deaf, blind, and mute with an impaired sense of smell. With only a sense of touch that could be relied upon, Ms. Bridgeman proved educable. There's a picture of her threading a needle with her tongue as Helen later learnt to do. The rest of the story is common knowledge and remembered somewhat idealistically in THE MIRACLE WORKER. Helen Keller had lost her two most important sensory modalities and with it most of her ability to communicate. Her most precious sensory asset was touch that would become her bridge to the outside world. Through a system of tactile communication she was meticulously taught to express herself, read and write with others' help, and write she did, very eloquently.

Helen Keller was nearly *deafferented*. She very nearly lost sensory (afferent) contact with her environment[Ψ] , but is not quite comparable because in such situations many persons theorize at least that the unaffected sensations, touch, smell, awareness of one's own body become even more acute. One might think that some persons who are not of a strong constitution perhaps older folks, might begin to hallucinate or at least retreat into their own world. Helen Keller became an eloquent spokesperson for the disabled, but more than that she could appreciate the tenuous relationship we have with our physical environment. Her religious faith is interesting in this regard. Her religion like that of the father of William and Henry James was developed out of the voluminous writings of one

Emanuel Swedenborg who wrote of material sensations in face of an inner intuitive faith and a gentler, less deterministic, brand of Christianity. The conflict between sensation of the material world and faith is incorporated into the following in which she comments that an inner faculty:

"that brings distant objects within the cognizance of the blind so that even the stars seem to be at our very door. This sense relates me to the spiritual world. It surveys the limited experience I gain from an imperfect touch world, and presents it to my mind for spiritualization. This sense reveals the Divine to the human in me, it forms a bond between the earth and the great beyond, between now and eternity, between God and man. It is speculative, intuitive, reminiscent. There is not only an objective physical world, but also an objective spiritual world."[97]

With real world sensation impaired you are less bound to matter. The process is liberating in a sense. As their connection with physical reality becomes more tenuous, imagination takes hold. It's the same when we close our eyes to become less bound to the material world and entering a fabricated alternate existence, the mental world of dreams and freedom of expression. Many artists and writers maintain that they do their best work in seclusion. They are more inventive when shielded from reality. Edgar Degas was blind and isolated in his later years. We all know of the example of Beethoven, practically deaf in his later life. Music is composed abstractly in the pure mental sphere analogous to pure mathematical thought. The relationship of music and math as pure isolated systems that may be disconnected from experience has been frequently noted. Edward Rothstein in his recent book Emblems of the Mind makes this point:

"These moments of illumination -- finding relationships and hearing links where none existed before--are known to all who have been students of mathematics and music, but it would be hard to say just what is learned or how. That illumination also seems to have little relationship to life experience. In both disciplines the extent of one's understanding seems strongly determined by 'gifts'--gifts of temperament ad insight, ability to coordinate concepts (and in music, to coordinate finger and breath with concepts). The "prodigy" is a figure almost native to music, mathematics, and other activities like chess--dependent less upon experience with the world than with insight into a seemingly closed, abstract universe. For example, Felix Mendelssohn had, u the age of seventeen, composed a dozen symphonies and the famous *Midsummer Night's Dream* overture. Mozart's abilities at an even earlier age caused him to be paraded around Europe by his ambitious father. Despite early abilities, of course, musicians may require experience to mature into great artists; but mathematicians do not require even that. The mathematician Carl Friedreich Gauss corrected his father's calculations before he was three years old and jested that he knew arithmetic before he could speak. Mathematics is a "young man's game" as the mathematician G.H. Hardy put it (and few enough women have gone into it)). Hardy pointed out that Galois died at twenty-one, Abel at twenty-seven, Ramanujan at thirty-three, Riemann at forty--all having achieved near immortal Stature in the history of mathematics." [98]

## ISOLATION:

Music, mathematics and chess are in their own ways pure isolated systems where real life practical experience makes very little impact. These systems of thought are pristine and isolated from the real world hence embracable by a child's mind. The brain of a prodigy likely does have anatomic connections, synapses that prepare it, yield aptitudes for certain specific endeavors. No where is this more apparent than in fields of thought, music, math, chess, that do not require much contact with the real world, These are pure systems unto themselves. It is as if the brain is structurally the link between experience and invention is broken. The sphere of pure thought is sometimes better unhinged from ordinary material reality. In his book SEARCHING FOR BOBBY FISCHER Fred Waitzkin which is really about his own son, a child chess prodigy gives rare insight into how a young child with a natural aptitude is drawn magnetically into a field of endeavor almost as a key into a lock or a hand in a glove. In chess and in mathematics it is not unusual at all for a child to manifest his talent, spontaneously outstripping adults. Child prodigies typically are drawn to their field of genius despite any attempts on the part of their adult parents to discourage them. The father was also drawn to chess at an early age but his son was beating him regularly before the age of seven[99]. Sensory experience is sometimes necessary for invention but sensory deprivation may also be an asset as it was at times for Beethoven. Skilled composers don't usually use auditory feedback that ordinary persons would depend on if they were to set down music. To those with such musical skill an inner voice gains precedence. Beethoven, at the height of his powers became deaf and withdrawn, retreating into an inner world of intense abstraction so that some of his greatest works were produced in the torment of his later life, the last quartets and his Ninth Symphony.[+] Of Beethoven's later years David Ewen writes:

"When the spirit seized him, and he had to put down on paper the visions that tortured him, Beethoven was seized by what he once described as "raptus." This happened to him when he wrote Missa Solemnis. "In the living room, behind a locked door, " described his friend Schindler, "we heard the master singing, howling, stamping. After we had been listening a long time to this almost awful scene, and were about to go away, the door opened and Beethoven stood before us with distorted features, calculated to excite fears...Never, it may be said, did so great an art work see its creation under more adverse circumstances."

Here we see the other side of exuberant pain in which intense and prolonged input beats a learning pathway in the brain. The opposite, also a potent force physiologically, is lack of input, denervation hypersensitivity, which increases activity in susceptible neurons. The Associative areas of the cortex react to a lack of sensory input. Some deaf persons are observed to crave auditory input, so much that they create it. In the case of Beethoven, his latter compositions are the culmination of his art in the third or greatest period of composition. But inadequacies in other areas of his life, namely his lack of a stable relationship with a woman and other human relationships, clearly drove his art. Artists such as Dostoyevsky and Solzhenitsen and even people imprisoned over long periods often find isolation and sensory deprivation, the lack of distractions and dependence on one's own imagination, advantageous for creation. Terry Waite the emissary of the Archbishop of Canterbury held for 1763 days in Lebanon, passed the time in captivity much of it in solitary confinement writing his autobiography which perhaps might never have been written under more pleasant circumstances[100].

Denervation hypersensitivity is a neurophysiologic mechanism for such creativity in isolation from the mundane or real world. Which can actually interfere with the creative abilities. At times productivity increases in direct proportion of to disconnection from the reality, when associative areas of the brain are allowed to function without interference from sensory input. The deepest areas of human speculation are advantaged by the disconnection from reality. How closely is creativity tied to perception? This is how Einstein puts it quite eloquently:

"There exists a passion for comprehension, just a there exists a passion for music. That passion is rather common in children, but gets lost in most people later on. Without this passion, there would be neither mathematics nor natural science. Time and again the passion for understanding has led to the illusion that man is able to comprehend the objective world rationally, by pure thought, without any empirical foundations - in short , by metaphysics. I believe that every true theorist is a kind of tamed metaphysicist, no matter how pure a "positivist" he may fancy himself. The metaphysicist believes that the logically simple is also the real. The tamed metaphysicist believes that not all that is logically simple is embodied in experienced reality, but that the totality of all sensory experience can be "comprehended" on the basis of a conceptual system built on premises of great simplicity. The skeptic will say that this is a "miracle creed." Admittedly so, but it is a miracle creed which has been borne out to an amazing extent by the development of science."[101]

This "passion" which Einstein rightly mentions is more obvious in youth, is natural and has to do with brain anatomy and physiology (for "passion" read "nature"). In certain remote regions of human endeavor Pure logic and reason are very likely built in to the structure of the associative areas of the brain. One way of rephrasing the above is that associative regions of cortex may continue to function connected to or disconnected from the sensory areas and that at times associative processes work to advantage in isolation from sensory ("empirical" or "positivist" ) regions.

Lastly, here's an observation I have made as a non-musician. Beethoven was obsessed I think, taken over by relatively simple themes and melodies which would consume him and this one can see in his music. These it would appear, went round and round in his head, creating an explosion of emotion. Everyone's familiar with the rather simple theme in the "Ode to Joy" and how it builds through repetition in the finale of the Ninth Symphony. A similar theme had been in Beethoven's brain for years before that expressed in his *Fantasia* written considerably earlier. Most of us mortals may hum or whistle a tune and then leave it alone. We can let go of a simple melody after a while. There's the wonderful "Sanctus" of his Missa Solemnis too, led by the violin and soloists that develops into a thing of beauty. And this violin theme is very similar to the main theme of the first movement of his violin concerto. His music is full of such examples. Beethoven was easily obsessed and consumed by simple themes in much of his music, some of which almost seem never ending. Indeed it seems to me that once a good idea came into his mind, he had a problem getting it out. Less interference or distraction from outside sounds and noises might contribute but such obsession, it would appear is a necessary but not sufficient condition for creativity, a pretext for high concentration of effort. You can say the same about other music. Just a couple of examples: Janacek's iterations of two or three notes and of course, Tchaikovsky's emotional crescendos built on simple "answering" or repetition of themes by one part of the orchestra, then another until an emotional apotheosis is attained. Given that creation is in many cases an obsession, it is reasonable to ask whether the rest of us too easily give up on a theme without seeing its true intrinsic beauty or without being able to develop it fully.

Lastly, here's an observation I have made as a non-musician. Beethoven was obsessed I think, taken over by relatively simple themes and melodies which would consume him and this one can see in his music. These it would appear, went round and round in his head, creating an explosion of emotion. Everyone's familiar with the rather simple theme in the "Ode to Joy" and how it builds through repetition in the finale of the Ninth Symphony. A similar theme had been in Beethoven's brain for years before that expressed in his *Fantasia* written considerably earlier. Most of us mortals may hum or whistle a tune and then leave it alone. We can let go of a simple melody after a while. There's the wonderful "Sanctus" of his Missa Solemnis too, led by the violin and soloists that develops into a thing of beauty. And this violin theme is very similar to the main theme of the first movement of his violin concerto. His music is full of such examples. Beethoven was easily obsessed and consumed by simple themes in much of his music, some of which almost seem never ending. Indeed it seems to me that once a good idea came into his mind, he had a problem getting it out. Less interference or distraction from outside sounds and noises might contribute but such obsession, it would appear is a necessary but not sufficient condition for creativity, a pretext for high concentration of effort. You can say the same about other music. Just a couple of examples: Janacek's iterations of two or three notes and of course, Tchaikovsky's emotional crescendos built on simple "answering" or repetition of themes by one part of the orchestra, then another until an emotional apotheosis is attained. Given that creation is in many cases an obsession, it is reasonable to ask whether the rest of us too easily give up on a theme without seeing its true intrinsic beauty or without being able to develop it fully.

We have seen that on a physiological basis, sensory deprivation is analogous to denervation hypersensitivity. It may increase one's response to residual sensory stimuli in such a way as to result in an exuberant response, an exuberantly painful response or more often flowering of the imagination or even cause a retreat into one's inner world and hallucination. Alternately, one may learn to confabulate, in other words to make something that is not out of an incomplete picture that is there and there may be one cause of an hallucination or illusion. How is it that we can see something that is not really there? Why do we see something that does not exist?

Receiving stations stand ready to receive input whether there is any input or not. These receiving stations may be disconnected from their inputs and hence unaware that input is curtailed. Anosognosia, unawareness of a disease or problem, is the result. Being unaware that you are blind (Anton's syndrome) is but one example but there are other related problems that come from the sudden end of basic sensory input in the brain. Unusual visual phenomena occur in migraine, and persistence of visual imagery or palinopsia that occur because again the receiving neurons are still there to receive input even though they may be receiving nothing. A visual afterimage persists long after it is gone for some people who have had damage to the occipital lobe. Ordinarily an object shown in the blind visual field say on the right side with a right hemianopsia will persist in the preserved visual field. Patients with occipital lobe damage frequently make more of an image than is actually there. This happens a lot when a small part of their vision is preserved. They may see a certain form, say that of a person in their affected visual field and then when they turn their eyes in a way that they can more easily see, they find out the person is not there. Most of them don't seem to mind this problem too much, but they mention it.

## MIND SENSES WHAT THE EYE CANNOT SEE:

You can understand how we see what is not there. It is harder to explain how we can see what our visual systems were not designed to see. I was teaching my third grade son about the color spectrum. I happened to mention that the spectrum we see is only a very small part of the waves that are out there. Ultraviolet and infrared waves that border the visible spectrum are invisible to the naked eye, but the really interesting wavelengths are so far from visible light and don't generate heat so that we can't sense them at all, radio waves, x-rays and gamma rays. Trying to explain to a curious eight year old how we know about things our bodies can't sense, that our eyes were not designed to see, is a job.

One way we know these certain invisible waves are there is through instruments designed to pick them up, not the least of which is our table radio and television set, that an eight year old can understand. They also fit into our theories about what light is. If we have a wave of a certain wavelength, why shouldn't there be waves of a shorter or longer wavelength? But there is something quite profound here. It took tens of thousands of years of human history to understand the meaning of the light that impinges on our eyes, by this I mean to appreciate visible light as just a small part of the electromagnetic spectrum. Actually we learned about what light really is only recently and of course we still have a lot to learn about light.

We all know what some of these insights were starting perhaps when Isaac Newton's prism which separated out all the colors from white light, then to the speculations about the wave nature of light, that the various colors differed from each other only in terms of wavelength and frequency, later the discovery of invisible portions of the electromagnetic spectrum, the invention of various instruments, radios, Geiger counters, x-ray detectors, that could respond to invisible electromagnetic waves, finally to instruments and detectors that could generate images from parts of the spectrum that we could not see. Every time you take an x-ray, from a simple chest x-ray to a CAT scan, you are making an image using invisible rays, and astronomers make radio frequency, IR , gamma ray images from doubly invisible objects. Not only are they too far away and too small in our sky for our eyes to see, but they also radiate energy within invisible parts of the electromagnetic spectrum. Quasars are invisible and far away. The most distant quasar is felt to be up to 13 Billion light years away[102], which means we are receiving electromagnetic radiation (here in radio frequencies) old enough to have been sent shortly after the birth of the universe. Their energy comes to us from the farthest reaches of the universe and a time only shortly after the birth of the cosmos. Certain very distant galaxies that powerful telescopes can sense help to tell us about the early history of the cosmos. In looking at them, we are peering into the farthest reaches of time and space. This is because they are billions of light years away and hence the electromagnetic energy that we sense is extraordinarily old, in some cases billions of years old, so we see these quasars and galaxies as they were when their energy set out on its journey to us. These very distant objects have passively sent messages to us, but we have had to invent great instruments for seeing.

When the Hubble telescope went up in 1990 the first thing that came to my mind was the biblical Tower of Babel and man's hubris in making something that would reach into realms he had no business going or knowing anything about. "Whoever ponders on four things, it were better for him if he had not come into the world: what is above, what is below, what was before time, and what will be in the hereafter." (Mishnah) When the Hubble telescope malfunctioned I was certain it was due to some divine intervention. Are we exploring where we have no right to go? We are breaking out of our limiting structures in many areas simultaneously. This applies to biology as well as physics. Should we be redesigning animals and plants for our own purposes? In every instance so far, just as we are convinced we've reached and ultimate understanding, we discover that our line of sight is limited and there is always a lot more to learn.[103] Near the turn of the century the man who measured the speed of light, Albert Michaelson said in a fit of hubris, that physics had discovered the basic laws of the universe[104], expressing his disappointment that there was little more to do. A few short years after, Einstein published on the special theory of relativity unifying our ideas of time and space. It seemed to me that we might just be exploring a realm God might not want us to know about. that maybe there was a reason why the Hubble telescope had failed to work. We were building things to stretch into a world where we didn't belong. Or, maybe I was just kidding myself into thinking such exploration was more fundamental or basic than it actually is. We always turn out to be more ignorant than we give ourselves credit for!

A lot of books about neurology describe how we see what we can see. But they fall short in describing those things that are way beyond the capacity of our visual systems to see, but somehow we manage to see anyway. To me this is the most profound part of that way we see. Every time we look at an x-ray image, hear a radio, use ultraviolet light, we step beyond our biological capacities into the realm of human invention. The vision we inherit is miraculous enough. There are worlds beyond that we've discovered well beyond the reach of our biological systems and their advanced design. Did you ever wonder why we can see well beyond what our organism was designed to see? This is accomplished at higher abstract levels in the association cortex of the brain, that is those areas not involved primarily with basic sensory reception or muscle movements, in other words thought and motivation which is hardly taken into account in various neurology or neurobiology texts. Because the function of these areas of the brain isn't as readily "visible" or easily accessible to examination, that is the workings of the association cortex are less manifest in behavioral observation or testing for sensory reception, our conception of how association cortices works is vague. But we have a strong hint at the function of these brain areas when we become aware of all we have accomplished that is well beyond the bounds of our basic sensory and motor apparatuses. What I mean here is that it's fine to describe how we walk and maintain balance. We can run also at a certain speed and over a certain terrain that is far surpassed by many other animals. Few men can run a mile in as little as four minutes under the best of circumstances. This is our biological endowment. On the other hand, at this stage in the twentieth century machines that increase this capacity are abundant, from all wheel drive vehicles that let us surpass the fastest predators of the African plain to automobiles and airplanes. We've gone well beyond our biological endowments and this is on a very practical level and we've done so utilizing our association cortex, whose machinations have been little described physiologically. It's fine for us to follow the electrical impulses carrying visual information to the brain, but much more difficult to trace the thought processes physiologically connected with the understanding of light as electromagnetic radiation or the design of instruments that increase our traveling speed.

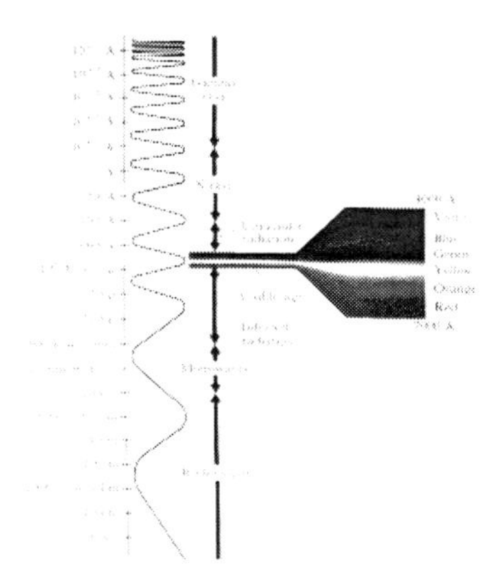


**Figure 42: The visible spectrum accounts for only a very small portion of the electromagnetic waves that we are aware of and utilize every day. We are not stultified by the eye's abilities.**

Visual impulses, after arrival at the visual or primary sensory cortex in the occipital lobe, spread rapidly to secondary and tertiary sensory cortices then after that, ramify widely. As these impulses spread they are harder to define and follow and the pathway is less stereotyped and varies from subject to subject, particularly if the person is involved in a complex task involving multiple brain areas. We learn this, not primarily through lesioning experiments that can only show us how certain capacities become limited when brain regions are removed, but more through physiologic experiments. The PET scan SPECT, and certain MRI[Ψ] techniques now allow us to visualize increased metabolism in specific brain regions, over time and in response to specific stimuli. And different persons, different groups of people, men vs. Women for example, or normals vs. Stroke patients, may employ different anatomic strategies in order to accomplish the same task. One study showed men and women using different halves of the brain to perform the same nonsense syllable rhyming task. Men and women were equally skilled at the task but men tended to use mostly their left hemisphere whereas women employed both the left and right hemisphere indicating as is well known that cerebral dominance for language is expressed more strongly in men than it is in women.

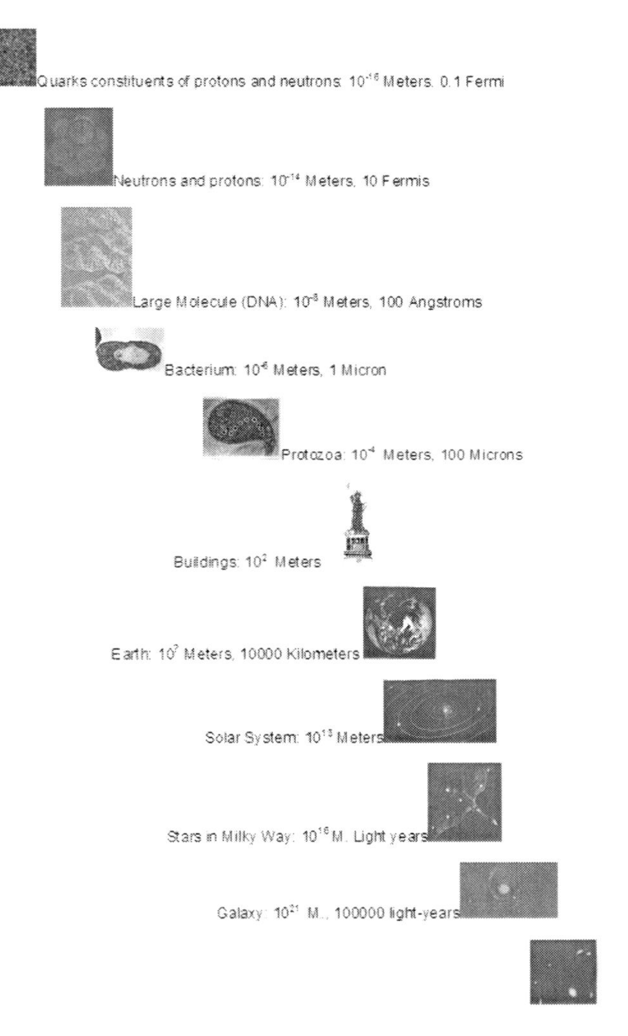


**Figure 43: Man's vision extends well beyond his biological limitations into micro and macro worlds[105].**

So the question of how our brain and mind functions becomes even more fundamental. Given that we've already achieved a level of function that far surpasses our biological endowments as far a perception, and motor function are concerned, that we now live in a situation resembling little our original biological beginnings perhaps as a ground primate in a grassland in Africa, are biological, physiological neurological concepts going to prove to be adequate to explain the whole range of human behavior, or should we perhaps be looking at something more?

A lot of our abilities are no longer in our brain or in our bodies. We've *externalized* our abilities and hence much of ourselves. Much of ourselves is non-biological, but not material either; what we are is much harder to define. Our common memory is stored as data and is manipulated and manipulates and affects us through instruments of our own design, computers as well as other recording machines. Just as our machines extend our maximal velocity from a mile in 4 minutes, memory is extended with magnetic and optical storage devices, our ability to calculate which is slow and unreliable is magnified thousands of times. Biology ceases to define us for our abilities are extended by external tools. The most that can be said is that the best brain takes account of some of this, certainly it can't appreciate or apprehend all of the information we generate, but then may serve some executive function or other. Ultimately, we think, we are in control. Our abilities are leveraged so through external contrivances that we see much more than one would have thought we ever could see given our visual system, that is attuned only to a very small part of the electromagnetic spectrum at only limited intensities. We see so much farther, smaller, larger than we should be able to.

Mankind has long ago left his biology in the dust. That's precisely how we differ from all other animals that are forever bound by their biological endowments. Biology does not constrain us. Our biological instruments can see red green and blue, and vastest natural vistas of only a few miles, yet we've figured out how to see the entire electromagnetic spectrum and to the edges of the cosmos. This is not just a testimony to the genius of human kind. We can pat ourselves on the back for having come so far in our development. It is more than that, plain evidence that there is a driving force propelling us to achieve what is beyond our biological inheritance. You have to ask, whether or not this will to go beyond what we is biologically determined or does it come from another less-defined source. Has human will attained a life of its own? We do not have a definite answer to this question since such a will is unprecedented in non-human animals. Perhaps we've hit upon something that biology alone cannot account for.

From our limited view of a nucleus of material reality of our day to day we have expanded our vista into the realm of the very small. We take for granted now seeing things beyond our original capacity to see. In the micro-world we can see a Paramecium or bacterium using an ordinary light microscope, or a virus particle utilizing electron microscopy, and even the structure of a large molecule with x-ray diffraction techniques. All of these instruments have extended our horizons. In even smaller realms we conceptualize the constituents of atoms, protons, neutrons electrons and constituents of protons and neutrons, the quark that is an abstract mathematical concept, but yet a basic building block of matter. We learn that our material world is comprised primarily of empty space. And utilizing these mental tools we have begun to speculate about the beginnings of things the origin and the end of the universe, extending this horizon even further, well beyond the beginnings and ends of our own lifetimes. Indeed every time we open a history book we are in a sense gazing beyond our own natural span of years. What other animal can lay a claim of understanding of time before and after his own lifespan?

## A REALITY MORE VAST THAN OUR IMAGINGS:

The other side of this is our looking out to the stars into a cosmos vaster than our imaginations. Only a few hundred years ago we speculated that heaven and hell lay just over and just under the surface of our earth, that we could happen into these imaginary realms by just traveling the surface of our planet. Now we peer out into the heavens using earth bound optical radio and gamma and x-ray telescopes and orbiting instruments trying to discover the ends of the universe and perhaps other life forms like ourselves. How far can we see? Because of the big bang the farthest objects we can examine are receding from us at the fastest speed. Generally objects at the greatest distance from us are traveling away at enormous speed. Our horizon that part of the universe we can see from our vantage point is limited. If the universe is 10 to 12 billion years old, as it is felt to be, and we are seeing an object a billion light years away, the light (more properly electromagnetic energy - we may be looking at energy outside the "visible" part of the electromagnetic spectrum) 5 billion years old. By the time we receive this light, the object in question will have moved at such high velocity much farther away and changed a great deal. Indeed it may not even still exist. And the light we are seeing was sent before the formation of our own planet some 4.5 billion years ago, sent before there was even any life to appreciate it. How far can we see? Using these facts, the speed of light and of recession of heavenly bodies, knowing our capacity to receive the faint signals from distant objects you can make certain calculations but it is clear the information that we get from the farthest reaches of universe is billions of years old (because that is how long it takes for it to arrive here) and does give us a "picture" how things were then but not how they look today. Recall that when this light was originally sent, there was no life here on earth, no earth or sun. Our horizon is no where close to the edge of the universe, or near the beginning of time (if we take into consideration the age of light that we see at vast distances then we can apprehend the fundamental equivalence of time and distance.)

According to our current understandings our visual horizon is not infinite, it is roughly calculable. In general terms we can see the distance that light is able to travel since the universe began. We can't see further than x light years, x being the age of the universe, what ever that age is and it is not definitely known (perhaps around 12 billion years.) A light year, the distance that light can travel in one year, is about 6 Trillion miles. It's almost a given that information can never be transferred faster than the speed of light. As we peer at objects further and further away we can see things as they were longer and longer ago. Since the universe, like us, is only getting older and older, it stands to reason that as it ages, our horizon gets wider and wider; we can see further, so that over time imperceptibly, more and more distant objects rise over our horizon. By this time, these objects are so far away, that they are flying away from us at a speed, close to the speed of light. If you could only see an object 10 or 12 billion light years away, then you could see things as they were at the beginning of time just after the big bang itself!! But it doesn't work that way of course. Since galaxies are flying away from us the light dispatched from the most distant of galaxies was a lot closer to us at the time it was made and the velocity of recession has to be taken into calculations of arrival time in light years. There are finite limits of our sight, but they are not determined by the structure of our eyes or our nervous system. The laws of physics limit our vision.

The most important implication of Einstein's relativity theory early in the twentieth century is that time and space are essentially equivalent concepts. No where is this point made more strongly than in consideration of visual horizons. Time and space may be taken as dimensions on a graph and are essentially equivalent from the standpoint of mathematics. Someone can locate me at this very moment as I sit a certain place in a physical universe by specifying my coordinates in a Cartesian system of axes over the continuum of space and time. This is graphically represented below, where time flows from the bottom to the top of the paper and is thus on the vertical axis and space is seen in two of its dimensions represented by a plane in a diagram that occurs at a certain time t. As time passes represented along the axis from the bottom to the top of the paper, space occurs as a series of planes extending along its own axes. One can locate an event in space and in time.

Nothing can travel faster than the speed of light. If an event is given as a point at a certain time t, then the distance that light travels over a future time higher up on the graph over the vertical axis can now be represented as a circle. Light cannot travel more than this distance over the period of time from t to t1 so it is impossible to see further than the distance represented by the circle. As time passes and light is able to travel farther and farther, the ever expanding light circles describe a cone. Our visual horizon has a mathematical limit as represented by this cone. An event outside the horizon cone cannot influence or be influenced by the event as depicted on the graphical diagram[Ψ] . This is depicted in the figure by the separation of horizons of observers A and B. Never their twain shall meet.

You can represent a life on such a diagram with a limited capacity to see, limited only by the horizon as dictated by the speed of light which is an absolute, only the image of a life is a great deal fuzzier. Instead of an event, which starts out as a point, a life extends slightly farther over the vertical axis of time and of course, we humans physically extend a certain distance out into space as well but better represented by the extent of our visual horizons rather than the width of our bodies. Nevertheless the principle is the same. A life is a contiguous extension of perception, a vision over space and time axes, and may be graphically represented in same way as a physical event over the space-time continuum. To find an ancestor or descendent one needs only to travel to their space-time coordinates, limited only by our technical ability to accomplish this and the absolute speed of light.

One's own life, the beginning and end of perception, can be revisited by a form of travel. Space and time coordinates are mathematically and physically quite equivalent. Physical laws make no distinction here. Now it is true as humans that we perceive space and time in a way that is different than their mathematical representation. Time flows in our perception whereas space does not. Events are perceived to be caused by previous events so that time is unidirectional whereas space is not. Time will only flow for us from the bottom to the top of the page in our current representation and never the other way around. This is often referred to as the arrow of time. Space has no such arrow.

This presents one with an interesting conundrum for going according to mathematical representations, time is treated fundamentally different than the way we consciously think of it. In math, the arrow of time can conceivable flow in either direction. Going back in time is not a problem at all. Physical processes and events are reversible. Why time flows ever forward from the mathematical perspective has only to do with the second law of thermodynamics which states that over time in a closed system, entropy or disorder increases, but this matter seems to me to be almost an insignificant footnote a limited explanation, given as almost an afterthought because some explanation is necessary to explain the forward arrow of time. Otherwise as stated, space and time are exactly similar types of dimensions on physical axes.

To humans, time is fundamentally much more different than space coordinates. It seems to me this bears on birth and death. While we extend as a certain quantity of matter over limited spatial dimensions, our lives extend over time in a fundamentally different manner marked by a beginning, which is perhaps our conception or possibly our birth and an end, which is death. It is not possible to revisit a life or an event in a life in the real world by going back in time. This could violate causality, which seems to be also an absolute. To travel through space, on the other hand is not a problem. One can revisit a place but not a life.

Since time is relatively equivalent to space in mathematical terms, the math, it seems to me, is at variance with common perception of time. Pure math seems inadequate in its handling of the concept of time despite the limited provisos of the second law of thermodynamics, which by my own common sense (take that for what it's worth), do not seem to be a strong enough reflection of the human perception of time flows. It seems to me a mathematical statement of time flux needs to be more consistent with human experience. Why is this consideration so important? Therein may lie an explanation of the mystery of life and death[106]!!

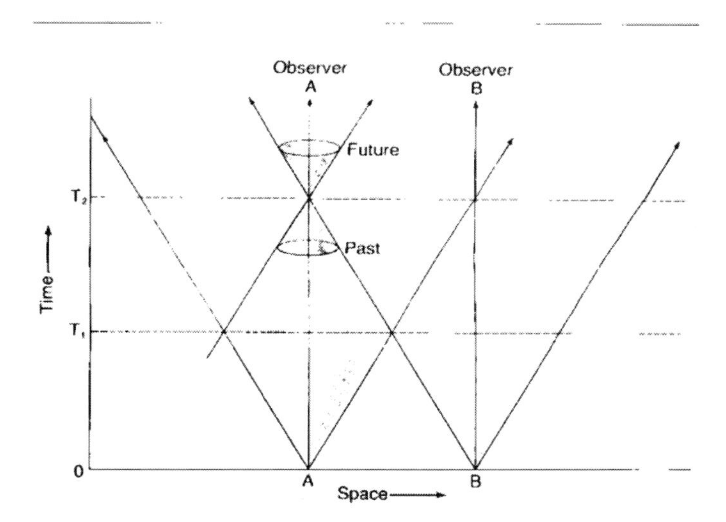


**Figure 44[107]: Astronomical distance horizons. As time we can see more objects because their light has enough time to reach us. Time, meaning the age of the universe, is on the vertical axis and we are able to see over longer distances.**

## MUSICAL LITERACY:

Legend has it that the young Mozart was capable of hearing a very long piece of music just once and then could reproduce the piece exactly. After that, he could perhaps set it down on paper with the utmost fidelity. Of course those of us who enjoy music at all, and that is most of us, commit certain songs and melodies to memory. But there are two different styles of musical memory, different strategies used by the musically literate, those who can read and write music, and most others who are musically illiterate[108]. It is not at all surprising that literate musicians, trained more classically, have a tendency to recall music literally, that is they often reproduce melodies and harmonies with precision, note for note. In other words they have very highly trained and competent musical memories. But after hearing a melody enough times, most people will start to be able to hum, sing or whistle the tune. Most people after going to a concert or a Broadway show, that they like, end up singing or humming some of the tunes, but if we are not literate, chances are we won't reproduce the music exactly. Instead, what they pick up is the general shape of the melody, perhaps the alteration of going up and going down the scale particularly at a point in the piece that we like, but we can't except in few cases, tell you exactly what you just heard.

Neurological tests indicate that musical illiterates use mostly their right or non-dominant (non-language) hemisphere to function musically, but literate trained folks depend on linguistic processes in order to memorize and make music. In other words they tend to have a precise picture of a piece in terms of notes, rhythm and chords as if it were down in notation much as if you could memorize a Shakespeare soliloquy by seeing the print in your mind's eye. This goes against the popular belief that music resides or is taken care of by the more artistic or emotional right brain while math and scientific pursuits are handled by the left or language or analytical half of the brain. The real truth is that complex composite tasks such as musical writing and appreciation involve large areas in both cerebral hemispheres.

It's a shame that such a large popular literature has evolved purporting to teach people how they need to develop their full potential and use both sides of the brain. in daily pursuits they absolutely are using both cerebral hemispheres. in any case, quite famous musicians have who had strokes and other lesions have been looked at [109] In some cases, classically trained literate musicians have been crippled by lesions to the left hemisphere. They simply could not analyze, listen to or write music linguistically using their left hemisphere, and their lofty strata of musicianship surely involved intimately almost their whole brain as would be expected. Brain lesions that affect musical function produce defects affectionately called amusias which like speech deficits can be analyzed into types expressive (inability to produce) and receptive (inability to understand) amusia. [Ψ] Some musicians with limited lesions in one or the other hemisphere are able to carry out musical functions anyway. Perhaps less demanding musical situations with simpler popular music or musical tasks might involve less of their brains and they would have been able to get away with functioning despite their cerebral lesions.

## WRITING HISTORY:

There are two distinct strategies in recording events. One is record them literally with complete accuracy. You can use writing instruments or machines, computers or calculators. This is what, in popular parlance is done with the left hemisphere of the brain which is more analytical and closely tied to written and aural language, especially literacy, what I would call a book-keeper's or accountant's reality. The other strategy is to recollect the general impression or shape of reality, more of an artistic recollection perhaps incorporating emotion but this is a less faithful, more fanciful recollection as well, a far less exact rendition but closer to the way that most of us record reality, by inexact impressions and subject to mental editing. Julian Jaynes makes the interesting point early on that bringing memories into consciousness and recording them, we are not reproducing actual experience.

"Or introspect on when you last went swimming: I suspect you have an image of a seashore, lake, or pool which is largely a retrospection, but when it comes to yourself swimming, lo! like Nijinski in his dance, you are seeing yourself swim, something that you have never observed at all! There is precious little of the actual sensations of swimming, the particular waterline across your face, the feel of the water against your skin, or to what extent your eyes were underwater as you turned your head to breathe. Similarly, if you think of the last time you slept out of doors, went skating, or -- if all else fails-- did something that you regretted in public, you tend not to see, hear, or feel things as you actually experienced them, but rather to recreate them in objective terms, seeing yourself in a setting as if you were somebody else. Looking back into memory, then, is a great deal invention, seeing yourself as others see you. Memory is the medium of the must-have been. though I have no doubt that in any of these instances you could by inference invent a subjective view of the experience, even with the conviction that it was the actual memory."[110]

Reminiscence is editing altering and fitting in to one's rubric. We don't actually record the whole of actual literal experience on a blank film the way a camera does. We extract certain details and give them back subjectively. Literal photographic renditions are a novelty in art and literature. But you always have to ask how much should art mirror reality.

## Realism, Distortion and Creativity:

Sometimes it can be harder to present reality than to invent it which you can appreciate seeing realist paintings of shiney cars or sugar dispensers. Whatever you can capture with a photograph you have difficulty reproducing with a brush. When artist succeed at these difficult tasks we find it amazing. You may surmise that mimicking reality with precision utilizing a brush is no trivial matter but every bit as hard as creating an impressionistic work. From the purely evolutionary standpoint one can see the advantage of being able to memorize then perhaps reproduce a carbon copy of the real world in order to be able to respond to it. Planning a motor response or strategy of response is best done when we have an exact idea of what is there. But as we have already seen, neither perception nor memories are that accurate. Perceptions are not exact reproductions and memories are subject to editing. You have to wonder why? Perhaps some internal misrepresentation may be more adaptive than literal reality.

Two further processes, distortion of reality and creativity are even harder to understand from the point of evolution and adaptation. If an animal or person distorts or changes reality, it makes it that much harder to plan and manipulate in an adaptive way. How do we know there is distortion in perception? For one thing there are so many different strategies for seeing. The spider with eight eyes has a much different strategy for recording visual events than man. Undoubtedly he's designed to make primitive responses to simple elements in his visual world with more or less hard-wired direct connections from his eyes to muscles that can make a change whereas we humans are more attuned to take in the big picture rather than to respond to small elements of a scene. Even among people though, when a lot of people witness the same event the accounts disagree. They can't all be right of course and this non-reproducibility of perception is a sign that there is distortion.

## CREATIVITY:

What of creativity? Few artists seek to make an exact copy of a real image though that is one approach to art, more akin to photography. But a photograph would include none of the message of the artist, nothing of his inner self. Granted, the photographer selects his images, seeking to photograph something worthwhile and thinks nothing of altering his photographic subject or waiting for just the right moment to take a photograph. The painter, on the other hand, does not need to manipulate reality except in his mind's eye and is free to distort images at will. Distortion may just be a matter of style as it was at the time of El Greco's Saint Jerome. El Greco had no particular perceptual deficit. He was merely painting his subjects as was the custom of his time, as an elongated figure. Still this distortion as maladaptive as it is makes a work of art. It is a change that biology and evolution just cannot explain.

The origins of abstract art are not to be found in biology that can barely explain what we do see and how our vision adapts us to the practical world. A record of reality, no doubt, serves a useful purpose, such as an exact internal map of territory and internal picture of prey and other animals in one's own species. It is easy to understand how an internal mental map might help an animal. Predator and prey would both have a distinct advantage knowing the lay of the land and obstacles to be encountered in a chase. The usefulness of pure imagination is more difficult to defend. The biggest problem is that imagined distorted pictures of reality are inaccurate and may actually be maladaptive. Much better to have an accurate picture of reality than an inaccurate one, all else being equal. This is at first glance.

Two further processes, distortion of reality and creativity are even harder to understand from the point of evolution and adaptation. If an animal or person distorts or changes reality, it makes it that much harder to plan and manipulate in an adaptive way. How do we know there is distortion in perception? For one thing there are so many different strategies for seeing. The spider with eight eyes has a much different strategy for recording visual events than man. Undoubtedly he's designed to make primitive responses to simple elements in his visual world with more or less hard-wired direct connections from his eyes to muscles that can make a change whereas we humans are more attuned to take in the big picture rather than to respond to small elements of a scene. Even among people though, when a lot of people witness the same event the accounts disagree. They can't all be right of course and this non-reproducibility of perception is a sign that there is distortion.

Yet it is also apparent that the imaginative faculty confers a biological advantage. If you can imagine your habitat, then you may keep it in your mind's eye, creating different perspectives than are obvious from you specific vantage point, perhaps see it in a different way, make plans, even gain competence surprising your prey or your enemies. It does confer an advantage not only to be able to see your environment, but to be able to flip images and see them from a different perspective, to further manipulate these images in your mind. With writing or the computer, these manipulations multiply a hundredfold.

One could ask, of what possible use could abstraction or unreality be from the purely biological point of view? How can evolution possibly explain imagination, particularly unreal or surreal imagination? Look at the Picasso below. We see the face as we never see it with a front and profile view simultaneously. This is not altered perception but a manipulation of reality that we find novel. To be able to manipulate in the abstract sense, instead of having to try things out in the real world, is valuable. That is what computer modeling is about. If you wonder about the effects of global warming, factories spewing carbon dioxide and other greenhouse gases into the atmosphere, you don't have to wait to see it in the real world. You can model it, use tools to expand memory and manipulative abilities, then make changes in virtual reality. It gives you a chance to test your theories out before applying them. A football coach can try out a play on paper before bringing it into play. This manipulation is not restricted to visual imagery. Apart from the purely adaptive advantages that such manipulation confers, it serves a sociological function increasing cohesiveness of the group.

The same could be said about exact recollection of events. There is always a conflict between the accurate telling of a story and imagination. The Greeks defeat the Trojans and burn down the city. This is done through force of arms and cunning. But through the telling and retelling of the story it is embellished and a conflict between mortals becomes a struggle between the Gods, killing becomes a heroic act done for vengeance or some other higher purpose, imagination becomes myth and over the years the story becomes more interesting and universal in its appeal. For a group this process increases cohesion and group identity where little of it existed before. Man being a social animal such creativity can be viewed as a biological adaptation.

Consider the earliest known examples of cave painting found as long as 23000 years ago in Southern France and Northern Spain at such Paleolithic sites as Lascaux and Altimira. For our Cro-Magnon progenitor hunters these were intimate images in which hunted animals were much more intimately and accurately portrayed than were humans. Some of the paintings seem to be telling a story or illustrating a myth perhaps of a mighty ancestor. In one image the hunter is downed by an enormous speared animal his entrails spilling on the ground[111]. We witness the fierce fight of a legendary hero in the upper Paleolithic age of hunting and gathering. Sculpture tended to be primarily sexual, accentuated with prominent genitalia, buttocks and breasts. What function did all of this artwork serve? We might theorize that almost for the first time man was able to manipulate mental images of his prey. Cro-Magnon kept his prey, his living, in his mind's eye. Primitive art evidences premeditation and planning, functions of the human frontal lobe. This image painting is mental rehearsal about he act of hunting. Imaging and rehearsal has the same quality as the work that is part of our dream journey in the middle of the night. Mental rehearsal will ultimately enhance performance and expresses anxieties about possible failure and other dangers. Our Cro-Magnon ancestors went to bed thinking of their day and work. Their visions accomplished for them the very same thing as ours do today.

But there is much more to the story. Cave painting is almost always situated in inaccessible parts of caves (aren't our own dreams similarly inaccessible?) . To get to them, one needs to be able to ford underground streams and slip through dangerous craggy narrow dark passageways, an adventure into the depths of the psyche and the traditions of the past. Abstract and practical concepts are conveyed on cave walls through imagery. Religion may have been an adventitious outgrowth of a basic mental imagery with pictures deliberately set deep inside his caves, beyond the region of habitation. Or perhaps they were placed in these remote regions for posterity so that in a sense, these cave painters were painting for us rather than for their contemporaries who might destroy their images. The artwork placed in deep inaccessible regions is suddenly made more secure, far removed from daily traffic and at the same time becomes secret, remote and mysterious which always makes the symbols all the more exclusive and powerful. Consider images made in utter darkness except for artificial torchlight quite possibly in the middle of the night. Then the work would really be part of dreamscape, not practical reality. Painting animals confers upon them a certain immortality, a way of thanking them for making life possible. Perhaps religion began this simply thanking the very animals that provided sustenance, food and skins then only later imagining some external force that created these hunted animals. These primitives did not take without saying thank you, but even at that time wondered about the universal questions that we ask today. Why they were there, and how were they able to make a living from their surroundings and even about the permanence of death. These deeply situated images would have worked for initiation rites that introduce to the maturing initiate the permanent and transcendent myths and heroes of his heritage, to teach not only the abstract elements but also practicalities of survival both economic and sexual. Some of these areas have multiple footprints and may have been a communal spot where dancing ceremonies occurred in a remote location far from dangers but also in secret regions that were difficult to approach. The more primitive Neanderthals seem to give evidence of the first burials in human history although some finds have been discounted. It makes on wonder about when man became conscious of his own mortality. [112]

**Figure 45: Image from Lascaux cave[113].and Venus figure[114]., Picasso "Woman with Long Hair" [115] (1936) with typical simultaneous front and side perspective, showing visual distortion.**

To distort and to invent to imagine as well as to see. All of this serves a certain adaptive purpose. No one knows the origin of music. An awful waste of effort to create rhythms, melodies, and harmonies, time and effort that could be used to build shelter, to hunt animals, make weapons to defend against enemies etc. Yet enormous effort seems to have gone into the production of organized sound. Why? One might think hearing should have developed for the purpose of signaling, detecting enemies and prey, yet from simple auditory reception the facilities of speech, writing and music eventually developed. The greatest argument for music is again, that it is an instrument for social cohesion, the tribal war and hunting dances, rituals of marriage and birth and other communal events. The hardest concept for me to grasp though when it comes to art and imagery whether it be visual or auditory, is distinguishing between the practical and the abstract. It is easy to postulate that art and music began for some practical purpose, and even in the most advanced condition still serves some practical purpose such as social cohesion. Yet at the same time it takes on a life of its own, which is abstract and disconnected from reality, more from the dream world of night than the practical world of daylight.

## DREAMS AS REHEARSAL :

The origin of dreams, imaginings is the same as the origin of visual and aural art. It is nocturnal. Visualization perhaps started as a form of rehearsal giving the opportunity to try out certain strategies perhaps in hunting, fighting and other fields of endeavor in a different arena so as to avoid the high stakes game of reality. Today we've developed the tools of imagination to a high degree in computer simulations. We can test stresses on buildings and bridges in bearing weight and in simulated storms, test out materials for human joint replacement undergoing various stresses, model weather and seismic systems, test automobiles before production and avoid destruction and expense using computer models. At the same time we're experiencing the same evolution as our forebears did in Lascaux cave. The imaginings take on a life of their own and begin to devolve their own mysterious purpose as we use computer simulations to create an alter reality. In the far reaches of our offices and homes we may enjoy and live in a dreamlike alter reality born of darkness, not the bright light of day.

One might even propose that as the imaginary facility was cultivated in society and specialists evolved who could make a living in a civilized milieu, women were as attracted or even more attracted to men who could make a living with mental creations as they were to more intrepid men who make a living by hunting or with their hands. Could it be that the brain of a man was as attractive to the opposite sex as his body and may have given certain individuals a reproductive advantage? If we take this argument farther one might propose that a different kind of evolution is going on today as there must be many women who actually prefer mental function to brawn and thus natural selection is occurring under what may be viewed as unnatural that is civilized circumstances.

Mating us often viewed as a matter of female choice. Since males have little investment in their offspring, they are less involved in upbringing and do not have to maintain pregnancy for nine months in the case of humans they can afford to be less finicky the choice of a mate. Females, having much more investment in the relationship, must choose well. But it isn't generally noticed how much of this choice is made by males themselves well before the female comes onto the scene. Females, particularly viviparous ones, such as humans, hold onto their young during gestation, must nourish them and can only be involved with a few offspring at a time. Hence their investment in small numbers of offspring is great and their fecundity will critically depend on their choice of a mate. They will want the most physically fit male or perhaps the most successful or dominant male. What most of us do not appreciate is that systems of dominance have evolved which immediately inform eligible females which male has achieved dominant status. Some baboons and other primates have established readily identifiable hierarchies in which the dominant male gets to copulate the most times with females in heat. Other less dominant males remain in the periphery of the group. Homosexuality may have evolved as part of a readily visible dominance scheme, homosexual males staying in the periphery and never getting a chance to copulate.

Birds such as the pied flycatcher have evolved so-called Lek systems. The eligible female does not come into a group scouting for a suitable male. Instead the males have already decided who will be elected to do most or all of the mating, the other males in the group performing some other function. A male who copulates so many times is not going to be a very effective father, so that the strategy in these groups would appear to be that male dominance, determined beforehand strictly among males, is the paramount consideration in offspring fitness and some other means of taking care of young offspring, has to be worked out. One can readily draw conclusions about males in human societies. Don't they posture male against male, to determine some dominance structure well before the arrival of any female? Presumably, she will respond to superficial trappings of male dominance, perhaps some sign of power, say position, income, power, clothes, car and make her choice. Such sports as football, and war especially seem to serve a function of jockeying for power purely among men that may explain why both athletic events and war are strongly connected with aggressive, abundant and exuberant sex.

The biological explanations go very far but fall short of explaining what has ultimately developed in civilization with regard to the visual and auditory arts. Does our biology define what is art and what is not art, the world of the beautiful? I think it does but not entirely. We all are subject to the immediate effects of near universals, what are called in ethology, *releasers*. This is the moving target to the duckling who will bond with it as its mother, the red flag of the stickleback fish, redwinged blackbird and bull, the stridulation of the cricket, all of these sensory experiences trigger specific behaviors. For men, the voluptuous female breast or buttocks is a releaser, so are lips and eyes of a certain variety and the conception of what is sexually attractive is very much the same particularly within a given range or ethnic group.

On the other hand advanced concepts of beauty or certainly not biological or bioethological. To a physicist beauty and truth are close relatives. At the very least beauty seems to be a requirement for truth in the sense that the truth is beautiful[Φ] . What does something need, particularly a mathematical equation, in order to be beautiful? Beauty has a classical (almost Grecian) form which includes some symmetry of design, simplicity of conception and a minimum of preconceived notions. In the lingo of mathematics means a small number of axioms and constants which are viewed as fudge factors for physicists in an effort to ram or pigeonhole an equation or theory to make it fit with reality. The classic example is the pre-Copernican Ptolomaic picture of the solar system consisting of rotating spheres with the earth at the center. For a long while it was possible to describe the motion of the planets and stars in this system. Only it needed so many "epicycles" and other fudge factors to accommodate movement in the heavens, a person had to be an expert to figure everything out an predict planetary motion. Despite the best efforts of the church which felt it had to conform to an erroneous notion of bible lore that place the earth at the center of the universe, the Ptolomaic theory had eventually to be scrapped in favor of theories of Copernicus, Galileo and Kepler. Why? The heliocentric picture was intrinsically more beautiful, simplistic and did not need fudge factors and anyway proved to be a far more accurate predictor of the motion of heavenly bodies. Once the intrinsic beauty in a system is apprehended we are transfixed to me. We are willing to follow our theories and mathematical conceptions to the ends of the earth, to the outer limits of credulity. Hence we have our "string" theories, the notion that physical forces such as gravity may be described in terms of a vibrating string, and science and science fiction models about billions of alternate universes and wormholes into alternate universes, time travel becomes possible and we may think of out own cosmos in ten or more dimensions, (some of these unobservable) or we have the fantastic notion unbelievable on its surface that the universe, the entire universe, exploded out from a so-called singularity, out of a volume smaller than a pinhead. Why do scientists believe this even though it is contrary to common sense? They are transfixed by the beauty of mathematics. Its what the psychoanalysts call cathexis, attachment, desire. Once attracted sexually, you are more than willing to ignore little flaws such as eccentricities or birthmarks. In fact these obstacles make you further cathect the object of your desire. For mathematics this beauty resides in simplicity and conception of design and everything in our current environment appears to act in conformance with its basic principles.

## THE BEAUTIFUL:

But we meander off from the track. Conceptions of beauty range from the purely biological which has something to do with human releasers, well beyond this to the more abstract artistic conceptions which are both natural and unnatural in design, way beyond, to the purely abstract and way out (or at least this way to me) conceptions of pure mathematics. The point is that conceptions of beauty (and thus perhaps of truth) go well beyond any biological parameters. Or do they? Psychoanalytical ideas, regardless of whether or not you agree with them, have gone a long way in finding the meaning and motivation in art. Works of art are made to express some inner motivation or desire and psychoanalysis largely based on the analysis of works of literature and art, has helped in this conception. When we seek to understand a work of art, one of our goals is to understand the motivation that brought the artist to create the work. Much of this motivation is biological. In writing, sculpting, painting, dancing, the motivation comes initially from passion. A famous example is the Symphonie Fantastique of Hector Berlioz. This spectacular work was energized by an attraction Berlioz for an actress who at first, ignored him. But then when someone excites you sexually, it's not uncommon to hear music. Maybe this is a form of visual to auditory synesthesia some of the sensory experience perceived, the rest of it in our imaginations, the entire experience drawn from widely separated regions of the brain. Each separate brain area takes care of its own modality whether this be visual, auditory,tactile etc. To be able, not only to hear, but then relate the experience to other senses, increases the emotive impact of a work. Gustav Mahler would write extensive programmatic notes for his symphonies, but then deny reject the basic programmatic nature of the music. He didn't wish to limit the meaning of the work, yet he would cut it off from a combined sensory experience that could only intensify its effect. This intensity is later brought together by some as yet to be described co-ordinating principle or executive.

The only reasonable formula to be arrived at here is that in music and in art, biology lays the basis for primitive forms that serve a legitimate biological function. Esthetics, what is attractive, and repellent, is directly related to biological function. We're programmed to like what will be good for us, the mates that are most likely to successfully transmit our genetic endowment, supposedly with foods that will be most nourishing for us, and to be repelled by what is dangerous or harmful, from predators and waste. What is harder to explain is what makes us go on to create whole paintings, tapestries, plays, dances. How are these pure biological functions? Biology is the basis, but the brain is exuberant in its creative power.

Basic biology of the beautiful is what is sexually appealing and includes the releasers of ethology and all efforts to increase fitness, the passing on of genetic traits to one's offspring. Conceptions of beauty range well beyond the biological imperative, into the realm of the artistic (art for art's sake) and then to the abstract even getting confused at times with what is true (epistemology), not just in the sense of the naked realism found in works like the painting above of the sugar dispenser in a restaurant, but all the way into alternate conceptions of beauty deriving from imagination and creativity, a beauty that need not resemble tangible reality. What about this abstract notion of beauty??

Realism and the bio-ethological useful aspects of beauty are part of man-in-nature. On the other hand more advanced conception of beauty, the artistic and abstract ideas and going even further the nexus of beauty and truth, place us above nature to an extent. Advances of civilization that allow us to take part in pursuits not directly connected with survival and procreation allow us to see our selves as being above nature. There are the contrary arguments: even super-natural man and all his conceptions are basic expressions of instinct and thus are a part of man-in-nature scenario but at the same time, he is to some degree above it. His conception of truth and beauty today ranges far beyond an accurate and useful picture of his physical world. The question will always be asked about whether man is in nature or above it. The basis of his perception is natural but his imaginative capacities go far beyond nature and adapt to civilization as a thing in itself rising above nature.

In summary our visual apparatus may be taken as a metaphor for all sensory experience. We see much more than by rights we should be able to see given this apparatus. We are well equipped to make our living in this world from the biological perspective. We are good predators and are proficient in avoiding our enemies as well. We even have built in protections from such harming influences as the damaging rays of the sun.

What our peripheral sensory equipment picks up falls far short of what we are capable of appreciating which is far beyond our immediate and even remote biological needs. For example the biological imperative has little to do with understanding the atom, composing a symphony or erecting great art museums. Seeing infra-red waves or receiving radio-frequency energy from distant stars is far from what our eyes and visual apparatus was designed for. Our vision is far more encephalized than in any other species. More than that, the brain invents new perspectives, new visions, even new worlds.

## TOTALITY OF EXPERIENCE, E:

This is a far cry from other animals and plants. For example, there are raging debates about plants and little worms that "respond to light". For plants there is a tropism that makes them grow automatically toward light and some flowers bloom with daylight and close at night, for the worms a few small light sensitive cells that help them to distinguish darkness inside the earth from daylight found on the earth's surface. Do these plants and animals see by virtue of their response to light stimuli? They may have a behavioral response to sensory stimuli without having a specific receptive organ. A certain response may be linked to a specific percept. They have afferent and efferent limbs here that make a rudimentary connection between a stimulus and response, but no appreciation of the visual stimulus, no associative element. By contrast consider the relative perceptual deficit that we humans have a gap between what is perceived and our innate perceptual abilities that is taken care of by our associative or cognitive brain, not the primary sensory cortices for vision, hearing, touch and so forth. Added to this is our invention of capacities outside of the brain such as books and computers. These catapult human capacities far beyond anything expected given our biological endowment. Yet data and algorithms for manipulating information still remain accessible to our intellect creating a new role for the brain as a springboard for human capacities. If the totality of sensory experience is E (for experience), the primary sensory data received by afferent regions of brain designated as P, (for simple perception) and the rest of experience contributed by associative and cognitive areas of the brain and other instruments controlled by the brain as A (for associative capacities inside and outside the nervous system including books, computers and other instuments) then:

$$E=P+A.$$

Alternatively stated to accentuate the fact that what is experienced is far more than what can be received by our peripheral sensory organs which pick up far less than our understanding of the world:

$$E-P=A$$

While there may be no very good way to quantitate any of the letters in this expression, one certainly gets the impression that what is actually known and experienced by us at this late phase of the Twentieth century is by now only given to us in small part by P and that A plays much greater role in our total experience, in other words:

$$A>>P$$

Also much of A is non-biological in that it lies outside the brain and is dependent on the brain's invention of instruments of recording and is accessible by interactions between humans. Whereas one could picture in primitive societies or earlier in human history P may have been closer to or even much surpassed A. For most plants and animals A apparently = 0. The invention of writing and mathematics most certainly increased A enormously as has the high-speed computer, the physical understanding of light and magnetism, and countless other inventions.

Unlike other animals, we have radiated into far-flung biological habitats world wide not on account of heritable biological characters but because our brain allows us to invent and adapt to different environments from sealskin coat of the Innuit to the African loincloth, from the igloo to the jungle hut, under water and on the land, our inventions allow us to make a life in widely different environments. This is a far different strategy for radiation into other habitats than all other animals. There is some interaction between biological adaptation and habitat radiation. Animals must take steps to evolve through variation, then can radiate. A cat who is a little different than his fellows and has more fur, may be able to survive in a slightly colder climate, one that can survive for a few more days without food from a kill, may take his family into a zone with fewer wildebeest. Thus ordinarily heritable physical change occurs first, fueled by variation. This allows the animal to radiate into different environments and to make a living there. The raw material of physical differences that allows variation and species radiation into different environments is genetic variation which is increased by gene mixing from sexual reproduction, the origin of sex in animals, or at least this is part of the accepted explanation. Human adaptation is different. Men are able to live in virtually all habits because of invention that is an endowment of the brain, his intellect. He changes his clothes rather than fur. He can get rid of his own hair if he wishes.

Beyond the adaptation of invention is imagination and our vision into outer space. Sooner or later we will radiate into our line of sight, to the moon or to Mars and inhabit other worlds. This will be the biggest biological event since the Devonian period, some 300 million years ago in which the Crossopterigian fish[#] grew fins resembling feet and developed air breathing lungs so that they could survive for the first time out of the water. Only rarely in evolution have living things made adaptations of such magnitude taking animals life from the sea to mud to dry land which was becoming abundant at that time. We will experience a similar revolution in our conquest of gravity that confines us to our earthly habitat. Even today man is the only species with a worldwide habitat. But the day will surely come for us to transport our food and our atmosphere or find a way to create a habitable place in distant regions of our solar system and there is nothing to stop us from radiating into the seas which were home to our progenitors.

We only rarely remark about the miraculous extent our vision and how it outstrips biological capacities. The whole notion has sort of crept up on us over the centuries, but this vision widens every day. During the short period of our growth in vision and radiation into wider and wider habitats our biology has remained remarkably stagnant. And without it's changing any more our minds, not alteration in any physical characteristic, has launched us into incredibly variable habitats far wider than any other individual type of organism has been able to adapt to. The brain is what initiates continued growth and the success or our species.

## MORE THAN THE EYE CAN SEE:

Not too many people are aware how our gaze has increased only very recently. It is true on some levels things have changed for the worse. Few of us go out into the dark night anymore to apprehend the wonders of the heavens. If we do, ubiquitous incandescent lights light adapt our eyes since all of us spend most of our time indoors. When we do get out, most of us live in populated cities with abundant light and industrial pollution that obscures all but the brightest first magnitude stars. For most of us only very rarely, if ever, do we view the true splendor of the night sky, a lot people never do. Our ancestors, by contrast, had long hours almost every night to marvel the sky and to watch it, noting the changing vista over seasons and the wandering planets and cycles of the moon. Primitive men constructed legendary models of this motion of which only few survive such as Stonehenge and Chichenitzta which seem wondrous to us but are actually not so surprising considering how closely men and women were tied to the sky. Monuments and written works have taught us that humans even our "primitive" and prehistoric ancestors looked up at the sky at night and wondered about the repetition of the seasons, movements of the planets to speculate about forces larger than themselves. Yet until 70 or so years ago, no one knew how vast the universe really is[#] . In 1923 Edwin Hubble utilizing the 2.54 meter telescope high on Mt. Wilson was able to pick out individual stars of our neighboring Andromeda galaxy, now estimated to be 2.2 million light-years away. Before then Andromeda was only another smudge, a spiral shaped nebula. But this relatively recent discovery had all kinds of implications of cosmic proportions. It meant that Andromeda was an entire galaxy of stars similar to our own Milky Way, containing tens of billions of stars. Many of these stars undoubtedly have planetary systems of their own much like our sun, too small and dark to be seen with our earth-bound telescopes. The greatest likelihood is that some of these planets are home to intelligent life forms. More strikingly, we now know there are hundreds of billions of even more far-flung galaxies each containing billions of stars. Current reckonings indicate that there is an almost infinite number of unseen and unseeable galaxies beyond our visual horizon. These galaxies are simply too far away for their radiant energy in the form of visible or non-visible electromagnetic waves, to reach the earth within the timeframe of the age of the cosmos.[#] It's also hard to fathom that only since 1838 has man had an idea how far were even the nearest stars. In that year Bessel was able to estimate the distance of a close star within the Milky way, 61 Cigni, at about 11 light years. He did so utilizing parallax a method of triangulation used by surveyors, only in this case the bottom of a right triangle is the distance from the earth to the sun, 93 million miles. The very closest stars, alpha and Proxima centauri among them, about 4 million light years away, may be estimated from this method of triangulation. The size of the earth can be estimated by this method utilizing the sun as a fixed distant point that it is on earth scales keeping time and season constant. On that basis one can figure how much the sun angle changes when one walks a known very far distance and so figure the curvature of the earth. The earth's distance to the sun may be calculated on geometric principles as soon as the earth's diameter is known but this was accomplished only in the 1672 by Cassini, relatively recently in modern history. So distances can be estimated from triangulation by a certain bootstrap method. But for most stars that are much more distant and for galaxies, this method fails because on astronomical scales the distance from earth to the sun is extremely insignificant. The distance from earth to other stars in our galaxy may be obtained from an estimate of their absolute brightness and then calculated from their apparent brightness from earth. Distances of galaxies may be figured from their velocity of recession calculated by a Doppler red shift. Utilizing simple formulas astronomers take advantage of a train whistle phenomenon. As the train travels away, the pitch of the whistle appears lower and the speed of the train can be calculated from the drop in pitch. In the same way, the speed of recession of a star can be gotten from a red shift in its color. We now have a universe of star containing galaxies. We know that these galaxies have been flying away from each other ever since the origin of this universe some 15 billion years ago in the Big Bang. Our line of sight has expanded more than exponentially since the first yeas or our momentous century, and we have an idea that there may be many more galaxies of stars than we will ever be capable of seeing the visible horizon.

"Do you sense a creator, World?
Seek him beyond the canopy of the Stars!
beyond the canopy of the stars, he must live..."
-Schiller and Beetoven, "Ode to Joy", Ninth Symphony

Who is the God of such an enormous universe? The concept of a God who split the Red Sea, hovered over the tabernacle, who would send his only son to a small insignificant planet, such a *limited* God is as anachronistic for us as idolatry was for Abraham 4000 years ago. A lot of people maintain that science is irrelevant for religion. Science, knowledge, must fundamentally alter our point of view. Otherwise we are like some pig-headed person who refuses to be swayed by the facts. Here is just one example.

Given our new appreciation of the vastness of space and time how do we see God?

In the biblical account of the Golden Calf Moses is absent for a while receiving the Law and Aaron is left in charge. The people, feeling insecure, and unready to accept completely an abstract immaterial God, pool all their resources, throw in all their gold, to create the Golden Calf, a rather generic material deity. Moses returns, shocked at the events, smashes the tablets of the Law, and not surprisingly, the idolaters are eliminated. In a fascinating direct encounter, the Deity argues directly with the mere mortal, Moses. He wishes to annihilate every one of this "stiff-necked people". But Moses pleads for the Israelites. How would it look, Moses asks, to the Egyptians who have only just been taught a lesson, and to posterity if the very people whom God had protected had been destroyed? It wouldn't seem that God was very powerful at all, and by implication, people should think of worshipping other Gods, who might bring them better fortune. So goes the logic of religion of that ancient time. And besides, Moses offers to erase his own name from Holy books. As a mere human person cannot be expected to prevail upon God and alter God's opinion, it is not reasonable to accept here that Moses' arguments actually prevailed against God's. We have to accept this story as more of a commentary about Moses' inestimable character as the greatest prophet. In fact, God ups the ante, offering to make Moses the father of a totally new people, quite possibly testing Moses' resolve at this point. But Moses will have none of it. No, when we read this account today, we have no choice but to accept all that happens including the Golden Calf and the destruction of the first set of commandments as being part of a grand pre-ordained design which Moses, even in his greatness, does not alter, yet the story is testimony to the greatness of Moses as well as God.. We also see that at that critical time, with Moses away in the desert, the people need to have a deity, especially at that time some kind of a solid material entity. The people were not entirely ready to accept a spiritual essence, choosing in stead to have something tangible, a God that could be seen, a God closer to that of the Egyptians and other surrounding peoples.

**Figure 46: Michaelangelo's Moses: On his second descent from Sinai, Moses is transformed with miraculous beams of light[116]**

•

Then a most disturbing thing occurs. Quite in the middle of all this betrayal, anger, retribution, noting that he is still favored, Moses asks to look upon God's essence himself. Moses requests to look upon God! Hannah Arendt makes the interesting point that according to tradition God is heard and not seen[117]. This is no accident. Firstly the visual sense is a lot more precise than hearing. Auditory cortex is much closer to the emotion laden areas of the brain, the limbic system deep to the temporal lobe, than visual areas. As a rule visual information is more analytical, hearing more emotional and mystical, hallucinatory experiences far more often are auditory rather than visual, all perfectly fitting for religious experience. Moses wants reassurance that God's essence will still be with his people but Moses was asking for even more a greater degree of perception than anyone had ever achieved as he says, "Show me now your Glory." Incredibly God acquiesces, but Moses must be shielded from seeing the face or front of God. To protect him from the awful vision God puts him in a cleft in a rock, Moses' eyes shielded by God, lest he die. The material calf God is here contrasted with the infinite mysterious spirit God of Abraham. Moses, his greatest prophet, will see only the back or perceive God passing, not the Divine Countenance. Even more remarkably, Moses himself is transformed by this vision so that his own appearance is frightening. His skin of his face became radiant, his hair wizened. People feared to approach Moses so much that he had to wear a mask on his face with horns or more rightly beams of light. Moses had been powerfully transformed by his vision.

Ancient biblical descriptions seem to be divided between God's immanence and His Greatness. Some would use this as evidence of multiple authors throughout Jewish history. Others would be content to point to both qualities. It's a little like complementary but contradictory descriptions of particles and waves in physics. God vied with Egyptian deities as a sort of national God and protector, hovered over the tabernacle, and yet was a sort of huge, ferocious and terrible being unknowable even to the greatest of the prophets. The Bible blends percepts, immanence and transcendence, and appeals to primitive as well as advanced ideas, ancient and modern about what God is.

In our own time our vision of God has to be influenced by new discoveries. It is not true that science does not as some argue, influence religion or as others would have it, that science obliterates religion. The God of our Twenty-First Century universe, even if He exists only in the minds of some people, needs to be unprecedentedly more vast and abstract than previous conceptions, more intangible. Science, knowledge, has direct bearing on modern concepts of God. Science is a tool for the apprehension of The Deity.

Copernicus and Galileo may have expelled us from the center of our cosmos. In those days our field of view was only a tiny fraction of what it is today. In our own Century we have discovered ourselves to be at the center of something infinitely more vast. As we gaze out to the stars and galaxies what we see looking in any one direction is what exists in any other, the universe is homogeneous and to us symmetrical. When we look to the left we see about as many galaxies flying away at about the same speed as when we look to the right. And the background radiation from the original Big Bang explosion is everywhere the same, no matter in which direction we turn our antenna to pick this radiation up. Our gaze into the very large is supplemented by our appreciation of the infinitely small, both realms limitless. We can see now a great deal farther into billions of years of time that come before us and which will be here after we are gone. The thing is we're again right in the center again of a maelstrom, vaster than ever imagined 100 years ago. Where are we in this vast sea? We can construct Cartesian axes of space and time and put our physical spatial dimensions on this graph and the length of a human life in time. Compared with the largeness and smallness of everything else our extent in space-time is but a dot. What about our line of sight? What we can see and apprehend that is a great deal for some of us more extensive. This graph gives us a new notion of life and death. Where we begin and end in time, physically, is insignificant as you begin to appreciate that it matters little about the length of time we are on the earth just as our physical size is insignificant. What really matters is how far we can see what we apprehend in our short life. Who can peer into the ends of the cosmos past the realms of life and death? Who will know that come before and what will be in the hereafter? Vision.

# Chapter 4

## ICONOCLASM

"If we evaluate a human being as a type, we need not take the individual case into account, and that is so convenient. It is as convenient as evaluating an automobile by its make or body type. If you drive a certain make of car, you know where you stand. If you know the brand of a typewriter, you know what to expect of it. You can even select our breed of dog in this way; a poodle will have certain inclinations and certain traits, a wolfhound will have others. Only in the case of man is this not so. Man alone is not determined by his origins; his behavior cannot be calculated from the type. The reckoning will not come out even; there is always a remainder. This remainder is the freedom of man to escape the conditioning factors of type. Man begins to be human only where he has the freedom to oppose bondage to a type. For only there, in freedom, is his being--being responsible; only there "is" man authentically, for only there is man "authentic,". The more standardized a machine is, the better it is; but the more standardized a person is, the more submerged he is in his race, class, or characterological type, the more he conforms to the average--the more inferior he is from the ethical standpoint." [118]

-Viktor Frankl

# ICONOCLASM

From Whence Comes Motivation?

Every age has its own idols, its "isms" for us, Mechan-ism or scient-ism. This is a systematized belief, that anything understood deeply enough can be explained in mechanistic terms and is no different from any other belief ingrained at an early age. Mental events comprise a subset of all phenomena, by no means unusual, and thus can be reduced to physical processes that take place primarily in the brain. In its deepest sense this belief system implies that pure thought is reduced to events that take place in matter. Experience is mechanical and material. But foundations grounded in matter in the palpable as opposed to the invisible, are beginning to crumble. There is evidence that we find these old paradigms inadequate to begin with. For instance, in the United States, the richest and most powerful nation in history, depression and a profound disappointment with life is endemic, especially in affluent neighborhoods. This shows material comfort and spiritual fulfillment are different. Is it not significant to find so many persons suffering from a profound anxiety, ennui with life, that we function zombie-like without a direction or purpose? A rise in psychosomatic illness signals deeper pathology. This does not mean that the basic method of reasoning is to blame, only those cherished structures no longer suffice.

What makes you the person that you are, determines your proclivities and behavior including reaction to positive and adverse stimuli, level of happiness, productivity and whether or not you have emotional problems? If we only knew! How much could we profit from such knowledge! What of the age-old questions of nature, the way persons are built, how we are vs. nurture, what our immediate environment makes of us? Are people basically good or evil? Do we have only a thin veneer of civility and control with evil kept constantly in check (We certainly get that impression in Yugoslavia today and looking back through history!) Why are some people more ethical than others? Why do some persons care too much about ethics and morals while others care not at all? In the past these were philosophical even metaphysical questions; it was OK for everyone to have their own personal opinion in the absence or knowledge, but not anymore. Science is slowly pushing back the limits of philosophical debate and answering many of these questions, and frequently surprising intuition.

Despite all of our scientific advances we have only a rudimentary idea about basic human motivation. It is all we can do to vaguely state that motivation may reside somewhere in the frontal lobes. The most advanced research has defined, only to a limited extent some of the influences that shape decisions in terms of anatomical substrate and perhaps more usefully, certain neurotransmitters as we shall see below. But where is the kernel or core of human will? When you decide to do something as simple as moving your finger, where is the ultimate source, anatomically of this decision? We do not know! Our tests, functional MRI or PET scans merely point out activated areas of brain that are implicated in such actions; they may show a pattern of activation that varies from subject to subject and between situations, but they will never show the ultimate source of even the most rudimentary decision. They do not show us from whence derives motivation, especially the drive to change one's world or the direction of one's life.

Two persons raised the same way, even growing up in the same household, can turn out different. No one has been able to predict how someone will turn out by looking at his or her environment, which frustrates social engineers. There is a lot of debate about what constitutes a healthy upbringing. Bad persons can be born under ideal circumstances and in many cases persons show tremendous resilience despite., or even because, of their being dealt a bad hand. There is little question that in our society we tend to trust the highly born. Most of our heroes hail from patrician lineages and we scoff at the lowly born. It's the self-made person who we truly should admire, but we don't trust him. He is a nouveau riche. We seek a purebred highbrow who has had everything handed to him. Yet we all know of persons lowly born, who have achieved a great deal (even if this is not the usual case) and there is perhaps more social mobility in American society than most places around the world. Most of us don't think very much of Rousseau's concept of the noble savage.

We all know someone who grew up under adverse circumstances having to struggle against seemingly insuperable odds, and yet have achieved material, educational or ethical milestones beyond imagining, and of others whose lives were a disappointment despite their having all of the advantages of material comfort and close nuclear family. We enjoy citing these examples because they seem to prove that some persons have an inner fire while few others have any burning passions inside. Examples abound of persons born to limited means who become rich or famous by virtue of burning ambition. Bill Gates, the Intel guy, Clare Boothe Luce, Einstein and Freud, Faraday, and a host of others were born poor and to unknown families.

When it comes to human behavior facile rules, break down, not always, but occasionally, and enough to make all the difference. This slight infrequent breaking of rules, this difference or residual is just enough to stake our claim in the world; it's what makes us a little difference and the same stuff is what makes our civilization, our whole consciousness advance. It's swimming against the tide. It's not doing what is manifestly or immediately advantageous, not secreting behavior that is predictable, but choosing instead to perform unpredictable acts, separate from the mold of contingencies and the people who pull all the strings, iconoclasm, by which we mean not

Momentum= Mass X Velocity, but rather,

Force = Mass X Acceleration.

Sometimes, perhaps not that often, a person does the unexpected, fails to go with the flow, but instead, exerts a force on his world which in turn, changes his course or his outer world. The paradigm, or rather its exact opposite, that illustrates the concept, is

Parkinson's disease, a disorder of the human motor system. The victim of Parkinson's is at his basic worst, a non-initiator of action. He is inertly immobile. If he needs to stand or begin to walk, his greatest problem is actually starting out. Once he gets moving he can continue at a constant velocity. In fact, he has difficulty stopping, and turning will cause him to fall. His body at rest remains at rest, in motion, remains in motion at constant velocity, unable to accelerate in any way, that is change velocity, in physical terms, apply force, in the way we all take for granted in our daily lives. This force is rooted in the human will, motivation, the desire to exert a force and bring about change.

Parkinson's disease is caused by Dopamine deficiency in the basal ganglia of the brain, specifically in the substantia nigra or black substance which is less black in Parkinson's brains because there is also less melanin, a metabolite of Dopamine. In his popular book, *Awakenings,* Oliver Sacks showed how there is spectrum from the Dopamine deficiency of Parkinson disease to Dopamine excess brought about by excessive Dopamine replacement therapy. Patients may run the gamut from immobile inertia, to aggression, excessive movement, and hypersexuality depending their Dopamine levels. Thus Dopamine is the transmitter of more than simple motor action acceleration or force. There is evidence connecting Dopamine with basic motivational processes, an inner fire, pleasure seeking, motivation.

Cocaine increases brain Dopamine by blocking Dopamine transporter, the protein that transports Dopamine from the synapse, back into the neuron, where Dopamine ceases to be active. Every neurotransmitter released into the synapse needs to have some method for its deactivation lest it continue to work forever and cease to be a phasic messenger. For Dopamine this is the Dopamine transporter. Certain persons are more apt to become dependent on substances such as cocaine, may even crave stimulants, because of an inborn deficiency in their Dopamine systems. This anhedonia hypothesis, meaning the absence of pleasure, also known as reward deficiency, is arguably caused by many factors.

Perhaps a person is constantly beaten down, made to feel that his own life is worthless, or grows up deprived. He will become emotionally stunted, hypoactive, will have what is frequently noted in depressed patients, psychomotor retardation. Or, he might have to an inborn dopamine deficiency, some learned or environmental influence that reduces Dopamine effect, or probably more likely, if the process is genetic, a polymorphism in Dopamine or other neurochemical receptors. He might then crave substances that provide him some relief from his deficiency. Addictions run in families, particularly alcoholism, which is rare among some ethnic and even racial groups, highly common in others, but also obesity and certainly other drug addictions and compulsive behaviors, such as gambling that seem to afflict certain kindreds over many generations. We all know that cravings are multifactorial, that there are environmental influences to be sure, but could a certain susceptibility for pleasure giving substances, whether they be foods, alcohol or drugs, be genetic?

The D2 Dopamine receptor subtype is implicated in this process. Experimental Rats try to stimulate certain nuclei in the brain, that contain Dopamine, especially the nucleus Accumbens. Alcohol, Cocaine, and nicotine, increase the effect of Dopamine in the brain. The ventral tegmental area of the midbrain also has Dopamine secreting neurons that project to the nucleus accumbens. Other transmitters, serotonin, enkephalins GABA also get into the act, stimulating or inhibiting the ventral tegmental Dopamine containing neurons. The final common pathway though is the nucleus accumbens that has D2 dopamine receptors. Does an inherited alteration in Dopamine D2 receptors predispose a person toward alcholism and cocaine addiction, pathological gambling or other cravings?

Evidence is contradictory at this point but there is some support in work by Blum et al.[119] Have found an increased incidence of a certain gene type in alcoholics designated as A1 allele, which associates with the D2 Dopamine receptor. These associations are fairly weak if they exist at all, but are more convincing in more severe cases of alcoholism. For example, if the A1 subtype has a frequency of about 20% in normal Caucasian control group, there may be around a 45% incidence among Caucasian alcoholics, not a very convincing association, but enough magnitude to be statistically significant. D2 receptor subtypes seem to account for some susceptibility to alcoholism and also the A1 polymorphism connects with other disorders, motor tics, compulsions, attention deficit disorder and autism[120]. Others have found that alcoholics, as opposed to the rest of us, are simply not as sensitive to the effects of alcohol, that is may require higher doses, more drinks to become uncoordinated, high, or have changes on their EEG. In other words, future alcoholics may have an inherited decreased susceptibility to the acute effects of alcohol.

What if alcoholism and other cravings were found to be true genetic based disease states? Does that absolve the alcoholic from responsibility for his own behavior? A more important practical consideration is that understanding drug cravings, which may have to do with an ascending tracts in the brainstem and Dopamine transmission, an internal reward system, may one day help us to find medical treatments for all sorts of cravings and addictions.

To me, the interesting part has to do with the weakness of the data. The statistical strength of Dopamine receptor subtypes in ethanolism is fairly minimal and even there, some groups reach contradictory conclusions. Under the best of circumstances it is far from possible to draw firm conclusions about a genetic or any organic basis for any specific human craving.

The best model of craving is Tourette Syndrome which is a study of the irresistible urge. In Tourette's the patient has an urge to perform certain acts, ordinarily simple tics or mannerisms such as forceful eye blinks and shoulder shrugs, but also more involved activities throat clearings, vocalizations and other utterances, sometimes expletives, "coprolalia" ( from Greek, literally fecal babbling). Some persons have copropraxias (praxis, to do) and needing to touch themselves for example, even in public. The

Touretter will be able to suppress his irresistible urge for a while, but the urge will always be expressed. He will try hold out until he is alone, but the act will always get done and there is no way to stop it. Not surprisingly, Touretters have various compulsions of all types, even to ritualistic behaviors and obsessive-compulsive disorders sometimes gambling. Dopamine is the major transmitter implicated in Tourette syndrome, as drugs that block Dopamine, the anti-psychotic drugs, are by far the most effective tic-suppressors, but other medicines, especially Serotonin-increasing antidepressants, seem to be most effective for compulsions. These medicines are effective at treating many secondary symptoms such as an addiction to gambling. An organic disease such as Tourette's is the strongest argument in favor of the notion that humans are subject to predetermined irresistible urges that cannot be controlled, also that one may use organic means, namely drugs, to control these urges. But Tourette's is a disease which says little about health and superlative ability.

In fact most information about neurotransmitters and psychiatric disease, the brain, behavior connection, is generated about what we know about the effects of certain drugs. Typically certain classes of medicines are found to be helpful for certain conditions, sometimes quite by accident, then it is up to researchers to find explanations for why this might be so. Thus isoniazid used to treat tuberculosis, was found, quite by accident, to diminish depression and cause in some, susceptible persons, a form of mania or excitation. From this observation, the first generation of antidepressants, known as tricyclics developped, amitriptylene and imipramine. These drugs were only later found to affect the transmitter brain norepinephrine. Tourette's was at one time thought to be a purely psychiatric disease. Today, many of us doubt that such a thing exists as a psychiatric disease because of known physical concomitants. The first hint that Tourtette's is actually neurological came from empirical observations about the effects of Haloperidol on the tics in this disorder. From clinical this observation, a Dopamine hypothesis about the causation of the disease arose. Why? Haloperidol was known to exert its effect, primarily through dopamine blockade.

## Iconoclasm Defined:

Igor Stravinsky was an early twentieth century iconoclast. His *Rite of Spring* explodes with energy, rhythms so earthy and complex Stravinsky had difficulty figuring out how to put them to paper. This was a heady time, in the days before the Russian Revolution, the Great War, in the years between Einstein's Special (1905) and General (1916) Relativity, the victory of Quantum mechanics over a Newtonian Physics. *Le Sacre* raised more than a stir at its premier in 1913. By that time Stravinsky had already introduced two wonderful earlier ballets, *Petrushka* (1911) and *Firebird* (1910) whose music was influenced by his brilliant teacher Rimsky-Korsakov. But *Le Sacre* was different. Its offenses were sufficiently revolutionary, too much for his audience to take. The premier touched off one of the more celebrated riots in music history. It is hard for most people to appreciate how sometimes great art must offend, not fit neatly into the blueprint of previous works. Here Stravinsky literally helped create modernism. The ears and eyes of the musical public were not quite ready for this, but that is what must happen, once in a while, if we are to make any advance. It is emblematic in art.

Le Sacre is about a pagan ritual that welcomes Spring with the worship of the earth. The theme is a ritual in which a chosen sacrificial virgin whirls and dances herself to death under the watchful eye of sage elders. The music crashes through the gates breaking conventional romantic musical idols along a path of conquest. Great music and art is most of the time is simply validating and uplifting. Yet if Stravinsky's sound is appealing, and though explosive and primitively advanced rhythm, it does not sound all that unusual to us today. We have heard so much since then that we hardly know what all the fuss was about back then. Its appeal is not to emotions primarily which is better done with melody than rhythm. Still the music finds new patterns of activation within the cortex, new patterns of arousal and mental exercise. Ironic that a break with the romantic past in the early twentieth century depicts pagan ritual – iconoclasm in the service of idolatry. Le Sacre may break idols, but the earthy primitivism of matter still wins out over mind.

The archetypal iconoclast was Abraham. The Abraham in Biblical accounts, broke all of the idols of his father and idol merchant, Terah, and then left the ancient land of Ur to change history and become the father of nations. People would pay good money for these idols thinking them to be gods. "Why did you destroy all my idols?" Abraham's father cried, (so the legend goes) for they were more than gods, they were Terah's livelihood. "That giant statue over there who is bigger than the others, destroyed all the other gods and then himself." Abraham is said to have retorted sardonically. Then and there his father realized how stupid was attribution human will and motivation to inanimate objects and began to worship the spirit-god as did his son.

Iconoclasm. There are echoes of Abraham. The convention was to worship inanimate material objects, themselves the products of human hands "Why do you worship statues that can't do or feel anything? How can these objects possibly have any control over your life?" you hear Abraham say. Chances are his father sympathized with him. Undoubtedly Terah sympathized with his son, but didn't have the personal fortitude to leave his life behind, or perhaps he was by then too old. Chances are he merely made his living selling the dumb statues but didn't believe in them. Abraham's insistence on a non-material spirit who created and controlled the universe was the first iconoclasm. in history. That is not to deny that to this day we worship physical objects.

People still need to see something tangible and real and churches have made peace with this human clamoring with its extensive iconography that a lot of people require in their religious worship. When you consider it, this dependence on an icon is one of the most powerful and pervasive forces. Seeing is believing and it's hard to believe that something spiritual, numinous, is actually there.

Idol-breakers accentuate humanness. They swim against the tide, eschew the conventional. they arouse hatred and alienate. This doesn't mean that the outrageous is always good or valuable, nor that art needs to be outrageous in order to be good. Without form, or meaning, expression turns crude, vulgar. Formless meaninglessness is what we have in rap and rock music of a drug influenced new age. Recall that a whole political movement revolved about the sacrilegious Andres Serrano's "Piss Christ," a photograph of a crucifix immersed in the artist's urine. This drew criticism from Senator Jesse Helms that nearly did away with the National Endowment for the Arts. Innumerable articles were written about the subject. Finally the Supreme Court got into the act and supported a standard of decency as prerequisite to federal arts funding. Many persons recoiling from this outrageous art that offends sensibilities, have used this for a pretext to condemn the entire art world. Yet it is in the very fiber of art to swim against the tide, to challenge convention with the outrageous. Sometimes this simply offends and does so without substance with little regard for form let alone our deepest feelings. Such amorphous garbage does not express anything. Rather it tramples over us.

Adolescence is in part about the same rebellion, a declaration of one's own identity, as if to shout to the world, "I've arrived! I'm here and in your face!" It's the source of a youthful enthusiasm that to adults maddens and insults, but it is what an individual identity is about. At least in adolescence you can say that you are on the road to something.

David Gelernter in his book *1939*, *The Lost World of the Fair*[121] gives us a poignant look of youthful enthusiasm and optimism, a post-depression world in the throes of a world war when one would think there was little reason for such optimism. Gelernter presents a young architect and his own unconventional visions for playgrounds, schools and buildings. A building as a vision can positively change lives and society and its institutions. Creativity begins well before you have been infused with the conventional, while you are learning in school, as you are flushed with the new, and a host of things you have yet to learn. After you have reached a state of high competence in a field, you begin to see things as others do. You lose that youthful creative spirit. Having established a line of thought in your youth it is usual that you are trapped in it, you will spend the rest of your years in that area. For one thing, you just have too much invested.

Unless, you are fortunate enough to be able to do something totally different in later life, a phenomenon well described in Betty Friedan's *Fountain of Age* and in Gail Sheehy's *Passages.*

## Enter Biology:

Medical science has uncovered a large number of biological processes that profoundly affect personality and behavior. As more data accumulates we are finding that behaviors and internal personality structure are controlled by innate, genetic internal elements. These come less from Western literature and more from biology and chemistry. Controlling for environmental factors, intellectual, musical and athletic abilities are largely inherited. It seems to be no accident that abilities as well as diseases and disabilities run in families. It was more than the musical environment that made musicians in generations of Bachs, Mozarts and Beethovens. In retrospect, it is hard to believe how much we were taken in by arguments of the strong environmentalists anyway. All of us know of examples of children driven by some inner force though never trained in the field of music medicine or law, and never encouraged to enter into it were driven in a certain direction anyway. What determines such a drive is more than a mere talent. It is motivation and proclivity. Society functions best when it allows people to make their own choices.

Most of us tend to gravitate into the fields for which we are the most suited. Society would do best by not restricting this choice and basically leaving people to pursue their own dreams. One example is Richard Wagner a towering figure in music the greatest composer by far of German opera. Friedrich Wagner, a minor police official who died 6 months after Richard's birth, was probably not his father. Richard was most likely the illegitimate son of Ludwig Geyer, an actor and singer who married his mother a year after Friedrich's death and provided all important exposure to the theater while absolutely forbidding him to have anything to do with music. These facts are even more ironic since Wagner, noted for his later anti-Jewish sentiments was probably the son of Geyer a Jew*. But his paternity would go a long way in explaining his proclivity for both literature and music and his grand scheme to unify the arts in one magnificent operatic structure. Richard studied music on the sly and Geyer died in 1821 at any event. Hearing Beethoven's Fidelio in 1829 was something of an epiphany for Wagner that awakened his own latent interest in music. Before that point his artistic interests were mainly literary.

More and more we are hearing about spectacular discoveries bearing on biological links to behavior. A genetic defect in MAOA, monamine Oxidase A, which is the enzyme that deactivates Serotonin in the brain was created in mice. These mice with an up to nine-fold increased brain serotonin are more tremulous as pups and more aggressive adults, exhibiting increased grasping in a clumsy attempt to mate, irritability, and a tendency to bite and attack. Two years earlier a Dutch kindred lacking MAOA was discovered with overly aggressive with males who would rape, commit arson, and brandish knives, try to run down an employer with a car. These men were also found to have elevated brain levels of Serotonin due to a deficiency in the Serotonin metabolizing enzyme MAOA.[122] We know personality and behavior can be altered in using drugs that increase or decrease levels of Serotonin and other chemicals, so why should veterinarians be hesitant to use the same chemicals in animals that are known to alter emotion and behavior in humans? Many of them do will give your pets with emotional disorders doses of antidepressants and tranquilizers such as Prozac and Valium.

Even more biological ammunition comes from ethology which looks at the innate behavior patterns of animals. Animals inherit much more than physical characteristics such as eye and hair color. Complex patterns of behavior are also inherited and have profound effects on maturation and mating. Some speciation is determined more on an inherited behavioral basis than any physical quality. Certain simple physical characteristics act as releasers of behavior. The famous example given by Konrad Lorenz a pioneer in this area of study is the propensity, expressed over a limited period shortly after hatching, of ducklings to follow a moving walking animal. Almost always this animal turns out to be the mother of the hatchlings but if a human should intervene and begin to walk away during this critical period shortly after birth then the ducklings will continue to follow the human who will be treated as the ducklings' mother. This is imprinting. What is required is an inborn pattern of behavior, wired into the central nervous system, and a releaser that brings out this behavior often in a critical period. Earlier in chapter one the releaser of the male stickleback fish which is another classic example, was mentioned. The best example of imprinting pointing to the interaction of environment and inherited behavioral patterns is birdsong. Most birds are born with a simple limited repertoire of song patterns. These basic patterns are later blossom into their full complexity and diversity during a critical period of learning that occurs early in life. A bird's song is endowed with a local species-specific dialect that is superimposed on simpler patterns and learned through contact with older adults who have fully developed song. The male utilizes song mainly to attract a female of the same species as part of a mating ritual. If he is raised in isolation from other males, the very basic innate unembellished song pattern is what remains. Therefore the song of the male bird is largely inherited and partly learned as well. Not only the basic song apparatus and wherewithal to produce the song, the vocal and neural apparatus, but also the proclivity to learn more complex song patterns during a specific period of the bird's early life. Song, as much as any physical character, ensures that females only of the same species are attracted for mating purposes. Perhaps other closely related females might look almost exactly the same and mating might possibly occur, purely on the basis of morphology or appearance. But the specific male song, and heritable differences in actual mating behavior, help ensure that this will not occur. Male aggressive behavior is controlled by the same factors. The birdsong is sexual and has both attractant and aggressive repercussions. Other males, hearing conspecific birdsong that is part of the competitive mating ritual, if done in their own territory, might be inclined to attack so that a physical fight will occur. More often the song is a declaration keeping competing males away from females and out of the singing male's territory. The male bird will sing more vehemently if deprived of sex or sexually excited. Aggression will be more vehement as well. While in birds with more complex brains some specific elements have yet to be worked out neurophysiologically, it has been possible to show in certain stridulating insects such as grasshoppers, that inherited song dialects, also used to attract females, translate into firing patterns of single neurons tuned to respond (fire) to exact frequencies produced. Thus there is a certain inherited lock and key when it comes to bird and insect sound production, a correspondence between the motor output, influenced as it is by internal states, and sensory response.

It's also impossible to escape generalization of these same principles to human language dialects. Presumably with us, language is more a matter of learning with minimal genetic component (excluding the basic abilities to speak and to respond to language) but the effect is the same. Mostly by learning humans are "tuned" to respond to specific languages and dialects and mating and other social interactions are much more likely to occur among persons who are genetically, racially and linguistically similar. Our analogy also extends well beyond language over the whole gamut of ethnicity with its own biological and cultural and linguistic components. Ethnic linguistic variation, which is an evolutionary tree of sorts, largely parallels genetic differences. Human racial varieties parallel ethnic and linguistic ones.

Considering that almost all organisms have evolved methods for kin recognition[123] ranging from elaborate olfactory systems in paper wasps, tadpoles and sweat bees, to location of eggs in certain birds we should note that many of our ethnic concerns are biologically determined. Humans go out of their way not to breed with close family members. That would be a disadvantage in two ways. Recessive traits, many of them injurious, would be expressed doubly by mating with a close relative[Φ]. Also mating with very close relatives would limit the store of genetic variation, so that making populations so homogeneous they would not have the basic fuel required to evolve. The fuel of evolution is individual variation. Organisms evolve when variations in individuals affect fitness.

On the other hand evolutionary forces limit out of group mating. This is true especially in complex social animals such as humans. Each large group or kindred is also a separate evolutionary experiment and there is pressure to mate within a group (ordinarily for man a common language or ethnic group) and fierce competition for survival between groups as well. The survival of the individual's genes is married to that of his group. If the group is large enough, there will be adequate genetic variation to fuel adaptation. History describes kindreds conquered, displaced, obliterated. There are records of such total destruction since the invention of writing. When you consider it, it seems that language may have developed entirely to separate out different ethnic groups or, the biblical descriptions of such processes through marriage, and war are very likely highly accurate. Groups segregate after separation of a founding member into separate family groups or kindreds then continue to grow apart. Two related kindreds that separated only few generations ago may later make war against each other much as sons of Isaac and Ishmael who had Abraham as their father or Jacob and Esau (founder of the Edomites) who were both sons of Isaac. What this means is that we need come to some accommodation between mating with relatives that are too close (as exhibited by incest laws) and those who are too distant (the mechanisms of language, religion and war are instrumental here.)

Ethology has shown that behaviors are inherited just as much as morphologic characters. Behavior is the efferent side. On the afferent, receptive side it is very likely that we inherit a propensity to perceive the world in a certain way. It is certain that a whole logic is wired into the brain and that the origin of this logic or wiring system that shapes our perception is in evolution. We turned out this to view the world in our own specific way as humans because evolutionarily that is what was most adaptive; that is what increased our fitness in the world. All of us wonder about what is truth, what determines out perception of truth debates which we have today but which were fully shaped by enlightenment philosophers by Hobbes, Hume, Berkeley and Kant. The debate in epistemology, the origin of truth, how we know what we know, whether what we think is true bears any relationship to an objective reality, revolves about how our world is formed by the interaction of the inner workings of our mind and perceptions and how this fits into a pattern that is our world view. Is there a mental form or pattern into which we fit our perception or a blank slate into which we place individual raw perceptions to eventually make a mental model. The striking thing for me, is that all of these philosophical debates were formed before Darwin whose greatest lesson for us is that the form of our mind, and brain arose from nature and very likely reflects nature. Every perception is molded by and placed in context with a mental pattern that is inherent. In the brain, and the mind, every perception has its place and context in a mental system comprised of inherent logic and emotion. That system is a product of, and must reflect, nature, otherwise it would never be instrumental in helping us to adapt to our own environment. Thus science has materially contributed to the philosophical debate about truth. When these men argued in the seventeenth century they had not the tools to properly close their debate having had almost no knowledge about the form and function of the brain which is the organ of perception, or its biological origins from nature.[‡] What these men lacked was an understanding of the critical link between the knower and the known. Not appreciating the nature of this link, they rightly questioned humankind's understanding of reality, whether understanding was true or not. My contention is what they did not understand is how the brain and hence the mind were born of nature and the universe and was therefore linked with it. Knowledge is thus critically linked with the knower, for the knower has adapted practically to truths within the universe. Presumably mankind continues to compete successfully with other organisms because his view of the world is at the very least adaptive, if not accurate. Perception does not arise in a vacuum, it is not created willy-nilly from the outside or above but arises empirically through daily encounters with an unforgiving environment. When you see rightly, your dealings are bound to be a success; misperceive and you could have a disaster. Man, as part of nature, will most likely be an accurate manipulator and systematizer.

So the third aspect of the scientific examination of personality and behavior, after analysis, and behaviorism, is biology. And here we have made the greatest strides in the latter part of the twentieth century. Biology definitely explains a lot. Major objections to Darwin and evolutionism, far and away the most powerful biological theory before the revolution in molecular genetics, that we are in the throes of at present, have come from radical religionists, and they are armed with fundamental misunderstandings about a Christian and English speaking literal interpretations of their Bibles and little, if any, appreciation of other areas of enquiry.

Scientists have failed to inquire about whether it is possible for a mechanism of free will to be built into a biological system. Could it be that the brain is designed to be a repository of free-will? This would be contrary to our understanding of all other mechanical systems to be sure, but is a theoretical possibility that should not be overlooked. Of course it is contrary to our understanding of other mechanical systems and that is what makes it so interesting. We have alluded to examples of other systems, scientifically described, whose behaviors cannot be readily predicted. These include descriptions of sub-atomic quantum behavior where the uncertainty principle applies. Since we cannot know at time $t_0$ the position and momentum of an elementary particle at the same time, the behavior of such elementary particles at time $t_1$ cannot entirely be predicted. Also, in recent years, science has explored chaos theory which describes the relatively unpredictable, yet still deterministic in the wide sense behavior of such areas as weather prediction and the stock market, even EEG and EKG non-rhythmic patterns. It is even more probable that free-will will fall into the purview of chaos. Another alternative is that of new areas of inquiry about self-organization order out of chaos theories as described by Ilya Prigogene and Stuart Kauffman[124] for example. Or, we may be forced to admit the possibility of free-will built into neural systems via scientific principles that have yet to be defined.

Psychologists would like to be able to predict behavior and prognosticate on abilities. If nothing else this would legitimize their science by giving it predictive power. What if one could develop a test that would accurately predict school performance or find the best recruits for officer candidate school? Imagine if one could tell who among us are most likely to commit a crime. Parole boards could know which inmates to keep and which to set free. Our penal system would be turned on its head. For the first time one could come up with rational arguments, a high probability that a certain personality type would be at risk to commit a crime, to preemptively incarcerate or neutralize a criminal prior to his actually committing a heinous act.[ϕ] . What if you knew which persons were more reliable, which ones had a tendency to steal, and the proclivities of all candidates, then you would have an idea what person could be hired and which to turn down. The Holy Grail of psychology is finding instruments that can *predict* future behavior. The history of such attempts is full of failure because behavior is willed and not determined.

The best that you might hope for is to be able to define abilities, proclivities, and more vaguely on emotions and motivation. You could make a statement about various influences then, but would be unable to predict behavior with any measure of certainty. A probabilistic model is all you could hope for. With those tools you might furtively enter the realm of prediction, knowing depending on the design of your test you may not be accurately assessing all you intended to measure.

In a certain sense it may not matter very much. For example consider the much-vaunted SAT test used for high school students. The Education and Testing Service has tried to design a test that can predict college performance, to provide an objective means to compare high school juniors from many different schools and backgrounds. We all know that a student with high grades from a mediocre high school may not be as academically gifted as one with lower grades from a more demanding or competitive class. The SAT seeks to correct for these deficiencies. The SAT test is very subject to a practice effect which means scores improve considerably on repeated testing because the questions from exam to exam are not that different. Consequently the SAT's and other standardized tests have started a mini-industry of courses that essentially defeat these exams.

What about IQ tests which measure intelligence and seek to stratify persons into prognostic categories predictive of school and work performance? It is shocking how little IQ testing is questioned in the lay and academic media especially when you consider the checkered history of the tests and how they have been used by pseudo scientists of all stripes in the past. In the early part of the Twentieth Century these instruments, developed by such pioneers as Binet and Terman, administered to military recruits found a low general level of intelligence, somewhere around the 13 year old age level. IQ tests administered to new immigrants on Ellis Island yielded even more frightening results. Many could not tell you the month or year or copy a figure implying a high incidence of mental subnormality greater than 80% for some groups of Eastern Europeans. It wasn't long before such tests became the principal argument for the eugenics movement seeking to limit immigration from oppressed and indigent populations in Southern and Eastern Europe on the basis of their certifiably inferior intelligence. Allowing large numbers of these dirty and ignorant partisans would encourage mixture of inferior stock into our more purely Anglo and Nordic gene pool and end up lowering the intellectual endowment of future generations of Americans. So went the argument. When within one or two generations these dirty poor stupid and immigrant families raised the intellectual standards of public schools and colleges and were picking up Nobel prizes and other high academic honors, the fallacious arguments had been forgotten and new excuses invented for keeping immigrants and inferior groups out of the United States. Laws were passed that blocked immigrants from entering the U.S. between the wars. No one knows how many souls were later slaughtered in the second World War and the holocaust on the basis of these fallacious judgments.[125] The same arguments have been made continually and in our own time in fact very recently. The well-publicized book THE BELL CURVE makes very much the same point. Persons perform on IQ tests very much along racial lines and so, the argument goes, the wisest use of limited resources would be to cut your losses, and not emphasize teaching of abstract topics to persons ill-equipped to learn them. This was a position essentially endorsed in a position paper later published in the Wall Street Journal whose many signatories included IQ mavens of all stripes. Nothing in The Bell Curve explains why, much to the dismay and puzzlement of the testers, IQ has steadily been slowly rising from 1918 to 1989 consistently in constant IQ points to the tune of about 24 points. A person scoring at the 90th percentile 100 years ago on the progressive matrices test, one of the least culturally loaded subtests, would end up in just the fifth percentile today. This is the so-called “Flynn Effect” named for James R. Flynn who first noticed it when studying intelligence in the U.S. Military, but the effect is quite robust, seeming to extend over the decades to just about every different I.Q. test.[126] Just one major area of fallout is the estimates of dementia in the elderly who come from a period when I.Q. scores were considerably lower. If elderly I.Q.’s are compared with those of younger folks, you would be looking at the Flynn effect not any measurement of dementia. For that one needs to dig out measures of these own folk’s past performance. While achievement test scores decrease or at best remain stagnant, I.Q. scores are slowly increasing and no one knows why. I.Q. is supposedly mostly a genetic endowment, not very culturally determined and should remain constant over generations but doesn’t. Is our intelligence actually increasing perhaps through better nutrition, more contact with electronic devices and more abstract concepts?. Almost certainly not. There is most likely some systematic slow change in the means of administration of the test that introduces artifact, perhaps a similar artifact to that which stratifies the races in such treatises as The Bell Curve. No measuring device is perfect and this one may be subject to the biases of the measurer. And the same tests are used to classify and prognosticate on your own children, even though the instruments claiming to measure intelligence actually do no such thing. Even correlation of test scores with academic performance is poor.

Current research indicates that about 50% of variation in human IQ is inherited. Common techniques used to explore this interesting area include correlation coefficients of certain testees, identical and fraternal twins raised together and apart, biological parents and their children whether adopted or raised by them. The upshot of such research is that mental tests correlate more strongly with inheritance than upbringing. Your IQ is much more likely to correlate with your biological sibling or parent especially an identical twin, than adoptive parents or siblings. Heredity is more important than environment. Not surprisingly, mental abilities also correlate fairly strongly with each other. Spatial and verbal abilities while being handled to some extent in separate areas of the brain, correlate well with each other for the most part.

Starting from the seed of the IQ test, a whole branch of psychology has developed around the use of tests and instruments that seek to measure aptitudes and emotions. These consist of hundreds of tests and inventories many of which are all too familiar since we have all taken many of them. The simplest are the test administered by teachers and we all know how imperfectly such exams actually measure mastery of subjects we have taken in school. After you come out of an exam aren't you always questioning whether you had a fair or and unfair test? In many cases you have not been tested fairly but at least a teacher's exam does get you to study. The closest thing to an IQ tests are exams of "scholastic Aptitude" given to high school students in an effort to prognosticate on their abilities. You can improve your performance markedly on such exams through practice and a mini-industry has evolved just because students do improve their scores utilizing these courses and the test largely determines entrance into competitive schools. If you can improve your scores then the word aptitude is a misnomer, but at the very least, the test cannot be quantifying any innate academic ability and the exam is quite easily defeated. The problem is that there is no very good way of comparing students of diverse background and from different schools. Psychological testing has evolved into a whole specialty.

Many people are not aware how many different tests there are. Tests measure abilities and proclivities of all types. Interests, personality types, emotions, linguistic mathematical musical abilities, character, honesty, frontal parietal lobe functions, are all measurable with these instruments. Tests are not used only to admit a student to a certain college but to help quantify brain injury for court cases or to help decide whether or not to hire a prospective candidate. As on all measures you accept a certain limit of accuracy. Schools and employers are well aware that in trying to find the best candidates some of the best students and employees will be excluded while other poor candidates will be accepted and are willing to take their chances that on the whole, in most cases they will find the fittest candidate. Besides, in case anyone inquires (or sues) they can always show that all comers no matter what their race or ethnic background were subjected to the same "objective" measures. The most common instruments used to measure brain function are IQ tests especially the Wechsler scales and Wechsler memory scales, MMPI, Halsted Reitan and Luria Nebraska batteries, Beck depression inventories etc.[127]

A few cognitive deficits are specific and can be localized to a given region of cerebral cortex. They localize to areas mostly because of lesion experiments that show when a certain area of the brain is cut out then these functions are impaired implying that the affected area must somehow be connected with performing the function. Contrary to popular belief, decreased function via lesioning does not necessarily infer that the area in question is responsible for the function. One situation is termed Gerstmann syndrome after the psychiatrist who first noted a curious combination of problems. The subject with Gerstmann's is unable to write, recognize his own fingers, calculate, and tell right from left. He has in order of appearance above, agraphia, finger agnosia, acalculia, right-left disorientation. The lesion is in the angular gyrus of the dominant (usually left) parietal lobe. All of these problems, taken individually, can be caused by anatomical defects elsewhere in the brain, but when they occur in combination, they almost always point to a problem in the left angular gyrus or an area near it. This doesn't mean that the angular gyrus performs these functions, far from it. However it is one area that is necessary for the performance of these individual tasks. Why?

The angular gyrus processes data arriving from many sources, the visual cortex, auditory areas etc. It helps to further process this data. It is one of the areas of the brain collating and operating on information from diverse sources. Some of the tasks are actually composites multiple subtasks. Let's take the problem of acalculia. Before you are able to use and manipulate numbers you have to be able to recognize them properly through either an auditory or a visual pathway usually, i.e. you must be able to receive numbers and the problem. Then you must understand the operations, addition, subtraction etc. that you are expected to perform. You have to have some language function in order to know what to do to operate on the numbers. This language function is the same as what is taken care of by the left hemisphere of the brain and as such persons with some types of aphasia (see previous sections) may have a number language defect or a type of "aphasic acalculia". Such problems in linguistic interpretation ordinarily localize to the dominant cerebral hemisphere.

What if you are confronted with a complex problem whose answer you don't know by rote such as addition of more than two or three columns with carrying and place holding? Then if you have a spatial deficit such as could be caused by a lesion on the right side of the brain you could get into trouble. Hence such problems involving extended computation and place holding are trouble for those persons having right hemisphere deficits. Other folks have more a pervasive problem with calculations that do not involve place holding or language function. Those people ordinarily may have more extensive bilateral defects such as occur with Alzheimer's disease and have "anarithmetria." Therefore as we have found with some other cognitive deficits such as naming, arithmetic computation, calculation, is a composite function that is not entirely localizing since so many steps are necessary for it to be carried out correctly by the brain. But where it occurs in conjunction with other defects as listed above, it localizes to the left angular gyrus [128](picture) Complex mental tasks, even those as seemingly simple as naming and calculation involve many areas of brain that are all necessary for the completion of the act. Each of these areas do not perform a function that is sufficient in and of itself, nor is it specifically involved in doing the function in question, say to name or to calculate. An area of brain, if lesioned, may impair a certain function. Yet it is rarely possible to connect that particular cognitive function specifically to that area. The most you can say is that particular area is necessary for the performance of that function, not that it alone is totally responsible for performing it.

Cognitive functions are characteristically composite in this way, that is, they involve putting together simple tasks into a complex whole. The brain usually parses out each simple function to a specific zone. Therefore each of these possibly disparate zones may in fact be responsible for a deficit that we see when we examine the patient. The classic examples are with calculation as shown above where a simple spatial problem such as space holding may impair the function and that is localized to the parietal lobe, or there may be a problem translating numbers linguistically that has to do with the language areas of the brain and either of these two simple deficits is sufficient to cause the larger deficit of acalculia because each is a necessary prerequisite to calculations. Another classic example is with naming, i.e. anomia. In order to name an object you have to have some visual recognition which implicates the visual areas, ie the occipital lobe and connecting parietal lobe, and you have to somehow connect the language area of the brain with the visualized object, this involves white matter connections from the visual to language areas, and even after the language to visual connection is made, you still have to find the right word which involves in the language area choosing from a vocabulary repertoire of nouns from memory banks and then in Broca's area of the brain, making the sound, either saying it or writing it. If any of these links is broken, the subject will seem to have an anomia. The clinician's skill is in further localizing the deficit which is almost always done on the basis of accompanying deficits. For example a subject with anomia who has a visual deficit will have his lesion localized to the parietal or occipital lobes whereas an anomia that associates with a decreased speech production in general is located in the frontal lobe, near Broca's area. Hence for brain localization we depend on the concurrence of more than one, hopefully two or even three deficits. A single task especially a composite task, often involves disparate areas that are each responsible for simpler functions but each is a link in the chain which is broken, a lesion in any of a number or areas will result in the single deficit.

Psychologists and neurologists too typically find certain deficits on behavioral test or on examination. On the basis of these problems they try to find a diseased area of the brain. A patient with a Broca (motor) aphasia is supposed to have a problem in the inferior posterior frontal lobe (Broca's area). Usually we are correct when we try to localize problems in this way. Aphasias or brain-related disorders in handling language, are the easiest cognitive deficits of all to precisely localize. With a proper analysis of the specific dysfunction, the clinical impression of a lesion correlate very well with what is seen on a CT or an MRI scan. In a few instances we may find a patient with a Broca type aphasia does not have a lesion where we expect it, in Broca's area but somewhere else, often in the deeper part of the left hemisphere or even in the thalamus and we are often wrong, because there are a number of other brain areas that are necessary for the performance of the function and any of these might be responsible. However we do a lot better when we identify a *combination* of problems as these localize to specific areas of brain most precisely. What if we have a Broca Aphasia and right sided weakness? Then we know where the problem must be, in the Left frontal lobe. It's a lot like localizing with one coordinate compared with triangulation which is a lot more precise. For this reason one might in fact achieve *deficit triangulation.* A lesion in the brain is most like a radio transmission that is best localized along more than one axis. These axes in space are the specific deficits. Even a seemingly simple classic deficit such as naming turns out to be a composite function involving a number of brain regions.

It is most fascinating to compare changes in personality with alterations in the brain. Here we have to keep in mind how personality is on even a higher level of abstraction than even cognitive cerebral functions. Not only is the topic in question composite, culling together multiple brain areas but also there are much more subtle problems in recognition of a personality change. We are talking about seeing human patterns of response or characteristic style of response that each of us has to specific sets of like situations. How does a person characteristically respond to stress, for example? Does he worry and perhaps become immobile and withdraw, or is he a person who takes action. This requires at the very least, repetitive periods of observation before a conclusion can be drawn, or perhaps at the very least, self report.

One of the biggest problems with the materialist paradigm is the following: Why don't people always behave in a way that is manifestly advantageous to their self-interest? Why do they persist with behaviors that are injurious to their interests? It is the stubborn tendency of persons to act against what is manifestly advantageous that thwarts attempts at social engineering and makes behavior unpredictable. Most of us, as a general rule, will conform to rewards and punishments of our society. Young persons see that lots of money is to be made in business. Therefore, our colleges are enrolling business majors who expect to make a killing when they graduate. When repressed by an immoral and corrupt dictatorship or crime boss, 99% of people will follow the rules. But eventually, evil will be thwarted by the few who are courageous enough to act against their own immediate manifest self-interest. How do we account for this phenomenon?

This poses a terrible problem for those who seek to control others by pulling strings and manipulating rewards and punishments. Not everyone will act as you expect them to. There may be no better way to preserve the language of a small minority than to forbid its people from speaking it, as the English did in the case of Scottish Gallic still spoken today, no better way to preserve a minority religion than the unfriendly suppression of its practice. Umberto Eco in his book, FOUCAULT'S PENDULUM writes, tongue-in-cheek, of the Templars, ruthlessly repressed by Philip IV of France in the fourteenth century, its leaders tortured into confession and burnt at the stake, yet whose descendants exist to this day as a sort of underground, precisely because of this repression. In short, what we have identified is a rebellious spirit, unexpected behavior among some, by no means all, repressed people, which makes life difficult for the despot or manipulator, makes behavior that much more difficult to predict, and in some ways defines what makes us human.

All kinds of persons seek to predict your behavior and to profit from their predictions. Schools and teachers stratify their students, in an effort to teach more to the quicker ones and less to the slower ones. Presumably it helps to concentrate on students that that they feel have more ability, but it also explains their failure to reach the others. No one has proven their prognostications to be correct. Marketers of products are betting they can create a demand for consumer goods. Employers want productivity and the government demands a high level of behavioral conformance to maintain the peace but also for less laudable enterprises such as collecting taxes and fighting wars. Politicians are trying to find ways to control you, whether they be dictatorial and are seeking means to get the populace to conform to their rule, or trying to get elected which they frequently do by stretching a wet finger to the winds to figure out which opinions will elect them at the moment. Men who bet on sports want to know which team is most likely to win the Super Bowl and or which horse will take the race and what the odds are. Wall Streeters need to know which way the herd is stampeding, mostly to run in the opposite direction, and profit from the result. There's very little doubt that those persons who can predict behavior, or better, control behavior, stand to benefit enormously. The only thing is that daily, all of these educators, pundits, politicians, sports fans, entrepreneurs are proven dismally wrong. The essence of human behavior is unpredictability. Whereas you more or less know what to expect from a machine, the way persons react, frequently surprises.

That makes persons somewhat less subject to prognostications and manipulations. In Russia, thousands of women are beaten by their husbands. Almost all of them take the torment, humiliation and physical assaults in silence. As in many places, women are treated as property. But one woman, not a woman of means, is regularly visited by her alcoholic husband from whom she is now separated, because he has a right to see his seven year old son. She lives in a tiny flat in Moscow. This lady behaved differently than all the rest. She started a hot line and an organization of battered women, a clearing house and a support for women of all social classes who are abused by men. There are thousands of women in her shoes, but only this woman dared to behave in an unexpected manner. And in India, Africa, the Middle East there are courageous women also. Bear in mind they constitute only a very tiny fraction compared to the ones who suffer in silence. Despotism and abuse usually work except in a small number of exceptional instances. But this makes all the difference. There are those, who have begun to stand up, for their own dignity and the dignity of their sisters. In 1944 Hanna Senesch, a 23 year old Jewish girl, inexplicably parachuted into Yugoslavia in the vain hope of joining the Hungarian underground. Of course she was aware of what would happen to her if she got caught by the Nazis. Hungarian underground partisans, learning she was Jewish, did promptly turn her over the Gestapo who executed her. What possesses a person to defy all the odds in this way? It is the very same quality that makes behavior unpredictable. Not everyone in every circumstance will act for their own benefit. Some persons are moved by another force, the very substance that confounds those who seek to control and manipulate others. Evil doers are usually right; most people will conform to the will of the power structure. That some will not conform, that is what make life interesting, and it renews faith in humanity.

## Character:

On the other hand is a behavioral determinant that you can more or less count on. This is the person's character or ego structure. It consists of habitual behaviors that are somewhat mechanistic in their predictability and automatic nature, also of beliefs, points of view, that are part of personality structure. Thus behavior is composed of two contradictory elements, the unpredictable non-mechanical side and the characterological constant side.

At first glance it seems that character is the more mechanistic and determined part of human nature and not subject to human will and motivation. The proper understanding of character and personality does make behavior more predictable though to truly know a person, to get past their masks and facades, is no mean task. But it is an error to suppose that the unpredictable side of behavior is the part subject to free will. We are powerfully motivated to change our nature and character as when we are transformed by work, by education or efforts to improve our habits or ethical response. On the other hand what appear on the surface to be random willful behaviors usually aren't. Persons without character fail to show up for work or commit random acts of murderous violence, performed under the influence of base tendencies, resulting from forces other than free will. The supposed free-wheeling murderer or crook is not acting as much with his free will, as collapsing before his baser instincts. To fashion a character that others can depend on, that will give a reliable response, this takes motivation, commitment, and executive function, a higher more pervasive will. One builds character through premeditation, motivation and will-power. This means that people can count on us to show up, give our best performance, do our best work and also that our behavior is to some extent predictable, but in doing so we are exercising a stronger human will than the slothful forsaker of work or person who commits the random act of violence. I saw a man who took a bullet to the face which went through one optic nerve causing blindness. He heard a lady scream and decided to help. The attacker, a would-be rapist, shot him in the face. Who was exercising his will at that time, the rapist or the man of character? On the surface the rapist's behavior was not constrained by social mores. The rescuer, on the other hand, was a man of conscience and character ready to put his own well-being on the line for another person. The rapist's behavior was determined by base instincts, whereas the rescuer's was made by active suppression of those same animal traits, the mastery of fear, for example, and the motivation to achieve a higher good. The character that had been formed over the years of his life proved itself over a few minutes of human action. This is a matter of great confusion in our culture, determining who is free. Is it the person, on the surface, constrained by ethical conduct and character, or someone who is free to act on his baser drives? Who is free?

The answer unequivocally is the ethical person. This is a paradox though only a superficial one, that we readily fall for if we fail to think things through. By dint of years character development and personal sacrifice, someone may finally be able to master immediate desires, to sublimate for a higher, less immediate reward never dreamed of by the person of little discipline. It's the base part of our person, the pure biology that shackles us.

Recent work has shown that murderers have a high incidence of brain damage and are deficient in frontal lobe function. When tested by neurological examiners or looked at with MRI and other imaging techniques there is a very high incidence of frontal lobe dysfunction. More violent types also frequently have a history of violence in their own life particularly as children[129]. An intact frontal lobe acts primarily through inhibition. This we have seen in earlier chapters that explored the ideas behind neurological control. It is a general principle in the nervous system that certain motor paradigms are hard wired into lower centers, then this motor repertoire is mostly inhibited or at best chosen from a set of lower motor paradigms by higher centers. The deep tendon reflex and the withdrawal reflex that reside in the spinal cord, are examples of this type of mechanism controlled by higher brain centers. Damage these, and you have a motor system out of control, one that works automatically by its own accord, and not adaptively. We will see that we expect much more from the frontal lobes of the brain than inhibition however. The frontal lobes initiate and plan complex motor activity.

It's a given that what is basic to you will eventually be expressed. This is not just on the biological level and there are many examples. Srinivasa Ramanujan, who is compared with the greatest of mathematicians in history, was born impoverished in India. Some inner force, more than pure talent, drew him to mathematics and he absorbed the material in the few great mathematics texts that were then at his disposal. At a young age he began reciting formulas including the value of pi to many places (Later methods of solving for pi was to be just one of the tasks he became most famous for) and at the age of seven started in a high school. But he failed college exams at two separate institutions due to total immersion in mathematics and took an insignificant post as a clerk until he came to the attention of superiors and almost by accident, caught the attention of the mathematician G.H. Hardy at Cambridge University and the two men collaborated for a number of years. Ramanujan died at the young age of 32, most likely of Tuberculosis, but notebooks he left behind, are a treasure of mathematical material and are being edited and published. His work has been applied to theories bearing on the physical origins of the universe and in theoretical physics and his abilities have been compared with the greatest of mathematical minds[130] in fact, Albert Einstein's background is in some ways similar in that his innate talent, not recognized by most of his teachers eventually shone through, as if on its own accord, almost as if his deep thoughts could not help but to express themselves and be noted by others[131]. The point is that despite any or all disadvantages, inner talent comes through, as if it can not keep from being expressed. The innate nature of talent is a characteristic of endeavors in mathematics, music in particular. In no other area of human endeavor is talent as a basic quality of the individual, so apparent.

It is so easy to make the generalization that what is innate and basic will eventually come through. Even though the experimental evidence is lacking, it is reasonable to ask, how it is possible to describe a basic self, and having done so what do we do with the information? Life as a process can be seen as a constant battle between what we are, and a buffeting and erosion of what really matters. Most of us aren't resilient enough not to make certain compromises and eventually what we are, our own personalities gets worn down. Most of the erosion takes place on the side of the individual person. Outside forces change us inside. Yet some other persons, relatively few of them, are capable of chiseling away at their surroundings and make a great change seemingly through sheer force of will. Those persons are larger than life to the rest of us. Albert Einstein was unwilling to make compromises with what he knew to be true, that nothing could travel faster than the speed of light. He was willing to live with the consequences even if they contradicted with common sense. If consequences caused time to dilate or to contract, he would still insist on what he knew to be correct. Semmelweis dealt with derision of his colleagues when he suggested that childbirth fever may be spread from patient to patient because examiners weren't washing their hands. His heroic insistence on what he knew to be true in face of ostracism by his colleagues cost him his mental health, but saved many women's lives.

Most of us are too willing to compromise in almost every instance with what we know to be true. For material gain or success, we sacrifice integrity, a bit at a time. Perhaps it's for a little bit higher income, or more security for our family or the short-sighted approval of a superior at a job, in gradual small pieces, imperceptible steps, we chip little by little at our real selves, until what we know to be true suddenly isn't important anymore. One day we wake up and there is no integrity left. This is the story of all but a few person's lives. The real price of "success" is how much you really want it, the currency used to pay for it is dissolution of the self.

The frontal lobe is an extension of the efferent and motor limb in the same way as are the basal ganglia, but functions at an even higher, more abstract, level in the task of motivation and planning and strategy. It will be recalled that the basal ganglia are needed for intact detailed motor function, provide acceleration that comes from force, a change in the inertia of constant velocity. The frontal lobes may be viewed as an appendage to basal ganglion function in some ways. To make a motor movement each system, the frontal lobes, the basal ganglia, the cerebellum, makes their own individual contributions. These integrate to create fine movements. As far as frontal lobe abilities we all know persons who excel in motivation and the force of will and others who will be taken up by forces that surround them. By sheer will, exert a force a change in their environment. Most of us just go with the flow of life and rarely have the chance, or I should say, take the chance, to alter it.

Just trying to get to what you are, how you are basically built is a slippery process. Some things like religion are inculcated at a very early age and so probably form some of this basic structure. Very early on most of us have a good idea about our religious origins and the culture we belong to. Jewish young persons frequently intermarry. I have seen a number of cases in which a spouse, who is Christian, is made to take on the Jewish faith and to raise the couple's children as Jews. This comes about not because of any doctrine that is intrinsically Jewish, but because Jews, being so much the minority, are jealous about the preservation of their faith. Nevertheless attraction between young people of different backgrounds and faiths occurs naturally and, if you study the racial and ethnic characteristics of almost all of us, you can appreciate to what great extent this ethnic mixing has occurred throughout history. No one is racially pure. Genetic studies will inevitably give the lie to virtually all claims of racial purity. Certainly no one is racially Jewish, (and there are Jews of all races) or German or Black for that matter, just look around. But in any case, religion serves as a crucible for the notion of giving up what is learned at an early age. What about those instances where a person, out of love and desire to marry, is forced to give up his or her religion and identity? In many cases Jewish families, due to persecution and having to work against enormous odds and sacrifice that comes with being Jewish, decided to throw in the towel, to give it all up. Examples are many among famous persons in history, Benjamin Disraeli, Karl Marx, Felix Mendelsohn, Gustav Mahler, Madeline Albright to name just a very few are from converso families. These persons, all nominally Christian, must frequently discover something of enduring, from their former persona, perhaps a remnant of a family tradition, love for social justice, preoccupation with something larger than their immediate mundane life or even sympathy for an oppressed minority group. Something basic remains. You can take the man out of the country but...

We all know of people who have been forced to subsume their identity for various reasons. Perhaps a wife on marriage will throw her training and career to the wind, subsuming her entire identity, including her religion, folding it into those of her husband. She may even take on a new life in a different culture. Throughout her early marriage she may be deeply involved in bringing up children and very busy with other tasks. She may perceive very little at first. But sooner or later the children are going to grow up and out of the house, and she will be less busy and she may suddenly notice how much of her self has been lost. It happens that the institution of marriage and even dating is very much like the process above described. At the same time we mate, we subsume a larger of smaller piece of our identity into that of the union, and, to a great extent, our partner. Most of us agree that theoretically marriage should amplify rather than squelch one's individuality. As things work out, this positive outcome rarely occurs.

Giving up of a piece of oneself, what is basic to our own personality structure inculcated at an early age, will eventually raise great anxieties. Anxiety turns up in a number of different guises, expressed in a variety of ways, subtly at first almost imperceptibly but eventually makes its appearance. This is seen in some free-floating cases that we eventually diagnose as anxiety disorders, what used to be called free floating anxiety and what is called today, panic attack disorder. Not surprisingly, these problems are especially common in women and occur I think as almost a fear of death which of course exactly what the problem is, the demise of the person, the death of the self.

For psychiatrists the personality, formed at an early age, is enduring almost by definition. Personality disorders, though not as overtly pathological as other forms of psychiatric disorder, are considered untreatable. Admittedly certain personality traits cause great misery and are maladaptive for the person afflicted, changing and even ruining lives. The sociopathic, histrionic, compulsive or "borderline" traits may be life crippling, but these are almost unalterable conditions. Better to be an overt neurotic, more obviously sick, but easier to treat.

The brain mediates personality, for it is simply a more enduring aspect of mental life. Character is mediated by the cerebral cortex in particular. Cerebral cortex is composed of numerous interacting, often conflicting elements that are mutually influential. This is unlike our discussions of other parts of the nervous system.

## The Network:

We have observed that as a general rule the nervous system is hierarchical, with higher centers controlling, modulating or inhibiting, lower more hard-wired responses. Not so in the cortex whose various parts interdigitate and interact. The cerebral cortex functions as a sort of committee structure. Anybody who has been on any committees knows how they work. The more members of the committee to spread responsibility out, the less gets done. Some members of committees assume relatively greater, some lesser influence. Usually committees meet endlessly, accomplish nothing, unless there is some executive to take control and commit it to act., yet as we shall see, the greatest problem will be in trying to determine how we have the illusion, if that is what it is, of some unitary control or aim in the entire structure.

In consideration of a neurological approach to personality we have to come back to the lesion experiment. Lesions and their effects give us a powerful and at the same time, incomplete look, at the workings of the brain. It 's important to remember that if a lesion results in absence of a given function that merely shows that the part of the brain is necessary to that function, not that the lesioned area performs a given function, but that it is a necessary part of that apparatus for carrying it out. Lesion Broca's area and the subject will be speechless for awhile, mute, then a non-fluent difficult slow speech will emerge and frustration at not being able to use language to get one's ideas across. Your first impulse is to conclude that Broca's area in the left frontal lobe, is responsible for making speech. That impulse should be resisted. As it turns out other lesions far-removed from the left frontal lobe will also produce similar findings. This means that the subject needs Broca's area in order to make speech, that it is necessary, but not that this is the function that Broca's area provides. It does not make Broca's area a sufficient producer of speech or language. As it turns out, Broca's area intimately interacts with other parts of the brain which work together in the production of language. These include Wernicke's area in the temporal lobe, the supplementary motor area and other regions which form a sort of network of activity. What we are talking about is moving from a lesion model of brain function, equivalent to a modular view of function, to the conception of a network or committee like function. Heretofore, prior to the consideration of the cerebral cortex in and of itself in our previous chapters, an hierarchical modular view was very useful and generally an accurate approximation of neural function. For the cortex considered on its own, this view will no longer suffice.

M-Marsel Mesulam made this point eloquently in a ground-breaking article, "Large-Scale Neurocognitive Networks and Distributed Processing for Attention, Language, and Memory[132]". The brain described as a computational machine is non-sequential for the most part, massively parallel in structure. This point was made in the first chapter, "Inside the Neuron", in pointing out how the brain differs from most modern computers. Slowly computer science has evolved to the point where Silicon-based thinking machines resemble brain functions and even anatomy. Cognitive science has benefited from cross-pollination between computer and brain specialists.

How does a modular construct differ from a network anyway? You can pretty much disable most desktop computers by removing any specific part or module. It almost doesn't matter what the part is. Remove a board, or a disk drive, a monitor, the power supply, in other words any of a variety of modules, and the thing will stop in its tracks. We've already pointed out how the brain is different. You can removed an awful lot, even obliterate a hemisphere, and the person will make every effort to get around a deficit. All the parts interact, functions are widely distributed, and this goes especially for higher cerebral functions. Taking memory as an example, certain very important structures, the hippocampus, the mammillary bodies, limbic system and others, are responsible initially for laying down a memory engram. Destroy any of these paired structures on both sides and the person will suffer a severe impairment in learning anything new. What about old memories? These will be unaffected, because, they spread out, become very widely distributed over multiple cortical areas over time. They are over-determined. Thus old memories are extremely resistant to erasure, are far less vulnerable, than are new memories that are just now being laid down.

Thus we've evolved a more mature picture of cerebral function as we look straight at the cortex, a sort of network model. The single lesion will impair function, but only because every cortical area interacts as part of a network with other areas. Such basic functions as memory, language, focusing of attention, depend on neural networks.

## Attention:

The process of attentional focus will illustrate this point. Over the years a number of neurological conditions have been described in which there is an alteration of attentional focus. As it turns our attentional processes pertain to such a wide array of disorders as: attention deficit disorder, schizophrenia, encephalopathy that happens inside hospitals, strokes which cause "neglect" of one side of the body and so-called disconnection syndromes, hyper and hypovigilant states, obsessive compulsive and anxiety disorders including panic attacks and flash-backs that occur in post-traumatic stress syndrome in which it is impossible to get a person to unfocus attention just to name a few. These can all be crippling conditions in which a neural basis is slowly being elucidated. If we can come up the cause of a condition, we are one step closer to altering its course.

Attention is determined not by any single brain structure, but through interaction of way stations. Probably the oldest observation having to do with attention has to do with the parietal lobe. Clinical neurologists long ago noted a defect called "neglect" that was especially prominent in patients suffering from a right parietal brain injury or stroke, something seen almost daily on the wards by practicing neurologists. A stroke affecting the right parietal lobe may cause the patient to ignore his left side. First of all data is not getting in to the cortical areas that subserve sensory function from the left body because these reside in the right parietal cortex, data about the position of a limb, sensation and so forth. A person may be so ignorant of his body, that he may come to the conclusion someone else is lying in bed with him not his own left arm and left leg. He may at first lose all contact and knowledge of his own left side as if it is alien to him belonging to someone else. And if this same region is weak or paralyzed, as it often is if the stroke involves more frontal areas as well, he will report that everything is normal denying all weakness, so-called anosognosia, unawareness of disease, a type of neglect mentioned previously. In extreme cases the poor patient will have no knowledge and be unable to focus on the entire left side of the room, will fail to acknowledge a person approaching him from his left, fail to eat food on his tray left of the midline, etc. Studies have shown that the right parietal lobe, is implicated in a spatial sensory ordering of one's world. The parietal lobe, posterior or behind the central sulcus, performs this primarily sensory spatial function.

What of the motor component? This has been connected with the frontal eye fields, which are known to be necessary to direct visual attention to objects on the opposite side of the visual field, left frontal eye field throwing the eyes, directing visual attention, to the right, right frontal eye fields directing attention to the left[φ] . The frontal eye fields located in the inferior part of the frontal convexity, direct more than visual attention to the opposite side. A lesion placed there will result in motor inattention. A third, somewhat more mysterious area that is necessary for attention is the cingulate gyrus, referred to previously as part of the emotional part of the brain, the limbic system. Most probably the cingulate adds some emotional or motivational value to directed attention. So we have described three areas, the parietal lobe (area referred to as "PG", the frontal eye fields ("FEF") and the cingulate gyrus.) Destroy any of these three regions and a deficit in attention to the opposite side of one's world will result. They interact as a network[ψ] which is *not* tantamount to saying that any of these three areas functions as a separate module, that alone, any of these areas, PG, FEF, or Cingulate, is thought to perform the function of focusing attention. They all work as an interacting committee or network for performance of this function. PG and FEF most likely impose a spatial map on the outside world, at least the opposite half of the world, in the words of Mesulam, "...areas PG sculpts the attentional landscape, while the FEF and surrounding areas plan the strategy for navigating it." and, "...the cingulate component could introduce a value system into the perceptuomotor mapping of the extrapersonal space."

Understanding attention from a systems approach will provide the first step in describing conditions, mentioned above in which there is a pathological alteration in attentional abilities. Some conditions are more disabling than others. For example akinetic mutism features markedly decreased verbal output and movement, "abulia" total absence of drive and motivation. In the passive form the patient has the wherewithal to move, that is he is not paralyzed, nor is there a lesion that alters his ability to receive sensory input, yet he sits like a barnacle attached to the bottom of a boat, inert and unresponsive, his condition resembling quite a bit a sort of catatonic state. The akinetic mute may have a lesion in the medial frontal lobe, an area that attaches drive and motivation.

It's tempting to view Attention deficit and hyperactivity disorder as a frontal lobe deficiency. A functional imaging (PET) study showed the frontal lobes of adults who had been formerly diagnosed to have Attention Deficit Hyperactivity Disorder (ADHD) and had children with this diagnosis, are not as active as the frontal lobes of controls without that diagnosis[133]. This implies that the disorder may run in families, as a deficiency of frontal lobe activation, that persists into adulthood. Thus children aren't the only persons with pathological hypofunction of the frontal lobe that requires treatment. Adults have a similar disorder as well. This early 1990 study performed by NIMH researchers and published in the prestigious New England Journal of Medicine, made it into the news media and had great impact on thinking about the disorder, reasonably estimated to affect between 2 and 4 percent of school children. For the first time, an objective test had shown an abnormality in the brain, some functional pathology, supportive of an ADHD diagnosis. The study was done on adults because of ethical concerns about doing a test involving radioactivity in teens who could not give informed consent. But a 1993 follow-up study by the same lead author was barely noticed. Here he finally was able to do PET scans on ADHD teens. This study showed no difference in frontal lobe metabolism in patients vs. controls.

That data is sparse at this point has not stopped practitioners and patients from jumping in headlong and spawning a mini-industry in the diagnosis and treatment of adult attention deficit disorder. Patients are lining up to be "diagnosed" by specialists. Some of them read descriptions in advertising literature or in magazines and "discover" that this is the problem they've had all of their lives, only it was unrecognized by their teachers and doctors. Among schoolchildren, especially boys, large percentages get treated with Ritalin a stimulant medicine used to improve behavior and performance in a windfall for the manufacturer of this drug. If parents and adults who are queuing up to be diagnosed realized the full implication of the disorder perhaps they would be less anxious to be diagnosed!

ADHD is a conceptual descendant of the former "minimal brain dysfunction", MBD, a term that has now fallen by the wayside, but MBD is still a useful concept. The idea was that the condition was caused by some perinatal brain insult. The period of prenatal development, up and including birth, is a terribly vulnerable one for the brain. Genetic defects, alcohol and drug use by the prospective mom, infectious diseases, high risk situations, especially diabetes, and a host of other factors, adversely affect brain function. Severe perinatal cerebral injuries cause cerebral palsy, a condition caused by an injury at about the time of birth that remains for a lifetime. More subtle brain insults might cause "MBD". Instead of observing an obvious palsy or weakness there could be lesser signs such as a cognitive deficit or mild problems with coordination, for example what is termed mirroring of movements, the obligate automatic movement in one hand while the other is supposed to move alone. Also specific learning deficits were frequently diagnosed in such children while, others presumably mediated by regions of unaffected brain, were left intact. The bottom line is that although there were abnormalities and delays in brain development, there were not deficits pervasive enough to cause either a global decrease in mentation - full-blown developmental delay, or cerebral palsy.

A combination of learning deficits and trouble sustaining attention with behavioral problems made a diagnosis of MBD. Today's point of view with ADHD is a bit more restrictive, although it has to be admitted there is a very high frequency of learning disorders as well. The triad of *inattention, hyperactivity,* and *impulsivity* defines ADHD with a set of examples and criteria in and out of the classroom to confirm each of these subcategories. A few subjects may be inattentive and impulsive but not overtly hyperactive and still have the problem. Checklists for parents and teachers have evolved to make things appear objective. As it turns out if a child seems to be doing poorly at school but yet should be able to master the subject matter on the basis of properly done IQ and ability tests, ADHD is often thought of. Generalized emotional and motivational problems, sicknesses such as chronic illnesses and epilepsy, lead exposure, even sleep or perceptual problems, also need to be ruled out. Given the above a child functioning below his ability especially a disruptive moto-perpetuo, is very apt to get this diagnosis. Children with ADHD tend to be adventurous and physically uninhibited. They are the first to hang upside down in playgrounds, to jump off furniture, may be especially prone to injury, and if they don't have a gross developmental delay, may acquire gross or lower extremity motor skills *earlier* than their normal peers. One even gets the feeling that perhaps one day they might grow up to be brave intrepid warriors due to a lack of inhibition.

Often they do display what are often considered deficits of frontal lobe function. They tend to be uninhibited in social situations, to misbehave and disrupt their classes. Some are class clowns, stealing center stage from their quieter colleagues. They move about constantly and get distracted by any stimulus, being unable to concentrate on the task at hand, especially a reading task, and follow it through to completion. In milder cases the hyperactivity may be masked for a while in a doctor's office so that the diagnosis may be more difficult. Some display imitative behavior or what is called utilization behavior. If an object is present in the room, they won't be able to stop themselves from using it, even if they know it isn't theirs and they are not supposed to touch it. Another child might be more inhibited, but these kids will pick it up. Behavior is, in other words, less restricted by usual norms and mores that inhibit others. Ritalin and d-amphetamine, stimulants, calm these kids down and help focus them on the task at hand. With a higher degree of activation they are able to focus better and there is the paradox, because you wouldn't think of using a stimulant on a person who is hyperactive. The paradox is that a stimulant seems to calm them down and focus them. But it all is not so surprising really. Most of us have seen that a good rich cup of coffee will focus our attention and allow us to do our work, one reason why coffee is such a popular drink. Without it we are tired in the morning and our activity tends to be aimless and non-productive. Coffee rouses us into productivity and gives our work direction. Perhaps without coffee, our higher centers activate incompletely and we are aimless. Ritalin and amphetamine do just the same thing for these young kids. When a child has ADHD the effect is very dramatic, so much so that the teacher who hasn't been forewarned, will notice if the child skips but a single dose. In fact the acid test is an unequivocal response to medication, -does or does not the child respond to Ritalin - indicating that what we may be dealing with a defect in nerve transmission.

Simply put, ADHD is a defect in brain activation. Properly activated, the brain functions normally and the student concentrates, completes his assignments and does well in school. and he doesn't have to deal with the failure that would have occurred otherwise. The teachers are happy too. A diagnosis made. The child is classified, pigeonholed. He fits into someone's rubric or model. All is well. But we have a deficit in studies that show how these persons do longitudinally. How long does Ritalin need to be continued? Is ADHD a maturational defect that recovers? Is it then possible to stop medication for the child to live happily ever after? We don't know. No one has really looked into the matter.

Seeing a lot of these kids the subset in whom Ritalin "works", (What I really mean is that it helps, the student is better. Teachers and parents wouldn't be without it.) something far more pervasive affects these kids, some, not all, of them, and it is at least a deficit in what are classically considered to be frontal lobe functions, proper inhibition, social grace, planning for the future, delayed gratification (strategizing for some future instead of immediate satisfaction), patience, motivation - in short all of the things that we typically admire in and which mediate "success" at least in our own society. Some of these kids have all the ingredients built into their very nature of sociopathy. They want what they want now. There is little conscience or remorse. The concept of MBD in ADHD resurfaces in force.

The sociopath or common criminal is extremely interesting in that guilt and remorse seems is amputated from the personality (absent from the criminal's mental structure). Thus among persons convicted of rape, murder and violent crimes sentenced to death none express any remorse for their crime. This is a remarkable and universal finding frequently remarked upon by experts who deal with these personalities. It is not that they are unwilling to admit to the crime. Remorse is totally foreign to the sociopath's way of thinking.

Jesse Timmendequas sentenced to death of raping and murdering young Megan Kanka in New Jersey, and also with a record of similar previous crimes, got to speak freely in the penalty phase of his trial about his preying for victim and family, but asked that his life be spared so as to first comprehend the gravity of his crime. He should be kept alive for his own edification. He was only speaking the truth. The terrible gravity of his action was something he truly didn't appreciate. Since this represents an enduring permanent aspect of the sociopath physiology, there is good reason to neutralize these persons, by permanent imprisonment. They must never be paroled or released into society to inflict further harm on other citizens. Indeed the major function of the criminal justice system should never be conceived as revenge, but rather protection of citizens and maintenance of order. The penal system most importantly has an immune surveillance like mission, walling off the predatory invaders from the general mass of children and persons in society.

You can't reduce the problem of ADHD merely to a chemical deficit in most cases. The problem is more pervasive. It gets into areas that are hard to quantify such as directed sustained attention, motivation, planning, social intelligence, inhibitions, appreciation and fear of consequences of one's actions and generalized executive function. In many cases that first consultation for ADHD is but a harbinger of a lifelong maladaptation, at least for our own societal mores, may be the first sign of generalized sociopathy.

First ADHD does seem to cluster in families. This conclusion is drawn from clinical observations, as again, there are no very thorough studies. This does not necessarily mean that a problem is entirely genetic. A variety of physical and adaptive disadvantages derive from being raised under adverse social circumstances, toxin exposure perhaps even lead exposure, and infectious diseases, poor self-esteem and most importantly, a vacuum of values and passing down of maladaptive work and social habits.

Then ADHD is associated with other brain conditions, especially learning disorders of various types, such as dyslexia, but also Tourette syndrome. The correlation of ADHD with other neurological conditions indicates that the problem is physical, and that ADHD may be significant for a more widespread disorder of cerebral function.

In fact, though we may loosely classify ADHD as a type of frontal lobe hypofunction, we can't isolate such an integrated part of the brain in this way. We are indebted to such great neuroanatomists as Ramón y Cajal, Lorente de No, Walle Nauta and many others, men of patience and foresight, who felt the importance of their work and painstakingly mapped innumerable tracts and interconnections in the brain. Utilizing stains and making gross and microscopical observations, these men were able to show us how exactly how the cortex of the brain relates to lower centers, and how specifically different cortical regions with varying microscopic architecture, relate to each other. They laid the groundwork for the modular and network conceptualizations that we rely on heavily for understanding brain function. Great men and women of later generations, it is impossible to name them all, have followed in their footsteps and have confirmed using more modern techniques just about all of the earlier observations. The only thing is that over with time and further rendering and utilizing more sensitive techniques for example radioisotopes and antibody techniques, the interconnectivity of the brain is even more extensive than originally dreamed of or at least found by original observers. The most basic statement that can be made is that parts of the brain connect more widely with each other, ramify, much more than we all thought even just a few years ago.

The frontal lobe is in a unique position to serve as executive. It connects with virtually every important neurological structure. Connection with the limbic system (limbus for edge) lying below the cortex (see "Beginnings") are strong. Recall that the limbic system is involved with emotion and affective color, the hypothalamus and amygdala connected with high emotion and rage, and how these areas connect to the hippocampus and anterior thalamus and are heavily involved with memory formation. The frontal lobes interact strongly with the thalamus. In fact in many ways the entire cortex, but the frontal lobes most specifically, can be seen as a further development, an embellishment of the thalamus which acts as way-station to the brain. Finally the frontal lobes connect and interact strongly with other cortical surfaces, most particularly and interestingly the parietal lobes to which it connects via the superior longitudinal fasciculus[Ψ] . Such strong connection implies integration in a network and control and places the frontal lobes in a central executive position.

The "pre-frontal" areas of the brain, that is the part of the frontal lobe not having a precise function in movement, that portion that lies in front of the motor strip on the border of the central sulcus in man are the best developed of all other animals comprising some

30 per cent, of the brain more than in any other animal. This prefrontal region has an alternative definition, that volume of frontal lobe projected to by the medio-dorsal nucleus of the thalamus. (thalamus figure) because the thalamus projects widely over this whole area and to all of it.

In a sense, the cerebral cortex as well as the frontal lobe may be viewed as embellishments of the more ancient thalamus. The thalamus is a way station intimately applied to the cerebral hemispheres. All sensory input, somesthetic, touch, vibratory sensation, pain, hearing and vision project from the periphery via thalamus to the cortex. Efferent (outgoing or motor) function also needs to go through the thalamus on its way to the periphery.

The most posterior portion of the frontal lobe and the most ventral or inferior part have a specific purpose, motor control and olfactory reception, respectively. Cut out the very back of the frontal lobe in any animal including man, and the animal will be paralyzed, the inferior part of a man's left frontal lobe and he will be unable to speak. Similarly the sense of smell or primary olfactory cortex seems to be connected with the rear and interior of the frontal lobe where anatomically you can see nerve fibers coming from the olfactory nerve. These are the obvious functions of the frontal lobe but this leaves a very great mass of tissue in humans completely unaccounted for, the prefrontal cortex. The function of the prefrontal area is far from obvious and hints to its function have only been forthcoming in recent years. The defects caused by lesioning large sections of the frontal lobe are subtle and entirely human, connected with such soft areas as personality and executive function. Indeed, at a time when American neurologists and doctors were heavily involved with cutting out sometimes vast regions of the frontal lobes earlier in the twentieth century, many of them were surprised about how much tissue could be removed from human brains. The method used for frontal lobotomies at that time was to jam a sterile ice pick through the inner aspect of the orbits where the eyes are and wiggle it about knocking out a good portion of the medial frontal lobes, likely causing a good deal of bleeding and tissue loss. Sometimes electroshock therapy was the only anesthesia. As if this methodology were not surprising and frightening enough, experience showed that deaths were very rare, and even more surprising neurologic deficits were difficult to appreciate. Intellectual defects were not detected using standard instruments and personality changes, being hard to measure, went undetected. In the meantime this surgery had a therapeutic effect in that patient's behavior became more tractable and less resistant. It became possible to talk with the patient and communicate perhaps for the first time in a while. It appeared that large portions of the prefrontal lobe were not necessary at all for day to day function, large portions being lesioned but causing no profound deficit. What we are talking about here is the very obverse of the topic at the beginning of this chapter. What you find in lobotomized patients is not an obvious deficit[134]. Instead you notice that the patient suddenly becomes more conforming, tractable, less threatening, easier to manage. It's the kind of stereotyped conformity made famous in "One Flew Over the Cuckoo's Nest. That was not the original intent of this procedure at all.

It had to be admitted, as first noted by the Portuguese neurologist Egas Moniz in 1935, winner of Nobel prize, and pioneer in the operation called frontal leukotomy on psychiatric patients and others, that such lesioning had a therapeutic effect on severely affected patients with psychiatric disease, especially those afflicted with types of schizophrenia causing violent behavior or severe obsessions. These symptoms could be relieved even doing crude procedures that obliterated part of the frontal lobes even as in those days was the case if the lesioning was not visible and you could not do the surgery under guidance of the imaging techniques we have today such as CT or MRI scans which would be used decades later. The procedure decreased the need for forceful restraint in the days when there were no very effective drugs to combat certain mental illnesses. Recall that major tranquilizers such as Thorazine and antidepressants first appeared in the 1950's and antidepressants at about the same time, also minor tranquilizers such as Valium appeared later so that the only drugs in the psychiatrist's armimentarium were barbiturates like Phenobarbital, and Meprobamate or Miltown which were merely very sedating. These drugs offered nothing in the way of specific treatment of specific symptoms of schizophrenia, delusions and hallucinations.

When you took out even large portions of the frontal lobe, you had a more tractable patient, who, in the long run, was a lot happier, needed less restraint, need not struggle against his caretakers could now walk about in freedom and yet there was no measurable change in cognition his ability to talk and function, in fact he was improved! What a marked therapeutic effect such an organic approach to mental illness had and it made the psychiatric ward into a much more therapeutic milieu[Φ] . In the days before doctors had effective specific medication, such surgery was undoubtedly a godsend and it was very popular too, not only performed on many indigent wards of the state, for in those times patients tended to be cared for in huge state run mental hospitals, but even in a few famous cases involving well-to-do subjects such as Rosemary Kennedy, sister of the John Fitzgerald.

Every advance, even such a primitive advance by today's standards as frontal leukotomy, is a two-edged sword. Not every one who employed the operation was thinking of the patient. You can imagine how such surgery must have been misused especially in such regimes as the Stalinist Soviet Union and China, whose psychiatrists were controlled the State and made to treat political dissidents with their potent organic therapies which included electroshock, insulin shock, and undoubtedly the frontal lobotomy. How many political dissidents and other disorderlies such as artists, musicians poets etc. were treated with these modalities or perhaps just executed for their mental excesses we will perhaps never no even in this post-peristroika era. But it is also the case that these therapies have also been applied with the best of intentions, that is in a sincere effort to decrease constant struggle and human misery at a time where there was really very little else available to treat these unfortunate patients and their families.

But little by little some pieces of subtle information began to emerge. There were in fact changes in such subjects, not at first apparent using any psychological testing instrument, but present nonetheless. The first hint of such change was, of course, the therapeutic efficacy of the surgery which was done to make resistant even violent behavior more placid and tractable. Certain patients seemed to be struggling constantly and were prisoners of extreme emotional states. Many of these persons could be helped by amputating this troubling part of their personality.

After lesioning large areas of the frontal lobe, psychologists had a difficult time finding any deficits on their psychological test instruments. Certainly classic tests of cognitive function, namely IQ and achievement tests, turned out to be normal as did tests of verbal memory, concentration and planning. Few tests showed any abnormality at all, except one rather simple-minded test which is now frequently cited the Wisconsin Card sorting test. Yet in the crucible of real-life situations many patients with frontal lobe lesions turned out to be dismal failures. They could not judge their own behavior or objectify themselves and frequently were inappropriate in social situations. They could not hold a job, maintain social relationships or appreciate the big picture in their lives, perfectly content to live dependent on other person's largesse and not caring about not having their own identity. They were insensitive to other's feelings as well.

Psychological testing shares some methodological flaws with objectivist behaviorist psychology of which it is a relic. Psychologist's instruments test for external behavior, above all attempting to objectify human potential and personality, but ignore internal states which are valid study phenomena. The ignorance of internal states which early psychologists, particularly American psychologists, could not agree was a proper subject of scientific inquiry, is a major pitfall that still poisons the field of psychological assessment. I have often been impressed by psychologists reports sometimes involving eight or more hours of comprehensive "objective" testing. These are long impressive looking reports that can take up sheaves of paper, yet most fail to see the big picture and miss major deficits in patients. The subject may be a dismal failure in life and be unable to function, yet perform adequately on psychological profiles or a minor basically irrelevant deficit may be focused upon in the meantime missing the big picture. A physician's assessment is free to look at internal states and thus has a tremendous advantage in patient assessment. On the other hand, it has to be admitted, the doctor's assessment may not be quite as "objective". If, on the other hand you mean by objective that conclusions are can be agreed upon by multiple observers, psychologists who perform these psychological assessments, then alas, these reports suffer from the same deficits that physician reports do i.e. there is a great deal of disagreement between professionals.

This abandoning of internal states due to a bias against inquiry into non-behavioral elements in personality has been eloquently addressed from the philosophical perspective by John Searle in his book *The Rediscovery of The Mind.* In it he presents forceful arguments that are valid, pointing to our lack of attention to internal states[135] which are, to be sure, a proper study of for scientific inquiry. Admittedly internal states are more difficult to objectify or even to get at but that does not mean we should stop trying to understand. Perhaps our understanding needs to ascend to a new plateau, maybe we need new tools before we are able to understand internal states and phenomena, but let us keep trying.

Moreover it was already known that there were cases of frontal lobe damage that had caused real effects. Many authors had published cases on war and other injuries and profound alterations in personality not the least of which was the celebrated nineteenth century case of Phineas Gage who described by his physician Harlow in 1868. Phineas Gage was a previously highly responsible and with a "well-balanced mind" and fastidious and dependable personality who working at his job as a foreman on a railroad had a rod blasted through the inner part of medial frontal lobe through his upper orbit (eye) on the left side mostly. This produced a marked alteration in his personality. Whereas he had been a "God-fearing, family-loving, teetotaling, and scrupulous honest working man"

"The equilibrium or balance, so to speak, between his intellectual faculties and animal propensities seems to have been destroyed. He is fitful, irreverent, indulging at times in the grossest profanity, manifesting but little deference for his fellows, impatient or restraint or advice when it conflicts with his desires at times pertinaceously obstinate, yet capricious and vacillating, devising many plans of operation, which are no sooner arranged than they are abandoned in turn for others appearing more feasible. A child in his intellectual capacity and manifestations, he has the animal passions of a strong man..." [136]

The greatest likelihood Is that these effects of frontal lobe lesioning that are so well-known now were not even known to the Moniz or at least most of the original purveyors of frontal leukotomies and other similar procedures. Otherwise such widespread use of psychosurgery as it was then called might have given them a lot more pause. Brain areas inferior to the cortex in the human perform little integrative function. Lower areas are quite specialized leaving it to the cortex to put it all together. The Thalamus, in particular, receives inputs from lower centers. Sensory input from the spinal cord and cranial nerves arrives at thalamic way-stations, mainly the ventral posterolateral (VPL) nucleus. Motor inputs also converge on their own separate thalamic nuclei (ventral anterior, ventrolateral VA, VL) from many areas including but not exclusively from the spinal cord, cerebellum, and basal ganglia. You'd think that some final processing must occur within the thalamus, being that so many sensory and motor inputs converge upon it. Within the structure, however, motor sensory and other processes are kept separate. While there is extensive communication, (cross talk) between all of these modalities motor function, auditory and sensory, areas that converge upon thalamic nuclei, there is virtually no dialogue between these modalities, between motor and sensory inputs inside the thalamus itself. The thalamus is pretty large but not really large enough to perform this highly complex, multimodal integrative function. Therefore the great bulk of correlative and integrative work is left to the much greater developed cerebral cortices. Individual sensory and motor modalities are separate in the thalamus. Why?

Humans reserve this integrative function, interaction between sensory and sensory and motor, finally associative function, pretty strictly for the cerebral cortex. This is still another piece of our uniqueness puzzle. Integration is a cortical function. The cortex is that anatomic substrate for what is distinctly us. The cerebral cortex embellishes the thalamus the great way station complex of nuclei on which so much data converge. VPL will project to the primary sensory parietal cortex which will take things from there. Then in the cortex this sensory information will first ramify widely, extensively link, to other cortical regions and link with motor function as well. This means that up until the level of the cerebral cortex there is no solid singular integration of disparate neural functions that will define a single person. In some lower vertebrates there is very little cortex, there is very little beyond the thalamus. We see as a general rule in lower animals that higher integrative functions are performed in more peripheral areas of the nervous system, while in man such higher functions are reserved for more central regions. For example, in lower animals a lot of visual integration is performed right in the retina.

As far as the phylogenetically new prefrontal area is concerned, it is not silent. More recently we have only begun to appreciate some of its functions. In a general way, one could see that many patients subjected to prefrontal lobotomies seemed almost to be less human, less themselves, although this procedure often did serve to make their behavior more tractable. You could see that many of them seemed to be less spontaneous, more docile, even zombie-like. We see patients with frontal lobe lesions who seem to lack any spontaneity. Words such as aspontaneity, abulia, mutism, are used to describe them. Some of them have various tumors or strokes that affect this front part of the frontal lobe. Even processes that affect other areas of the brain primarily such as diffuse increased pressure in the head, closed head injury etc., seem to affect the frontal lobe preferentially. The anatomically newest area, or developmentally "highest" area seems to be the most vulnerable and likely to malfunction.

There is little initiative in these patients and like lifeless physical objects they are subject to the laws of inertia. It's hard to motivate them. They have no initiative. But once they start something, they can't seem to stop it either. They can't terminate a hand grasp. When you can finally get them to start doing something they have trouble ending the task. Ask one to draw a cross and he has to be cajoled. If you stick to your guns and finally get him to draw a cross, he'll keep doing it, in other words perseverate. Ask him to draw a circle and he may do so only continuing to put in a cross as part of the figure. This happens with damage to the convexity in the dorsolateral frontal lobe. It is interesting that this are is more or less contiguous with the motor areas of the brain, with a region in the frontal lobe known as the supplementary motor area. The supplementary motor area is basically contiguous with the still more primitive primary motor area, which we've seen has a more precise, motor homunculus on its surface. It's thought that motor movements are basically initiated in the supplementary motor areas but that more abstract motor plans and strategies come from the prefrontal area, also to some extent motivation and drive.

Lesions in prefrontal areas closer to the midline also occur, for example, in patients who have had rupture of blood into the brain from anterior communicating aneurysms. Here the problem is a mirror image to the aspontaneity referred to above. Patients are hyperspontaneous in a shallow way with a lesion in this orbitofrontal area. They have "oscillating affect" that alternates between shallow rage, depression or euphoria. They may be involved in such petty crimes due to social disinhibitions as shoplifting, sexual assault etc. Because of the seemingly contradictory ways that the dorsolateral as opposed to the orbitofrontal lobe seems to work, the eminent neurophysiologist Karl Pribram suggested that the dorsolateral frontal lobe activates whereas the orbitofrontal region inhibits[137].

The opposite of perseveration at least superficially is called Field Dependent Behavior which also occurs as part of a frontal lobe deficit. Here the person seems to take all his cues for behavior from the environment. If he sees a cup he will have an irresistible urge to drink from it even though it is empty. I've noticed many children with attention deficit disorder seem to pick things up in the office, for example, a stethoscope, tuning fork, or hammer and use it no matter who tells them to stop. This is essentially the same as so called "utilization behavior" [138]. While perseveration seems to signify behavioral stability, field dependent behavior evidences instability, both relatively higher orders of behavior subserved by the heretofore clinically "silent" prefrontal area.

We've established that so called "higher" or more advanced brain areas are evolutionary embellishments of lower regions, that such areas developed later and develop later in the embryo where evolutionary steps are recalled and restaged. The cortex embellishes of the thalamus, the supplementary motor area embellishes the primary motor area, the prefrontal area the frontal area, as each lower region proves inadequate for the needs of the animal appearing later on the evolutionary scale. The most recently appearing areas are hardest to "grasp" objectively, but they are the areas that make us human.

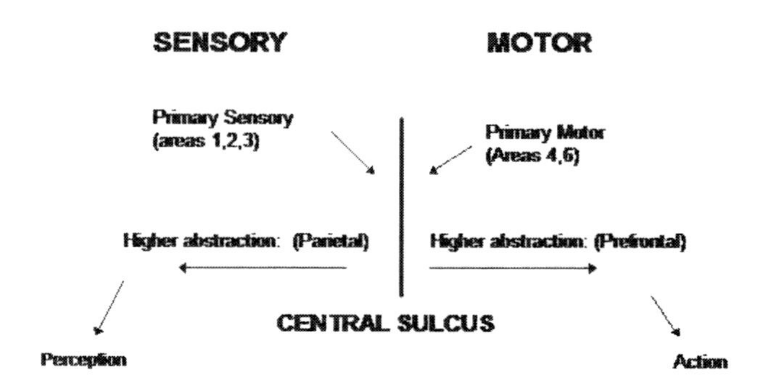


**Figure 47: Depicting the Central Sulcus flanked by primary motor and sensory cortices. As one travels away from primary cortices, there is a higher level of abstract processing.**

As the rear portion of the frontal lobe is responsible for initiating motor activity, the cells comprising the area bordering on the central sulcus are designated as the primary motor area. As I talked about in chapter electrical stimulation of this region would result in movements or combinations of movements in a limb on the opposite side of the body. In the 1930's Wilder Penfield at the Montreal Neurological Institute, was interested in excising areas of the brain that were thought to be responsible for generating epileptic seizure discharges. In striving to avoid lesioning the language areas of the brain, he began utilizing the technique of electrical stimulation of the cortex in an awake patient. Surprisingly he found that he could map out a little albeit slightly deformed caricature of one side of a person over the surface of the motor cortex, a figure now called an homunculus. Stimulating an area would often cause a specific movement on the opposite side of the patient. The area representing the feet was at top side of the brain, the head toward the bottom, rather like an upside down man hanging by his feet. Certain areas of the body were over represented such as the face, hands and lips, the very areas requiring precise motor control, while other areas got short shrift for cortical space.

On an electroencephalogram certain electrical rhythms over regions of the cortex can be seen that go away with motion of the limb on the opposite side of the body that these regions control. In fact, even the thought of motion will obliterate these so- called Mu rhythms which are thought to represent a kind of motor running or idling of certain cerebral areas that are ready and waiting for action.

A similar homunculus could be constructed along the posterior edge of the central sulcus for sensation corresponding to the primary sensory cortex. Secondary motor and sensory areas also exist in the frontal and parietal lobes representing larger less specific body areas and more complex movements and sensations. Finally the surrounding areas also segregated as to sensory vs. motor function deal with the initiation and action and reception of sensation on still a more highly integrated and abstract level. These are the association areas of the brain.

Touch, pain and position sensation may be organized according to the locus on the body stimulated, just as movements involve single or multiple body areas. For other sensory modalities such somatotopic organization is inappropriate, i.e. for auditory processing and the sensations of taste, smell and vision. For the auditory system in the area designated as primary auditory cortex that resides in the temporal lobe, reception is organized according to specific tones i.e. is "tonotopic". Certain cells within the primary auditory cortex react to sounds of a given pitch. The organ of Corti in the ear is organized in a similar manner with the apex of the organ resonant to lower pitches and the base vibrating more to higher ones, so that one can easily see that axons projecting to the brain conserve the same essential organization. Other brain areas that surround primary auditory cortex bring about a more integrated appreciation of sound. Some of these areas for example if stimulated, may cause a person the recall a certain melody or even a memory attached to a musical theme.

What we see from the above is that there are areas of the cortex that are homogeneous with respect to sensory modalities, be they hearing, vision, or touch sensation. Within these specific regions, located in the temporal lobe for hearing, the occipital lobe for vision, the parietal lobe for touch sensation, and so forth, is further specific organization. The best example is somatotopic organization of body parts in the parietal lobe that subserves touch and position sensation, the so-called sensory homunculus. There is a similar topical localization in the temporal lobe, as we have seen, for specific tones, and in the occipital lobe for parts of one's visual field. All of these cortices have surrounding, secondary receptive areas that interpret more abstract attributes of sensation, for example put words and sequence meanings and melodies for sound, interpret movement for vision and so forth. They are handling a higher level of perceptual abstraction. Other regions of cortex even further removed, help interpret this data contextually, often along with information that is cross or hetero-modal, involving sight and sound, for example. Information is first uni-modal, involving one sense, then interpreted at a higher level of abstraction usually not far removed from the primary sensory region, then redigested at far-flung locales over the cortex. Full apperception or experience involves committee like cross talk between far flung cortical regions abuzz with stimulation and interpretation.

These areas have to be connected. The full range of cortical interconnections, the job of white matter tracts that interconnect cortical regions, has just begun to be appreciated. It's easy to see that the view of the brain or cortex as merely a system of modules, without full appreciation of the interaction of these modules to perform function is entirely inaccurate and incomplete. It's fair to ask what happens when these connections fail.

This failure to connect is not uncommon and was explored most specifically by Norman Geschwind who wrote extensively about so-called "disconnection syndromes" over 30 years ago. There are many fascinating examples, some of which we have explored in earlier chapters. One particularly instructive one is the disorder "Alexia without agraphia", a specific problem where a person loses the previously acquired ability to read, even though still able to write. These patients can understand words presented to them aurally. Some can read a word spelled out to them on their hand. They can spell and write, sometimes fluently. What they can't do is read and interpret visually, a written word or sentence syntax. The disorder is a disconnection. The visual areas of the brain in the occipital lobe, can see a complete word, but the visual information is prevented from getting to the language-interpreting area of the brain in the left temporal lobe. The fibers that go from the occipital to the temporal lobe are destroyed, most often by a stroke in that area. Patients who have alexia without agraphia, have a lesion in the left occipital area of the brain and also the connecting fibers from their right occipital lobe that connect the visual brain with the left temporal lobe. This is a thick white matter bundle known the splenium of the corpus callosum. As such they cannot see to the right, have a right hemianopia (see "vision") but also they are unable to read, not because they cannot see the letters, but due to an inability to get this information to the brain's language module.

Thus we have one of many examples of defects of cross talking areas of cortex. Information from one region or module, must reach other modules if the brain is to function as a whole. Other examples we have explored previously are anosognosias, unawareness of diseases or deficits. The language module surrounding the left Sylvian fissure is disconnected, therefore unaware, of a specific difficulty or deficit and thus is unable to "report" to the examiner that there is anything wrong. Sometimes this unawareness is quite striking as when a person is completely blind yet does not report blindness. The subject instead invents answers to visual questions; he confabulates. This is known as Anton's syndrome. Similarly a memory defect causes confabulation, invention of wrong answers, when language module is disconnected and thus unaware of a deficit in what is known as Korsakov's syndrome. There is ignorance of one's left side or neglect when the right parietal lobe is affected by disease, and so forth. All of these deficits result from disconnections, mostly isolation from the language or reporting area of the brain.

The frontal lobe is what it is primarily because of its connections. The frontal lobe connects directly to the limbic system which it overlies. Recall the limbic system relates to emotional coloring of memories and experience. It has rich connections to the frontal lobes which overlie it. All the other cerebral lobes, enjoy vigorous connections as well which means that the frontal lobe has heteromodal or multi-sensory associations. As we have seen diffusely projecting nuclei within the thalamus such as the dorsomedial nucleus also connect with the frontal lobe. The ascending reticular system responsible for activating the entire cortex and also the diffusely projecting cholinergic nucleus of basalis of Meynert has rich frontal lobe connections all for a reason. In many cases frontal lobe dysfunction comes not from destruction of parts of the frontal lobe, but results mainly from dissociation with other cortical and subcortical regions, comes from disconnection[139]. When you look at the brain, the first thing that impresses you is the massive size of white matter areas in relation to the gray matter, which is a thin ribbon overlying it. This tells us that the business of the brain is more the connection of concepts and data than initiation. Look at a map of the United States or any civilized country. The first thing that impresses you is an elaborate system of roads, the white matter of cartography, the towns, cities, and megalopolises are the neurons and groups of neurons or nuclei in the brain, relatively unimpressive in a bird's-eye view. The frontal lobe as much by its topography and connectedness as through initiator of action plays its own unique role in integrative executive function.

Understanding of whole brain function has evolved from a lesional or modular construct, through a committee structure as described above with a description of networks of elaborately interacting modules communicating through an interstate highway system of white matter tracts. The modules work at all levels of sophistication and function as we have seen, gradually integrating more and more data at higher abstract levels, areas of cortex loosely handling greater levels of abstraction as one travels further from the primary sensory and motor cortices.

Very loosely speaking the frontal lobes are felt to bring about the "responsible" elements in our personality. A person with good frontal lobe function is a serious planner, initiator, worker, saver, one who can follow through with an elaborate scheme and is not afraid to work and sweat for long periods to accomplish his goal. He will always follow through. He's a bull-dog as much as his personal plans are concerned in that he will never let go until he accomplishes what he wants. This single minded individual is the entrepreneur, the leader in our own society. As a general rule, he has qualities that are seen as positive and adaptive, at least in American culture. The opposite situation, which we have conjectured might be due to a lesion in the frontal lobe, might be the person with ADHD who has trouble maintaining his focus, accomplishing and following through on his task, is easily distracted by any stimulus that comes along.

As far as the frontal lobes are concerned, various authors have described their own particular versions of what is meant by *hypo*, diminished, vs. *hyper*, exaggerated *frontality*. According to one very reasonable point of view schizophrenia may represent the extreme of *hypofrontality*. Schizophrenics are rarely the neatly groomed shirt and tie type of person. They are sloppy and unfocussed. They rarely are able to function in life or social tasks, to plan and stay with a specific goal and work it through. Schizophrenics tend to move from job to job, from superficial relationship to superficial relationship. Two of the primary diagnostic features of schizophrenia as presented by Eugen Bleuler a turn of the century psychiatrist who gave the disease it name, still widely quoted are *apathy* and *ambivalence*. The opposite of our social mover or entrepreneur. They fail in social and material endeavors. On a more basic level, schizophrenia represents a loss of or weak development of ego structure, poor recognition and appreciation of self with profound lack or reality testing.

According to this view, if schizophrenics represent one pole, namely the hypofrontal pole in this scheme, then the opposite or hyperfrontal pole is the obsessive-compulsive disorder (OCD). The major characteristic of persons with OCD is that they can't let go. They are interminable worriers. Before leaving the house they have to check the oven, not once but over again, then the locks, then look through their pocketbooks to make sure they have their money. These poor folks worry so much they can never get out. They may be victims of certain ritual behaviors that they cannot control. For instance they may have to wash their hands over and over again because of concern over germs. They may never be able to accomplish anything because of their disorder.

This construct, schizophrenia on one extreme, OCD on the other, receives its own support from PET scan data, at least interpreted superficially. As can be seen in, the schizophrenic has lower frontal lobe activity, the OCD higher frontal activity than normal. There is more relevant information bolstering this construct. OCD is successfully treated in many cases with medications that increase Serotonin in the brain, the so-called SSRI's specific Serotonin uptake inhibitors. Newer drugs used to treat schizophrenia work primarily by blocking serotonin, especially working according to the model of the novel drug Clozepine which revolutionized treatment of schizophrenia for some patients. This implies that schizophrenia in part involves an overaction of serotonin or serotonin receptors and that OCD may involve a certain underaction of serotonin. The SSRI's on the model of Prozac, are extremely useful for many neurotic disorders characterized by over-worry, increased fear and anxiety such as panic attack disorder and some forms of depression, also Tourette's syndrome which at least as far as this particular characteristic is concerned, is similar to OCD. Psychiatrists like to use many of these drugs together as all patients present with a gamish of paralyzing psychiatric problems. They need to be vigilant so far as the function of serotonin in the brain in managing their patients.

Then our businessman or entrepreneur, the productive person, the mover or shaker of our society needs to be removed from our model or spectrum of frontal lobe function. I jumped to my own conclusions prematurely. Entrepreneurs have some but not all of the characteristics of hyperfrontality, at least some of them do. What do they lack, compared to our OCD, or panic disorder patients? Why don't most entrepreneurs require Prozac or something similar? As a general rule they don't suffer from much *anxiety*. These productive folks are fully able to get off the dime once they make a decision, may easily decide whether to fish or cut bait, not so our OCD subjects, as we have seen. In fact, more productive folks, once they decide in an informed way, can just do it or, as Yogi Berra said, "you can't hit and think at the same time." They follow through with their decision with a minimum of doubt or regret. Not surprisingly folks who do and produce will not fall along a spectrum of pathology. The pathological models as above described, do not refer to them.

As I have indicated it is common for ADHD, sometimes thought of as a manifestation of hypofrontality, may often co-occur with Tourette's, related to OCD, a hyperfrontal disorder. This should remind us that our model suffers from being to simple, generic or global. There are conflicting portions of the frontal lobe, for example midline versus lateral surfaces that are responsible for differing aspects of frontal lobe function. The midline portion of the frontal lobes, is more closely connected with the emotional or limbic part of the brain and so is connected with basic motivation. The more well developed and evolutionarily newer lateral frontal lobe, on the other hand, is more involved with the actual intellectual aspects of strategy and planning, reasoning and flexibility, with directed attention, order and the actual sequencing of action. So we may conclude that the frontal lobe concerns itself with sequencing, executive control, future memory, drive (especially the midline of the frontal lobe) and self-awareness. These are all hard commodities to appreciate and measure on psychological test batteries.

So-called "reduplicative amnesia" is a curious though transient disorder of psychological function. It sometimes occurs in the healing phase of head injuries and is related to frontal lobe damage. Here the injured person correctly identifies his surroundings, so that it is clear he can register and recall memories and is functioning verbally. However he insists that his surroundings have been duplicated in a different place, that is not really where he is. A related situation called "Capgras Syndrome", is the insistence that acquaintances or family members look and talk like exactly as they should, but are actually different people reduplcated somewhere else. "Yes, this lady looks and talks exactly like my wife but doctor, it isn't her!!" Somehow it appears the affected person hasn't the wherewithal to check out reality, or perhaps has an uneasy feeling of unreality that comes with the partial dissolution of the self or what may loosely be called ego structure. In the Freudian sense, it is the ego or self that is involved with truth or reality checking.

We are beginning to define abundant conflicts of this kind in our model built on disturbances seen among patients. Parts of the frontal lobe counterbalance each other and add to the emotional and intellectual matrix of consciousness, the frontal lobe may also be seen modulate countervailing influences of the parietal lobe. This has been extensively described by Lhermitte and others[140]. Lhermitte described patients with a so-called *environmental dependency syn-*

*drome* who had, as above described, an urge to touch and use objects in a room, to be distracted by all kinds of stimuli not relevant to their situation of tasks, ignoring generally accepted norms of behavior. These patients had frontal lobe deficits or lesions. The common denominator, as described in some distractible subjects with ADHD as above, is that such persons are unable to keep an appropriate distance, unable to control their desire to approach objects, indeed have trouble even objectifying objects. They approach objects and use them in an uncontrolled manner, unable to keep themselves from being distracted for a larger purpose. Lesions of the frontal lobe promote this type of inappropriate approach to objects. Lesions of the parietal lobe of the brain, on the other hand, promote inappropriate maintenance of distance as we have seen, so-called neglect. The model here is that the frontal lobe promotes distance in conflict with the parietal lobe which promotes approach. Lesion the parietal lobe and the frontal lobe has free reign. The person objectifies himself. Lesion the frontal lobe and he can't keep himself away.

Very similar models are constructed for other lobes of the brain. As we have seen lesions of the amygdala, the almond-shaped nucleus of the temporal lobe, on both sides will cause a KlÜver-Bucy syndrome. Persons or monkeys with KlÜver-Bucy syndrome are unaggressive, excessively docile, hypersexual - they will approach almost any inappropriate object in a sexual way, and hyperphagic. Some affected folks will eat vaginal creams or their own pubic hair. The amygdala as we have seen, has extensive connections with the hypothalamus and other emotion generating parts of the brain as an integral part of the limbic system. In a way patients with lesions here may be said to be *hypotemporal*. The opposite, what may be called *hypertemporality* happens in the so-called Geschwind Syndrome. The temporal lobe is often the site of abundant electrical seizure discharges which propagate through the amygdalas inside the temporal lobe. The amygdalas, being in the pathway of an electrical discharge, may themselves become electrically active sometimes pathologically so. The amygdala will start to generate its own electrical activity. This is a process called kindling. An electrical discharge is like a spreading forest fire which lights adjacent structures in the brain. Pretty soon these electrically active areas light up on their own, may start their own fire, in other words. As the amygdala becomes more active, the subject acquires high emotion which he connects with almost anything. He may be overly serious or morose, aggressive and hostile (in opposition to emotional docility as in the Kluver-Bucy syndrome). Such persons are often grandiose, very frequently hyper-religious for they see great cosmic significance in everyday events, proselytizing everyone in sight. In our region in Central Pennsylvania where religion is an important part of daily life, I tend to see persons express this inflated sense of emotional significance in religious ways. So the direction of expression may have something to do with one's environment. Many persons with temporal lobe seizures are hyper-graphic. They want to write everything down, even their most mundane thoughts. As they attach emotion to words, poetry is a prime outlet many times. This Geschwind syndrome sometimes in more sometimes in less extreme forms, is very often observed in persons who have seizures that come from the temporal lobes. As KlÜver-Bucy is hypotemporal, Geschwind is hypertemporal[141].

What I am developing here, which is a bit more involved than a simple committee model of brain function, is what I call a "*push-pull model*" that describes conflicting influences of various groups of neurons within and between lobes of the brain resulting in some final state that is a sort of sum or compromise between elements in the network. One problem as we shall see is that this model derives from appreciation of pathology in the brain and as such may not be totally applicable when we consider the personality structure of normal or even better-than-normal functioning individuals.

**Table 7: A "Push-Pull" conflictual model of brain modular function (see text)**

| LOBE | HYPO-FUNCTION | HYPER-FUNCTION |
|---|---|---|
| FRONTAL | ADHD, SOCIOPATHY | ENTREPRENEUR |
| | SCHIZOPHRENIA | OCD, PANIC |
| FRONTAL (VS PARIETAL) | Environmental dependency, "Utilization" | Neglect, Objectification, Distance |
| TEMPORAL | KlÜver-Bucy | Geschwind |

In Chapter one I presented an anatomical view of consciousness. The cerebral cortex, reticular activating system, thalamus, limbic system, each added their own singular contribution to the brain product we loosely called "consciousness". Recall that the reticular activating system would keep the cortex awake, the cortex itself lent meaning to percepts and actions, while the limbic system aided in memory and emotional coloring etc. Each area would contribute on its own, and influence other areas (much of this influence was reciprocal) until the whole brain put out its final product. Now, describing in more detail the workings of the cerebral cortex, it is possible to add to this committee model, which still pertains, a push-pull summation model of conflicting influences.

Its very tempting to generalize from pathological states and apply these principles to real people. Simply put, according to the above model, if the temporal lobe and particular the amygdala had its way with your personality, took over, you would be a very serious person indeed. All of a sudden you would wish to attach cosmic significance to every small event. Suddenly you would get religion. Sprinkle in some influence from a hyperactive frontal lobe and a rigid system of rituals would be added to the mix. We have our evangelist for whom nothing is more important than serving God, who needs to pray x number of times a day, follow religious precepts to the letter, who can't understand the sinner who does not take religion as seriously, has seen the light and is unable to accept any other religion. The more stereotyped, the more ritualistic the behavior, the higher the comfort level of this person though a high level of comfort is never achieved. Every religion has its own specific algorithms and here I don't mean to single any of them out. The archetype is mnemonic device of rosary beads, but multiple repetitions of the Amidah, the 18 benedictions said at least 3 times a day in the Jewish faith is another great example. These may be considered almost mantras which give great comfort to persons who follow without thinking, previously laid-out paths. The point is that for certain personalities, and this is admittedly an oversimplification, a formulation, the temporal lobe has taken over the personality. These persons are hypertemporal. Every practicing neurologist knows what I mean when I describe such persons. We know many of them from our practice. Persons with seizures emanating from the temporal lobe tend to have such personalities, i.e. are “temporal lobish”, in common parlance. The theory is that excessive electrical discharges in the temporal lobe indicate increased activity there. To make matters more complicated, this theory is flawed. Though temporal lobe epileptics tend to have this personality, most of them don't. For the ones who do, controlling their seizures has little to no influence on their personality. I have seen persons who first acquire these disabling personality characteristics after a temporal lobectomy, the removal of a temporal lobe, an operation done with great success to control their epilepsy. That in itself is fascinating, that an operation will demonstrably cause, almost in a flash, a profound personality transformation.

Its really something to find the rare patient who has electrical seizure discharges within the temporal lobe, who will express emotions as part of their seizure. Some will laugh uncontrollably, (“gelastic” seizures), some cry. I know one woman with tuberous sclerosis, a genetic disease with abnormal benign growths or "tubers" deep in the brain, who cries and cries when she has an epileptic seizure. One tuberous growth is in a temporal lobe. Dostoyevsky had temporal lobe epilepsy and had many of the characteristics described. His seizures had an aura of high emotion, ecstasy, his literary themes were highly religious. He certainly took himself very seriously and had a compulsion to write. Would one consider him then hypergraphic? He would continue to write under the most adverse personal circumstances. The only reason why we read and admire him is because of his genius and the obvious power of his insights as opposed to many other persons compelled to write everything down despite their more pedestrian abilities. But it makes you think about what is truly is the meaning of what we hold to be important or even sacred.

Is it fair to apply these known anatomical influences as they have been defined, to all of us? How much do the same formulations apply to the ordinary personality and to men of genius who have no known brain pathology? By this calculus each of us is the mere sum of modular influences in a committee structure, and resultant of contradictory push-pull influences. This defines our basic self, our baseline personality.

Having said this, we need a model also for environmental influence. In recent times we've begun to understand how some of these might work. Lately articles have been written about how playing Mozart might have a positive effect on learning[142]. Students who listened to a Mozart sonata were able to score higher on spatial reasoning tasks given to them than a control group. What's more, exposing them to the minimalist music of Philip Glass had no such effect. So it was not music itself, but a particular pattern of stimulation that had an effect of increasing brain power.

Immediately this information was applied in a public educational program for disadvantaged kids in a New York City neighborhood. How would listening to Mozart improve academic performance? One way is by appealing to logical patterns of firing within the brain. We have observed that the auditory cortex in the temporal lobe has a tonotopic organization. Listen to a melody an the succession of notes stimulates specific neurons in a pattern over the primary auditory cortex. Next, adjacent areas of cortex participate in higher abstract level interpretations. Suppose you should make these musical patterns excessively simple, and meaningless, appealing to the basest of patterns in the brain, or, even worse, that you play, instead of popular music, sounds with no pattern, dissonant, or having no melody or chord progression. Then the cortex isn't stimulated in any way that would promote intrinsic patterned responses. But suppose you present to the brain the opposite, some advanced patterns of tones and chord progressions over time that are familiar, follow intrinsic patterns of cerebral excitations. Such thinking would be a long-shot logically, but presumably, Mozart, a childhood genius, was able to access the brain's most basic intrinsic firing patterns when he played and composed music at such a young age, such built-in progressions of tones that are basic to patterns of firing built into our own cerebral cortices. Mozart's music is thus a "match" for our own intrinsic cerebrotopic firing patterns built in to brain anatomy and thus a form of mental exercise. Mozartian music patterns may be a fit like a key in a lock with intrinsic cerebral neuronal patterns of activation much in the way that male bird song of fits with female receptive patterns of brain activation in the same species and thus produces excitation (the desire to copulate). Given what we know about how such patterns ramify, stimulate other adjacent and non-adjacent areas of the brain, and how electric signals reverberate throughout the brain's structure, pretty soon the music will be appealing to, bringing to the fore, the brain's own basic logic or wiring patterns. Before Mozart, having listened to the usual unpatterned noise that we are usually exposed to, maybe these logical cerebral circuits had fallen into disuse, or were never appealed to at all. After listening to Mozart, these unused portions of the brain are dusted off, taken for a test drive, and are now ready to go. Pretty soon we are able to solve mathematical problems, to see spatial constructs that we were not prepared to appreciate before. When we say a that a great composer appeals to what is basic in us maybe we're talking about more than emotion. Being presented with a musical template congruent with cortical firing patterns brings about an unprecedented level of comfort, and at the same time excites the brain, whets one's appetite, for further thought and stimulation. No wonder why I like Mozart so much. I've always felt good music appeals to what is most basic inside of myself.

Which makes you wonder what a person is doing when you observe him doing nothing? Suppose he is awake. Is his cerebral electrical activity patterned or unpatterned? As you observe him sitting in front of you, you've no idea whether a thought, or a melody, or a picture is in his head, or whether he is thinking of nothing. Think about this in your own case. What is happening for you most of the time when you are mentally idling? Most likely the answers are as different as varied personalities and temporary mental states. Some people like to turn off when they relax. They have little taste for patterned thought, music or visualization. Others have an unrelenting need to think and perform mental gymnastics. In all cases the awake brain, when unengaged in productive (motor or efferent) activity is idling. The brain is an electrically active organ that never stops except rarely under very pathological circumstances, as when you are under deep anesthesia. The ignition is always in the "on" position. The only question is whether activity is patterned or unpatterned firing and about the complexity of firing patterns.

By playing Mozart you are stimulating, tickling, intrinsic brain firing patterns. You are pulling up memories that previously were buried and unavailable as presented in Beginnings the chapter on memory, where memory engrams were pulled out of oblivion by "handles" attached to them up to the surface ready to be accessed in the process of recall. These memories include logical patterns used in learning and problem solving. Thus Mozart helps the brain to idle logically and pulls abilities, previously buried, to the surface, making them accessible.

Brain disorders alter the firing patterns within the brain and the effects are more enduring and usually permanent. These disorders reinforce the argument that the final common pathway for all thought and behavior is the wetware of the brain (Table 2). A whole host of disorders may alter the personality and sometimes are first noticed with a personality change. Some of these affect the brain diffusely and merely cause dementia starting with different patterns as with Alzheimer's disease which typically begins either with memory loss or language function problems or Pick's disease. Other degenerations such as Creutzfeldt-Jakob disease begin variably and the mode of presentation depends on which neurons are attacked first. Creutzfeldt-Jakob disease is caused by a new type of organism, the so-called prion or proteinaceous infectious agent, the only reproducing biological entity that doesn't contain nucleic acids. It is the same as the mad cow disease of common parlance, that affected British cows and a few humans otherwise known as a "slow virus" since symptoms may not be manifest until decades after infection. In one of the commonest forms of Creutzfeldt-Jakob disease, the cerebellum the part of the brain controlling coordination and balance, gets hit first. Only later is there a change in mental function and the personality. However, patients with Creutzfeldt-Jakob disease are frequently encountered on the psychiatric ward of a hospital. In the typical case an elderly person who has no history of psychiatric disease, presents with severe depression or schizophrenia. Now these disorders most of the time start at a younger age and when they do occur for the first time in the elderly in a person who apparently is not prone to develop them, this raises a red flag of concern that what the psychiatrist may really be dealing with, is a brain and not a psychiatric disease. Your first attack of depression or schizophrenia should not happen when you are old. On the other hand, Creutzfeldt-Jakob disease almost always occurs in the elderly.

Another very interesting case of personality altering diseases is a kind of neurosyphilis, an entity called "general paresis of the insane" (GPI) in which there is an active syphilis infection over a very long period of time, a meningitis affecting the coverings of the brain. At one time up to 4-10% patients hospitalized in chronic psychiatric facilities had GPI but now it is rarely seem thanks to penicillin and other antibiotics. At one time, syphilis caused GPI was held out as the archetype of a mental disorder caused by a physical process, but no more since it is now so rare that few clinicians have any experience with it. Still GPI is still a sort of model of an illness with a known biology and pathology which causes severe mental change, not just generalized deterioration in mental processes. This is because so much of the course of events is well-known and documented. The psychiatric changes go hand in hand with pathology in the brain to such a degree as to make an extremely potent case for physical processes being the main cause of mental affliction. Some small proportion of affected individuals start with megalomania and grandiosity, having a king or ruler complex and there seems to be a specific tendency of GPI to cause that particular problem, though other processes that affect brain function diffusely can cause severe paranoia as well. Idi Amin the Ugandan dictator, was said to have suffered from GPI. Before falling from power Amin murdered tens of thousands of his fellow countrymen, many personally, as he apparently had sadistic tendencies. He is said to have eaten the heart of his wife who he had executed, and performed a variety of bizarre acts. At the peak of his power he allied himself with the PLO which gave rise to the interesting raid at Entebbe, and was much honored by the United Nations facts that are even more bizarre considering that nearly everyone was aware of his intense cruelty and bizarre behavior[143]. Though to my knowledge Amin's affliction was never proved at least publicly, this raises a question about the unusual behaviors of a number of famous personalities whose profiles fall far from the usual norms. Perhaps craziness is in part what makes these persons so unusual and is responsible at least in part for some of these folk's success. In recent years, brain syphilis has been very rare though some cases have again been associated with the immune suppression that happens with HIV disease.

The only problem with GPI is that it is mostly an historical disease with which today's clinicians have little contact. However the pathology is well-described. Syphilis is caused by Treponema pallidum, a spirochete which is a close cousin to biologically and in its biological effects on the brain and other organs as the popular Lyme disease organism Borrelia burgdorferi. Central nervous system syphilis is a disease that develops over a long period of time, taking years ordinarily to cause its full effect. It starts by causing a chronic meningitis, an inflammation to be found in the spinal fluid that bathes the brain and spinal cord and affects the meninges which cover these structures. Apart from finding signs of a syphilis infection in the blood, a positive RPR test, the best way to document the effect in the nervous system is by a spinal tap that allows examination of the spinal fluid. The spinal fluid shows evidence of inflammation in the form of increased protein content including the antibody portion, gamma globulins also increased inflammatory white blood cells (lymphocytes) in the fluid. This is the definition of the term meningitis, signs of inflammation one can see in the spinal fluid. The gamma globulin is interesting. The antibodies are directed against the syphilis spirochete. Clones of IgG antibody actually react with Treponema pallidum, so this is one of the body's defense mechanisms against the bug, which persists despite the poor patient's best efforts to eliminate it. Because the infection lasts years the meninges become thickened and the syphilis treponeme actually ends up getting the best of these defense mechanisms invading the brain tissue itself. Brain derived inflammatory cells known as microglia are present in brain tissue, but neurons are lost and eventually there is loss of valuable brain volume as the infection takes over, the treponemes continue to be present right in the brain tissue and a chronic infection ensues ending in alteration of the personality as we have seen as well as later general dementia and dissolution of mental processes and finally death. The pathological and microscopical processes parallel mental and cognitive changes to such an extent in syphilis that it can be used as a model for such changes. At its worst, Lyme disease seems to be a milder process, but the changes are essentially the same, eventual invasion of the nervous system causing chronic meningitis affecting the spinal fluid and brain coverings, followed much later by actual invasion of the sanctus sanctorum the brain itself in a process that is very much like what used to be seen in patients with GPI. Lyme disease affects peripheral nerves and cranial nerves more than does syphilis yet its Central nervous system effects are rather less pronounced. This is a difficult call since there are no cases today where patients are untouched by multiple antibiotics which while they may not cure the disease entirely in rare instances still likely lessen natural processes. Still Lyme disease is more catholic in its tastes than syphilis seeming to affect just about any part of the central nervous system, cranial nerves especially the facial nerve causing paralysis of facial muscles, the spinal cord and peripheral nerves. On the other hand the personality changes apart from fatigue are not as vividly described as in the old cases of GPI. Lyme is perhaps more protean in its clinical manifestations than syphilis and a lot of personality changes have been connected with Lyme infection especially various forms of chronic fatigue with very poor documentation for this effect. Depression and all kinds of personality alterations have been connected with Lyme meningoencephalitis in popular literature and among patient support groups. Yet very few claims have been documented objectively. At one time various Lyme treatment centers had sprung up throughout the country treating persons who complained of symptoms empirically but with little to no documentation of any disease process in most cases.

A host of other diseases infectious and non-infectious, infect the brain in their late stages and often cause mental changes. Other forms of infection, killed off by antibiotics in the rest of the body, may hide out in the spinal fluid partly excluded from antibiotic penetration by the blood-brain barrier, an unusual combination of tight capillary endothelium, basement membranes, and astrocytes that exclude unwelcome molecules from the central nervous system (described in chapter one). The BBB may block antibiotic penetration into the CNS as well. So it is possible for certain dangerous organisms to hide out as for example, the pneumococcus bacterium which lives in our mouths and nasal passages close to the spinal fluid and is the most frequent bacterial invader of the adult CNS.

More tragically perhaps, certain cancers also hide out in the central nervous system well after they have been cleared and cured from the rest of the body by strong chemotherapy. You can cure a cancer everywhere. There may be no signs of it for years sometimes, yet just one or a few tumor cells may have lodged themselves in the CNS, once in protected by the BBB where the chemotherapy agents cannot get at them, only to cause a recurrence of the tumor right in the brain when you discover it. This has happened in childhood leukemias especially where the whole process is cured as far as the body is concerned, only to later find that some cells have hidden in the brain, out of the reach of drugs. Other tumors such as melanoma and breast cancer may behave to a much lesser extent, in the same way. The best methods of treatment pre-empt CNS invasion with early preventive radiation treatment and specific aimed radiation beam techniques such as the gamma knife if such tumors appear. Tumors may spread to affect parts of the brain or themselves spread microscopically into the spinal fluid causing a tumor meningitis, carcinomatosis. Generally personality changes aren't noticed first in these conditions, just the specific effects dictated by the part of the nervous system invaded.

Illnesses currently classified as psychiatric are, in reality neurologic that is they are brain and not mental disorders. The mental symptoms are secondary to brain changes. This is particularly true for psychoses, the affective (emotional) disorders such as depression and bipolar (manic-depressive disorder) and schizophrenia, so-called thought disorders, but is undoubtedly true for neuroses and personality traits and disorders as well. Theoretically these disorders should be treated by a neurologist or brain specialist, not a psychiatrist primarily except to the extent that the psychiatrist has organic orientation and training (he or she is in fact a physician first and foremost which is good) or if not organically oriented may at least address psycho-social fallout which is secondarily affected by the primary brain disease. My point of view is sounds rather shrill and extreme yet is borne out by various separate lines of evidence, viz.:

1. It has not been possible to define an environment conducive to the development of schizophrenia. Schizophrenics come from all kinds of environments. Even taking into account that schizophrenics tend to be raised under less afflu-

ent circumstances that they tend to come from disadvantaged socio-economic groups, there are still a number born into the wealthiest families. The disorder tends to cluster in families with a 10 to 11 times greater risk in first-degree relatives of patients (siblings, children, parents) than the general population (lifetime risk as high as 1%) and if you are the identical twin of a diagnosed schizophrenic, you have a much better than even chance of having the disorder yourself which implicates a substantial genetic component. The clustering in lower socioeconomic groups illustrates that all other things being equal, schizophrenics and those with schizophrenic characteristics are less successful in our society which is not at all surprising given characteristics we've already explored. All attempts at trying to define the kind of environment conducive to the development of schizophrenia have failed. Such theoretic constructs as schizophenogenic mother for example have had never been supported by the evidence.

2. Schizophrenia represents a pervasive and all encompassing pattern of mental dysfunction which is extremely disabling. Subjects have difficulty following logical progression of thoughts and instead reasoning is contrary to common convention thus conclusions tend to be false. One recent example. Lyndon Johnson was president when Martin Luther King was shot. Johnson was a powerful man who had access to a lot of information. Ergo Johnson must have known about plans to assassinate King and is thus responsible for the King assassination. Since conclusions derived from these erroneous logical processes are unlikely to be correct, persons are not likely to be successful in most endeavors. Another common example is flow of thought by loose sound or "clang" associations as occurs in some forms of art. e.g. from spine, swine, supine, behin(d), mine, entwine etc, . Schizophrenics have thought progressions that are all but impossible for a normal conventional thinker to follow, hence they are observed to have looseness of associations,

what is manifest as tangentiality or circumstantiality sometimes much worse in their conversation. Often there is complete breakdown of mental processes so far as these can be ascertained aurally or in writing.

3. There is no evidence that any kind of talk or non-organic derived psychotherapy is of benefit for the cardinal symptom complex of schizophrenia. Psychotherapy and social interventions may allow improvement in one's marital, employment or benefit status, or may put the patient into a milieu where he will be able to better function yet seems to have no bearing on the basic disease process. Physical interventions on the other hand, especially drugs, alter the cardinal manifestations of the disease, not just via sedation, drugs do tend to be sedating, but by reduction of positive and negative symptoms, illogical thought associations, delusions, hallucinations, paranoia, and social withdrawal and generalized decompensation.

4. Schizophrenia may be the result of a number of different processes but they are all brain processes. Early attempts to find a common microscopic pathology failed. There is no specific degenerative change you can observe under the microscope. Modern imaging techniques have found common abnormalities particularly whole brain atrophy, enlargement of intracranial spaces, ventricles and sulci is very common in schizophrenics though by all means not always present and is a finding even when patients are compared with a non-affected identical twin. Certain areas of the brain have been found to have reduced volume or hypofunction. These have included variously in different studies, areas of the temporal lobe, and some PET studies show hypofunction of the frontal lobes. The problem in characterizing physical features of the disease underlines the fact that schizophrenia is not one disease, but the endpoint of many processes, cerebral insults, environmental, genetic but above all, physical that affect the brain and in-

> terfere behaviorally with thinking so as to cause looseness of associations and erroneous reasoning, delusions hallucinations, withdrawal, and other manifestations that spur clinicians to make this diagnosis. Many of the disease processes listed above, encephalitis, syphilis, porphyria, Creutzfeldt-Jakob disease among them, may lead to the diagnosis of schizophrenia, though once any of these underlying processes is defined, once one is aware of a specific cause, that is in fact listed as the disease even though some of the behavioral manifestations may be similar.

Functional neuroimaging (PET) scanning has allowed researchers[144] to monitor, moment to moment some positive symptoms of schizophrenia, such as auditory hallucinations. From such studies one sees not only the decreased activity in the frontal lobes chronically, but activation of such areas as temporal lobes, during an auditory hallucination. Speech areas of the brain also take part in auditory hallucinations.

**Table 8: Some Diseases presenting with personality change.**

| Disease | Cause | Findings | Examples |
|---|---|---|---|
| Syphilis/Lyme disease | spirochete | paranoia, dementia, neurological focal changes | Idi Amin<br>Robt Schumann,<br>Friedrich Nietzsche |
| Pick's disease | degeneration | dementia, language disturbance | |
| encephalitis (Herpes) | Virus, e.g. Herpes | seizures, change in level of cons. Focal signs | |
| Alcohol | death of brain cells, other injuries | neuronal loss, effects of trauma | |
| HIV disease | Retrovirus | secondary infection, direct viral invasion, focal or generalized findings | |
| limbic encephalopathy | tumor antibody that cross-reacts with brain cells | altered emotions, dementia | |
| Alzheimer disease | amyloid accumulation | memory or language disturbance, then dementia | Ronald Reagan<br>Arthur Fiedler |
| Porphyria | inborn enzyme deficiency | episodic madness and melancholy | King George III of England |
| Schizophrenia | various | "thought disorder" | |
| Wilson's disease | enzyme deficiency responsible for carrying Copper | liver flap<br>movement disorder<br>hemolysis, liver failure | |

Another interesting case in point is the familial disease Acute Intermittent Porphyria. This is a disorder that ran, as hemophilia did, within some of the Royal families of Europe whose most famous protagonist was George III the Hannoverian king of England during the period of the American Revolution. Although at first appearing to be an able ruler, in later life, he seems to have suffered from more and more protracted periods of insanity, now attributed to porphyria, which causes episodes of delirium and abdominal colic. Porphyria is a disease induced by a simple inability to break down hemoglobin as one should due to a defective enzyme. Hence persons with the disease build up toxic by-products that poison the brain and peripheral nerves. Certain foods, drugs and other substances drive individuals into acute attacks. Between attacks or bouts with the disease, subjects are for the most part, normal. The affects of George's disease on history in particular the separation of the American colonies, is probably overstated.

Herpes Encephalitis causes personality and emotional change that progress rapidly over days. The virus lives in neurons in a dormant state but probably gets into the brain due to its close proximity in such structures as the trigeminal ganglion or olfactory nerve as it tends to attack the medial frontal and temporal lobe. Listlessness, depression, disorientation, headache neck stiffness are followed by fever and convulsions by which time it is often too late to expect a compete recovery. Nowadays we have effective drugs such as acyclovir block viral reproduction. The trick is that Herpes encephalitis has to be suspected and diagnosed very early since without treatment viruses will proliferate in the brain which will be irreversibly damaged or the patient may die. Some patients have been cured but only after the virus has damaged both temporal lobes, resulting in a devastating recent memory defect. The subject is in limbo of disorientation unable to lay down new memories. Depending on how extreme the situation is, he may be disoriented living life entirely in the past, or some engrams may be able to get through and be stored resulting in a partial state of confusion and disorientation.

Limbic encephalopathy is a rare disorder associated with a common cancer, small cell carcinoma of the lung. It is not due to the spread of the cancer, but rather a substance elaborated by the cancer itself into the blood, perhaps an antibody called Hu, that is usually found in the blood of such patients. Hu is a brain protein constituent and this disease seems to be caused by an antibody against that protein. They most often suffer from memory deficit for as we have seen, the limbic system is closely associated with emotion and memory, but there can also be profound emotional and personality changes as this process affects the emotion expressing areas of the brain.

As time passes goes on more focal and specific disorders will be defined anatomically and functionally. This will allow more explicit statements about functional anatomy of the brain and how groups of neurons interact to form a complete personality structure. Susan Swedo and colleagues[145] have found a fascinating group on children who, in response to repeated Strept throat infections, have a complete obsessive-compulsive ritualistic syndrome. Their personalities change when they are reinfected, often going back to normal between infections. Some children have been treated successfully with plasmapheresis and other therapies that alter immunity. In some cases enlargement of the caudate nucleus in the brain and other structures has been associated with the temporary profound personality change. Hypertrophy of the caudate may be evidence for increased function within that area of the brain overwhelming other inhibitory regions. Movements such as chorea, may be associated with these personality changes as well which remit with therapies aimed at ameliorating the structural changes in that disease. If certain regions of the brain mediate alterations in diseased states, then it is reasonable to conclude that one's character is determined by a crosstalk anatomical brain loci.

Generally the conditions described above result in an obliteration of function of one or another part of the brain, which means that they may be considered pathological conditions. It is reasonable to presume that "normal" behavioral variations, in other words personality, is explained along the same lines. For example in America we live in a quite religious society, where the incidence of belief in God, and church attendance is very high. Yet you meet people who display a whole spectrum of beliefs no matter what religion they have, some whose belief in God and absolute ethical principles# as laid down by their religion is virtually absent, others in whom it has high importance and dominates their life. Part of this comes from upbringing and social background, of course, but it is rather striking that some persons are so basically more metaphysical and philosophical than others, involved with "bigger questions" of life and death, order in the universe and so forth, while others, care not at all over such things. Some person's lives need to be organized, dictated by others in authority, ritualized, while others disdain ritual and dictates of higher authority. Some of us need to follow exactly other person's rules, others of us do not. Given the observations of a spectrum of patients pathologically ensnared in ritual (Geschwind), others who are willy-nilly victims of their moment to moment desires (KlÜver-Bucy) and care not a whit for rules and ritual, undoubtedly this variation reflects differences in brain physiology. We know of environmental and situational factors in personality development. However, if personality is refers to patterns of relatively fixed and predictable behaviors in an individual these patterns are determined by differences in the brain.

Given a push-pull set-up and conflicting influences of various parts of the brain, characteristics may be described along organic scales of hyper and hypofunction of brain regions or modules. Thus it makes eminent sense to build an organic personality profile. Why should be confine considerations of organicity to obvious pathological states such as schizophrenia, Tourette syndrome, epilepsy and the like? These conditions are merely emblematic of other personality characteristics, perhaps extremes. All of us carry elements of these supposedly pathological states. Disease states may thus be useful in delimiting normal variances. Brain physiology and pathology sheds a certain light on the inner workings "normal" personality. With tongue in cheek I present my own fanciful conception of a brain derived (organic) personality profile. One day it is hoped arm with ever more sophisticated knowledge derived from anatomical and physiological probes, tests may be devised to plumb the depths of these conceptions, in much the same way that the Minnesota Multiphasic Personality Inventory (MMPI) graphs personality profiles on a series of scales on the basis of self report true false or other questionnaire. While a semi-projective test like MMPI, is based on purely psychological constructs, a new profile, based on knowledge of brain physiology would seem to have even more promise.

# THE BRAIN-DERIVED (PHYSIOLOGIC) MUTLI-AXIAL PERSONALITY PROFILE:

## (P-MAPP)

## Continuum:Spectrum:

I. Temporal Lobe:

| Hypo (KlÜver-Bucy) | | Hyper (Geschwind) |
|---|---|---|
| Sex Editor | non-selective | over-selective |
| Food Editor | | |
| Seriousness | non-serious | over-serious |
| Religiosity | | (religious) |
| Emotion | | |
| Hypergraphia (Script Editor) | No | Yes |
| Aggressivity | Low | High |

II. Frontal Lobe:

| | Hypo | Hyper |
|---|---|---|
| Inhibition | vagrant | Nerd |
| Anxiety | Sociopathy | Panic |
| Guilt | ADHD | Depression |
| | Schizophrenia | |
| Inhibition | Low | High |
| Imitation | High | Low |

| | | |
|---|---|---|
| Objectivity | Low | High |
| (Parietal Lobe) | Sensory | Motor |

III. Left vs. Right Hemisphere

| | | |
|---|---|---|
| High Left | | High right |
| Precision | Verbal (mathematical) | Artistic (imprecise) |
| | Auditory | Visual |
| | Speech | Song (expression, prosody) |

IV. Subcortical Structures:

| | | |
|---|---|---|
| | Low | High |
| | Motor artist | Athlete |
| Afferent | Entrepreneur | Artist |
| | Associative | Philosopher |

Consider the practical implications. Take the food and the sex editor. Some among us are truly omnivores thinking very little of what they eat and are willing to experience almost anything. With no constraints at all, some persons, children or mental patients typically, may put almost anything into their mouth, perhaps hair or toiletries as well as food (known as "Pica" for woodpecker-The biggest concern over the years has been children who may eat peeling lead paint from tenement walls but pica refers to all manner of unnatural non-food consumption). Bezoars are composite indigestible objects found on x-ray sitting in the stomach such as balls of hair or seeds or even wood or metal. I've seen patients pick up and eat hair or creams and so-forth. On the opposite end of the spectrum are persons extremely selective with their diets, a selectivity which ranges from rational to irrational over-selectivity, some of it highly rationalized. There are the vegans and vegetarians whose dietary habits are followed almost with religious fervor, that is any slight impurity the inclusion of animal matter will cause great anxiety. A lot of these persons feel it is morally wrong to kill and eat an animal. Strict diets may be followed for religious reasons such as Kosher diets used to separate those who practice from less faithful and demarcate Jews from non-Jews, also a host of religious proscriptions of other faiths. These folks are fastidiously choosy with their diets. Exactly the same holds for sex. Some persons are restricted to very select persons of one sex, others aren't apparently so choosy as far as objects of attraction or sexual habits and may copulate with animals or inanimate objects. Ambivalence to the object class of sexual or appetite attractants is neither a matter of hypersexuality or hyperphagia. These persons don't merely have increased appetite for sex or food, so much as they are just not choosy about the object of their affection or appetite. They do appear to be hyperphagic, hypersexed and exhibit little hygiene or cleanliness, and at least some of them tend to have a one track mind so far as their desires are concerned, so their may be at least some element of obsession[Φ] . The distinction is facultative and fastidious in just the same way that a bacterium may be facultative, able to survive in an aerobic or non-aerobic environment or fastidious, that is a pure anaerobe or aerobe, only able to survive in an oxygen-rich milieu or entirely without oxygen. A bisexual may have homosexual inclinations, perhaps, but can have heterosexual relations and he is thus facultative and differs from more fastidious homo or heterosexual colleagues. The consumer of kosher foods or vegetarian, is more fastidious than his non-kosher and non-vegetarian neighbors. The obligate monogamous heterosexual, the Kosher person, the vegan are thus in their own way all highly temporal (referring to the temporal lobe) while the omnivore Gourmet, and bisexual are perhaps the Kluver-Bucys of food and sex.

Generally, but by no means always, other characteristics on the temporal lobe spectrum such as hypergraphia, the need to write things down, overall seriousness, and religiosity covary, but these separate but related characteristics are also independent and deserve their own scale. Many hypersexual subjects are also hypergraphic for instance, which means they tend to tell us in writing about their varied exploits. By the same token It is possible to conceive of an overall scale of *temporality,* encompassing food and sex choosiness, religiosity, seriousness, graphophilia.

The same may be said about the frontal lobe. As mentioned, some frontal lobe effects conflict with the parietal lobe in accordance with Alajouanine above, especially the character of overall "subjectivity" versus "objectivity." Anxiety and certain types of compulsivity may relate to an overall hyperfrontal personality, whereas the far less caring sociopath would seem to be hypofrontal. A lot of this remains speculative but is part of a physiological personality profile which emphasizes that many personality characteristics are brain mediated. Thus these constructs are based on physiological data and thus reality, subject to modification with new discovery and less to theoretical armchair speculation.

Other subcortical scales apply. Some persons are kinetic doers as distinguished from primarily afferent feelers, entrepreneur-athlete on the one hand, versus the artist. Others have cognitive and theoretical orientation such as priests and philosophers who seem to be taken over by association cortices, as opposed to primarily afferent or efferent subjects. Some persons, by their own natures, are meant to have less contact with the practical world. They function as armchair speculators, or worse, are isolated from reality altogether within the ivy-covered walls of the university or the cloistered convent. The practice of medicine is, for me, a spectacular example of a real world oriented profession. As a medical resident initiate the idea is to teach by massed practice. You're fully immersed in blood and guts and stool, human suffering and disease day and night in training. I feel privileged to have had this experience. Entering medical school it's traditional for you to be introduced to a cadaver. This is more than symbolic. You get more than experience of life and death and disease. You are immersed in it. There is nothing like it. Quite the opposite of being sheltered from the world, if you want to experience it first hand, there is not a lot that can be compared with medicine. So from this standpoint the monk is the opposite of the physician. Of course, we are all mixtures of these cerebral influences and variations. Why confine these physiological considerations to pathological states when it is highly probable that they are responsible for normal variations in personality?

For any individual it should be possible to derive a score on all of these scales that defines their personality profile, what may be termed the brain-derived (because it is derived from a knowledge of brain mechanisms) or simply, Physiological Multi-axial Personality Profile, the P-MAPP. The idea is not so much that such scales may be exploited and used, but that much of a personality is derived from purely physical or brain mechanisms. This would be a combination of simple and composite scales. The sociopath, the person with ADHD, the "marine" or soldier of fortune[ϕ], roller coaster rider, drug addict, who thinks less perhaps of the possibilities of dire consequences of behaviors, would be contrasted with the highly anxious panic attack victim, the highly inhibited nerd with loads of pens and pocket protector, the social planner, the map maker the philosopher, who obsess on laying out a course of action before hand, sometimes to the extent of immobility on a composite hypo to hyper frontal scale composed of sub-elements of obsession, planning, anxiety scales, united by an overall scale of low versus high inhibition.

This kind of instrument, as with other psychological tests that are less based upon physiology, has all kinds of applications. One can define criminal types using this technology, but as I've pointed out previously, the best predictor or future behavior and the most empirical, is past behavior, one's personal history. I don't believe that personal historical data is used as much as it should be as a preventive measure. One's behavior, as it turns out, does not depart from certain previously established norms. As far as perpetration of a crime and indeed in all other fields or endeavor the two components to prediction of future behavior are proclivity, previously expressed talents and preferences, and the behavioral repertoire, skills and past experiences that enable certain behaviors to be performed. A man may occasionally raise his hand to his girlfriend or child. Having done this a number of times without paying any penalty, his behavior could inexorably creep into the realm of the slightly more outrageous, perhaps bruising a hapless family member. The next time he may break a bone or aggress against a non-family member. The point is that each time his new behavior is not totally different from past expeditions into the realm or brutality. Yet incredibly, our legal system seeks to keep historical data from discovery in order to protect the legal rights of the suspect. From the scientific standpoint this is tantamount to throwing away the best information we have not only about guilt or innocence but even more importantly in trying to prevent future tragedies. A psychological instrument, one rooted in physiology, will render at best a reasonable approximation of expectations of future behavior and performances which is likely to be inferior to a survey of past historical behavioral output. When it comes to doing what we'd like to do with psychological tests, measure the fitness of persons for work, to be admitted into academic programs, estimate the future probability of such diverse acts as completing a course of study, performing on a job and even committing crimes, all testing is woefully inadequate. The inadequacy of prediction is due to a residuum of volition. It is theoretically possible, though admittedly this doesn't happen very often, for person to suddenly wake up and decide he will change his life, cease acting the way he did before to do something fundamentally different. Even then of course abundant remnants of a previous behavioral style will creep into a new repertoire or paradigm, but the point is that psychological profiles and past history may be thrown to the winds.

With further research the push-pull or crosstalk anatomical model of the personality articulated here will become ever more pervasive and precise. The model presented above is used for argument, but is much too simple. Even on the basis of current evidence the concept of hyper or hypo frontality or temporality is much too crude. Personality characteristics are increasingly being defined by specific loci or parts of the frontal and temporal lobe. Persons with pervasive anxiety in our society show increased activity in the amygdala a structure that gives emotional valence to stimuli and especially certain parts of this structure, the central nucleus and in areas that closely connect to the medial prefrontal cortex, only a small part of the frontal lobe. Anxiety accompanies neurochemical events in these areas of the brain. Serotonin and hormones in the brain mediate these events. Hence it is possible to state unequivocally today, not only that activity in certain brain regions determines personality but that various drugs can be used to alter personality and character.

Yet there is a sense, that any psychological scale, even based on physiology or more accurately pathophysiology, physiology rooted in knowledge of disease states is inadequate, not only because, as we have seen, personal historical information probably supersedes, is a far better predictor of future behavior, but for other reasons that are harder to define. There is some residuum of behaviors and human actions that is unpredictable. Persons don't always act the way you expect them to. Often they go far beyond one's expectations so that one has the sense that any pschological or historical instrument will always prove inadequate to its purpose.

In one's behavioral schema some mental activities and possibilities are, instantly rejected or, legitimized. This instantaneous rejection or acceptance I conceptualize with the term editor. And it is in just the same way that other actions, be they dietary, sexual, and all of our behaviors and activities of daily life are accepted or rejected. No question about it, some of us are more uptight than others. For some smoking a cigarette, eating raw fish, insects and crustaceans, having incestuous sex relations or of the same sex are all more than taboo, they are unthinkable, outside of our personal lexicon of thoughts and behavioral repertoire. These ideas are wholly edited out, rejected, denied admission into our thought processes. Those persons with a powerful editor of this type given control over mental considerations, on the one hand are felt to be uptight, constricted, to possess a lot of guilt, a powerful superego, according to previous systems of thought which utilize some of the same ideas. More than that, one may be beyond guilt, that is certain thoughts behaviors are so automatically excluded that they cease to occur, never get admitted into consciousness at all, hence guilt becomes at once unnecessary. Why do we need to worry if the perverse can simply be avoided?

The editor: Turning off the logic machine

The editor is a firewall between the two worlds we all inhabit, Cerebus between real earth and the netherworld. There's an entire universe of unedited thought processes, the vast majority of which is considered to be junk, or worse, perversity. The unthinkable, the perverse, occurs in all of us. Mostly these thoughts are excluded from the light of day, as if these terrible thoughts do not occur, which in fact, they do. A lot of this thinking is not so terrible. Some is simply illogical or contrary to practical experience in the real world. As we age the realm of the unthinkable gets larger, meaning that rejection occurs more instantaneously with experience. I can feel this happening in myself. When I was younger, I was often dogmatic, but had less practical knowledge and I was more willing to examine other possibilities which today I reject out of hand. I have a much firmer notion of what I'll accept and what is unacceptable.

We inhabit two worlds. All of us function in our workaday world. We make a living and feed our families, concern ourselves with continuing existence, with practical immediate mundane matters. This is our world by day. Our main focus has to be secular practicalities, which as we have seen is a source of great dissatisfaction a feeling of lack of wholeness completion that is the lot of modern material man. But whether acknowledged or not, there is another simultaneous existence, one more nocturnal, supernatural, mythical, religious. This is a separate subterranean intellect, the world of fantasy. Each datum, each contact or memory engram may divide simultaneously into two sub-particles that go in opposite directions. One will remain in consciousness, in the world of applicable logical practicality while each datum impinging on our being may have simultaneously an alternate illogical valence. Everything has both a logical and an emotional effect, is interpreted on the level of practical reality and fantasy. Some engrams may simply go in one direction or another but often enough they split. For physicists this is nearly analogous to a photon being simultaneously a particle and a wave, to Schrodinger's cat being at the same time dead and alive its existence described as a probabilistic wave function, or, the split of an elementary particle into two virtual simultaneous states.

The dream is emblematic of this process. Dreams happen at night, out of the glare of daylight, and logical processes. Dream images are less precise, less practical than daytime apparitions. Dreams associations are sub-logical, they are freer than waking associations. In order to explore the meaning of a dream you need to engage sub-logical processes that I talked about above including but not limited to, mythic emotional association, pure clang or poetic association, linguistic accidents or puns, predicate reasoning, connection with bodily functions such as sexuality and urination. Typical dreams utilize incompletely processed memory engrams from the previous day or immediate experience as raw material for their subject matter. The dream is helpful for final processing as assignment of meaning of immediate experiences. For example I'm often bombarded with of shards of information that I cannot possibly process at a particular moment. I receive piles of mail which I will typically review over one or two minutes in rush while a patient is getting ready to be examined in the next room. I open it next to a garbage can which means I throw most of it out immediately. Occasionally a piece I have discarded comes back to haunt me later on and I think to myself, maybe I threw out something prematurely. The interesting thing is that out of a whole large pile, I may receive one or two pieces of significance, perhaps a magazine has rejected a paper I'd written, or a course that I'd really like to attend or something of that nature. I know if something like this comes my way I just can't deal with it at that time. I'm expected in the next room in one minute. What I do is mentally "file" data to be processed at a later time. Right now, I just have to think about something else. The full-blown emotional fallout of events of the day, decisions about which pieces of information to keep and which to discard, filing and other tasks will be handled as the day wears on perhaps on a subconscious level but also may well occur that very night in a dream. We have physiological theories relating to dreams and memory espoused by Crick and others. Dreams are viewed as a sorting device for memories helping us decide which engrams to keep and which to discard. Our brains are only so large the theory goes. Much of the data we take in has to be discarded. Animals who don't dream are unable to discard data from that day and need have a disproportionately large cortex to store all that un-needed information. The example given is the ant-eating Echidna, a type of primitive vestige of egg laying mammals, which doesn't seem to have REM sleep and thus has an enormous cortex for the size of the animal. (As we have seen in pictures of dolphin brains, other mammals all cortex was not created equal and mammals with large brains lack the advanced cellular cortical architecture found in humans which is multilayered and loaded with synapses). In any event we now have centuries of research on dreams that proves that dreaming involves far more processing of data than simple sorting or rejection of engrams, that dream work involves an awful lot more than engram acceptance and rejection functions. That is fascinating about dreams is that in them we accept associations, connections that in waking life would be rejected immediately[146]. Thus dream contents are connected one to another in far more ways than logic or strict semantic processes would allow. It's analogous to having far more nexes, synapses between cells, as we have seen in the brain, that there are so many types of synapses, chemical and electrical, perhaps synapse types that have yet to be discovered. There are hosts of such possibilities for the connections between ideas as well or dream elements, yet our waking logical processes exclude all but the most logical at least most of the time. This exclusionary or logical, linguistic function which helps us through most of our normal waking day, is the *thought editor*, which is turned off in dreaming and relatively as we shall see in a host of other mental processes - the dream is merely an archetype of this process. Our mental support is two legged, the by day logical semantic leg, the world of the material and mundane, and the nocturnal or dark more numinous or mysterious world of dreaming and fantasy.

Dream work is but one example, probably the most obvious, of mental processing that allows ideas to fly in under the radar of the logical thought editor. It is characteristic of early childhood that fantasy is not distinguished from reality. Parents and teachers are frustrated with the "lies" of young children for whom wishful thoughts are reality. As these kids grow up, suddenly they cease to lie (in most cases). Numerous studies have shown how suggestible young children are which creates a nightmare in courts of law. Studies by Ceci and a host of others have proven children to be highly suggestible. Ceci was easily able to alter young children's views of reality fully convincing them that fabricated events had occurred merely by suggesting that these happened in questions[147]. Stories of sexual abuse involving caretakers and parents are often as not fabrications of police, social workers and other professional questioners. Though much parental sexual abuse undoubtedly does happen such material has been used extensively by parents in custody battles but is notoriously unreliable. Children so often confuse their own dreams and fantasies with reality. Many normal children have pretend friends and act in every way as if they truly exist. In elderly folks too, dream and fantasy material creeps into their perceptions of reality. They may dream that a deceased parent or sibling is in the room and while awake, act as if this is fact. Delusions about an elderly spouse's infidelity are extremely common, perhaps stemming from insecurities of younger years which occurs in paranoia connected with dementia and involutional melancholia. In the very young logical judgment and reality testing are not yet fully developed, while in the old judgment is impaired by brain disease. And delirium or encephalopathy impair one's faculty for distinguishing fact from wishful fiction though in these pathological states, aging, dementia, disease, delusional systems tend to be primitive and underdeveloped compared with fertile childhood fantasies.

Fantasy and invention are what the arts are made of. Associations are emotional and illogical, often outrageous and fly in the face of reason but they work within the creative psyche. Connections here are out of the realm of the apparent or obvious for most ordinary persons. They work out of the range of consciousness. Using ordinary adult reasoning, for many of us it is still possible to "admit" this type of fantastic mental processing, but many persons, entirely exclude artistic illogical modes of thought. We see all degrees of admission and exclusion of alternate modes of thinking. Arguments in American congress with regard to public support of funding for the arts revolve about the inclusion of what is felt to be the ridiculous, illogical or experimental in the arts and how much of what seems to be patent absurdity ought to be supported or even tolerated. The very same considerations hold for religious thinking. I've often observed that for a religion to be successful, to take hold with a large enough audience, the stories and precepts need to be sufficiently illogical. If they are not illogical enough, perhaps not outright impossible or taxing incredulity, then there would be no room for blind faith which to many is the most satisfying part of religion, something logical scientific and mathematical reasoning does not provide. I can give many examples here but I hesitate to out of fear of offending sensibilities.

Myth and religion heavily utilize sub-logical associations. Indeed this explains their universality and success. Myth and religion need to be something other than logical. Otherwise they are cut from the common cloth of every day experience which is no fun at all. Religion wouldn't work for most persons except to give respite from the tyranny of science and logic. The Sabbath is so much more than a simple day of rest. It is an escape into a world ideas other than work and productivity.

When we use Jungian "free association" what we really mean is that we are freed from our repressive thought editor, to elevate anything that comes into our mind into ordinary consciousness. Associations are polymorphously perverse, methods of connections almost limitless and idiotypic, ranging from simple rhymes to jokes and puns, clang associations, similarities in appearance, human animal associations and influenced by personal experience.

What do all these processes, dreaming, fantasy, art, free association, poetry have in common? They are generators of new thoughts and emotions that occur before the level of rejection. They create possibilities. In the absence of a rejecting thought editor, ideas are limitless. This other world is thus a great generator of ideas which in the light of day would be thrown out, a remnant of the primitive and childlike mind, almost an internal random number generator of possibilities before they get rejected repressed out of hand.

The random generation of possibilities is common and necessary. Evolution works this way. Diversity in biology is the random generation of alternative forms which are then tried in the real world. The vast majority of these possibilities is, on the surface, ridiculous, too ridiculous to be considered. Most mutations have to result in the formation of an impaired protein product that will adversely affect survival. Some few mutations will improve the product. Even more important is the new combination that results in a slightly different phenotype every time an animal or plant comes to be. Each and every animal or plant is thus an individual experiment in reality which to a varying extent either succeeds or fails. As we have seen, so many animals and plants are sexual because sexuality is evolutionarily advantageous. Why? Sexual reproduction adds further raw material into this cauldron of diversity. If each organism were a clone of its parents, adaptation then radiation into other environments would be impossible. There is some evidence that under adverse circumstances, lack of water or nutrition or when organisms are under attack, for examples when bacteria are being killed off by antibiotics, that at that time they make every effort to increase temporarily their diversity so a to survive. Perhaps they will begin to reproduce again sexually, or enzymes that temporarily increase mutation rates may become active.

But at any event the scheme here is that organisms are first diverse then they find by practical experience, that they can compete in different conditions. In my housing development, squirrels have few predators and reproduce so as to overwhelm the carrying capacity of their environment. There is not enough space for all of them. The same is true for deer which live close to housing developments which have no non-human predators. This is the origin of Lyme disease which lives in deer ticks that parasitize humans and deer. Some of these squirrels are resourceful enough to live not in trees but inside attics and other man-made structures. Others haven't the "personality" inventiveness or other adaptive ability or perhaps are more dominant on the ground or in trees which squirrels more typically inhabit. The squirrel world is diverse (chac'un à son gout) hence more adaptive. By the same token, our human ancestors at various times in our history, in some groups sooner, among others, later, made a number of discoveries, the most important of which converted hunter gatherers to farmers. They discovered domestication of animals and plants allowing humans to extract a living for greater numbers of people on a smaller amount of land. The discovery of agriculture and domestication came first, the practical application second. But the very store of raw material of ideas which was able to escape detection by rejecting radar or the thought editor of hunter-gathering lifestyle created the possibility, galvanized later, into the practicality of agriculture and rise of cities, that we call civilization. As I'd pointed out earlier, bio-genetic adaptation has crossed into the realm of mental adaptation or the generation of ideas and this mode of preadaptation reaches a peak in humans.[Ψ] It is a commercial for pure scientific research. You don't usually know what practical benefit you'll derive from studying something until you actually begin to seriously learn about it.

Pharmaceutical houses are just now waking up to the same strategy in discovery of new drugs. The first step, a rate-limiting step, turns out to be the rapid generation of novel compounds which only later be tested for a specific effect. The greatest obstacle being the generation of a large enough number of new chemical possibilities[148]. One of the major advances in drug research is the use of robots to generate compounds with start with some basic characteristic then can be tested further for safety and efficacy, such as a ligand that can bind to one or another specific receptor. For example, a research group may need to generate a class of compounds (the more members the merrier) which bind to a certain receptor. Later on these compounds can be weeded out according to their stability and safety profile, strength of binding etc. and the vast majority of candidates can then be rejected until an ideal candidate is found. This is not unlike the crucible of evolution which occurs in the real world. First there is variation and lots of it. Only later on will all these varieties be tried. Imagination is, in essence sort of a random number or raw materials generator, whose elements must not be, at least at first, prematurely rejected. If there is some way to generate virtual compounds without actually synthesizing them then comparing their three dimensional stereo-isomeric structure so as to make some statement about expected therapeutic result so much the better. This is the function of a computer model.

Material disseminated over the Internet falls into roughly the same category. Most of it is untried. The process is not like getting it into a magazine or book which has to pass under to watchful eyes of an editor who decides what lives and what dies. Instead a lot of material is out there for all to see. Supposed news and gossip items are floating around with no one to confirm their validity. A small amount of material may be true and some may be worthwhile. But it is all out there for a certain public to see, a way to test its mettle. This is a mass of raw material, the subconscious of the publishing world.

Out of this disorganized morass comes a very small amount of worthwhile material that will stand up in the real world. Creativity, even scientific creativity, is based on reaching down into this giant welter of possibility. One example which is frequently cited is Kekule's discovery of the ring structure of Benzene, a pivotal insight that underlies the structure of many organic molecules. Appreciation of the 6 carbon benzene ring required a paradigm shift, away from the conventional linear array of carbon atoms which scientists had pictured carbon to occur up until the end of the 19th century[∅] . This insight, Kekule claimed, came first to him in a dream. In fact there are many examples of scientific insights that first came to the discoverer in dream form. It may come as a surprise that dreams provide profound insight not only for prophets and seers who uncover principles which are then accepted as a matter of religious faith or doctrine, but for scientists who while simultaneously recognizing the truth or insight of their discoveries, then subject these to the crucible of the scientific test. Theories uncovered in this way will be tested. Presumably the fittest theories will survive. But the imaginative dreaming facility provides the grist, the raw material, for scientific mill. The advantages are presumably that ideas are not prematurely rejected by a condemning thought editor but are brought to the light of day where they can be tested.

Just as the fitness of every organism will be tested in the real world, alternative mental states provide raw varieties of ideas that may be tested in consciousness, if they are ever brought to the surface and command the attention for the conscious brain. It is worth considering how these alternative mental states develop in all of us. Consider that developed normal adults, we have at least three highly distinct mental states that are rigorously separated from each other. We experience slow wave sleep, REM or dream sleep and waking activity, all of which are distinguished on the EEG and which you can see also with other physiological parameters recorded as when a polysomnogram or sleep physiological study. These separate states develop in infancy to crystallize into approximate adult form in early childhood. Of course children and adults, daydream at which time they are electrophysiologically awake but ordinarily involved in wishful thinking or fantasy that is usually somehow satisfying and a way of passing one's time when bored. The hypnotized subject too, is restfully awake. Electrophysiologically the hypnotized person has not entered an altered state and this does not resemble sleep at all despite the name which incorporates the Greek root hypnos meaning sleep.

The first electrical brain activity that develops in prematures which can be appreciated with electrodes place over the scalp is an uncoordinated burst suppression pattern. Long periods of virtual total electrical silence or quiescence are punctured with bursts of high amplitude brain activity that looks almost like an epileptic seizure and this occurs in a seeming random pattern. The first nascent electrical brain activity is escaping from the newly formed brain as it begins to function. Activity comes first, regulation and control second.

Every mother knows that by 18-20 weeks of gestation, there are periods of physical activity in the fetus alternating with inactivity. At first, these are so undifferentiated that it is difficult to correlate with changes in electrical brain activity. If you look at an EEG of a premature, you will not be able to tell whether he is apparently, behaviorally awake or asleep. Newborns sleep about 70% of the time while as adults we spend less than a third of our time asleep. By about 28-30 weeks of gestation sleep is broken fundamentally into active and quiet sleep, the forerunner of REM vs. Slow wave sleep. In active sleep you can see that respiration, pulse, eye and body movements are irregular. In a full-term (40 week) newborn, it is easier to distinguish active REM sleep from quiet sleep but there are periods of indeterminate sleep by then as well which it is harder to categorize. The work of distinct separation of states of consciousness is not yet done. A form of quiet sleep on the EEG termed tracé alternant is only seen from about 32-40 weeks. This is replaced by the sleep spindles characteristic of mature slow wave sleep that replaces infantile quiet sleep early after birth[149]. Slow wave sleep will still later differentiate into the four prolonged distinct stages that we have in the adult, very early in life. In adults there are four stages of slow wave sleep and shorter period of REM or dream sleep. The active sleep precursor to REM sleep decreases in relative sleep time very early in life. While infants may fall directly into REM or dream sleep, this rarely to never happens in a normal adult. Always an adult goes through what seem to be meticulously orchestrated slow wave sleep stages before entering

REM dream sleep. It may be 60-90 minutes before REM sleep occurs as you first drop off to sleep at night.

Part of maturity is the stricter definition and separation of altered states of awareness. Slow wave sleep is a firewall in essence between two worlds or two states of being, dream or REM sleep and ordinary conscious awareness (waking). This separation is so well developed that it is reasonable to ask why it is so necessary at all. Add to that the fact that memories of dream experience are so rarely recalled while awake that one must postulate that there is some active process of forgetting of dream material. The separation of these different states is so well made for most of us that it must serve an adaptive purpose.

The facility for dreaming, fantasy is seemingly present in many advanced mammals. Dogs and cats dream and have REM sleep. Perhaps they are fantasizing about a chase or catching prey, they seem to have episodes during sleep that look like a simulated hunt. Aristotle and Maimonides a twelfth century Aristotelian, both remarked that Animals have the facility for dreams and fantasy. Fully conscious awareness though, reality testing, was held to distinguish humans from animals. Waking consciousness is the pride of humanity, not that it too, isn't present, but in much more rudimentary fashion in other animals, especially mammals. There is electrical precedence and analogous state change on the EEG in other animals to be sure.

So in sleep we have one example of the alternative states, the conflict of reality of waking conscious life by day, with the phantasmagorical REM or dream sleep at night when things aren't so clearly visible. One shouldn't give autocratic thought too tight a reign on all of thought processes. As we age this capacity for fantasy, the generation of new ideas, withers. As we all know there is less REM or dream sleep with increasing age. This is emblematic of a more universal process I think. The crustiness of aging, mental rigidity, is no more than the rejection of alternative modes of thought. As soon as you begin to just know out of hand what is to be rejected and what accepted, your thought processes become more Parkinsonian. Free association, fantasy, day and night dreaming, play, suggestion, are all generators of new ideas. Admittedly It's a difficult balancing act we have between the two alternate worlds because there has to be some means editor in altered states, the editor is turned off and the brain is allowed to function without fear of rejection of ideas.

| Unconscious | Conscious |
|---|---|
| Fiction | Non-fiction |
| Poetry | Prose |
| Art | Science |
| Fantasy | Reality |
| Night | Day |
| Dreams | Conscious wakefulness |
| Play | Work |
| Stars | Sun |
| Free, Clang Association | Logic |
| Myth, Religion | Technology |
| Ideas | Matter |

**Table 9: The dichotomy of awareness.**

The dichotomy of awareness is much the same as separating primitive from modern modes of thought. As I have tried to point out in the chapter on memory, rather than throw old forms away outright, these remain a part of us. As sentient living beings we differ from machines in that the past is part of our current essence, history is important, the past is not discarded. It remains inside us.

The following table is adapted from a book on Dreams by Anthony Davis, but in form it is quite analogous to the table above, even without going into detail about some of the specific elements. The primordial mind is still a very important part of modern person's thought process. It has not been thrown away just replaced by a manifest thought paradigm. Some explanatory notes are essential. In man's acquisition of knowledge, the eating of the apple in the Garden of Eden, the change from hunter gatherer status to agriculture, the ego is gradually separated from the self, creating an ego-self axis which expands as man's abilities and knowledge base expand. There is a separation from the natural primordial self so that the entire structure becomes more complicated. One's view of oneself is expanded by marriage, community and other social factors. All of the other considerations in the table are self-explanatory. In primordial Ptolomaic terms, man is the center of the universe, but ever since Copernicus we've been aware that we are not the center of the Cosmos or even the Solar system. In nature we've matured from being subject and part of it, to at least partly, manipulators and beneficiaries of its largesse. With the advent of agriculture, our view of the flow of time was largely circular and seasonal, but in modern times we view the flow of time as synonymous with progress, so that time has acquired a linear quality. Our modern reality is "show-me" scientific enmeshed in matter and less in spiritual imaginings of the past. Life is viewed as finite, with eternal life reduced to the status of wishful thinking. Myth and ritual are expendable as religion takes a back seat to science and values are no longer absolute or god-given to the modern scientific mind, they must be reproved and changed.

| | PRIMORDIAL MIND | MODERN MIND |
|---|---|---|
| **Ego-Self Axis** | Short and compact | long and attenuated |
| **Cosmic location** | central | peripheral |
| **Relationship to nature** | subjective participation 'participation mystique' | separation, objectivity |
| **Concept of time** | rhythmic and circular | progressive and linear |
| **Reality** | the world of the spirit | the world of matter |
| **Moral values** | absolute | relative |
| **Life** | eternal | finite |
| **Attitude to myth and ritual** | essential | futile |

**Table 10: Phenomenological differences between primordial and modern minds[150].**

Primordial thought patterns are still a part of our modern reality. There is certainly a persistent element of these primitive thought structures in much the same way as the dark or nighttime side of our mental processes influence consciousness. Proponents of more primitive ritualistic thinking make a valid point. For example, we've seen in recent years an outcry against moral relativism in favor of less well-developed god-given rigid rules. On many levels persons would be better off following absolute precepts. The beauty of human consciousness as opposed to mechanical objects, is the persistence of historical or past modes of thought even as newer paradigms and strategies take command of the mind. Old modes are not just thrown away. On the contrary, they remain part of us.

Biologist Ernst Mayr defines the useful concept of *teleonomy* from teleos meaning purpose or end driven as in teleology. Biologists so often speak, of animal behaviors teleologically, as though serving a purpose or intention such as bird's migration south in the Wintertime to escape lack of food and harsh circumstances in the north. As we have seen the predator and prey behaviors of even the simplest one-celled organisms seem to be goal-directed. Mayr contends that these behaviors are determined to a program carried out encrypted in our genes and passed down through the generations. Seeming intention is the unfolding and carrying out of a program determined by genetics.

At some point this genetic program is applied to the specific situation the animal is in. A program of fight or flight is intrinsic to every deer chased by a predator. Hearing a noise or rustling in the bushes, the animal will alert. The nervous system will respond, adrenaline will be released set the heart pounding. Ears and eyes will be tuned to the stimulus requiring further investigation perhaps or the animal might simply flee. In doing so it has to respond to differences in the lay of the land, the placement of trees, not to mention the surprise moves and starts of the chasing predator animal, so the response is not simply reflexive or stereotyped. Other built in programs such as those calling for the avoidance of obstacles are called up in the deer who is now in an alerted state as they are in the predator as well which may also be experiencing an orienting reaction and adrenaline rush though with somewhat different result. Are the animals' behavioral responses merely complex reflexes, that is complexes of pre-packaged programs called up in specific instances that enhance survival? As Mayr writes:

"Organisms are unique at the molecular level because they have a mechanism for the storage of historically acquired information, while inanimate matter does not. Perhaps the was an intermediate condition at the time of the origin of life, but for the last three billion years or more this distinction between living an nonliving matter has been complete. All organisms possess a historically evolved genetic program, coded in the DNA of the nucleus (or the RNA in some viruses). Nothing comparable exists in the inanimate world, except in man-made machines. The presence of this program gives organisms a peculiar duality, consisting of the genotype and a phenotype. The genotype (unchanged in its components except for occasional mutations) is handed on from generation to generation, but, owing to recombination in ever new variations. In interaction with the environment, the genotype controls the production of the phenotype, that is, the visible organism which we encounter and study.

The genotype (genetic program) is the product of a history that goes back to the origin of life, and thus it incorporates the "experiences" of all ancestors.... It is this which makes organisms historical phenomena. The genotype also endows them with the capacity for goal-directed (teleonomic) processes and activities, a capacity totally absent in the inanimate world." [151]

But Mayr would make the observation that in animals with complex brains, especially in man, the programs referred to range further from simple genetic plans. Our brains are large, complex and plastic. The human's childhood and dependency are prolonged in advanced societies sometimes well into child-bearing years, as is adulthood and old age, so that learning, wisdom and culture may be passed down. Many of us are not allowed to practice our profession independently until well into our 30's. We are as children until then, immature and dependent in our art on our still older mentors. This is the major adaptive feature in higher animals, especially in humans and even more so in highly functioning humans. Superimposed upon our genetic or biological endowment is a cognitive endowment as well. Many times, but not always, it is the learned component that supersedes biological influence as when you're hungry but wait to eat until you get done with work, or you're angry, but do not attack, or refrain from sexual satisfaction with anyone or everyone whom you find sexually attractive, or perform an altruistic act that is contrary to your own personal interest. As you become more competent, you so often rely on what you've learned to function and refrain from acting on your first impulse or intuition which is so often erroneous or counterproductive. Thus the brain, or rather information or programs incorporated in brain function have gotten the better of biological endowment or impulse.

Even birds with their notoriously small brains function behave not according to built in genetic programs alone, but need to learn from their fellows. Fully-formed bird song is not a product of genetics only, but partly determined by learned behavioral repertoire. Songbirds acquire a local dialect from their fellow mentor older birds. In humans though this process, like others, becomes so complex and highly developed, is almost hypertrophied because of the sheer size and complex circuitry of our brains. Somehow, especially for us, Mayr's programmatic teleotomy seems insufficient to explain the richness or behavior and social interaction.

A purely biological point of view, especially biology overly reliant on mechanical principles, does not begin to account for the wide range of human feelings and behaviors. While even a mechanistic view of behavior may be somewhat useful, it cuts off a complete view of how humans function. It gives a partial view of humanity without accounting for a fuller whole. This is why biological science and psychology cannot account for or predict future behavior, because always there is a residuum. A residuum better expressed in the history of the organism, personal and collective history.

Perhaps this is why we are unable to come up with biological explanations for behaviors which should, by rights, be explained biologically. The neurobiologist Simon Levay has written extensively on homosexuality. In his book Queer Science he resorts to sociology, anthropology, history to explain homosexual behavior. This is due the inadequacy of endocrine and brain related explanations for this complex of behaviors. No neuroanatomic, genetic or endocrine explanation determines homosexuality and psychological explanations or inadequate as well.

It is better to acknowledge that complex behaviors will never be explained completely with biology because we have already wandered too far from our biological origens. Reasoning has to come from a realm other than the scientific or mechanistic. Full appreciation of science, a reverence for the scintific method should incorporate an appreciation of science's limitations.

Scientific understandings leave us persons without a will or soul. Biologists and evolutionists, if they believe that consciousness exists at all, will ask why it exists. How is consciousness an evolutionary adaptation? John Searle in *The Rediscovery of the Mind*, makes the observation that many scientists, frustrated by their inability to explain or understand consciousness using purely the scientific method, deny that consciousness exists. The scientific method depends on firmly reproducible objective data. Whatever you are studying should look exactly the same, no matter who the observer is. The identity of the scientist is unimportant. If an observation is scientifically valid, all scientists will get the same result. That is what makes the concept of consciousness so slippery. By nature it is felt and appreciated subjectively. Try as we might, it is near impossible to get at someone else's mind. We have tools that help us transmit thought and feelings, writing, language, music, and art yet one person's experience is still not readily transmitted to someone else.

Consciousness for Searle is a real phenomenon that cannot be ignored. By its nature it is subjective and this challenges the scientific notions of wholly objective knowledge. Consciousness depends on reports of subjective experience to define it phenomenologically.

Consciousness is unique in this regard. While undoubtedly yours and my consciousness has common features, neither of us can quite experience the consciousness of the other and another persons consciousness cannot be described in purely objective and readily reproducible terms. Consciousness, in a word, seems to be, at least for the time being, non-transferable from one person to the next. For a scientist to reject the existence of consciousness because it is impossible to objectify is an easy out, but it does not represent truth. Considering all of this consciousness seems impossible to objectify, and document which leads some scientists to suspect that it does not exist at all. Yet the notion that there is no such thing as consciousness, which all of us experience seems he absurd.

Consciousness is not a mere byproduct or epiphenomenon of advanced cerebral cortical activity, but functions as a command-control mechanism for rapid adaptation. Conscious will takes precedence over genetic or learned teleonomic programs, though it is undoubtedly superimposed upon them. Conscious activity arises out of a conspiratorial effect of neural modules that interconnect to make experience.. These involve motor planning modules of the frontal lobe, memory and emotional modules contributed by limbic structures which color conscious experience, correlative sensory data derived from the parietal and occipital areas and arousal mechanisms controlled by the brainstem reticular activating system among others. All of these are necessary elements of consciousness and we have examined the consequences of lesions and disconnections to assess their individual contributions. Conscious knowledge will end in the prey more efficiently fleeing his predator incorporating strategies for escape that are far and away more efficacious than ordinary reflex responses under these circumstances. We have seen but one example of this with our account of the ameba, and just as efficient strategies of predators. In humans we witness long term planning mechanisms that overwhelm reflex and simple behavioral mechanisms, for example the desire for immediate reward, with considerations for long-term gain, namely, delayed gratification. Planning mechanisms have become extremely elaborate in our case since the advent of writing and computers, allowing us to design and build complex structures as disparate as skyscrapers and symphonies. Advanced schemas including codified ethics and laws that are passed down as surely as genetic information from one generation to the next define societies and groups. These blueprints allow some social groups to compete more or less successfully against others. This modifies the unit of evolution from the individual stipulated in biology, passed from one generation to the next genetically, to his social group with information inherited using the tool of language and writing. Automatic behavioral mechanisms simply don't suffice to account for advanced this kind of behavior, which adequately illustrates the usefulness of consciousness that is bound to a powerful efferent will. The larger brain with greater numbers of synaptic connections allows for, is a necessary condition, a platform for the launching of will and desire, more advanced machinery for planning, a larger store of memory, experience. Thus we will have learning to fall back upon, and all other things being equal, will compete successfully against other animals with more meager cerebral resources[ϕ] .

More than supplying the raw ability to learn and pass down information, the larger brain is now seen as an instrument for consciousness, which makes humans far more plastic and adaptable. The invention of consciousness and its efferent extension, free will, has to be seen as being instrumental in bringing about the obvious success of the human experiment.

Consciousness began as a biological construct, surviving because of evolutionary advantage. It is formed out of the conspiratorial primordia of various brain modules. That being the case, it has spun off free will. Biology, evolution, mechanism has gotten far more than it's bargained for or even what may have reasonably have been predicted, in this process. It's ended in the great bulk of human creation being extra-corporeal even extra-cerebral, that is, cars and machines to increase our speed, grand buildings that provide shelter and comfort, stone, paper and computers to record and build upon our thoughts and expand infinitely human memory along with thousands of other devices, lastly grand schemes and mental representations of meaning of life on earth.

All of this may at this point sound rather obvious, yet for the majority of us, notions of a mechanical automatic man making only predictable reflex behaviors dies hard. Some scientists continue to preach that persons do no more than express a certain limited variety of instincts and behavior programs. To them behavior is nothing more than the drives built into us and a certain limited repertoire of actions we've been taught.

Thus we are left with two distinct alternative images, alternative forms. On the one hand is executive personhood fully developed but difficult to explain utilizing current scientific concepts and terminology. On this level we have to accept our ignorance our inability to explain everything. It's analogous really to our not having a firm grasp of physics, the beginnings of time and the creation of the universe. On the other hand we go on contentedly explaining human behavior on the basis of current mechanistic and material constructs. This latter course is hubris, abundant self-confidence in our current state of knowledge about human thought and behavior. Where does this latter course lead? It leads directly to Frankenstein, to zombies, doppelgängers, shadow men, persons without dreams, consciousness and will.

I had an unpleasant experience of being misjudged and abused by a functionary in state government. In the scheme of things it was rather trivial, but I took it very seriously at the time and I felt helpless, because in this situation, there was no one I could appeal to. This bureaucrat had essentially no oversight in her job. One morning I awoke to the vision of two lines of persons. One line was for people who would be sent to the gas chamber. On the other line were persons chosen for temporary reprieve. I thought about the identity of people on the lines, and who were making life and death decisions. Those with power were bureaucrats, worse, thugs with authority. The heinousness of the Nazi crime is magnified by the fact that the victims were far more human than their handlers. Victims were the highly educated, inventors, artists, poets, engineers, entrepreneurs, physicians, musicians, scientists whose life or death was in the hands of criminals. It was Hannah Arendt writing of the

Eichmann trial who introduced the pivotal concept of the "banality of evil". Daniel Goldhagen[152] (*Hitler's Willing Executioners*") and many others, showed how this was true by empirically confirming the obvious fact that the average person through silent acquiescence or worse, was responsible for crimes and mass murder. We live in a world painful to think about, turned on its head, where the vulgar mobs of evildoers have power over the good and just. How many of us go into our jobs - we are authorities in our mini-worlds and judge with authority?

The bureaucrat or apparatchik left to his own devices is an automatic shadow-man, a doppelgänger exactly. To do well, to do the unexpected, not the spectacular for which you may expect to get immediate recognition, but to perform small acts of kindness, means not always working for your manifest immediate self-interest, but for something larger than yourself, for some ultimate reward perhaps, that is not in clear view. To perform the unexpected residual and improve a world full of cruelty that is the meaning of life.

Our world is not one of Zoroastrian symmetry between good and evil. Evil is common, banal. Good transcends evil. To do good is to go beyond expectations. To be good, maybe at times have people get angry at you, not appreciate your every motive, even though it is always comes from a sincere effort to understand and to help is believe it or not, better than being handsome smooth, slick, sophisticated or always popular.

Lots of ethnic traditions include of beings without souls. These are zombies, doppelgängers, Draculas, Frankensteins, and the homunculus or little man of Paracelcus the alchemist and golems[153]. These latter products of human as opposed to divine creations, arise solely out of matter and thus lack a soul, or will of their own. Lacking a higher consciousness they are beholden to their human maker. The legend of the golem (Hebrew for embryo) is the best developed in Talmudic, and Kabalistic legend and explored extensively in Yiddish literature, theatre and musical productions. According to Jewish legend humans are divine creation imbued with a *neshama* for soul. Adam (="earth") for the first 12 hours of life had a vitality alone, but only later was infused with a neshama. Some sages, on the other hand, having some secret knowledge, had the power to create a golem, or clay man, through their incantations, out of ordinary earth as if they were by virtue of superior wisdom, given enough of the power of God to create a living being but one with profound limitations. The golem was an imperfect creature with no will of his own, no neshama, but could perform, zombie-like, as an automaton, according to the will of his human creator. The golem could not speak. Tradition has it that the word *emet* for "truth", or knowledge, was stamped on the golem's forehead. You could deactivate or decommission a golem simply by rubbing out the first Hebrew letter of the word emet, the aleph, thus converting *emet* to *met* for "death". Its creators would dance about the inanimate clay creature reciting alphabetical letters or the secret Name of God bringing the golem to life. The incantations that originally created the golem could be recited in reverse, the dance being reversed as well, deactivating the creature. Thus a material creature was made to live and die out of the word, reliving the creation of the world where matter sprang from nothingness, and man was made from clay.

The human creators had an undeveloped embryonic simpleton monster on their hands, one with raw physical power that would grow beyond the man's ability to control it. Eventually the Golem would have to be destroyed. Many stories and plays were written about golems created to protect the Jewish neighborhoods from gentile oppressors who accused Jews of the Middle Ages of Blood Libels and other false crimes. The golem represented physical power brought about by the earthy materialist knowledge that was the province of man who had not the power which was the property of the divine presence alone, to make a total person having its own will, speech and soul.

As with many legends there is truth in the underlying ideas. Human knowledge has advanced to the extent that our own creations extend and enhance human capacities. The computer and calculator are tools that extend human abilities with speeds unrivaled by our brains. We can use drugs and transplant human organs to extend life for finite periods of time. Our machines carry us many times faster and with much less effort than natural means of locomotion (our own legs). With our knowledge of the human brain we are able to compartmentalize and modularize nervous system function and have, as we have seen, revolutionized our own picture of consciousness and human will.

But as of the present we have no idea how to create or animate willful sentient beings. We have mastered the machine language, the technicalities of medicine to a great extent but haven't yet created a being with its own strategy for existence even on the level of our ameba. Our scientists are the sages of the golem story. They have created their own golems, shadow men, doppelgängers but haven't even a vague idea about how to create and manipulate self-consciousness, awareness or its efferent component which is human will. With our knowledge as limited as it is, we have been unable to recreate life, certainly life which acts and thinks on its own accord. We have been successful at creating neither Carbon nor Silicon beings that act of their own free will.

The major question that we have to answer is where is our deficiency in understanding of ourselves. Will full understanding come eventually from more of the same, from brute force of applying mechanistic and materialist principles, or is it possible we are missing something less tangible, less material, in human and even animal nature? When we finally come to understand all of the chemistry, the electrophysiology, anatomy, will we then understand ourselves?

What are the limits of the scientific method?

# Chapter 5

## BEYOND BIOLOGY

And he tamed fire which, like some beast of prey,
Most terrible, but lovely, played beneath
The frown of man; and tortured to his will
Iron and gold, the slaves and signs of power,
And gems and poisons, and all subtlest forms
Hidden beneath the mountains and the waves,
He gave man speech, and speech created thought,
Which is the measure of the universe;
And Science struck the thrones of earth and heaven,
Which shook, but fell not; and the harmonious mind
Poured itself forth in all-prophetic song;
and music lifted up the listening spirit
Until it walked, exempt from mortal care,
Godlike, o'er the clear billows of sweet sound;
And human hands first mimicked and then mocked,
With moulded limbs more lovely than its own,
The human form, till marble grew divine;
And mothers, gazing, drank the love men see
Reflected in their race, behold, and perish.
He told the bidden power of herbs and springs,
And Disease drank and slept, Death grew like sleep.
He taught the implicated orbits woven
Of the wide-wandering stars; and how the sun
Changes his lair, and by what secret spell
The pale moon is transformed, when her broad eye

Gazes not on the interlunar sea:
He taught to rule, as life directs the limbs,
The tempest-winged chariots of the Ocean,
And the Celt knew the Indian, Cities then
Were built, and through their snow-like columns flowed
The warm winds, and the azure aether shone,
And the blue sea and shadowy hills were seen.
Such the alleviations of his state,
Prometheus gave to man, for which he hangs
Withering, in destined pain: but who rains down
Evil, the immedicable plague, which, while
Man looks on his creation like a God
And sees that it is glorious, drives him on,
The wreck of his own will, the scorn of earth, The outcast, the abandoned, the alone?
Not Jove: while yet his frown shook Heaven, ay, when
His adversary from adamantine chains
Cursed him, he trembled like a slave…
Who is his master? Is he too a slave?

From **Prometheus Unbound** by Percy B. Shelley

"The experience of the last 150 years has shown us that life is subject to the same laws of nature as inanimate matter. Nor is there any evidence in a grand design in the origin or the evolution of life. There are well-known problems in the description of consciousness in terms of the working of the brain. They arise because we each have special knowledge of our own consciousness that does not come to us from the senses. In principle, no obstacle stands in the way of explaining the *behavior* of other people in terms of neurology and physiology and, ultimately, in terms of physics and history. When we have succeeded in this endeavor, we should find that part of the explanation is a program of neural activity that we will recognize as corresponding to our own consciousness.But as much as we would like to take a unified view of nature , we keep encountering a stubborn duality in the role of intelligent life in the universe, as both subject and student. We see this even in the deepest level of quantum physics,

-Steven Weinberg Life in the Universe,
*Scienific American* 271(4)p 47 October 1994"[154]

The Titan Prometheus ("forethought") fashioned man and brought fire from Mount Olympus. His brother Epimetheus profligately gifted all the other animals, fleet legs for running, flight for birds and so on, but ran out of attributes by the time he got to man. Prometheus made man upright so he could occasionally look upward at the stars, and taught man technology, astronomy the arts and language to separate men from the animals. Man was thus made, by virtue of his know-how, in the image of the gods. Prometheus had done his work so well that mankind threatened the gods. Zeus had Prometheus chained to a rock in the Caucasus. By day a vulture would tear at his liver that regrew at night, ensuring endless torment. Zeus would have freed Prometheus. But Prometheus was a rebel who refused to give up information which Zeus craved. Hercules heroically set Prometheus free. Romantic poets like Shelley and Byron were inspired by the Titans, and Prometheus in particular, as figures who resolutely stood up to the imperial powers of conventional gods as led by Zeus.

In Genesis, Adam and Eve are created in a pristine blissful ignorance, all needs, eternal life provided for in the womb-like Garden of Eden. Eve takes knowledge for herself, encouraged by the evil rebel Satan. In fact God created a world, beings and circumstances that made these events inevitable, and therefore in His own omniscience, He certainly planned to make things go just this way. Man, even in Genesis, was intended to have knowledge and all its consequences. If in biblical tradition knowledge is Satanic in Greek lore, it is Promethian, heroic. Both agree that man, in acquiring knowledge and thought becomes god-like. Surely experience teaches us that knowledge has good and evil consequences.

In former generations man's level of expertise grew almost linearly. There were distinct limits in our ability to know. Men in other times were bound to the earth. They knew nothing of what they saw as they stared into the sky, the true extent of the cosmos, of other galaxies just our own Milky Way, that contain hundreds of billions of stars. Earth's gravity limited them too. Only recently have men been able to fly and even to occasionally escape earth's gravitational field. And knowledge was very limited in other ways, about our own makeup, and how it might be changed, about the basics of biology and the biochemistry of genes and how genes can be altered. Also knowledge was once province of the relatively privileged few, while today it is democratized. Everyone with a computer and this will be just about everyone, will feed at the trough of knowledge as libraries as we know them, become a thing of the past. The quantity of knowledge more than ever, is growing exponentially, explosively. And so mankind is escaping all limits all bounds, as in the past, but the likes of which, quantitatively we have never witnessed before. On the brink of the twenty-first century we have, ladies and gentlemen, Prometheus Unbound, not freedom only, that would be relatively trivial, but knowledge and possibilities beyond bound.

What will happen when we truly are aware that we've escaped finally the limits and gravity of the earth, that we gaze upon the true magnificence of the cosmos, that once and for all we've thrown off the fetters of our own biological limitations? No one knows, nor can predict.

In previous chapters I explored biological and neurological mechanisms that lay the foundation for our humanness. As useful as these mechanisms are, we have found them wanting. Section one dealt with various states of consciousness, unconscious and altered states of awareness. Consciousness is a composite function involving various specialized modules tied together by an executive. The anatomical location of this kernel-integrating unit is unknown at present and the executive function cannot be described given our current level of knowledge. A complete accounting of the miracle of consciousness will require a basic revolution in our mode of thinking. In section two we looked at vision as a model for sensory perception. Again we had the advantage that more is known about the physiology of vision than any other sense. In Section three I explored memory, as the simplest mental faculty, would be more difficult to describe, given our current state of knowledge. Memory is far more complex than one would imagine, as it is linked with other cognitive abilities. Finally the fourth section dealt with executive function as an extension of motor or efferent function. In all cases there was a lot of very useful knowledge, yet there was the sense that mechanistic explanations were far from adequate. The deeper we got into physical explanations for neural function, the more we had to learn. Any scientist would acknowledge that they can't explain a lot of phenomena, but most would maintain that this is only due to the deplorable state of our current knowledge, that given time and enough hard work, “brute force” utilizing the scientific method, scientists will slowly tease out the answers to all of our deepest mysteries. Science is a temptress at once beguiling us as layers of knowledge are peeled away, but when it comes to understanding of ourselves our fondest wish may never be fulfilled or at least our deepest questions will not be solved solely by the scientific method. As far as humans are concerned there is more than meets the material eye.

Scientists are in the best position to appreciate the limitations of science. Layman or non-scientists see science from afar. Wowed by practical technological advancements and having limited understanding of scientific techniques and methods, they tend to view science as an all powerful extraordinary undertaking. Scientists on the other hand, aren't anxious to give up their admiration even majesty and mystique. With the deepest respect for the scientific method, I have come to the conclusion that there is far more to human potential than will ever be revealed by science.

This chapter we will make a more advanced argument. Here we will ignore the fact that scientific explanations are necessarily incomplete. We will begin to observe that humankind has already ventured past biological bounds. Previous chapters have scanned abilities and connected them with neurological and physical principles. Mental capacities are deeply embedded or immanent in biology. But biology does not explain all. Some capacities transcend biological mechanisms, typically the works of modern man in recent centuries and especially of recent years. This may be the very reason that these capacities have yet to be appreciated for what they are, beyond or transcending biology. One of the pivotal concepts of theology is an immanent and transcendent God by which is meant that the spirit of God is at the same time deeply situated in the world and hovering above it. Leaving theology aside here, I borrow this useful concept for the idea of humankind, embedded in (immanent) and at the same time, transcendent upon biology, science, matter, mechanics.

A reasonable question is what drives a species to explore new territory. Answering this question gives us a better handle on human exploration, the desire to explore possibilities outside of the current range. Is exploration of surroundings beyond our own a purely biological imperative which we fulfill so eminently merely because of the size of our brain and use of superior mental tools also provided by biology, or is there something more to it, a striving that is outside over and above what is provided by biology?

To answer this we need to examine what keeps an organism in a particular habitat and what are the forces that drive it to explore other environments, from whence comes wanderlust? A given species of plant will survive and compete best at a particular altitude within the dense lush growth of a rain forest. It is well adapted there. Where there is abundant speciation, this range of habitat can be extremely narrow. On the other hand when competition is limited the range can be quite large. Pines and maples compete successfully within a certain latitude describing a particular temperature and humidity ranges or alternately within a given range of elevation above sea level. It is not as if maple trees have not dropped seeds above or below this level, only that other species compete more successfully because of more or less cold or heat or water adaptation or due to different relationships with animal fauna.

Hominid forebears of humans adapted away from apelike ancestors fit for life in trees. Hominids favored the savanna and later the desert as bipeds. Pelves became shorter and wider, the lumbar or lower spine longer, the arms shorter all in preparation for a bipedal existence rather than a knuckle walking. According to new evidence, it was not a new hominid line that separated from the rest of the apes, chimps, gorillas and orangs between five and seven millions of years ago. Rather, chimp and human lines formed their own branch for a time separating from other apes. It is said humans share 98 percent of genes with the chimpanzee, our closest living relative, but this depends on the method of reckoning for what constitutes an identical gene. There are striking similarities and differences between the two species not to mention the most obvious, that our brain is about 1350 cc. Compared to the chimp's 400. A bigger brain of course comes into the world through its mother's pelvis. She needs to be able to walk as well as to give birth. Eve's curse for eating of the tree of knowledge of good and evil "I will greatly multiply thy pain and thy travail; in pain shalt thou bring forth children" is literally due to bringing forth children with large enough brains apprehend this knowledge. Our offspring matriculate in the real world. Since a pelvis capable of a reasonable ambulation can give birth to a neonatal head with a brain of not more than about 350 cc. and the adult brain of 1350 cc is some four times this size, the other necessary accommodation to the excessive size of the adult human brain is a prolonged dependent childhood, fostering the brain's physical as well as mental development[155].

Men, apes and indeed all species fill their own ecological niche. Time saw a progression of proto-human forms such as the Australopithecines, still later Homo habilis, Homo erectus and finally Neanderthals and modern man, Homo sapiens sapiens. After Australopithicus, the genus Homo separated on its own evolutionary branch from other apes. The first upright hominids were more adapted to life on a sparsely treed grass plain with their upright posture and built better for running distances than brachiating through trees. Erectus' brain size was in the range of about 900-1100 cc. Their tool technology named for specific sites of discovery as 'Oldowan' for the Olduvai Gorge or the only slightly more advanced 'Acheulian' for St. Acheul finds in France. A major difference in these is the size of sharp stone tools, larger axes in the latter replacing simple choppers. A debating point is whether it was merely tool-making technology, essentially static and producing few, perhaps 10 of so variants over a period of perhaps 1.2 million years, that co-evolved with increasing brain size or if it was something else, the change from foods made of primarily vegetable matter to a more catholic or varied taste for meats and a resulting more complex savage but at the same time cooperative society of specialists necessitated by a hunter gatherer existence, or even the development of some spoken language even which necessitated brain growth and hence made a more plastic and response to environmental stresses possible. One can legitimately ask, from whence derives the first furtive rangings into the growth of the intellect. From earlier chapters we know the brain as an organ of afferent, associative and efferent (e.g. tool-making) function, so it is easy to see that tool making or efferent function can only be part of the story behind increasing brain size and complexity. Intellect is a composite function of brain modules. It was not, as some theorists suggest, merely the freeing of human hands, by the upright stance of men, though that is related to the growth of the motor and sensory convexities of the brain which deal specifically with the hands. (Recall the hands on the homunculus of the brain are represented by the cerebral convexities and not in the midline which represents the feet.) we aren't the only organisms by far to stand on two feet and have free hands. Dinosaurs and ostriches have free upper extremities as well but very small brains.

Somewhere along the course of human evolution, the familiar menstrual cycle replaced the standard ape estrus cycle. This meant women were sexually receptive most of the time rather than only about twice a year during a brief period of fertility. And the enlargement of the male penis and testicles, adapted for increasingly frequent copulations. Sex was a bigger deal. These changes sponsored far more enduring cultural ties, perhaps long-term commitments, and stable family relations (either monogamous or polygynous). The growth of cultural relations in general was probably the most potent stimulant for brain growth. The development of culture implies a meaningful tradition is passed down and is in a communal repository to be built upon and improved. Homo Erectus dating from as long ago as 1.8 million years ago has been found as far afield as Java Indonesia. How did these proto-humans travel over water except via some floating craft that were used as tools for exploration and adventure? And how would it be possible for one proto-human to produce such crafts except by building on the experience of previous intrepid explorers?

Current evidence is that at least the later forms particularly the Neanderthals, successfully radiated all over the map, through the Levant to Europe and Asia out of Africa.. Parts of Neanderthal skeletons are found throughout the Near East, Europe and Asia though the lion's share of Neanderthal sights are in Germany where they were first discovered, France, and throughout Europe. It now appears Neanderthal with his thicker build, constituted a variant of the human form built for colder weather. There is a great debate as to whether thick boned Neanderthals are our direct ancestors or whether if Neanderthals existed today, some of us might even be inclined to breed with them as fellow humans, if they might be considered so closely related as to be of the species Homo sapiens, in other words. Neanderthals were thick-boned humans, with enlarged muscle insertions obvious on their bones, fostering the stereotype of large muscled cave dwelling brutes with thick widened brows.

But they were human in a lot of respects. Neanderthals were extant from about 200,000 to 35,000 years ago. Their brains were slightly larger than ours up to the 1500cc range, though we have no knowledge of their brain's microscopical anatomy or synaptic relations between cells and we can only conjecture even about the size of individual lobes, Broca's and Wernicke's areas involved in language and therefore social function . They were tool makers more advanced than H. erectus, founders of the Mousterian (for Le Moustier on the Dordogne) tool-making culture of 60 or so different tools, an exponential increase in tool making variation over their Homo erectus predecessors. There is reasonable evidence that they buried their dead in special ceremonies, pollen remains of flowers and animals to provide food for the deceased have been found with Neanderthal remains indicating some adornment of a ritually buried person, so Neanderthals would certainly have wondered, as we do, about the meaning of life as they looked up at the stars and conjectured about an afterlife and were aware of the limitations of time. I'd venture to say that few, if any, animal dwells upon its own death, as Neanderthals most probably did. No one knows whether Neanderthals, classic "cave men" spoke or not or how advanced were Neanderthal societies. We suspect that their wonder was in fact expressed through some primitive language, though there is some tentative evidence based on the shape of the base of Neanderthal skulls that they could not have generated as many phonemes as modern men are capable of. Their rites and dances, teachings about the hunt, human relationships, marriage and the family do not survive with their bones.

As much as Neanderthals were thick boned and thick muscled with big brows, modern humans are thin boned or gracile. Our skeleton and body are much lighter; there is more marrow, cancellous or latticed bone and less heavy calcium with smaller muscles. The earliest modern humans, are recognized by their gracile, thin-boned less brutish skeletons, especially a thinner brow ridge, due to the fact that the skeleton is what is left of them. Well before the invention of writing, some form of culture probably spread out of Africa within the last 200 thousand years along with a more human biological form. The vocal cords were now lower, the larynx longer, making more advanced phonation possible but at the expense of more difficult deglutition. The time frame for the emergence of modern sapiens is a very tentative estimate based partly on the rate of gene mutations mostly of mitochondrial DNA. Techniques for comparing DNA have been with us only over the couple of decades. Mitochondria, the only organelles with their own DNA outside the nucleus, are part of the protoplasm of the cell, hence the ovum, passed down to all of us only through our mother, maker of the egg cell. When an ovum is fertilized, the successful sperm deposits merely the nuclear genetic material, the rest of the stuff in the zygote's cytoplasm comes from mom. If you are to trace changes in DNA genetic material it is much easier to keep track if you know for a fact it comes from one lineage, namely only through one's mother and that through her mother.

On the basis of expectations of only a finite number of mutations in a small circle of 37 genes that is part of mitochondrial DNA over tens of thousands of years of human generations you have an internal clock in a sense which ticks in divisions of time, so many mutations per 10 thousands of years. Observing DNA taken from current humans all over the world, one can compare the variation in DNA which could only have accumulated through mutation, and draw important conclusions, about the age of last radiation of humans from their starting point wherever that might be to all other habitats. The two important facts are that the amount of variation between mitochondrial DNA between modern humans would seem to place the date of modern human worldwide migration fairly recent within the last 200,000 years. Another very surprising fact is that the mitochondrial DNA of indigenous Africans varies much more than the racially diverse DNA of all other locations -there is more variability in Africa itself than at all other locales around the world which would seem to imply that the races are much more related to each other than indigenous Africans. The divergence that are obvious to us and define the differences in humanity all over the globe amount to much ado about nothing from a biological perspective. Intra-racial is greater than inter-racial genetic variation. This further implies that modern humankind radiated out of Africa then spread all over the globe. Perhaps we had a common female ancestor whom we can conveniently call Eve, mother of us all, in geologically speaking, the not too distant past.

Most probably there were multiple separate radiations of proto-humans over the Old World, showing again that successful animals tend to radiate geographically, to increase their habitat. Homo erectus, Neanderthals and modern gracile humans radiated out of their original habitats separately. There is some debate as to whether we are descended from Neanderthals who lived over a wide range of the old world but we now know that the age of Neanderthals and modern humans overlapped by at least a few thousand years. Neanderthals and modern humans lived side by side. The thinner boned less robust more modern humans very probably out-competed the larger brutes. Modern humans may well have been far more fierce and competitive and killed the Neanderthals off. Why were moderns more successful? The best theory is a combination of smarts and more advanced social structures. More to pass down. Intellect beats brutishness. But let us not take this too far. It couldn't have been the meek inheriting the earth. We know from recorded history about ruthlessness and cruelty in our own species. Cro magnon may have eaten Neanderthal for lunch.

Enter gracile man, maker of bone tools of many varieties, bead and stone necklaces, Lascaux cave, user of wood and of fire, cave dweller and human art only some of which done for a practical end. Modern man radiated throughout the known world. 40,000 years ago they went as far as Australia. 15,000 years ago forebears of Amerindians crossed the Bering Strait into Alaska. Perhaps they walked over an icy bridge. What made them move at times from relative comfort out into the cold uninhabited wilderness, making the transition from what is the same to the different. What makes humans and other animals go where no one has gone before?

There are competing forces at work. Firstly there is crowding and competition. We may reproduce until there is no more carrying capacity in the environment. This holds for inefficient hunter-gatherers in particular. A given habitat is capable of supporting only a small number of persons occupying the pinnacle of the food chain. Ineffective hunting methods with limited implements may allow capture of only the most vulnerable prey, the slow the sick and the old of animal species that are themselves overpopulating their habitat. Competition for food and mates may drive away members of the community who will be forced to make their way in a remote location. The bulk of adult males may wish to strike out on their own, forming an new family unit, founding a new people because of, or in spite of, the competition. The great bulk of geographic radiation results from the experiment that each individual of a species ultimately is. Each person is like an individual seed that falls from a tree. He will land on a habitat and either survive or not survive, bear fruit or wither, much as does the seed of a tree (perhaps with a higher probability of survival than a plant seed).

Animal or plant, all organisms naturally radiate to fill their niche. They are always testing the limits of that niche. Where there are many species, as in a tropical rain forest, competing for lush resources but trying to make a living in the same way, the size of the ecological niche will necessarily be very small, but out on the tundra where less species compete, the niche will be geographically larger. This is exactly the same situation as competition in an economy between start-up companies[156]. The company in an economy is analogous to the individual in a ecosystem or possibly a small social group, or better, a species. A crowded field will produce many superspecialists, a phenomenon which we see in computer or bio-tech fields in America, the tropical rain forest of economies[157].

Economic speciation is intense and furious. But one species has arisen which has unfair advantage over all others. This is man. He is an omnivore and supremely adaptable to all conditions. He may be seen to compete with other carnivorous species at the apex of the food chain such as tigers, lions, and hyenas, yet his diet is varied and level of invention without bound. In latter times humans will have learned to domesticate and raise the animals they consume and eliminate the need for hunting and gathering altogether making the city, civilization, possible. As we have seen the major biological competitors for mankind nowadays are not other carnivorous species but bacteria and insects which challenge us by their pure level of bio-adaptation and short generation time that matches the brain's rate of bringing ideas to fruition. [158].

Humans are unique as a single highly sapient species that fills an enormous range of habitats and are distributed worldwide. Though he may rarely choose to do so, a Scandinavian may easily breed with a Hottentot. More importantly, as vividly illustrated by Jared Diamond[159] in his book *The Third Chimpanzee* although some human groups such as the Australian aborigines may have been isolated and have had no exposure to modern culture and technology, their intellect will allow them to integrate into modern society within a short time and even to fly our jet planes or operate machinery. Admittedly it may take them generations to fathom the meaning of the new technology they have easily mastered, which is the reason for so much war and strife in countries recently given their own autonomy with decolonization but they may take advantage of technological advances nonetheless. We are one species, the differences among us are literally skin deep. Genetic studies show that we have separated and differentiated only quite recently in geologic time. Think of what might happen if the world were peopled with two or more sapient species especially if both were as self-serving and bellicose as our own. Or is it inevitable perhaps that sapiens should exist as a single biological entity? It is quite possible that once we had close rivals, Neanderthals or others that modern humans took out of the competition. Or perhaps circumstances, a long chain of accidental circumstances led to our descent are so unique on earth, or even in the universe. This is the stuff of the anthropic cosmological principle[Ψ] . We are unique.

The natural tendency is for any species to radiate, to expand its geographic and ecological base. This will make possible an increase in population and make the species less vulnerable as well. A wide base is more difficult to push or topple over. Species spread out over a larger geographic area unless pushed out by other competing species or if they are simply unfit for a different habitat or if a boundary such as a large mountain of body of water prevents spread. Perhaps mankind, able to compete in many directions is sort of a jack of all trades while some other specialists may out-compete mankind in some limited sphere of endeavor. Human genius overcomes most habitat limitations. We are able to survive in the tundra or the Arctic as well as on the Savanna. Geographic obstacles don't stop us anymore. We haven't spent long periods underwater and few of us make a permanent dwelling under the seas. Someday we dream about living on other planets perhaps planets around other stars but this seems a long way off. In the meantime we are testing the carrying capacity of the limited space we occupy on earth, as population is expected to top 8 billion by 2020.

Humans, by virtue of their ability to adapt to environments without having to make biological changes, with the aid of the brain in other words that makes them infinitely more adaptable, were able to survive in all kinds of climates and habitats. They are widely successful because of not being limited by their biology alone. Then there is the wanderlust, which is an intangible, the sense of wonder about what lies over the next hill or the further horizon, curiosity breeds success.

We can only theorize about what gave one group an advantage over the other. They migrated or radiated separately. When it came to Robust or heavier boned humans exemplified by the Neanderthals, they had their own migration and were widely successful in Europe and Asia, over a much longer period than whatever we are aware of in our proud later gracile human history which only spans a few thousand years. When you think about it, nature was remarkably patient in our development, considering that life has existed on our planet for maybe 3.5 billions of years, hominids diverged from apes 5 to 7 million years ago, the most advanced hominids for a long time, the Neanderthals and their thick boned relatives prospered 200,000 to maybe 35,000 years ago. Over this long period of hominid development, there is little left for our examination except a surprisingly small variety of tools and some primitive burial sites. Then there were modern humans who were around for 40,000 or so years. Their tool making technology utilized bones, ivory and materials other than stone (finally) and we begin to see some cave drawing and other signs of abstract thought, later cuneiform and still much more recently writing with an alphabet starting about 3500 or so years ago. And it is only here in the twilight of our existence that we maintain, and I'm speaking of any extant culture now, that any of our gods saw fit to endow us with some code of ethical, moral or religious conduct, or actually declared themselves finally to men as wholly conscious sapient fellow beings. Awfully patient were these supreme beings, whichever of them we believe in, to wait all this time and appear to us just now, when we finally have the faculty of understanding. The latest human form, gracile attenuated, light, fragile man. A better more adaptive model, not because he's harder to break, but due to the fact that he can think better, learn faster, store information. Gracile man is a repository of information, more powerful than bone and sinew, information that can be passed down, superseding genetic information inside every living cell. Other men need not start experience from the beginning, so there is no need to reinvent the wheel. Finally it is possible to build upon the past to exist upon a ready-made foundation to stand upon the shoulders of giants. If he can write so much easier is it to lay a plan, to plan a society , build and edifice, do a painting, write a symphony rather than some hodgepodge of random ad-lib sound.

Over the last 200 or so thousands of years, forebears of modern men radiated into diverse climates very possibly emerging out of Africa, spreading the fertile crescent and thence into Asia and earlier than 50 thousand years ago, into Australia. This latter journey was undertaken quite possibly on makeshift floating structures resembling boats by inordinately courageous folks driven into discovery of new lands that would support them.

## Carbon and Silicon Wed:

With Chemistry being a distant memory and unnatural subject for most of us, few appreciate the implications of the elements Carbon and Silicon. Carbon and Silicon on the Periodic Table are in the same family of elements. Both have 4 electrons in their outer shell. Electrons don't haphazardly encircle atomic nuclei as planets orbit their sun. A star may have almost any number of planets of different sizes. At the level of the atom we are in the quantum world. In the ordinary macroscopic world Newton's laws are a good approximation of reality, not so in the atomic world of quantum physics. Electrons occupy a specific kind of orbit called a shell. Atomic shells distinguish themselves others by distance from the nucleus of the atom. The nucleus contains much heavier particles, neutrons and protons. There are rules of occupation for electrons. The first shell is the one closest to the nucleus and can hold 2 electrons. If this shell is completely full in an atom that has two positively charged protons and two negatively charged electrons, we have an atom of Helium. The atom with shells filled lives a satisfying but lonely life and will not bond with other atoms. This is true for so called noble gases, Helium is an example, which do not form a chemical bond. Other elements bond only in certain ways dictated by their outer electron shells. Carbon and Silicon are alike, members of family IVa. Carbon has the atomic number 6, Silicon 14, which is the number of positively charged protons that occupies the nucleus of the atom and positive charges need to be balanced by negatively charged electrons. Now it is rule that a given shell will contain a certain number of electrons. For the first shell closest to the nucleus that number is 2, for the second shell 8, for the third 18, each shell being successively farther from the nucleus of the atom. The inner shells fill first as electrons are added. For a shell number, 1, 2, 3, and so on, S, the number of electrons to fill that shell is $2S^2$.

You can easily see that that for Carbon with 6 electrons, four will end up in the outer shell which is the same as for Silicon with 14 Silicon has 3 shells in comparison with Carbon's 2. Silicon is a larger and heavier atom with electrons further removed. Silicon's chemical bonding behavior is very much the same as Carbon's simply because the number of electrons in the outer shell of an atom determines its behavior when it comes to forming a chemical bond. When two atoms bond, they share electrons. They do so because in and of themselves they have shells or orbitals (diminutive for orbit) that are incompletely filled with electrons. To anthropomophize, atoms are unsatisfied and need to associate with other atoms in order to achieve fulfillment, to fill their orbitals or shells. Noble gases such as Helium have perfectly filled electron shells and have no need to make an association with other atoms. They have a patrician and isolated existence. Other elements are unfulfilled in and of themselves. Many of these associate (bond) extensively. Carbon and Silicon are in the latter class.

In consideration of their unfilled outer electron shells one of the things both Carbon and Silicon frequently do is bond with four other atoms. In so doing Carbon or Silicon will share an electron with each. A perfect atom to bond with is another copy of itself, Carbon with Carbon or Silicon with Silicon each achieving the happy status of 8 electrons in its outer shell[Ψ] . Either Carbon or Silicon will form a perfect crystal by latticing in this way.

Carbon is a social element. It bonds with itself extensively and most importantly not as a crystal lattice structure as illustrated above but in chains of double and single bonds Carbon also associates extensively with other elements, particularly Oxygen, Nitrogen, Sulfur and so forth, even with metals such as iron in hemoglobin. Carbon's repertoire of covalent bond association is truly amazing. This is organic chemistry of course, the chemistry of life. It's important to realize that all of this, all life, is made possible by chemistry and physics, the quantum properties of the element Carbon.

If Silicon is the same as Carbon, then why is there not Silicon based organic chemistry, Silicon life? This is the stuff of science fiction. Maybe somewhere on our planet, perhaps on other worlds there might be organisms based on Silicon chemistry. Well, Silicon is not the same as Carbon. Silicon can form chains with itself but partly owing to the larger size of the silicon atom, Silicon-Silicon chains are unfortunately not very stable. By virtue of its being a larger atom and because of other properties of chemical bonding Silicon life does not appear to be possible. Just one example is in the way Silicon bonds with Oxygen. Carbon and oxygen associate extensively and in a variety of ways. The simplest relationship is Carbon dioxide, a gas whose properties are taken advantage extensively in organisms in energy storage and respiration. $CO_2$ is cycled by plants which fix it in sugars that store energy for almost all forms of life and produced in the respiration of animals and plants. $SiO_2$ , Silicon dioxide, is totally different. It is a solid, not fit for any respiratory cycle, a crystal, the formula for glass, sand, and quartz. Carbon dioxide has a double bond between Oxygen and Carbon, Silicon Dioxide a single bond and it forms another stable crystal lattice

The Silicon crystal lattice has other interesting properties that make it useful. It forms the basis the current semiconductor industry, the heart of computer electronics. As we have seen the Silicon crystal lattice is essentially the same as Carbon's. Life does not utilize Carbon crystals, but takes advantage of Carbon chains. Silicon crystals on the other hand lay the basis for the semiconductor industry. Silicon wafers may be 'doped' with impurities especially Boron, with just 3 electrons in its outer shell, and with Phosphorus, with 5. The impurities add an extra electron freely mobile within the structure in the case of Phosphorus doping making it negative (N-Type Silicon). Boron doping will create a deficiency of one electron giving a free positive charge (P-type) which given the motility of negatively charged electrons will create a semi-conductor, the basis of all computers.

The interesting part of all of this is that P and N type silicon are conjoined to replicate electronic components, diodes that allow current to flow in only one direction, and most importantly transistors, which act as electronic switches. Transistors acting as switches or gates are always either in an open or a closed state, allowing or disallowing the flow of current. This is the binary state as discussed in Chapter one of the computer which processes series of 1's and zeros in binary code, also to some extent neurons too, which at any time are either firing or not firing. Other impurities including Germanium, and other elements and even Carbon, can be added to the Silicon crystal lattice or wafer which may be used to make the wafer size even smaller and enhance electrical and storage properties. This aspect of materials science, the creation of a better Silicon device is explored mainly in U.S. and Japanese laboratories. It is extremely basic and important research which has the potential to, multiply storage capacity, increase miniaturization and speed of computer devices. The interesting thing is that Carbon and Silicon do not readily form stable crystals together on the same wafer device, though these limitations may someday be overcome.

Transistors replaced vacuum tubes, which served the same function, during the 1950's. I recall going with my dad to the drugstore to get vacuum tubes tested. We'd suspect a tube was blown when a table radio or television stopped working and a tube didn't glow. Usually we were right. When I was very young transistors replaced vacuum tubes. Transistors were a lot more reliable and never had to be replaced. Transistor radios replaced the larger table versions and we could take them to the beach. Most were advertised according to the number of transistors, 4, 6, 8 or 10. The more they had the better was the reception because transistors primarily served as amplifiers of the radio signal.

The very first computers were mechanical not electrical contrivances. They were bulky and could only do limited calculations due to errors in mechanically setting dials and gears and the inherent slowness in mechanical gear motions. Mechanical computers are analog devices with inherent inaccuracies. When you look at your analog watch with hands that point at the time, you can only estimate the exact time. Inaccuracies will multiply with a large number of arithmetic operations in any analog mechanical computer. The great mathematician Wilhelm Leibniz the co-inventor of Calculus, built a mechanical computing machine in 1672 , writing presciently, “For it is unworthy of excellent men to lose hours like slaves in the labor of calculation which could safely be relegated to anyone else if machines were used.” The first modern computers were conceived in the early 1940s used to figure artillery shell trajectories. The Mark I used electromechanical relays as on-off switches that opened and closed utilizing an electromagnet, a relay. The machine was useful but slow because of its mechanical nature. What was needed was a pure electrical device. The next advance was to replace mechanical switches with vacuum tubes. As we have seen the ENIAC computer used in the war effort to help develop the H-Bomb had 17,468 vacuum tubes and weighed 30 tons. The computer size decreased, speed, power, and economy increased with advancements of methodology and hardware. The invention of the transistor was a revolution. Transistors are much smaller, more reliable and generate less heat. Next came integrated circuits in the '70s and '80s and finally the Silicon microprocessor. Computer speed is measured in FLOP's , floating point operations per second. The Mark one could do one operation in 3 seconds working at the speed of .3 flops whereas the fastest modern computers work in the teraflop range (1 trillion flops). This increased speed derives roughly from product of the number of switches and the speed of switching. Computers with 10's of thousands of transistors became possible with the demise of the vacuum tube. But the biggest advance was the microchip made of Silicon on which you could place the equivalent of thousands of transistors in a small space. A simple flat transistor element may consist of a juxtaposition of N-P-N or P-N-P Silicon on a wafer. Complex circuits could be etched on wafers and mass produced. Parallel arrays of Silicon be used to create machines with incredible computing power[160]. Tiny light beams of different electromagnetic frequencies etch smaller and smaller patterns on modern Silicon wafers and there is a race to develop and use this technology to mass-produce smaller chips with the equivalent of millions of flat transistors.

We all know what modern computers can do. Younger persons can barely imagine life without them. Elementary mathematics curricula use calculators. Consequently some children are not able to calculate without a machine. But we all depend on computers in our daily lives. They extend and expand our abilities. Even if we can figure things out on our own, especially if we can, calculators do figures much faster and free our minds for more abstract tasks. For advanced students of mathematics the scientific calculator is commonplace and leverages abilities. Task specific machines are analogous to other machines which extend abilities and function as tools. A man can't run very fast but is able to travel faster than the speed of sound in an airplane. While our body limits us, we can only get our legs to propel us so fast, still we dream of making past these limitations and our will takes us so much farther in a lot of instances. A man is quite limited in his ability to dig foundations but this task is made amazingly easy with heavy equipment. The computer is no different. It's becoming an indispensable tool. Those who have it and have more powerful than those who don't and so computers proliferate. There are those with advantages of wealth and other comparisons of cognitive power. There are two kinds of persons, those who have computers and those who don't. The haves will outcompete the have-nots every time. Armed with computers humans learn faster and will undoubtedly out-compete those who don't in such diverse areas of endeavor as learning, business and war. Men with the most modern tools will generally perform better in the same way that an army with modern weapons, an air force with the latest equipment and so forth. The computer is a tool, similar to a backhoe or a car or jet plane but with a catch.

Computers work not in the mechanical world so much as to extend understanding. As long as they are simple single task machines such as calculators, and graphers or game-players, they simply extend human abilities just like any other machine. But a multi-task machines that can calculate, search libraries of information and retrieve data, process words, recognize and produce speech, all with great fluency and accuracy, start to look very human. The Silicon wafer has made all of this possible.

A lot of people have begun wonder about the computer's ability to think. Will it be possible in the future, perhaps by virtue of advanced processing and brute force, power, increased computation speed, increased number of Silicon transistor equivalents, parallel arrays of processors, for Silicon machines to "think" or even become conscious beings? Today computers function merely as extenders of human capacities, as slaves. They don't initiate thoughts or actions and seem unable to feel or experience. Computer characters that function as human equivalents are by now deeply embedded in American culture, from Hal in "2001" to R2D2, and Data in Star Trek. Spock was a descendant of computer in earlier science fiction genres. Computers seem not even to have the will to live, fear of death, that we know occur in the smallest brained birds or rodents, certainly not those of dogs or dolphins, but who can tell what will be in the future with the explosion in computer technologies? Is a complex biological Carbon-based structure with nucleotides, proteins and all the complex chemical structures that go into making a "living" organism a necessary ingredient to true consciousness or can this cognitive outcome be achieved differently? No one knows. Some of us suspect consciousness, whatever is meant by it, as sensed by us, will never be achieved by a lifeless Silicon based apparatus. Were we to make Silicon machines that thought and felt as we do, then we would be as gods creating as "second generation" of thinking initiating beings viz. "Let us make man in our own image" replicating biblical creation.

Would humans be able to exist alongside sapient silicon based computer beings endowed with their own free will? Inevitably each superior cognizant being would be challenge the other for dominance. Wars would be fought. One being would inevitably replace the other much as Cro Magnon replaced Neanderthal. And one would end up winning out, maybe wiping the other off the face of the planet. It would be a war between the elements Carbon and Silicon, the former sapient alive with blood flowing though its veins, the latter made of cold sand and glass. For the foreseeable future, Silicon serves Carbon. Silicon is a lifeless tool, incapable of strategizing, a planning without a destiny of its own.

As an interesting sidebar, it would seem, that it would be virtually impossible for two sapient species to exist side by side. Endowed with free will, each would inevitably protect its own interest and there would be wars for supremacy. This is what we've experienced throughout our own history, where intolerance inevitably occurs between races and nationalities. We've seen so many fights break out between sports fans from different cities, in competition between colleges, annihilation of populations and world wars. Smart aggressive beings will always crave dominance. What is the wisdom of actually seeking other intelligent beings from outer space as envisioned in the SETI project? If such intelligent beings exist, and this may be likely, they will inevitably be somewhat different than humans and we there is certain to be an altercation between them and us. Under ordinary circumstances, it is so difficult to see humans and aliens just sharing common wisdom. That scenario would seem more than a little naïve.

Human expertise in creating computers is advancing rapidly. Yet it would seem our ability to create willful sapient Silicon beings is just as much of a longshot as our finding aware and alive aliens in space. This is all for the best because such "Contact[161]" is far and away most likely to be destructive for both sides.

The picture I'm presenting is the computer as a tool, an unfortunately lifeless, car, plane, gun, backhoe, extender of human abilities. And it's striking that as circuitry advances, as computer chips get ever smaller and more powerful, that humans and computers begin to work as close proximity, hand in hand, that we have become ever more dependent and are living closer with our computers. This is an observation that seems rarely to be made, but it is what I call the marriage of Carbon and Silicon. What has made the intimate relationship between humans and computers possible?

First of all the tininess of the Silicon chip. Not too long ago they took up buildings or rooms, now mostly desktops or even smaller spaces. Secondly everybody has access to them, whereas in the not too distant past they were tools for the elite who where using them for a single purpose. Today's machines multi-tasked and small, portable most of the time, and all of us depend on them. We have an intimate relationship with them. There is one everywhere we work. For some disabled persons, computers replace damaged parts. Persons who cannot speak use computers to express themselves. Some of these folks can merely type messages. Others use machines that interpret impaired speech. Silicon devices can be made to stimulate the human cochlea. Here is the scheme. Surgeons implant a device with multiple electrodes into the human cochlea. The cochlea normally translates sound vibrations into electrical impulses which is what the brain can process. In persons with nerve deafness, this cannot occur. This array of electrodes placed in the cochlea makes possible the direct electrical stimulation of nerve cells. A tiny microphone is used. Microphones normally change sound into electrical energy but the microphone is connected to a computer that alters electrical data according to multiple paradigms so as to ease interpretation by the brain. Then a tiny radio device transmits this information to the implanted cochlear electrodes. The nerve-deaf subject, especially if he is a child, can learn to process this altered information and hear, maybe for the first time. The deaf are made to hear. This would not be possible if the computer were some giant device. As things are, the computer device is somewhat large, close to palm sized, but may be place in a pocket or belt. Such devices raise some alarm in the deaf community. Deaf persons sense that they are different, a special minority, a group unique unto themselves who communicate by signing. Few hearing persons know sign language so that the deaf can send private messages. Some of them aren't so sure they want this uniqueness taken away with an electrical device. But cochlear implants are not for every non-hearing person anyway.

A paraplegic patient can be made to walk. These are persons whose spinal cords are damaged by trauma for the most part, whose legs are paralyzed. Very simply, one places electrodes over large leg muscles. When an electrical impulse gets to the muscle, it will contract, moving the hip, knee or ankle. A Silicon computer device controls the succession and amplitude of the electrical impulse in very much the way that they are progress within the spinal cord first to anterior tibialis, then quadriceps then the gluteus maximus, then the gastrocnemii, then to the iliopsoas on the opposite side, and so forth to produce a functional gait. Someone who is wheelchair confined and highly motivated may be able to get up and stand, first of all, and then walk, usually short distances. Other persons use electrical stimulation to prevent muscle atrophy. In paraplegics lower extremity muscles languish unused and wither away. Electrically stimulated muscles are made to contract against resistance and can rebuild muscle mass. Here as with the cochlear implant, the Silicon device itself isn't actually implanted, but is carried about, outside the body.

A patient with intractable right hand tremor can have a pacemaker wire implanted in his left thalamus which helps control the hand. Tremors can obliterate normal functioning movements and turn a dominant hand into a useless liability. A person may be unable to feed or to write. The thalamus is a bundle of nuclei. Each nucleus in the thalamus is an anatomical grouping of neurons. The sinus node and a-v node of the heart each have an intrinsic rhythm and drive the rest of the heart to beat at a specific frequency or rate. Many thalamic cells also fire with an intrinsic rhythm at a specific frequency. As we saw earlier, this rhythmicity drives the rest of the brain determining the state of consciousness, whether we are asleep, awake or in REM sleep. Intrinsic thalamic rhythms also drive motor systems. In motor systems this driving rhythm causes a tremor under certain circumstances. In just the same way that a pacer may be placed in the heart, a neural thalamic pacer is used in the brain to overwhelm the pathologic rhythmically firing neurons that cause tremor. The device needs to be placed in thalamic VIM or ventral intermediate nucleus of the thalamus. The pacer can be turned on and off, and set from the outside with the use of an electromagnetic control. Just like a cardiac pacemaker, it is implanted, along with its wire, into the chest wall just under the skin. Alternatively this abnormal intrinsic rhythm is lesioned with a probe that will take out the VIM nucleus or via a gamma knife radiation. This is a simpler way of ending pathological rhythmicity that causes tremor.

Physicians literally "reach into" the brain every day. Some stupendous examples involve replacing poorly functioning brain cells with new ones. In Parkinson's disease immature (fetal) midbrain cells and also cells from the adrenal medulla that secrete Dopamine, the transmitter deficient in Parkinson disease, are implanted into the brain with some success. Therapeutic fetal brain cell transplants are being tried in such diverse disorders as stroke, which results in the death of brain cells, and in spinal trauma[162]. These techniques are being tried on a practical basis, without anyone even noticing their philosophical implications. Scientist clinicians are reaching into the brain to treat disease and improve function. In doing so they are taking furtive steps beyond the limitations of biology; they are unbinding the patient from his biological limitations. They provide the proof that the patient, and any person adds up to much more than his own biological limitations. Though of course the victim of Parkinson disease or stroke is unable to operate on himself, he is the beneficiary of a communal level of knowledge or expertise.

Inevitably similar techniques will be used someday to maximize physical and mental performance. On a rudimentary level, stimulants and hormones increase atheletic performance in competitions around the world every day. Drugs, antidepressants among them, improve function in day to day living by their chronic effect on intrinsic neurotransmitters. Silicon devices add precision in drug delivery. Precise fixed amount of drug is dripped into the spine using a technique that has become commonplace today. Baclofen is delivered directly into the cerebrospinal fluid to inhibit neurons and control spasticity or pathological tightness, resistance to motion in muscles. Spasticity often results from spinal injury from trauma, birth injury and such diseases as multiple sclerosis which prevent the inhibitory controlling impulse from higher areas of the nervous system from reaching the spinal cord. When you give large doses of Baclofen by mouth, which causes a lot of side effects, very little of it gets into the brain and spinal cord where it must work. It may be excluded by the blood-brain barrier. Given intrathecally, directly into the spinal fluid, the results can be stunning. Suddenly a person formerly unable to walk, can get up on their feet. Others in a bed-ridden contracted state are finally able to stretch out and move. Similar pumps are used to deliver narcotics to control intractable chronic pain, to dose insulin and other drugs requiring precise titration schedules. It means the difference between functional vs. non-functional life. Tiny Silicon wafer circuits have made all of this possible and for us to begin to ask where does the living Carbon based biological system end, and lifeless Silicon begin, proving again that we are far more than our biological machine which is our body allows us to be. Devices are born of a dream or a conception but are made into reality.

We all use keyboards and pointing devices such as the mouse and trackball, and joystick. But the problem is that in order to communicate with a machine which is separate from ourselves we need to translate our desires into a form the machine can understand. A lot of steps impede the direct transfer of information. This relationship needs to be more intimate. By this I mean that what we ultimately need to achieve is full incorporation of Silicon devices into our brains to communicate electrically with the relevant tissue and improve and expand upon normal function. When we use devices to overwhelm a disability, that is just practice for situations where such devices would really be useful, namely the elaboration and expansion of normal human capacities. As we have seen computers far surpass the human brain in two important areas, information storage (memory) and speed of calculation. Memory and retrieval of information is not trivial. We have seen that memory intimately ramifies with all other areas of cognition. Libraries of data are stored in smaller and smaller space. Soon optical systems may store data in three-dimensional arrays. They will make our compact discs that already store whole encyclopedias or shelves of books on one small disc, seem large by comparison. Information will also be infinitely more accessible, easy to call up with the aid of infinitely more mnemonic “handles” than even the living brain is able to provide, not only logical knowledge categories, but by spelling and phonemic and positional classification schemes.

Far more sensitive input devices will be necessary. Hopefully, cells and Silicon wafers will make direct electrical connections as both systems work via the transfer of positive and negative electrical charge. Presently devices are touch pads, joysticks and electrodes placed about the eyes that sense the field of gaze in eye movements. These may be useful for fighter pilots for faster communication with a precision machine and in virtual reality games. Electrodes can be placed on any electrically active surface to gain sensitive feedback, for example in muscle where the state of relaxation or contraction can be monitored by computer, analogous to proprioceptive input into the brain. This data can be used to control movements for athletic movement or to modify accelerations in machines used to carry people. A computer commanded by eye movements may make rapid communication possible for the first time in patients with the locked-in syndrome (See Chapter One) who are unable to move anything but their eyes and otherwise are awake but can’t communicate. Some persons with severe forms of cerebral palsy and acquired neurological diseases may benefit. Perhaps EEG activity and distribution may be used to communicate with Silicon devices that can easily recognize patterns and spatial distributions of brain electrical activity. Such devices may someday be used to stop abnormal electrical discharges such as epileptic seizures, but more importantly sense and control brain waking an sleep rhythms on long trips such as space journeys. These devices amplify command and control, the efferent side of neural function.

We use many devices on a routine basis to extend the afferent powers. Infrared and other detectors extend our sensitivity to signals and energy sources that as biological organisms we haven’t receptors for, electromagnetic energy outside the visual spectrum.

These devices are mostly used in war and for night vision as in hunting but detectors on earth and in space have multiplied our powers of observation. Astronomers observe today in all parts of the electromagnetic spectrum, infrared, at radio and gamma frequencies,. analyze and collate with the aid of computing machines. In the not too distant future we can expect more devices to directly stimulate nervous tissue, to expand sensory function in much the same way as the cochlear implant does, and enhance tactile and visual function as well. Predictably these may at first to treat the blind and sensorily deprived, but one day they will be used to extend the sensory function of normals. We will have Silicon devices surgically implanted, some directly into the skull and brain. These devices will extend our sensory abilities, improve motor function, increase our memory and augment cognitive function. They will be sensory, mnemonic, cognitive, and motor enhancers and extenders..

Of course, everyone depends on Silicon devices to extend capabilities. Silicon devices calculate, write, retrieve and store abundant information. In selected situations as miniaturization progresses these devices may be implanted. Carrying small notebook computer devices around with us as we do even today is tantamount to implantation anyway. Our relationship is growing closer to these digital assistants that a lot of us couldn't be without even if we ended up on a desert island. These communication devices extend our capacities. With cardiac pacemakers and implantable defibrillators[Φ] being so commonplace today, we joke that the cardiac patient implanted with such a device is indestructible, he is prevented from dying.

The argument that we humans are nothing more than our biological endowment succumbs then to the power of human invention. Expansions of human abilities and possibilities that are even now possible in the computer age. We are more than our biological endowment allows us to be, we have become more because we have dared to imagine. Our inventions have had us escape our own innate limitations. We have hatched out of our biocapsule.

Carbon and Silicon are ever more being tied together. They are married. The result is an exponential growth of human capacities. Computers have done far more than affect our daily existence. Subtly, slowly, they have altered our philosophy, have changed the way we see ourselves. Firstly, as I have been saying, we are in no way limited by our own body our biological endowment. That is a tremendous realization. More than that has changed. By some reckonings, the sum total of human experience that is a life can be boiled down to a huge quantum of information. This is a legitimate claim. The next time you talk to your spouse or an acquaintance, or any other person, what you see before you will fundamentally change from a clothed body and face that you recognize, something material, to a package of non-material ideas, perceptions, and actions, something immaterial in other words. What resides in your brain right this moment, the effects of all of the memories and experience which are part of your life, is expressed in the form of information.

Suppose we could set all the electrical activity in all of your brain cells at exactly settings that are there at this moment which we designate $t_0$, then recorded also the exact anatomy and array of synaptic connections, also the placement and composition of all cells and substances such as proteins and nucleotides within cells, this is just information, admittedly an awful lot of information, but the point is that it can be stored in machines designed to store mountains of data. If we have all of this data stored in a Silicon device, we have an essence then of what a person is at that specific time $t_0$ and presumably have a good handle on what is to happen at $t_1$[φ]. What's more an essence, expressed as information content, is separable from a biological entity not unlike a soul in religious parlance.

Perhaps more importantly the integrated consciousness that is a person and more will represented in a "second generation" inside a Silicon[ψ] device. Whenever our civilization has advanced far enough, and this will not be anytime soon, all information should be carried by vessels that are non-biological. Someday we will be fully developed as non-biological beings, capable of experience well beyond our current limitations, sensitive to vibrations along all areas of the spectrum, vision in simultaneously infinite projections, enhanced tactile sense and epiphanies of emotion such as those never experienced by any one with paltry biological endowments. What is important is that we will reach this stage, and I have every confidence that we in fact will, in spite of our limited abilities and because whoever or wherever we are, we have an inner drive to reach beyond our limitations, we crave more knowledge and we dream, we want always to reach out and experience more. We are material yes, but there is some essence that is not expressed in material terms. I would ask this question. Have advances in twentieth century science proved that we have a non-physical essence then, what is indeed a soul?

You might point out that in fact, this inner essence is already recordable, though not with precise fidelity, in the form of books and other recording devices that we leave behind, long after our body has disintegrated. You'd be right of course, we can always see a performance from a deceased actor, or listen to a recording of a singer or conductor at any time. The difference is that while we leave ideas, performances, pieces reminiscences of ourselves, after we are gone, we haven't yet evolved a technique to preserve and make immortal the human experience that built into every Carbon-based living human creature. We are witnessing a performance, experiencing ideas, but the person "larger than life", is no longer is experiencing or living the part. We've not yet evolved a way of preserving intact, consciousness.

If you still maintain that there is nothing to humankind except biology there is yet another means of attack on this assumption. Through mastery of genetics we will soon find this biology in and of itself is a moving target. Genetics too can be reduced to information transfer and manipulation. Genes are nothing but the biochemical transference of data. Most of the time this information is transferred *vertically* from one generation to the next but more and more, as we are learning, information gets transferred in other ways. As we have already pointed out, neither simple cell division nor reproduction is designed to replicate with complete fidelity, a critical point that is usually lost in basic texts on the subject. This happens despite the fact that on the molecular level DNA is copied with amazing accuracy and speed. Animals and plants reproduce, yes, but they do not make exact replicas of themselves. Children are not clones of their parents. That happens more because of a reshuffling of genetic material that occurs in mating, but occurs to some extent among organisms reproducing by simple fission as in bacteria and protozoa. Variation between individuals is the raw material for evolution and each individual is an experiment in survival and fitness. Another example as we have seen is in the embryo where, by rights, through simple DNA replication, there should result in a huge number of identical cells. Embryonic cells differentiate despite their identical genetic complement, because of differences in expression translation into proteins. The embryo begins as a zygote which as a single cell reproduces by faithfully copying DNA so that each new cell will have its own supply of an identical product. But as this cell division takes place, cells differentiate and specialize so giving rise by embryologic development to a whole organism, which often contains hundreds of different types of cells. Therefore even the simplest forms of biological reproduction are not one hundred percent faithful, and this residuum of difference is what makes biological reproduction a truly unique enterprise.

Genetics is the key to manipulation of life for our own purposes. It is hard to believe that the structure of DNA the molecule that carries the bulk of genetic information, was elucidated only in 1953. This is one of those seminal practical discoveries akin to Hubble's discovery in 1929 that M31, Andromeda, was not just a “nebula”, or cloud of gas inside our own Milky Way galaxy, but a separate galaxy that contains its own stars, thus implying that many of those smudges or nebulae, are actually other galaxies equivalent to our own. Many of these galaxies contain hundreds of billions stars. This single discovery miraculously expanded the field of human gaze. Suddenly God, if there was a God, was not God of the world, or of the Solar system, or even of the Milky Way, but the God of hundreds times hundreds of billions of solar systems, a god of a wider cosmos than had ever been imagined by any of our ancestors, and all of this at one fell swoop. The study of genetics opens up similar possibilities for us, only here we are gazing inside or ourselves. Few persons will ever realize the full implications of Hubble's discoveries just as very few can apprehend the secrets and full implications of the new genetics.

New discoveries in genetics will allow us for the first time to manipulate biological reality. The genetic code is the Periodic Table, the Grand Unified Theory, of Biology. Scientists are discovering that genetics is far more complex than they imagined, the more they discover the more complex the phenomenon their work becomes. As it turns out, if you know the all about the chemistry of reproduction of the DNA molecule you still understand very little about the myriad ways this technique is used in nature. Nature is extremely inventive. But there is not doubt that we will be able to design a different human form and pass down this change. That is why when I see a movie about men in the distant future, say the year 2100, I can be confident, men won't look or think like *that* the way they are always presented like current modern, humans. People will be fundamentally changed in the future, in form and function, of that we can be certain (that is if we don't do ourselves in before that, and I doubt that will be the case.) We will make these changes according to our own will and won't any longer be subject to biological serendipity.

Our basic identity at this stage is known. We know what a human is. There is little confusion about the basic form or shape of a human person that has two eyes, a head, a trunk flanked by two arms and two legs. But think one day what it will be like when we are able to change this form, as will some time inevitably be the case. Suppose someone takes the decision that some persons should be specialists at working at a desk and that these would be better shorn of their legs and perhaps with shorter arms and longer fingers to fit keyboards. Or perhaps there should be generations of professional basketball superstars for our entertainment, with small cerebra, large cerebella, and long legs and further that these features should be passed down through generations. A great confusion would ensue about what is the human form and at the very least about what specific physical characters make a person. As time goes on the human morphology will change dramatically and there are likely to be humans of many different varieties, endowed with a special function perhaps. These changes, for better or for worse, will be premeditated and planned. Living long enough, or coming back in a time machine we would not recognize our own descendants.

(Humans will undoubtedly be able to change their own form by genetic manipulation. At first this will start with the elimination of undesirable traits.

Mutations for such disorders as sickle cell anemia, cystic fibrosis, Tay-Sachs, Hurler's syndrome, Huntington disease will be plucked from the human genome or strands of DNA that correct for these disorders will be inserted. The consequences of performing these acts may not be so great, though there are effects we will fail to predict. Along these lines we will inevitably discover that certain traits, especially personality traits, are genetic. Homosexuality, alcoholism, even hypochondriasis, anxiety and panic disorders will be found to be monogenic or polygenic traits. Already scientists have identified certain genetic types that increase risk for various cancers, especially mutations in tumor suppressing genes, Alzheimer disease, atherosclerotic stroke and heart disease, diabetes, and immune disorders such as rheumatoid arthritis and multiple sclerosis. These traits can be tested for and will soon be used as fodder by medical and life insurers to exclude or rate prospective customers. But techniques for genetic manipulation, the elimination of such disorders or DNA strands that predispose to disorders will inevitably occur.

As if this weren't enough, along the way scientists will uncover certain characteristics that increase math or literary prowess, spatial ability or general intelligence. Traits for height, muscle development, speed, physical attractiveness will appear in scientific literature along with the technology to alter or mix such traits. Inevitably we will plan to raise super movie stars, soccer players, physicists or some combination of the above. Perhaps we will seek to eliminate psychological traits such as self-doubt, guilt, anxiety that might be conceived to get in the way of accomplishment. In some cultures then, particularly in cultures where there was sufficient wealth and expertise, the human form would change. By the year 2100 or so people would take on different characteristics. They'd be untrammelled by any anxiety depression or guilt. You would see more men over 6 feet tall, with large muscular biceps and thighs, bred in other words for increased sinew and intelligence. The marathon would be run in an hour and thirty minutes, and the brains would create unparalleled works of art and science. Perhaps professions would be decided at conception rather than allowing a person to gravitate into the way of life that he is inclined to.

This is a fanciful yet highly likely peek into the future. Very little will be left to chance. The future of the human race would proceed by design, by careful planning, in other words. What is likely to be the outcome of such planning?

Linkage is a basic genetic concept. Genes on a chromosome are either close to or far away from other genes. Recall that a gene is defined as a strand of DNA encoding for a single protein. In this manner traits are linked to each other which means in practical terms that they are likely to occur together because they are close on the same chromosome. As DNA is moved from one cell to another certain traits are very likely to be moved together at least at first, before techniques get precise enough. But this is a technical obstacle that will be overcome. Another more difficult problem that will not be solved so easily is that genetic traits, single genes, have multiple effects. The sickle cell trait doesn't just cause sickle cell disease. It protects against malaria, and the Tay-Sachs allele may well have done the same for tuberculosis. But even this is not a particularly difficult problem. However, protein gene products inevitably have multiple effects in many systems. If you want more or less skin pigmentation, you in an attempt to make sexy blondes or dark skinned virile males there will be fallout affecting other propensities as well. Suppose you find a trait giving rise to large muscular thighs and calves for future soccer players, you will find that many of the proteins giving rise to characteristics of muscle, also affect brain cells. In fact muscle proteins and enzymes have a lot in common with proteins inside the brain. Muscle and brain diseases have some common genetic characters. In other words you may alter mental characteristics in many ways that you didn't bargain for. Genetics is far more complex topic that is dreamed about in our philosophies. One single trait even in isolation will undoubtedly affect hosts of other genetic characters.

If we try to alter the natural order of things, we are inevitably going to get into some trouble. The best analogy I can think of though admittedly with something a great deal less complex, is the economy. In our world today we see various kinds of planned economies, and more or less natural laissez-faire type systems. The best example of a top-down planned economic structure is the Soviet type economy, now pretty much universally accepted as a dismal failure. Lenin said, in essence, let not the market place determine the production and distribution of goods, but have the state determine what will be best for everyone by design. Decision would best trickle down from the top. We, the politburo, will determine how many shoes will be necessary, tons of steel production, farm implements, beef and pork. Actually as things turned out the Russians ended up eating most of their horses and still starved by the millions, making int possible to state cynically why leave anything to chance when we can rely upon the human frontal lobe? One problem with this scheme is that various constituencies sought to gain personally from state decisions. But what really killed the system is that even in something so simple as an economy, there is an inestimable number of interactions to thwart any planning mechanism under the best of circumstances.

Indeed this is what happens to us every day in the practice of medicine. We are aiming to control the body's function by design. In the Intensive Care Unit we try to control fluid balance, heart rate, cardiac output, blood pressure, level of sedation - hosts of factors. The aim is to leave nothing to chance. Yet "chance" events, shock, cardiac arrest, cerebral hemorrhage among them, do occur and at such a rate that under most circumstances even with intensive nursing and monitoring, it is impossible to predict what will occur even a minute after an examination.

Which leads me to again ask the question: How do systems function better, by human design or by serendipity? In previous chapters we wondered whether insects and bacteria, which change randomly and rapidly adapt by natural processes would eventually win over humans endowed with intelligence. Complex systems with numerous rapidly interacting nodes, the economy, genetics, biology, nature, will be difficult perhaps impossible to master at any time even considering advancements in cognitive and computing powers that will come under our control. Note here that I have deliberately skirted the issues and fears in the area of bio-ethics that arise whenever we talk about manipulating genes. The most basic bio-ethical issue is will we in fact be better off after manipulation of our own genome.

The most powerful argument of all in plans for genetic design is that the human organism has already evolved via recombination of thousands genetic traits. Why haven't we evolved that perfect six foot stature with large thighs and perfectly functioning brains. I believe that looking at the whole system, where the human is biologically at this time, we will discover that the situation is close to optimal. We still have abundant genetic variation, the raw material for adaptation increase fitness and evolution. My guess is that even some genetic imperfections will be found to have widespread positive effects which have escaped our notice. Perhaps some very disruptive traits such as Huntington's disease can finally be eliminated. It is important to note that from the population geneticist's standpoint, even very injurious traits marking a person for severe disease, must have some adaptive fallout that confers an adaptive advantage, perhaps to the heterozygote, but an advantage nonetheless. Huntington families tend to produce more children and the disease affects persons after their child-bearing years, so the gene "survives" since it does not demonstrably decrease fitness. May there be other effects that accrue from eliminating this gene that we don't know about. There probably are.)

Genetics is a mechanism to transfer information vertically through generations, from parents to their offspring. Genetics comes from Greek roots meaning to produce or origin or birth. Nucleic acids as the biochemical substrate of genetics, have acquired a broader context merely as chemical information which under normal circumstances in the living organism, can amplify and reproduce. Genetic information is used more broadly than simple inter-generational transfer primarily in natural reproduction. While a particular set of genes defines a species or variety of organisms genes and groups of genes are readily transferable to other animals or plants. A particular gene encodes for a specific protein product. But the gene and the protein produced are readily donated to other organisms, in which case you may say we have produced a certain hybrid and you would be right, except a hybrid is formed by mixing nearly equal moieties of genes from each parent most of the time.

Today it is possible to transfer a gene from one species that has it to another that does not. Think of a gene as the instructional code in the language of a chain of nucleotide codons to produce a chain of amino acids that is assembled into a protein. Every cell of every animal and plant and even bacteria, contain the machinery to translate this genetic code, into real proteins. The information that defines all living organisms is in this universal genetic code. There is no reason why data from one organism can't be placed in any other given simple technology. Viruses do it and provide a model for information transfer. (Figure 9) The bacteriophage is a DNA virus that infects bacteria. It inserts viral DNA into an unsuspecting bacterium, which will then reproduce it and make new viral particles.[163]

The simplest examples of gene transfer thus occur in nature. A virus is made of a strand of DNA or RNA enclosed in a protein coat, in other words no more or less than a repository of genetic information, whose aim is to reproduce. You can argue, whether a virus is living or not, since it fulfills only one criterion of life, by simply reproducing. It does so by finding a mechanism for injecting genetic material inside of a cell, then enlisting the infected cell's machinery to reproduce the viral, as opposed to the cell's own, genetic material. The AIDS virus infects human lymphocytes, attaching to protein receptors, the CD4 protein which marks the cell surface, as a lymphocyte of a certain type. These surface receptors fit perfectly with the surface of the AIDS virus. Having attached itself to the lymphocyte cell membrane, the viral RNA is injected into the hapless lymphocyte and the lymphocyte's own cellular machinery reproduces HIV, an irony, since our immune system, of which the lymphocyte is a part, tries to protect us from invasion of foreign organisms.

In principle, a quantum of genetic information from any organism can be placed in any other to reproduce part of itself either as genetic information or used to make one organism's own proteins in another. Thus we have not simple vertical transmission of information, as we usually think in genetics, but horizontal transmission, from one cell to another. Over the last few years different techniques have developed to reproduce strands of DNA and to make gene products in large quantity. The most efficient is to move genes into bacteria that have an extremely short generation time, in order to make huge quantities of the gene protein product. Industry manufactures and even patents proteins for diverse purposes such as digesting oils and PCB's. Proteins are made courtesy of bacterial servants include medicines such as peptide hormones, growth hormone, insulin, clot-busting proteins such as t-PA used to treat heart attacks and strokes, Interferons and a large array of immune mediator signals. Specially bred plants incorporate genes giving them resistance to disease. Cloned sheep incorporating human genes produce protein product excreted in milk. It is now technically feasible to produce large amounts both of a novel gene and its product and this so routine that doctors use it to test for genes that implicate particular infectious agents to make diagnoses. Bacterial hosts allow workers to amplify tiny quantities of genetic material of the invading organism. If a small genetic fingerprint is there, amplification methods will detect it and a diagnosis of infection can be made. One method is called PCR the Polymerase Chain Reaction. This is simply a method to amplify a tiny quantity of DNA. DNA from one person is not identical with another person's DNA and each organism has its own segments of DNA that are unique to it as well, act as a fingerprint tht implicates that particular bacterium or virus in an infection, even if there is a vanishingly small amount of material to test. Crime labs use the very same methods. Human genetic markers are amplified and implicate criminals with relative certainty, provided adequate care is used a suspect's genetic material can be identified in blood, or semen. The perpetrator of a crime or the father of a child can be determined with a high degree of certainty.. This goes to show that once you discover something really basic, there will be thousands of different applications for the discovery.

Bacteria exchange genetic information frequently in nature. We have discovered this much to our amazement and dismay in examining antibiotic resistance. Bacteria may acquire resistence to antibiotics by making proteins that break down the antibiotic. Some of these protein making genes are encoded on plasmids, separate circular pieces of DNA that can be passed between bacteria, even between species. Some bacteria resist penicillin through use of a penicillinase enzyme that breaks the penicillin molecule down. Other bacteria in the same region, say within a given hospital or in the gut, can easily acquire the gene which produces penicillinase, thus resistance to penicillin. They can pick up this gene simply by coming to life in the same region as other dead or disintegrated bacteria and picking up as if taking up so much garbage or waste, the dead bacterium's plasmid. Alternately, the plasmid can be trasmitted as if by infection by a bacteriophage. Thirdly, this genetic information can pass between bacteria via conjugation. In any event, whenever you have a patient taking penicillin, bacteria that are penicillin resistant will be the only ones to survive and multiply, and you will thus be selecting for resistant bacteria. In fact this happens whenever someone decides to use an antibiotic. They are irrevokably changing a mini ecosystem, in a hospital or a gut or other environment, selecting for bugs that have acquired resistance, usually through genetic machinations, to the antibiotic. If you are one of those persons who goes to the doctor with your sinuses or urinary tract infection and for one reason or another getting various antibiotics, you are selecting each time for resistant organisms in your own body, an ecosystem filled with bacteria. You are not resistant to antibiotics as a lot of people may misapprehend, the bacteria in your body are resistant. This data is carried for long periods of time, in bacterial genes. Another thing is that bacteria are in constant competition in your body with other commensal or symbiotic organisms such as viruses and yeast. Kill the bacteria and yeast which are usually not affected by antibiotics, may get out of control, hence a perfusion of yeast vaginal infections in over-treated young women who always seem to be using one antibiotic or another. Little kids constantly under treatment for presumed otitis media ear infections, are unfortunately candidates for prolonged bacterial colonization, whereas had antibiotics been used judiciously, despite the over-weaning remonstrations of their neer-do-well parents that they be treated, this situation might never have developed. Humans are constantly doing battle with tiny bacterial organisms who have the advantage of a short generations time, and rapid development and exchange of genetic material which is transferred vertically, between parents and offspring and horizontally, between organisms. On the other side is design the inventiveness our human brains that can't seem to design antibiotic molecules fast enough.

The much touted Human Genome Project aims to map all human chromosomes and to relate perhaps one hundred thousand or so loci to specific human characters and diseases. What is not commonly realized is that each gene may have numerous variants when one looks at the whole human family. An amino acid is specified by a series of three nucleotide bases, e.g. CAG[Φ] , some genetic varieties may differ from others by as little as a single base, some may be altered at more than one site, and one interesting and important type of variation that was totally unknown until recently is the replication of a triplet or trinucleotide repeat. A series of three nucleotide bases is what specifies a particular amino acid in a protein chain of amino acids. As it turns out this trinucleotide repeat is a chain of nucleotides that gets longer with each succeeding generation. This turns out to be an important common mechanism for passing down neurological disease such as Huntington's disease, Freidreich's and a group of ataxias, familial diseases similar to ALS, so-called Lou Gerig's disease and Myotonic muscular dystrophy. The more repeats the inherited the more serious the disease and the earlier it becomes manifest as a general rule as has been found to be true for Huntington's disease and Myotonic dystrophy. Many years ago clinicians observed that not only does Huntington's disease run in families as a dominant trait, it also seems to affect persons in succeeding generations at an earlier and earlier age, a phenomenon known as anticipation. With Huntington's disease, the average age of onset is about 35. Those who have symptoms at an earlier age, have the illness more severely. Now we know why. The culpable trinucleotide repeats, get longer as they are reproduced in germ cells and passed through generations. This new genetic discovery is one that scientists never could have predicted by just knowing the basis of genetic chemistry as outlined such a short time ago by Watson and Crick.

But this advanced know how was unnecessary in the most ancient use of gene and many different mechanisms of heredity. Undoubtedly the first domesticators of animals and plants learnt how to mix varieties of plants and animals in order to gain more yield, hardier forms, and even alter the personality of the animal product. Dogs are probably the best example, beginning a long relationship with men that started some 12,000 years ago, are pictorially represented by the ancient Egyptians who counted them among their gods. We have segregated a single species Canis familiaris into about 400 different breeds, used for diverse human purposes ranging from companionship and protection to hunting and smelling out bombs and narcotics. Jacob in the Bible became a wealthy man when while working for Laban bred speckled sheep for himself which Laban had agreed to let him keep. Ancient men of agriculture bred and domesticated all manner of animals and plants for their own purposes, camels, oxen, horses, and cattle without the slightest notion of Mendelian laws of inheritance or modern biochemistry of DNA, certainly not a double helix molecule. Gregor Mendel with his peas outlined simple mathematical laws of genetic inheritance without having the slightest idea that we have today about the biochemistry and mechanisms of inheritance. Humans throughout history have been a part of biology, but at the same time, have always manipulated biological inheritance and that of all manner or biological organisms. Since this is part of what humans do and have always done, you may consider the science of genetics wholly natural, a product of human's superior mental capacity. This goes to show that we have used and designed biology for our own purposes well before the high technology of our own twentieth century.

Computers and our own genome are thus mere repositories and reproducers of information. In biology one may ask, what is the simplest unit of information transfer? Biology, biochemistry seems to like chains, polymers, concatenations of small numbers of simple units which end up doing everything that all organisms do. DNA is a chain of only four nucleotide bases which encode for some twenty different amino acids which begin to form proteins, also joined in chains. Amino acid chains may join each other and also add side chains and by the end of the process a protein will be acted upon by other protein enzymes to add side moieties such as carbohydrate residues onto the protein molecule at various amino acid sites. In the end the whole protein structure, composed of chains, side chains, and non-amino acid residues will have certain charges and bend and twist, forming a unique three dimensional stereoscopic structure. A particular protein may function in very much the same way having minor differences between species and also among various members of a single species; there may several equally effective, legitimate sub-varieties of a single protein, in other words in a single species. These will ultimately determine varieties or traits such as temperament or eye color, facial features and so forth so that all of us are unique though we have the same genes and chromosomes. In other instances the replacement of a single amino acid residue or misreading of a DNA strand ("mis-sense") may be caused by replacing a single nucleotide base, with another on a single DNA strand.

One of the most fascinating stories that is being uncovered at this time has to do with prions, which cause what are termed "slow-virus" diseases. These include Creutzfeldt-Jakob disease, new variant Creutzfeldt Jakob disease otherwise known as "mad-cow disease", Kuru, Gerstman-Straussler disease, fatal familial insomnia all severe fatal diseases that cause dementia, psychiatric changes and generalized deterioration in neurological function. These are the so-called "slow virus" diseases, or transmissible dementias, because they are infectious diseases under the right very restricted circumstances.

These terrible, fortunately rare disorders have been spread between humans, almost exclusively by direct transfer of nervous system tissue. Kuru was the first of these mysterious diseases to be extensively studied in pioneering work done by J. Carlton Gajdusek. Gajdusek's work showed that Kuru was transmitted by cannibalistic practices of the Fore tribe in New Guinea. They would eat human remains and women and children, the weakest members of the tribe, were left to eat the offal, remnants of brain and spinal cord. Years later they would come down with an inexorably progressive neurological disease the first would affect the cerebellum which controls balance, then the rest of the brain. When cannibalistic practices stopped, the disease became extremely rare. Another exceedingly rare nervous system affliction is Creutzfeldt-Jakob disease which has a worldwide incidence of about 1 per million. This progressive dementing disease will often cause death within a year or so and can spread between humans and between humans and animals especially by the direct transfer of nervous system tissue, most efficiently spread by transfer of nervous tissue as close as possible to the recipient brain. Does this sound impossible? It is not. About 10 or so percent of cases are familial or inherited, for example in a kindred of Libyans, the rest are sporadic or non-familial. Cases have been described that are iatrogenic, that is spread by medical practice, in which nervous tissue is transferred from one human to another. Neurosurgeons in the past, used cadaver dura mater, that is brain covering in some operations. Cases have occurred with cornea transplants, using depth EEG electrodes, and among persons given growth hormone injections for pituitary insufficiency, which was made of chopped up cadaver pituitary glands. This was before the days of recombinant growth hormone. But under ordinary circumstances the disease appears to be non-contagious, even between husbands and wives. The vast majority of persons with Creutzfeldt-Jakob disease fall between the age of 50 to 70, and even after being infected with brain or pituitary tissue it was estimated the disease would take many years, even decades to develop, hence the designation "slow virus" referring to an enormously long incubation period. Many patients seemed to have the onset of terrible psychiatric disease at an older age whereas they had no history of psychiatric disease while young, an unusual scenario, only to develop later, an overall deterioration in mental function, like Alzheimer disease except progressing to severe dementia and death at a much faster rate, myoclonus, sudden jerking of limbs, and an abnormal EEG pattern which would help doctors to recognize and diagnose this rare and terrible disease. The pathology of the disease, that is if you could find a pathologist who would handle the tissue for fear of getting the disease, consists of a "spongiform encephalopathy". Under the microscope one sees a coalescence of vacuoles or holes in brain tissue and a loss of neurons so that nervous tissue resembles a sponge. You could take this nerve tissue and inject it into the brain of an animal especially a monkey, and eventually the hapless creature would develop the same disease.

In late 1995 two cases of Creutzfeldt-Jakob disease were noted in British teenagers. There have been at least 15 cases of this "New Variant" or Bovine spongiform encephalopathy in all. The presentation is atypical in that it has much younger victims who tend to be seen first by psychiatrists and they did not have the typical EEG pattern. These persons are now thought to have acquired the disease by eating meat from British cows. The whole picture is not completely developed yet, but it appears that the problem arose soon after British cattle came down with a Bovine form of the disease, which was likely the result of these cows being fed offal or brain and spinal cord remnants from sheep. This was done as a nutritional supplement for cattle and comes from not wanting to waste any animal products from slaughtered sheep. Sheep may get a spongiform encephalopathy of their own called Scrapie, named for the tendency of afflicted sheep to scrape their backs against fences and other objects as a part of the behavioral manifestation of this brain disease. Then by a curious combination of circumstances cow brain was used by the industry as a binder for hamburgers and sausages, (to paraphrase Upton Sinclair on the meat-packing industry, they use everything but the squeal) until the practice was stopped in 1989 due to fear of bovine or cow disease being spread, but not before the disease spread to some humans. Thus the passage of this terrible disease appears to have been due to direct consumption of nervous system tissue in a most unnatural way from Scrapie affected sheep, to cows thus infected with bovine spongiform encephalopathy, to humans, some unusually young persons by the standards of naturally occurring disease[164].

Some might point out that the use of offal to feed cows, which are naturally vegetarian, and then the lacing of human food, with cow brain binders, is more than unnatural. It is anti-natural, an abomination. As we learn more about genetics and master a sort of techno-biology we will come upon more and more scenarios which offend our sensibilities. Some examples are in the field of infertility. Hiring young women for ten or twenty thousand dollars and using a healthy uterus to nurture a fertilized egg (zygote) from an infertile or just a wealthy couple which can't be bothered with a natural pregnancy and delivery is one example. It may be almost the same thing as the old practice of using a poorer woman's breasts as a wet nurse. But in this instance a baby develops moves, gets born and to a great extent the "birth mother" goes through the very human process of bonding, learns to love this baby. We have confusion and conflict, the stuff of legal battles. Artificial insemination though simpler and not quite as extreme, also creates problems. It's almost impossible to tell who are and who are not one' s relatives. You could easily marry your half sister without even being aware of it and barring stringent checks, abuses could well occur. Where is the guarantee that you are getting the sperm you thought you were getting? An interesting conundrum arose on one occasion where a physician, used his own sperm to fertilize a large group of unsuspecting women thus creating a large posterity for himself.

Even more bizarre are plans by physicist-entrepreneur Richard Seed, who announced late in 1997 in light of reports of the successful cloning of adult female sheep, Dolly by Edinburgh embryologist Dr. Ian Wilmut, his intention to clone humans. In order to make this project financially feasible, Seed intended to service infertile couples. The idea is to either remove the nucleus of an adult's cell, place it in an ovum whose own fertilized nuclear material has been removed, and to thus develop a perfect clone of an adult human, who would be an identical twin really, perhaps one generation removed, or, to remove the nucleus of the fertilized zygote of a couple, perhaps after this zygote has undergone a few cell divisions, in which case there would be a number of identical nuclei to be remove, and then to implant each of these nuclei, each with their own genetic material, into ova whose own nucleus had been taken out, thus creating an indeterminate number of identical twins. Some problems right off the bat are that we know that the genes and chromosomes of an adult person have been altered in many ways by all of the cell divisions that have occurred up until that time. Genetic expression of most of the genome has been blocked so that the individual cell can specialize in its function. But there is some evidence that genes age over a number of cell divisions. Genes likely change their chemical structure so that each cell line can only divide some fixed number of times. This process of specialization and curtailment of future cell divisions is one explanation for senescence. Undoubtedly dividing cells see an accumulation of mutations. These processes may explain in part why Wilmut was not successful in producing a clone of Dolly until after about 200 tries. However the success in a sheep which is, after all an advanced mammal, shows that human cloning is doable. Of course some genetic material is not in the nucleus, but the mitochondria which means that a minor amount of genetic material would be part of the ovum used as a vessel for the development of genetic nuclear material, so that in a small way the ovum carries at least some contribution from its egg mother. Presumably after this manipulation of heredity, a child could then have as many as three "biological" mothers, the ovum mother, female gamete mother, the birth mother, provider of uterine development, plus still a fourth mother who actually has the pleasure of taking charge of the child after birth

Seed uses Genesis as his rationale declaring that man is made in the image of God and he is taking an important step therefore to make man One with God, men manipulating biology will be God, for this is the destiny of man. That man thinks, means that he extends himself beyond his biological endowments and in that man is a free conscious agent, he resembles God. Men are made in God's image. But this is not the same as saying that Man is God, or at least it seems to me that all of this is so much hubris. If after all of our great scientific advancements you still need to raise antiquated metaphors from Genesis, you might go forward a few chapters to Tower of Babel legend, that teaches about the folly of men thinking their works bring them close to God. Our conundrum, which is not God's, is that we have always found, that while we may think we know so much, the more we know, the more we know we do not know. True knowledge only brings us closer to an appreciation of our ignorance and limitations.

As we learn more about embryology, we will more practical things that we can do. The protein products of genes such as *Lim1*[165] controls gradients of development, for example. *Lim1* is a gene that helps create the embryonic gradient that determines differentiation of the longitudinal head to tail axis in an embryo. Absent the protein product of this gene, the animal, toad, mouse, or human, will fail to develop a head end. Only the body will develop. Suppose a viable animal could be produced which lacks a head? This torso could be used to harvest all kinds of products, meats for one thing, could be produced as long as one could find a way to keep a torso alive. Industrial farmers could raise chickens and cows without heads which would alleviate concerns about husbandry or humane treatment of animals, for example overcrowding. Protein and other organic products and blood and organs could be harvested. The next step would be finding a way to preserve life in headless human embryos. This could save lives of headed humans as in implantation of organs and other parts and products. As we uncover the genetics of senescence it will inevitably become possible to keep certain human cell lines reproducing in perpetuity, if not to keep the whole human organism alive, as long as he or she is not destroyed by trauma or other unexpected event. Does any of this offend your sensibilities? The real question about any of this is on what level we are offended if at all. Where do we draw the line, decide what, if any of this is not allow.[166].

Such "ethical" dilemmas will be decided on the basis of subjective assessments that are essentially esthetic. That is, the headless animal/human embryo example is especially unesthetic therefore is not very likely to be accepted by the vast majority of ethicists no matter what the potential benefits. On the other hand the perceived benefits to be derived from a technique may well counterbalance esthetic sensibilities. Should we pursue these lines or research there will be negative consequences that nobody expected or bargained for. For example, in forming a live cloned baby from an adult human cell, we are very likely to see the effects of accumulated mutations, perhaps a much increased incidence of cancer and congenital malformations and various diseases in the cloned child. Up until now our children have been formed by the cell division of zygotes formed through the fusion of haploid gametes, one from each parent. The making of a new embryo from the genetic material of an adult diploid cell is dangerous. Hundreds of simple cell divisions that have occurred throughout the adult life may accumulate mutations and many other alterations in genetic material. In other words it is highly probable that gametogenesis has a sort of "cleansing" effect, nature's way of way of starting anew from scratch. Gametogenesis, the making of sperm and egg, may have other effects that researchers are just beginning to understand. For example in aging, DNA accumulates change in parts of the chromosome called telomeres on the ends of chromosome structure that may affect chromosome function in subsequent divisions. Chromosomes also form associations with various proteins, for example histones, which could carry over in a cloned situation. This is the kind of trouble you can get into when you think you know more than you actually do, the stuff of overconfidence, scientific hubris.

Surrogate motherhood is rather confusing and appalling as well, which may explain why it is not very popular. It's almost impossible to draw up a strict set of standards or criteria to meet all of the myriad possibilities here. The feeding of sheep meat to vegetarian cows, the lacing of hamburgers with offal and giving it to unsuspecting child hamburger eaters is unesthetic, abominable, unethical. Some people might say that in doing so, people have gotten what they deserve except that the innocent inevitably suffer.

Unfortunately our approach to the ethics of new science is not at all efficacious. We let politicians who have little or no knowledge and have still less understanding for ethics, determine policy, mostly by imposing moratoria. What politicians respond to is what happens to be popular at the moment. Successful politicians are trying to get re-elected. At least involving politicians enfranchises a lot more people, brings non-research scientists who may be more disinterested and less likely to respond to immediate reward at the expense of the big picture into some form of public ethical debate.

You can't put a moratorium on thought. To deny this is swimming against the tide. Someone perhaps clandestinely in this country or perhaps abroad where there are less constraints, will continue research. One cannot when all is said and done, stop progress. The desire to go foreword and continue to discover is an irresistible urge. Putting brakes exploration is a only finger in the dike. It will only increase the impulse for progress.

The bigger picture is that humanity cannot, in the long run be squelched or defeated. A woman I saw complained of awakening at night, full of sweat her heart pounding, in a state of panic. Our conversation seemed to shift almost instantly to an account of life at her new job. She had left school after a minor head injury to work full time for a computer firm in a phone-in customer service position. Her computer terminal keeps a running average of time spent on calls that come in. The time spent to take care of a customer phone call is calculated to be only 5.35 minutes. Her average was somewhere in the range of 6.55 minutes, which pulls down the record of her work group, which competes with other work groups at the same job. As she logs on and off her terminal, her breaks are also precisely timed and compared with other workers. Of course, some employers also keep video logs of goings-on in employee bathrooms as well as all computer activity at the employee's computer terminal even during off periods, though I had no idea if her particular employer was doing the same thing. Your boss, who has control of activities while you are in the office, also thinks he should know what you do on your off time, what drugs you take, whether you eat salads or steaks, smoke or do now smoke, have sex, and so forth, because some behaviors aren't good for your health and affect productivity. This reminded me of an HMO which installed doctors conveniently in airports and shopping malls, then videotaped them, in order to evaluate such physician quality measures as number of smiles per patient encounter and compare these numbers with results of patient customer satisfaction questionnaires. This information was useful, as the doctor who smiled less that average could later be nudged by management to turn out a larger number of smiles per minute. To managers of health care plans, medical care is a business which means that patient encounters, and operations, medical care, are mere widgets, no different than boxes of cereal. The aim is to give as little "product" possible and charge the highest prices (insurance costs), yet still maintain customer satisfaction. If you can make that box look bigger, with the aid of slick advertising and graphics, while actually putting a little less cereal in the box, and the customer fails to notice, that efficiency. Doctors who are still deluded with ideas of grandeur and think of their work as something other than a widget, bridle at this materialistic entrepreneurial view of healthcare. Despite these negative influences on our work, not a few of physicians continue to do what they have to, to do the best work for their patients. Entrepreneurs are material men who will attempt to control, to dehumanize. Managers who treat persons as commodities decapitate them. This is esthetically unappealing, abominable, immoral for most of us. As long as most persons can keep rampant materialism from sickening them, if they know what they are and what they are about, they can work steadily to reaffirm their humanity and that of others. Revulsion can be used for positive change. Humanity survives along with its head.

Twentieth Century history has shown that evil can triumph over the short run. The Third Reich lasted 12 or 13 years, the Evil Empire of the Soviet Union about 72. These political ideologies, whether coming from the right or left, have in common, the notion that humans are nothing more than matter, something that comes from a misinterpretation of modern science. Hitler and Stalin have left indelible images of mass graves, huge piles of human flesh, fitting remembrance for these abhorrent political systems.

These and other attempts to crush the human spirit, are short-lived in the scheme of things. No need to panic or worry. Good eventually will triumph, though we may see over the short run, the destruction of a lot of good people. Be strong, steadfast and work to improve and repair the world.

The mystery of Creutzfeldt-Jakob disease is far from being solved at this point, but the really interesting question is unprecedented in the annals of medicine, that is, what kind of agent is infectious yet at the same time a genetic disease? What possibly could be the explanation for this very unique phenomenon

The "infectious particle" is protein, and, its seems, contains no DNA or nucleic acid, in other words is not a biological organism, but more of a poison that induces a fatal change in its hapless host. This is designated a *prion* for proteinaceous infectious agent by Stanley Prusiner who continues to do extensive pioneering basic and important work, winner of the Nobel Prize for medicine in 1997. With these disorders it appears that merely specifying a precise concatenation of amino acids, by writing a certain sequence of nucleotide bases is not enough. The process is post-translational, that is, occurs after a specific amino acid sequence for the relevant protein has been specified by thr DNA template. The prion is a protein which has a specific three dimensional stereoscopic structure, designated PrP. PrP in the form $PrP^{SC}$ which is the disease causing form of PrP, sidles up to the natural form of this protein present in all of us, which is designated $PrP^{C}$ and causes a deformation of $PrP^{C}$ protein. The protein, thus permanently deformed, will sidle up against other copies of naturally occurring $PrP^{C}$ causing these molecules to deform to $PrP^{SC}$ as well, thus creating a chain reaction causing accumulation of the injurious neuron-destroying $PrP^{SC}$ and Creutzfeldt-Jakob and by similar mechanism the other spongiform encephalopathies. It's been shown that abnormal prion protein accumulates in specific loci within the brain, and that destruction of neurons correlates with accumulation of this abnormal protein. This is analogous to the formation of Amyloid, whose another accumulating protein destroying neurons causing the dementia of Alzheimer disease. In fact, prion proteins are capable of polymerization and coalescence into amyloid plaques. Hence, prion research may well apply to Alzheimer disease though amyloid protein is chemically distinct from Alzheimer beta amyloid. It is as if the brain auto-infects, the responsible agent spreading to adjacent parts of the brain in just the same way that an pus and organisms spread in any infectious process[167]. For the first time a protein is in essence a complete infectious cause or transmissible agent. This protein is partially insoluble and resistant to normal enzymes that break down proteins known as proteases, they are protease resistant, which makes them highly stable. The order of amino acids as specified by host DNA does not cause the disease except in the few instances where the disease appears to be genetically specified and to run in families.

The gene encoding PrP is on chromosome 20 in humans. In certain instances persons who differ in their own prion protein gene by as little as one amino acid specification sometimes by as little as a single nucleic acid, for example substitution of ATG for GTG at codon 129 for the PrP gene on chromosome 20 in both their chromosome copies will result in replacement of the amino acid Methionine with Valine and the subject will not be able to have his or her PrP deform to the pathological $PrP^{SC}$ and therefore will be unable to acquire Creutzfeldt-Jakob disease. For example all of the cases of New Variant Creutzfeldt-Jakob disease were homozygous for methionine at codon 129. The expression of other prion diseases also depends on changes in single amino acids at position 129 and other areas further up the line. A mutation at PrP codon 178 causes aspartic acid in the PrP to be replaced by asparagine. That mutation along with methionine specification at codon 129 are sufficient to cause a different genetic variant familial fatal insomnia. Fatal familial insomnia is a genetic human disease that has been transmitted to other animals. For some reason it appears to affect primarily the thalamus and first causes an alteration in sleep-wake cycles then secondarily progresses to dementia and death.

The distinction between inherited and acquired disease is blurred here. This has been one of the most controversial areas in psychology and development - which characters are acquired and which inherited. With transmissible dementias we have a situation for the first time in which even the mechanism of inheritance comes into question. Can certain characteristics be transmitted purely through protein transfer or does inheritance need to affect DNA or RNA? Prion like particles have been passed down in yeast cells which can inherited improperly folded protein which corrupts other protein in the cell. This process is self-perpetuating and is inherited by other yeast cells in a process of vertical as well as horizontal transmission without changing DNA or RNA[168].

Here we have for the first time an example of a genetic disease that is also "transmissible" in other words simultaneously infectious, albeit under very restricted circumstances. It is most efficiently transmitted to a host who is genetically susceptible, that is one who has methionine at PrP codon 129, by direct implantation of brain tissue. Like other infectious diseases, whether the disorder will be transmitted largely depends on the size of the inoculum. If you give a small dose, the animal or person is likely to escape infection. What happened to those poor young hamburger consuming British kids who acquired the new variant form of Creutzfeldt-Jakob disease is that they probably got a large dose of $PrP^{SC}$ from the cow brain binders and then this protein was absorbed. The protein seems to be absorbed largely unchanged despite the fact that in our gut we don't absorb intact proteins, but instead digest proteins into their component amino acids. Part of the explanation may be that proteases used to digest proteins may be ineffective here or maybe some protein is absorbed intact for example by engulfing immune cells that line our mouth and gut. Since this PrP resembles our own internal protein, it may not be susceptible to attack from the immune system which only recognizes and attacks strange proteins and that is how the stuff may gain entrance to the inner sanctum, even perhaps cross the blood-brain barrier as it must in order to cause disease. In fact our immune cells may provide the portal of entry for this lethal protein into the brain.

The discovery of proteinaceous infectious particles begs the question about what is the simplest particle of infectivity, the smallest amount of data that may be transferred which is at the same time self-propagating, thus self-serving , yet causes an alteration in the host.

The concept of information transfer unifies genetics and infectious disease. Infectious diseases such as Leprosy, Tuberculosis, Bubonic plague, Influenza or any of the other scourges of mankind, can be viewed in terms of mere information transfer. The infectious agent transfers instructions, information, into its host, and devises a way to utilize the resources of the host to reproduce. If the parasite is very good, that is if it has been at its game for a long time and has experienced millions of infectious transfers, both parasite and host will have had time to adapt to one another and the tendency is, over many generations together for the infection to slowly become less virulent. Viruses, bacteria etc. that cause the host to die swiftly such as Ebola , will have very little time to transfer from one victim to the next and won't be able to claim very many victims. The infected hosts will die before they can spread the dread disease. For the infectious agent there will always be an incentive, that is, one will reproduce much more successfully and widely if the host's reaction is slight, non-lethal, as chronic as possible, and if one can take advantage of the habits of the host to spread and reproduce the parasite and produce others of its own kind. Host and parasite over time adapt to one another, the host being selected just by the mechanism of surviving the infection, and there is a lot of pressure on the infecting agent to be less virulent. As time goes on, with host and parasite living together, they adapt sometimes more sometimes less happily to one another. The infecting agent becomes less like disease-bearing parasite. Infection and infectee begin to resemble symbiotes or commensals, in fact the line of demarcation between parasite and host and symbiote is blurred, as anyone with any knowledge of this subject knows.

Consider what has happened over the very few years we have been aware of HIV. HIV is a rapidly mutating virus. The first recognized cases were rapidly fatal, partially because when you have a new disease entity only the most egregious cases are the ones at first recognized but also most probably because the virus at that time was all the more aggressively virulent. The infectious agent took full advantage of the behavior of male homosexuals and i.v. drug users that allowed direct transfer of the virus through direct injection of body fluids into the bloodstream. Semen had access directly to the bloodstream because of the inadequacy of anal epithelium to resist unnatural use. Widespread use of needles was also an unnatural and unusual practice that allowed direct blood to blood contact and direct efficient injection of large viral inoculums. Early recognized cases had fulminent disease. Over a short space of time physicians learned how to treat AIDS and the disease evolved into a chronic form. The virus itself, may well have become less rapidly fatal. Don't forget that rapidly fatal fulminent infections tend to be selected against in favor of more chronic milder forms that stick around longer to allow the host to spread the organism to others. At risk groups hosts, i.v. drug users and male homosexuals may have evolved as well in a sort of survival of the fittest, those who were more susceptible dying early of the disease. But the really interesting thing is how technology, man's own design, has changed the character HIV disease and this is almost unprecedented. Physicians have devised means of treating unusual infections that AIDS patients acquire, but also combinations of drugs that include newer protease inhibitors. Now AIDS has been transformed from a fulminent to a very chronic disease allowing infectious hosts to survive almost to a normal life span. Now the host will be infectious for a long time. The only saving grace is that viral loads and titers may be very low in HIV carriers decreasing the likelihood of transmission, or infectivity.

The virus has taken full advantage of certain human behavioral practices. Despite what the media say and experts have warned us about, HIV remains rare in the U.S. and truly exceptional, among non drug abusers who practice heterosexual non-promiscuous mostly monogamous sex. By contrast, in North Africa even in the absence of advanced medical care which may well serve merely to extend interminably the period of infectivity of the disease, HIV affects a good proportion of total population of men, women and children. This is due to extreme promiscuity in men. Prostitution particularly has been implicated as well as other forms of unconventional (non-vaginal) sexual practices which are more likely to occur with a prostitute as opposed to a regular sexual partner.

Humans have had a long relationship with viruses of the Herpes family. We know this because these viruses survive within our tissues especially nervous tissue over a long life span. Studies have shown that the vast majority of Americans harbor different types of

Herpes viruses especially Herpes simplex, Herpes zoster the cause of chickenpox and shingles, and the Epstein-Barr virus that causes mononucleosis. The hallmarks of a mutually long relationship between virus and host are there. The vast majority of Americans are infected with these viruses while young. Most do not recall their primary exposure, which means that infection caused so few symptoms, it was not even noticed. That happens in the situation of all three of the above viruses. 80-95% of adult American harbor antibodies against the above Herpes viruses, which means we've been exposed and harbor viruses as well, but do not even recall having symptomatic disease. Most of us are immune to mononucleosis which is caused by the Epstein-Barr virus but very few of us recall having it. Virus and host have had a long time to adapt to one another. Since antibody can be detected over many decades in most of us, Herpes takes full advantage of our own machinery to reproduce. Perhaps it even gives us something in return for the bargain, that we are unaware of. This is done assymptomatically in most cases and only rarely will large amounts of the virus be reproduced and a person in the case of shingles will infect himself with his own internal Herpes zoster virus, or if extremely sick old or frail where immune surveillance is down, he may acquire virus from others. It can be said that Herpes family viruses are almost a part of us, less a parasite or disease, than a commensal. This is speculation but it is possible that humans have had such close relationships with parasites in our distant past, that their genetic instructions do indeed become part of our own genetic endowment, in other words that we today are infected with even older viruses whose infection is completely silent or undetectable, that our relation is so close that they've become a part of us, a neutrino of the parasite world.

A neutrino is a subatomic particle having almost no mass and no charge. Neutrinos are stealth particles. You cannot weigh them because they are so light, nor do they can they be detected by any device that measures charge, since they lack electrical charge. Yet some astronomers postulate that neutrinos are so numerous, they may constitute so-called “dark matter” in the universe that in aggregate may be the great bulk of mass or substance in the universe.

It is possible the great bulk of our genome is parasitic, yet undetectable as such due to the fact that these old parasites no longer cause disease. Disease is the major way we have of detecting infection. We may have been infected by numerous organisms that have since become a permanent part of the human genome, stealth particles that escape detection. Some of these have conferred certain advantages such as mitochondria that introduced new repertoire of energy utilization to early species and whose genetic specifications were incorporated permanently into the cell. It is theoretically possible that a good deal of genetic material now part of the machinery of the cell, actually originated from infection by other organisms rather than purely through adaptation and natural selection.

In the case of genetics, just as the study of space and information and computer theory, the more we discover, the more we learn what we do not know. We have abundant knowledge about the molecular biology of inheritance, but over the past 50 years or so since Watson and Crick's elucidation of the basic chemical machinery of inheritance more and more varying pathways of information transfer have been elucidated that need to be explored further. To think that because scientists can use bacteria to make genes or protein products or clone sheep from adult cells, that we are all powerful or have done anything but merely scratch the surface of genetic knowledge is pure hubris. The last few paragraphs give a limited view of how incomplete our knowledge base is. Distinguishing lines are completely blurred in the transfer of genetic information. What is the fundamental unit of matter used in the transfer of genetic information, DNA, RNA, protein in some instances? What is the basic unit of information transfer? Is it a single gene alone that seeks to propagate itself or a larger aggregate genetic product, a bunch of genes that have bet their futures together as members of a group that we call an organism? Prions, genes, chromosomes, plasmids, mitochondria, chloroplasts, nuclei, bacteria, protozoa have all been agents of transfer. What is the distinction if any between an infecting organism, a parasite and its host? Given the abundance of strategies for propagation after one's own kind, what is the distinction between an idea, a particle or datum and the material biological vessels used to promote and reproduce that idea? The distinction is analogous to matter and energy an idea freely mobile between a biological material state and a particle of information. Here we see in miniature a reaffirmation of perhaps the greatest discovery of the Twentieth Century the equivalence of matter and energy.

As scientists acquire more information our vision, if anything, blurs. Typical distinctions are indistinct. The fundamental relation between matter and ideas appears to be that matter serves as a temporary storage device or vessel for ideas but that as time goes on and we begin to master the process, matter and ideas will become freely interconvertible. For the moment, matter and ideas are indistinct. Both turn out to be an apparition or ghost, which as we try to grasp it, disintegrates in our hands. The more we learn the more we find out how much we do not know. Perhaps this is the fundamental secret of the universe that it is here for our pleasure, always just beguiling enough not to bore us, just large enough to be beyond our grasp, just barely in the range of wonder. Matter at best, and here we are talking about matter in the form of biological reality, is a temporary vessel for the propagation of ideas. At least at this early stage we have come to that realization. Perhaps the computer, our latest gadget for storage, manipulating and transmitting information may one day supersede the human brain. An alternate strategy would be to change properties of the brain itself to overcome its limitations. Humans may fundamentally alter our biology to increase mental and other capacities that are today totally out of our reach. To perfect our tools or to change our basic selves, or some combinations of the two, these are our options as we separate ourselves from our own limited biologically given capabilities and step beyond biology. We have this choice because at base we are composed of two sources of information, the biological or genetic and the computational or cerebral. For "lower" organisms this is expressed almost entirely in biological terms, while for us, this data is mostly computationally specified in terms of electrostatic charges perhaps in our brain, but also with dependence on extra-cerebral information repositories.

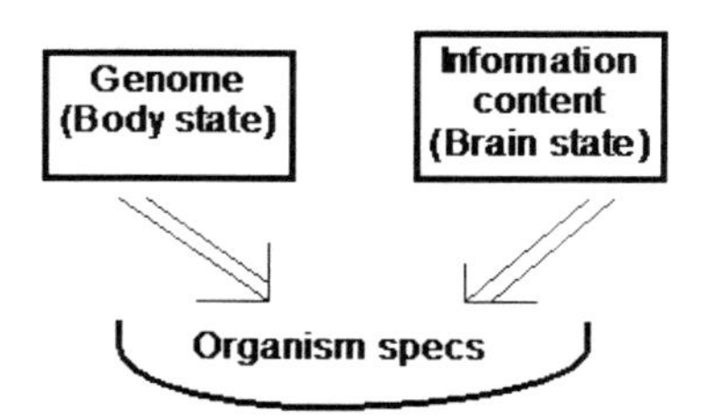


**Figure 48 The human condition. The total organism is specified by the genetic complement plus informational content of the brain.**

The idea here is if one died and were to come back, on what basis would we reconstruct the same individual, one who would not only exude the same response in a similar situation. But would sense and cognate, process i information in the same way. This newly reformulated person would be identical with the decedent in every respect and would in fact be that person. The two sources of information necessary for this specification are represented in the figure above.

Here is the cool position we are in: If we want, we have the technology to alter fundamentally our own form and anatomy. The danger is that we will change ourselves so much we will no longer be able to keep track of who we are. Suppose someone creates a new human with a bigger head and shorter arms, another proprietor makes persons with longer legs, bigger lung capacities and so on. After a while we'd accumulate so many varieties of the human form we'd lose track of what a human is. We would do this to compete with ourselves, to run faster, live longer, remember more, think better, experience life with greater intensity, etc. The other strategy is to keep form essentially the same, which would simplify accountability, keeping track of who we are (though I think we will hardly be able to resist making some "improvements" or at least eliminating diseases and disabilities) and utilize tools or our own invention - the best example is computers and Silicon devices, but cranes and trucks buildings and all civilized accouterments fall into the same category - and to work ever more closely with these machines, and still expand our capacities. The point is this goes to prove, that there is much more to us than is specified in biology. Even at this primitive stage of our existence, we are no longer confined by our biological endowments. It proves that biology does not limit us, our material nature does not even define us at all. There is far more to the human condition than is specified by biology. After we reach a certain point, perhaps getting rid of some imperfections and accidents, eliminating disability suffering and disease, the stuff that is done by medicine, and providing for basic needs food, comfort, jettisoning limitations, the human spirit will soar, showing once and for all, in case some of us remain unconvinced, that there is more to humankind than material existence.

This is a summary of who we are. At very bottom we are biological creatures of course. Our needs are basic to ensure survival. They are food, shelter, sustenance, freedom from disabling disease. If we are successful we reproduce our own kind, not an exact replica of ourselves but creatures that claim a resemblance to us, sons and daughters. In order to accomplish this most successfully, we use the tools of a family and more extended social structure. We seek to avoid death as much as possible. Death occurs anyway because without it, there could only be a vastly reduced number of experiments that we perceive as individuals of a kind, and without extensive experimentation with different individuals, there can be no adaptation, hence those of this certain kind would lose out ultimately in the game of fitness which involves adaptation to new environmental vicissitudes. Thus the paradox that organisms that will be successful need to die but individuals of a species as a general rule try with might and main to preserve their own life.

Biology is only the beginning. Biology is matter but without it we have nothing. Our anatomical and chemical form up until now, was the stable base from which persons were recognized and defined, and still are to some extent, considering our current state of knowledge. Our cells and our aggregate organism, is the repository, the growth capsule, the spore or protector of data, an abstract idea that is us. All at once this organism is us, but it has lately become only a launching platform which is necessary to allow human development and life. But at some point we are becoming bursting forth from the egg the biocapsule which has up until now formed and defined humankind.

Humans specifically, perhaps alone among all organisms, step beyond this basic biology. In this text we have come to consider the biological machinations as only a beginning, necessary but not sufficient for the nurturing of an idea. Humans have learnt through their history to step beyond this biology yet have heretofore but perhaps not forever, will use biological capacities as a springboard to greater accomplishment.

Long ago humans learned to take one step beyond the biological imperative, that is to want more than mere subsistence, wonder at the world. Mundane day to day existence and survival were one thing, but very early in human history there were those who insisted on something more than mere survival. These were the artists of Lascaux cave, the builders of homes and relationships and commitments. Some persons seemed to decide that there was something more than mundane day to day subsistence, the product of the hunt and material gain, that while at the same time fulfilling basic material needs, that they needed to take care of an inner sprit newly discovered. Surely our ancient ancestors dreamed as we do, about conquering death, the preservation or return of their own consciousness rather than its final end.

It is in this effort to step beyond the bounds of biology, dissatisfaction with mere survival but turning outward from ourselves, and inward too exploring uncharted regions in and out of ourselves, that was the start of self-expression, art. It began with an incautious curiosity and dissatisfaction with the status quo and meant taking risk, sometimes the risk of death, and a step away from the biological imperative. In the long run this tendency was instrumental in the survival of humankind even if it meant, as it inevitably did, short-term danger.

When the work of the day was done, when basic material survival was ensured, the artists of Lascaux cave risked life and stepped into another world of artistic production. What other explanation is there for their work except dissatisfaction with the daily routine and simple survival? Indeed what better explanation for the creation of a work of art or a symphony which has no obvious survival value a dissatisfaction with the ordinary and a need to escape the practical. We stand on two legs one in the real world but an second is firmly planted in an alter world of ideas and spirit. Both are necessary to support our basic human structure. In doing so we have taken the second step in our liberation from our biological nature, which is, after all, only part of us. Each new advance is a bootstrap that then accelerates further development. We have mastered our mechanical and physical limitations. Suddenly we are able to travel, store information, and transform and manipulate data as never before. Through it all comes all technology, music, arts, architecture, science and all intellectual adventures of mankind.

But there is yet a third phase that is in fact more transformative and it is almost a kind of Messianic era that we are entering today. It is the transformation of ourselves from our basic biological endowment. As we have seen computational devices allow us to store and manipulate data, and to sense as never before. These devices whether implanted or not, don't just enhance our cognitive capacity, but radically change who we are. Mind extenders increase our mental potential. And with this added potential we soon will develop new genetic techniques to change even that basic biological form. We will one day perhaps not recognize ourselves, our appearance will be materially altered by the machinations of an expanded mind.

Human life is a sheet of many dimensions. Every person who was is or will be can be found within as given set of space-time coordinates. Though Einstein is dead now it will always be possible to return to his own coordinates, at Princeton in 1942 and to find him, if we ever develop the technology to take advantage of this. There exists a giant address book that controls this information. Every life is suspended in this limited Cartesian system. Every person's life is a gossamer sheet extending as far as their eye can see, as far as their perceptual gaze, far out as their efferent grasp. To understand a language or apprehend music, play piano or design a building, then and there can be found in all these limiting extents among separate dimensions. Certain folks extend quite far in one or another of these dimensions in the area of cognitive function and consciousness and well as in space time. Others extend farther out on other planes, long life perhaps. In dying one should not worry for he is always there with some potentiality he ever had. If so the major task in life would be extending these potentialities, our multi-dimensional sphere. Each of us leaves something behind after we are gone, a child, our work, something written or created, that serves as a handle of limited extent from which another persons who carries on, can retrieve a memory engram analogous to the retrieval of memory inside the brain only here from some enormous repository of information in space and time. It should theoretically be possible at any point to go back to the appropriate coordinates that specify a life, and retrieve that particular existence again.

The cosmos is a giant fabric of n dimensions holding all that was, is and will be. Humans, within their limited capacity, strive to sense the contents of this fabric, like the way a phonograph needle, tape or computer disc magnetic head, or CD laser sensor plays contents over a limited extent of these storage devices, just as ribosomes read the sequence of nucleic acids on RNA. All of these situations have in common a limited capacity to experience at any time only the tiniest portion of the total contents of writings on the storage device. Religionists may call this cosmic fabric the Mind of God, but if there is a God, this is only the memory of the Mind of God, all the more impenetrable, unfathomable is the entire neural structure.

Every life is then an engram laid out in some gigantic neural structure. Each person hammers out for him or herself a structure of limited scope in perception, thought and action, a neural structure within another neural structure, the Mind of God. At any moment, this engram which is a life, may theoretically be retrieved, extracted by its memory handle. Each of us exists and ever will exist in the great eternal mental structure of our cosmos.

Given that we cannot be defined by a limiting form which give birth to us, our spirit, then is ever harder to grasp precisely who we actually are, our full essence. So much more is there to us than is delineated by material understandings. We have only to realize that we are far more than a machine, a mechanical contrivance. What we strive for, more than anything else, is to get our bearings in the midst of an exploding maelstrom of a cosmos. Where we live, knowledge built upon other knowledge expands exponentially, we have to be in a state of perpetual confusion, a divine diversion that commands our attention and ever holds interest if only because complete mastery and understanding always will be just beyond our grasp.

If this is true then it must be the idea rather than the receptacle device storing the idea that is the important element. The preservation of a living container, occupies us most, but is comparatively unimportant. A physical body is as a plastic compact disc is very nearly worthless without its proprietary contents. In just the same way the mind exists in an organism, a mere receptacle for information. Uncovering biological, chemical, electrical mechanisms gives us many tools for understanding ourselves, but describes life only at its most superficial level.

Mind and brain complementary understandings analogous to wave and particle descriptions in physics. Mind and brain, energy and matter, ideas and physiology are equivalent and interconvertible aspects of the same phenomena. Understand one aspect and your understanding is incomplete. If physics were only seen from the standpoint of particle interactions with no input from wave mechanics, knowledge would be one sided, incomplete. In just the same way the human condition is only incompletely apprehended through understandings of physical brain sciences. Hence we have complementarity. We may hope perhaps that one day scientists will unify complementary conceptions into a single unified theory, but this is not likely to happen soon and is likely to occur only after we alter some current fundamental misunderstanding.

This is the world of the future. As this book was obsessed with viewing the mind through the window of the brain, matter makes mind, matter makes ideas. If there is one lesson as we move into the twenty-first century, it is that the idea is the maker of matter. We are right now transitioning from the notion that matter makes the man to the idea as the essential element. We see this on a practical level as we experience even in our economy that persons who know how to manipulate things, are relegated to repetitive tasks are losing income relative to those who can manipulate symbols. Persons who manipulate symbols and words achieve success. Ideas have it over matter. The Mind has control over the brain. As time passes, in our economy, we are exiting the world of matter and enter to one of pure thought. We are breaking out the egg which bore us out of our biocapsule. As our efforts become more and more exuberant and adventurous, biology will be ever more surrounded, conquered, outflanked.

And our method for increasing adaptation to our world has changed strikingly in recent history. Formerly knowledge was revealed as from on high by a seer or a prophet. At present we depend on the scientific method to tease the facts out slowly, patiently and methodically. Science, discovery, not prophesy has provided the revelations for our age. Knowledge is won by dint of hard work, fastidious methodology. The work, sweat, sacrifice of discovery, making of man in our own image. Previously there seemed to be no purpose to the universe. That is the horse race. At present those who believe in purpose, in creation of an ordered cosmos at one instant are winning. How large is the cosmos? Large enough to keep our interest.

Here is a new optimism. Gone is despair and panic that results from wholly material man with fixed chemical composition, that there is little or nothing more to our essence than mechanistic relationships, the profoundly distressing and depressing point of view held by many of our contemporaries. The wholly material person derives from a misinterpretation of modern scientific data. We have good reason to hope, now with a proof that what we are will never be confined by material and machine, that we have crossed into the threshold of a brave new Promethian world- beyond biology.

Many of our most prominent thinkers, brilliant scientists, have sought to reduce us to our material constituents. They have used as their argument, advances in technology that have enabled us to study and understand more than we have ever grasped before. As I hope I have shown, they are defeated by their own discoveries. Great scientists such as E.O. Wilson, Francis Crick, among others, seek only to subsume all thought under an umbrella of biological causation. The argument goes something like this: The greatest inventions of the human mind, buildings, works of art, ethics, have come about merely as a product of a biological determinants, mere parts of a larger *biological imperative*.

Music? Well that came to be because of the need for the human family or tribe to form an emotional bond to help in the aim for adaptive societal cohesion. It's so easy to see how initiation, marriage, war ceremonies and other types of nodal tribal events were helped along by rhythmic tribal dances and mesmerizing repetitive patterns in sound which enlisted limbic or emotional parts of the human brain. Primitive art? The same mechanism. Primitive art illustrates techniques of the hunt, and inculcates respect for the hunted animal, which sacrifices its life for the survival of the tribal group. This served as an efficient means to hand down basic concepts from one generation to another. The groups that were more successful in doing so are the ones that survived, by virtue of their cohesion. Writing? Writing, indeed defines the boundaries of history and pre-history, and has been with us only for a few thousand years. This is biological too. It's always more adaptive to be able to record and pass down specific information.

Ethical principles derive from a biological imperative. Wilson and others have shown how the almost universal incest prohibitions make eminent sense from the biological perspective. Incest is maladaptive in that it decreases genetic variation and increases many fold the chances of having genetically defective offspring or even offspring with fatal genetic flaws. Just as incest taboos have arisen, and make excellent biological sense, so must all ethical principles be essentially from a biological adaptation. Good ethics increases fitness, adds to the probability of survival. No need to invoke concepts of a supernatural deity handing down laws that make eminent biological sense.

Great scientists more than the average mortal, are aware when they are reason on the basis of example. They know, or they should know that their examples will not suffice to describe a whole class of phenomena, particularly in the category of ethics. Reason by example is inductive as opposed to deductive logic. Induction presupposes that once you have uncovered a pattern in a few examples, the rest of the phenomena in the same category work by a similar mechanism. Hopefully you have come upon a pattern, but it just as possible that you have veered in the wrong direction. Reason by pattern of course, differs from deductive logic of stepwise implications.

Scientists. of all people should insist on filling in the blanks. A few examples may fail to establish a definite consistent pattern. Incest is an almost but not quite universally taboo. In Hawaii and ancient Egypt sibling marriages were encouraged typically in royal lineages. And a biological basis for ethical taboos is on even weaker ground in consideration of marriage law, which varies considerably from culture to culture. Some societies tolerate polygyny, even a few polyandry, some encourage levirate marriage in which it is incumbent upon a brother of a deceased man to marry his widow. Is it more adaptive to encourage polygyny in which few successful males sire a large percentage of offspring, or may this strategy only create discontent among the mass of male subjects, and end up being maladaptive in the sense that there will be a less variation with majority of offspring coming from few male progenitors[Ψ]? The genetic effects of monogamy vs. polygyny are roughly calculable. The most biologically advantageous system, one would expect, would apply across all cultures, but this is not what we see. One would expect the most potent arguments to be in the highly biological arena of sexual selection. Yet, the argument is weak even in this limited purview.

Other areas of ethics are far more unfathomable from a biological adaptation perspective. Laws about just treatment of helpless widows and children, all matters pertaining to altruism, religious rituals including sacrifice and hosts of other customs, the support and toleration of weak and unproductive members of society and those laws accepted as part of the social contract, ancient rules of war, the Geneva Convention are profoundly unbiological in that they are impossible to explain on an evolutionary basis. These laws may be accepted as "fair price" to pay for the object of societal cohesion but here we get into the very sticky areas of group vs individual selection, not accepted by all evolutionary biologists.

The unit of selection from the biological perspective is the individual. In biology the name of the game is to pass on your own genes. The organism is merely vehicle for the transmission of genes. When an animal performs an altruistic act, for example, warning other members of his group about the approach of a predator thus calling attention to himself or sacrificing his life for someone else's benefit as an soldier ant might do during an attack, or even better, standing up to a vicious animal in order to allow others to escape as is found even among primates, he is performing an act inimical to the passing on of his own genetic traits. Such an act will not be selected for except for a couple of considerations. If his altruism signifies that an equally altruistic act will be performed by others in the altruist's behalf, then there is a reciprocal relationship.

Alternatively, and this is by far the most frequent explanation, he may pass on his own genes providing other members of his group whom he has saved are very closely related, meaning that they share many of the same genes. One example is so-called kin selection. The ordinary animal shares an average of half his genes with first degree relatives, siblings, children and parents. If he sacrifices his life so that two children survive, he will more than break even, particularly if he has finished making offspring. The name of the game from the evolutionist's viewpoint is to pass on one's own genes. A gene for altruism will survive in a population if it helps the carrier pass on his genes. Otherwise the altruism gene is detrimental. Take a group of social insects where thousands of individuals share the same genotype. Nothing at all is lost with the sacrifice of an individual. Preordained sterility as often is possible as is known with workers of a hive. This is the calculus of biological altruism, which is not, as can plainly be seen, altruism at all. The natural altruistic act is performed for the benefit of the individual performing it. Otherwise most biologists dismiss the notion that the individual gains because of preservation of members of his group, so-called group selection. Evolution progresses for the benefit of the individual, not the group. The individual is the unit of selection.

In humans, and only among humans, things may well be fundamentally different. Throughout history large societies even cities and states have competed directly with each other. The consequences of losing a competition are often disastrous, particularly in war. Loss of a head to head combat most frequently ends in annihilation of the losing side and that often includes an entire city-state or nationality. At the very least men of the conquering tribe will inseminate women if they allow women to stay alive as an entire culture becomes subsumed in another. Certainly in human populations, though individual selection likely still predominates, group selection is important as well.

Thus group cohesion and all ethical principles that would promote the survival of the group hold little water when it comes to pure biological argumentation, while for humans things may be different. This being the case it is far more parsimonious from the scientific perspective to observe that basic ethics, that is the most primitive taboos, perhaps derive originally from a biological imperative, but from there on, ethics acquires a life of its own. When congress is deciding to change a law, primary considerations are whether the change is consistent with previous derived laws and principles or if the newly considered variation is not with these principles, not whether the change will be adaptive from the biological perspective. Not a few biologists might suggest that the primary consideration for ethics and laws ought to involve adaptive considerations, but all would agree that this would mean the primary thought in this instance would be for the fitness of the individual and this in turn would mean preserving the reproductive potential of the most fit among us, which is not what the operation of modern societies is about.

In fact, many of our laws invoke quite the opposite, protection of the weak and minorities among us. Laws in a modern society are (ideally) passed and enforced for the greatest good. One of the pressing issues in our own day, as in the past, is whether to accept immigrants and in what numbers and under which circumstances to accept them. Immigrants as a rule flee political and economic disadvantage. These underprivileged persons come to us starving, filthy, diseased, and undereducated. At first glance they are highly unappealing, that is until one realizes mostly by virtue of empirical (historical) evidence that immigration is the greatest engine of American diversity, most probably the main cause for our international hegemony. (Recall how the raw material of biological adaptation is also diversity.) Immigrants are no random sample of foreign populace. They among their compatriots are the ones who risked everything to, of their own accord, flee the status quo or persecution and economic disadvantage. Displaced persons are proven to be more productive than a native populace typically within one or two generations. Thus initial impressions are deceptive. More than that, the members of human societies are genetically diverse. Yet they compete, today more than ever, directly with each other, not only militarily in the past, but to a much greater extent in a global economy.

The difference between humans and animals is of course, the brain, which allows the maintenance of complex societies with their own geography, language, customs, ways of making a living, and defense. Human societies are cemented together by various cohesive elements, but the important point is how the fortunes of the individual are linked with every group member. And members of a societal group need not be as closely related genetically, need not be kin at all, in order to share a common lot. It is because of the brain that the units of inheritance are not only genes, but also common patterns culture, societal rules. All persons are a part of a vast societal framework, a collective consciousness. This information is reproduced and passed as surely as are genetic traits, from generation to generation.

Groups that espouse cohesiveness, productivity, freedom, respect for individual differences, a host of non-biologically based values, increase their own fitness. Groups that espouse a specific ethic or reason for being have the greatest chance for success. Those groups whose watchword is the fitness of the individual alone, selfishness, the immediate gratification of personal need, fall into the rubbish heap of history. Modern events prove this more actively than ever. Examples abound from Rwanda to Kossovo. What is right is not necessarily in keeping with a biological imperative. Ethics as in all other areas of human endeavor, acquires a life of its own.

Music too and all other art forms no doubt, originated as biological adaptation. But biology cannot explain a Beethoven symphony. The symphony was born in the limbic system or emotional mind of Beethoven or perhaps in a motivational area of his brain, the anterior cingulate. From there his cortex would have been instrumental in design of the ultimate structure of the symphony. But as I hope we have seen over again in preceding chapters, these brain structures have served as a platform or basis for invention. Without any one of these structures and many other brain constituents too, Beethoven would never have been able to write his symphony. But this says nothing about the final form of the work or its meaning, which comes not only from the brain, but something intangible in the man who wrote the music. That is to say without his brain Beethoven could not have written his music. The brain of Beethoven thus exerts somehow a permissive effect on his work, but does not determine the final form, which comes from something defying anatomical description deep within his being. That, in fact, is the story behind all invention. To state that the brain is necessary is a truth with a small t. In fact so are his kidneys, liver and so forth necessary, as but for them he would not have been alive to accomplish his work. To maintain that to know this brain anatomy and physiology in toto, would tell us how he did his work and all of the meaning behind it, is at this point in time a leap of faith.

The brain is and all its products are subject to the same rules as govern any complex mechanical contrivance. A good metaphor is the power plant of your car. At best this performance machine allows you to accelerate from 0 to 60 mph in a few seconds. Take out the battery, and the thing won't start. You conclude the battery starts the car. More precisely the battery is necessary to start the car, not that other parts don't participate in this effort and are not also necessary, least of which is the starter. Certain defects are fatal. If the oil pressure is down for a long enough period of time the engine will seize turning your car into lifeless scrap. In other situations the engine will continue to function, though at an impaired way for example with defects in the valves or carburetor. The car won't run smoothly and will lose power in passing and on hills. Maybe it will burn oil. But the thing will limp along and carry you for a long while. So it is for the brain which may have fatal or not so serious deficiencies. Punch a hole in the brain, create a lesion as by a stroke, and the organism, as a general rule continues to function. An accumulation of defects or a single serious event makes brain function incompatible with life.

Lesion the cingulate gyrus, and a composer will be unable to produce a meaningful work. The same holds for the left hemisphere. Both of these structures are necessary for the creative work of our composer. According to this line of reasoning, the logic of the lesion experiment, the intact normal functions in a manner indistinguishable from the lesioned animal. Both the cingulate and left hemisphere are necessary for musical composition. But necessity does not equal sufficiency. No one has proved the brain alone as flesh and blood is sufficient cause for high human creativity. Music, like ethics, may start in biology, then acquires a life of its own, separate from biological contingency.

In music, art, writing, architecture, ethics the first furtive steps of the initial journey probably did occur because of a biological need, but all of these endeavors and more, later acquired a fire of their own. Primitive music enhanced tribal ritual and cohesion. It's easy to picture it as an adaptation, in that tribes without music may have had more difficulty surviving than tribes that had it. Once having gotten started music developed as a thing in itself. That means that modern music plays many roles far removed from its original biologically adaptive function, which should be obvious even to a biologist. What are we to make of the development of musical notation and invention that culminates in Bach and Beethoven? Like the persons that produce them they were born of biology, but the entire future and form and not biologically determined. Biology set the process in motion, but kindled fires all about that are out of its control. How do we incorporate chamber music or great stages such as Lincoln Center into a biological scheme? Obviously modern music has acquired a reason for its own development as a thing in and of itself, a separate life. This separate life is not explained or even delimited by a biological imperative.

All branches of learning and accomplishment have gone in their own separate directions, branching from their biological origins. They've evolved into separate flames kindled by that original biological fire. All of these areas of endeavor began out of biological need, but at some point, in some cases very recently in our history, we have managed to break free of biological causation and limitation. The reason for modern advancements has little or nothing to do with original biological causation. All fields of intellectual endeavor worked by specialists who transmit and advance each individual field. The reasons for pursuing these separate endeavors now has little to do with survival and fecundity or reproduction, but more with individual self expression, societal advancement or enrichment, improving the quality of life. In essence by applying such knowledge we are reaching for a goal that transcends simple survival and reproduction. And in the end, as we have seen, we change ourselves.

What then is the role of the brain? The brain is a platform, making possible high human accomplishment. With the aid of a system of human invention and technology which bootstraps upon itself. The brain is merely a launch pad. The rocket attains escape velocity freeing itself from the limits of gravity. And as everyone knows, it is not the launch pad that determines the rocket's trajectory.

From the scientific perspective all fields of human endeavor will be found to have a physiological or anatomical, that is, a mechanistic cause. That cause resides in the brain. That the scientific method will reveal everything by brute force, that is the steady application of scientific method, over a long enough period of time, is an article of faith as much as is found in any religion.

From whence springs this freedom this creativity that we find in the intact functioning human? If not for the machinations or the white and gray matter of the brain, from where does this derive? With self-accusatory tongue in cheek all I can say is this comes from a certain Je ne sais quoi. No matter how wise we think ourselves to be, there is still a basic elemental constituent uncounted by any mechanistic theory. This is what some might call a non–material soul the origin of all motivation, of all thought feeling and action. Using current tools and limited by mechanical as opposed to spiritual considerations, I see little hope of ever localizing initiative in the substance of the intact functioning brain. The brain is a single platform housing only a small quantum of memory and thought. As human expertise advances each individual is less and less limited by the capacity of a single brain. As has been made abundantly clear by this time, the limit of awareness ranges far outside the confines of any single human head.

But what of perhaps some kind of communal instrument of thought? Indeed one expert may reach into the brain of another to alter or improve his function. The expert functions with knowledge shared between myriad practitioners of the art. A universal expertise, a communal knowledge transcends us all. The farther we wander in the journey of human understanding, the more we appreciate that now, more than ever how we have evolved into a form emancipated from the limits of our own body and brain. Humankind is as Prometheus unbound – unbound from our own biological beginnings.

Each of us, every human born, can be seen as a small node in giant pattern of human societal evolution and technological development, artistic expression. Each of us is born, takes his or her place, continues suspended in those coordinates in life and then in death. Awareness is a communal enterprise, each individual person no more than a node, a part of something larger.

"Hm-m," he said. "Lookie, Ma. I been all day an'all night hidin' alone. Guess who I been thinkin' about? Casy! He talked a lot. Use'ta bother me. Bot now I been thinkin' what he said, an' I can remember – all of it. Says one time he went out in the wilderness to find his own soul, an' he foun' he didn' have no soul that was his'n. Says he foun' he jus' got a little piece of a great big soul. Says a wilderness ain't no good, 'cause his little piece of a soul wasn't no good' less it was with the rest, an' was whole. Funny how I remember. Didn' think I was even listenin'. But I know now a fella ain't no good alone." –

John Steinbeck *The Grapes of Wrath* (Chapter 28)

A life lived is an indelible record in a huge space-time array that we call everything or simply, the cosmos. At any time, before or after the actual consummation of that life, it should theoretically be possible to extract that memory module, to replay a single life. The process is similar to a phonograph needle or laser lighting on a record or CD to replay its contents. As discussed, these "universal" memory engrams most likely are recorded and recalled holographically as are memories within the brain. Everything living and dead leaves a trail of its existence in space-time. These trails or engrams are ordinarily discoverable by use of the scientific method. They may be sensed by other means, through affect or subtly as through emotion, or sensed through intrinsic memory mechanisms, much as the emotion-charged memory of a loved person springs into our mind at times or subtly influences our behavior.

The difference between the atheist and religionist is teleological. The atheist and religionist would agree that events and lives leave a record or trail of discovery. The atheist sees these lives and events occurring without method or purpose. To the atheist the cosmos is a mechanical device. The religionist looks out into space and sees an Eternal Mind, with all of its constituents not the least of which is motivation, strategy and purpose. The religionist might admit that the cosmos seems to work mechanically, but it is a mechanical contrivance put together or set in motion by a higher being. Each person's mind, is a component of universal knowledge which in turn has a place in larger human history of life and understanding, a communal knowledge passed horizontally and vertically from person to person, community to community. The bigger picture the human essay, part of the Eternal Mind conveniently represented here not as some mechanical recording device, but symbolized by the entire human brain, signifying that these memories at the human, communal, historical, universal levels are multimodal. They require all of our pitifully limited human faculties, auditory, visual, and a lot more to appreciate their fullest splendor.

The rendering of the brain here of the Eternal, the depiction of the cosmic mind in the form of the human brain doesn't imply that the Eternal's mental machinery is as limited as the paltry contents of a single human skull. Instead it is meant to signify a certain completeness and purposefulness of being, a wholeness of thought, feeling, perception, cognition, volition that are composite functions of the brain with each of its lobes, brainstem, cerebellum and all of its other parts. The brain isn't everything, just a symbol. In our feeble attempt to grasp something infinitely larger than ourselves it helps to anthropomorphize it, at least as a first step in understanding.

In closing we should all be mindful that each and every one of us is a part of a much larger plan, that we are suspended for all eternity in this plan. It helps to keep this perspective even as we are harassed by the narrow limitations of our mundane daily existence at work and at home. This realization is the very meaning of freedom. In it we are suddenly transformed, unbound from grind of everyday life. We escape our own worldly concerns, as we jump out of our own skins, seeing how we transcend our own biological organism.

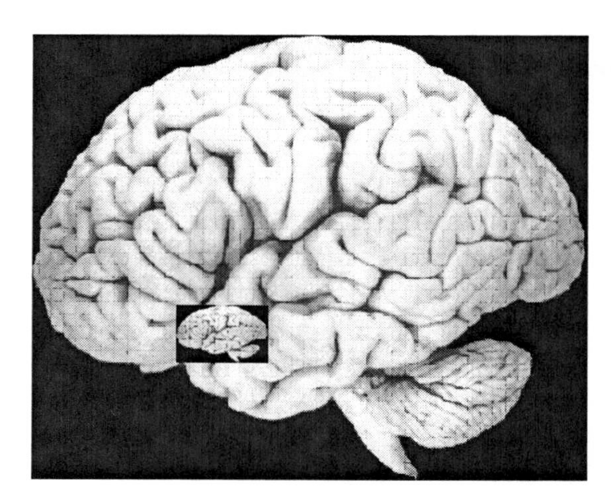

**Figure 49: A human mind remembered in the in the mind of the Eternal (not to scale).**

[169]**Figure 50**

# Bibliography

Ackerman, Diane. *A Natural History of the Senses.* New York: Random House, 1980

Ackerman MJ, Clapham DE. Ion Channels--Basic Science And Clinical Disease *New Engl J Med* 336:1577 (1997)

Adams. RD. Victor M. *Principles of Neurology* New York: McGraw-Hill,1993

Alajouanine, T, Aphasia and Artistic Realization. Brain 71:228-241 (1948)

Ambros IM, Zellner A, Roald B et al. "Role of Ploidy, Chromosome 1p, and Schwann Cells in the Maturation of Neuroblastoma" *New Engl J Med* 334 (23):1505-1511 6/6/96

Aminoff, MJ (Ed.) *Electrodiagnosis in Clinical Neurology (3rd Edition).*New York: Churchill Livingstone,1992

Annual Meeting of the American Academy of Neurology: "Update on the Neurology of Sleep" Course given at the 1997 Syllabus 146

Arendt, Hannah *The Life of the Mind* San Diego : Harcourt Brace & Company, 1978

Arey, LB, *Developmental Anatomy* revised 7th Edition Philadelphia: WB Saunders Company,1974

Adkanas D, Engle WK Sporadic Inclusion-Body Myositis and Hereditary Inclusion-Body Myopathies:Diseases of Oxidative Stress and Aging? *Archives of Neurology* 55:922-928,1998

Bach ME, Hawkins RD, Osman M, Kandel ER, Mayford M "Impairment of spatial but not contextual memory in CaMKII mutant mice with a selective loss of hippocampal LTP in the range of the theta frequency." *Cell* 1995 Jun 16;81(6):905-15

Barrett, William. *Death of the Soul: From Decartes to the Computer.* New York: Anchor Doubleday, 1986

Bartsch D, Ghirardi M, Skehel PA, Karl KA, Herder SP, Chen M, Bailey CH, Kandel ER "Aplysia CREB2 represses long-term facilitation: relief of repression converts transient facilitation into long-term functional and structural change." *Cell* 1995 Dec 15;83(6):979-92

Bear, DM, Fedio P Quantitative Analysis of Interictal Behavior in Temporal Lobe Epilepsy *Arch. Neurology* 1977;34:453-467

Behe, M *Darwin's Black Box* New York: Free Press, 1996

Benson, Frank. *The Neurology of Thinking*. New York:Oxford Univ Press, 1994

Berger RL 1990 "Nazi science -- the Dachau hypothermia experiments," *New Engl. J. Med.* 322 (20) (1990): 1435

Bever, TG & Chiarillo, RJ Cerebral Dominance in Musicians and Non-Musicians *Science* 185: 137-39 (1974)

Birge RR. Protein-Based Computers *Scientific American* 272(3):90-95. March 1995

Blum, K, Cull, J, Braverman, E., Comings, DE. (1996) Reward Deficiency Syndrome. *American Scientist* , 84,2,132-145

Blum K,Noble EP,Sheridan PJ, et al., (1990) Association of Human Dopamine D2 receptor gene in Alcoholism. *Journal of The American Medical Association* 263(15):2055-2060

Blum K, Sheridan, PJ, Wood RC, Braveman ER, Chen T, Comings DE. (1995) Dopamine $D_2$ Receptor Gene Variants: Association and Linkage Studies in Impulsive-Addictive-Compulsive Behavior. *Pharmacogenetics.* 5:121-141

Borwein, Jonathan M. and Borwein,Peter B. Ramanujan and PI. *Scientific American* 1989:112-117

Blake,Pamela Y, Pincus, Jonathan, Buckner, Cary, Neurologic Abnormalities in Murderers *Neurology* 1995;45:1641-1647.

Bolos, AM, Dean M, Lucas-Derse S. Et al., (1990) Population and Pedigree Studies Reveal a Lack of Association Between the Dopamine D2 Receptor Gene and Alcoholism. *Journal of the American Medical Association* 264:3156-3160

Brenneis CB. Belief and Suggestion in the Recovery of Memories of Childhood Sexual Abuse (Review) *Journal of the American Psychoanalytic Association.* 42(4):1027-53,1994.

Brodeur G. Schwann Cells as Antineuroblasoma Agents *New Engl J Med* 334(23):1537-8

Bronowski, J *Science and Human Values* New York:Harper & Row New York 1965

Brunner HG, Nelen M, Breakefield HH, Ropers BA, Van Oost A, *Science* 262:578 (1993)

Buchsbaum, Monte S, Spiegel-Cohen, Jacqueline, Wei, Tsechung Three Dimensional PET/MRI Images in OCD and Schizophrenia *CNS Spectrums* 1997;2(4):.26

Burenhult, Göran (ed) *The First Humans*, San Francisco: Harper "The Rise of Art" by same author P. 97-122, 1993

Capra Fritjof *The Tao of Physics.* Toronto: Bantam Books,1984

Carpenter, Malcolm B. *Core Text of Neuroanatomy* Baltimore: Williams and Wilkins Co., 1972

Carroll, Sean B., *Endlesss Forms Most Beautiful: The New Science of Evo Devo* New York, Norton and Company, 2006,

Cases O, Seif I, Grimsby J, Gaspar P, Chen K, Pournin S, Muller U, Aguet M, Babinet C, Shih JC, De Maeyer E, "Aggressive Behavior and Altered Amounts of Brain Serotonin and Norepinephrine in Mice Lacking MAOA", *Science* 268: 1763-66, June 23, 1995,

Casey D, Blitt (Eds.*): Monitoring in Anesthesia and Critical Care Medicine*.Chap. 16 Grantham CD, Hameroff SR, Monitoring Anesthetic Depth New York:Churchill Livingstone New York 1985

Ceci S.J. Bruck M. Suggestibility of the Child Witness A historical Review and Synthesis *Psychological Bulletin* 113(3):403-39, 1993 May

Ceci SJ, Loftus EF, Leichtman MD Bruck M. The Possible Role of Source Misattributions in the Creation of False Beliefs Among Preschoolers. *International Journal of Clinical & Experimental Hypnosis.* 42(4):304-320, 1994

Churchland, Paul M. *The Engine of Reason, The seat of the Soul* Cambridge, MA: MIT Press, 1995

Churchland PS *Neurophilosophy Toward a Unified Science of Mind/Brain* M.I.T. Press Cambridge, MA 1986 p.410

Churchland, PS, Sejnowski, TJ. *The Computational Brain* Cambridge, MA: MIT Press, 1992

Comings, DE (1998) The Molecular Genetics of Pathological Gambling *CNS Spectrums* 3(6), 20-37

Commella CL, Tanner CM, Ruzica RK. Polysomnographic Sleep Measures in Parkinson's Disease Patients with Treatment-induced Hallucinations" *Annals of Neurolgy* 1193 (34):710-714

Connolly TL (ed.) *Poems of Francis Thomson* Revised Edition, New York: Appleton-Century-Crofts, 1941

Crews, David. Animal Sexuality *Scientific Amercan* 270 (Jan. 1994):108-114

Crick,Francis *The Astonishing Hypothesis: The Scientific Search for the Soul*. New York: Charles Scribner's Sons,1994

Cross M, Kohrs K. *The New Milton Cross' Complete Stories of the Great Operas.* Garden City, New York: Doubleday & Company, 1955

Czeisler, CA and others. Suppression of Melatonin Secretion in Some Blind Patients by Exposure to Bright Light. *New Engl J of Med* 332(1): 6-11

Damasio, Antonio. *Decarte's Error*. New York: G.P. Putnam's Sons, 1994

DeArmond SJ, Mobley WC, DeMott DL, Barry RA, Beckstead JH, Prusiner SB. Changes in the Localization of Brain Prion Proteins During Scrapie Infection. *Neurology* 50: 1271-80, 1998

Dawkins, Richard *Climbing Mount Improbable* New York: WW Norton and Company, 1966

Degler CN, *In Search of Human Nature.* New York: Oxford Univ. Press, 1991

Dennett, DC *Consciousness Explained* Boston: Little, Brown and Company, 1991

DiLeo, Joseph H. *Child Development, Analysis and Synthesis*, New York:Brunner/Mazel, 1977

Diamond, Jared *The Third Chimpanzee* New York: Harper Collins, 1992

Diamond, Jared. *Guns, Germs, and Steel: The Fates of Human Societies.* New York: Norton, 1997

Duckett S, *The Pathology of the Aging Human Nervous System* Philadelphia: Lea & Febiger, 1991

Duke RC, Ojcius DM., Ding-E Young J Cell Suicide in Health and Disease. *Scientific American* 275 (6):80-87, 1996

Dulbecco, Renato *The Design of Life.* New Haven: *Yale* Univ. Press, 1987

Eco, Umberto *Foucault's Pendulum* New York: Ballantine Books Harcourt Brace Jovanovich, 1989

Einstein A "On The Generalized Theory of Gravitation" *The Laureates Anthology* Scientific American, 1990

Fogel, Barry S., Schiffer, Randolph B., Rao, Stephen M. (eds.) *Neuropsychiatry.* Baltimore: Williams & Wilkins, 1996

Frankel FH Discovering new memories in psychotherapy--childhood revisited, fantasy, or both? *N Engl J Med* 333(9):591-4 (1995)

Frankl, Viktor. *The Doctor and the Soul: From Psychoanalysis to Logotherapy* New York: Alfred A. Knopf, 1965

Fredrick LW, Baker, RH. *Astronomy.* New York: D. Van Nostrand Company, 1976

Freud A. Burlingham D. *Infants without Families.* New York: International University Press, 1944

Freud. S: *Character and Culture* "Edited by Philip Rieff, New York: Collier Books, 1963

Freud, S. *The Interpretation of Dreams* trans. James Strachey New York :Avon Books, 1965

Friedman, RC, Downey, JI. Homosexuality. New Engl J Med 1994. 331:923

Fuster, Joaquin M. "Frontal Lobe Syndromes" in: NEUROPSYCHIATRY, Barry S. Fogel, Randolph B. Schiffer, Stephen M. Rao (Eds.) Baltimore: Williams & Wilkins, 1996

Gardner E. *Fundamentals of Neurology: A Psychophysiological Approach*, 6th Ed. (Philadelphia, PA :W.B. Saunders and Company, 1975

Gardner, H *The Mind's New Science.* A History of the Cognitive Revolution. New York: Basic Books, 1985

Garrett L. *The Coming Plague*: *Newly Emerging Diseases in a World Out of Balance.* New York:Penguin Books, 1994

Gates W *The Road Ahead* New York: Viking Penguin, 1995

Gazzaniga MS *The Mind's Past* Berkeley, University of California Press, 1998

Gazziniga, MS (Ed.) *The Cognitive Neurosciences* Cambridge, MA: MIT Press, 1995

Gelernter D, *1939: The lost World of the Fair* New York: The Free Press, 1995

Gibbs WW. Gaining of Fat *Scientific American* August 1996

Goings, Ralph *Diner Still Life* 1977 Watercolor on paper George Sturman Gallery from *American Watercolors* by Christopher Finch New York: Abbeville Press, 1986

Goldberg, Elkhnan. "The Frontal Lobe" June 1 1990 lecture given NEUROPSYCHIATRY at Univ. Pittsburgh

Goldhagen, Daniel: *Hitler's Willing Executioners:* Ordinary Germans and the Holocaust. Alfred Knopf, New York 1996

Gould SJ The Evolution of Life on the Earth. *Scientific American* October 1994

Gould SJ, *The Mismeasure of Man* New York: WW Norton and Company, 1981.

Gregory, Richard L (ED.), *The Oxford Companion to the Mind* "Dreams in Ancient Greece" p.203 Oxford: Oxford Univ Press, 1987

Guttmann, Ludwig, and Guttman, Laurie. Axonal Channelopathies: An evolving Concept in the pathogenesis of Peripheral Nerve Disorders. *Neurology* 1996;47:18-21

Hartmann EA. Outline for a Theory on the Nature, and Functions of Dreaming: Dreaming . 6(2): 1966

Haywood, AM. Transmissible Spongiform Encephalopathies *New England Journal of Medicine* 337,1821-28, Dec. 18, 1997

Hebb,Donald. The Mind's Eye" *Psychology Today* ,1961

Heilman KM, Valenstein E (eds.) *Clinical Neuropsychology* New York: Oxford University Press, 1993

Hickman CD. *Integrated Principles of Zoology* Saint Louis :C.V. Mosby Company, 1970

Horgan "Get Smart, Take a Test" From *Science and the Citizen ,* Scientific American . 273,(5) November1995 page12-14

Huang, Z, Pruisiner SB, Cohen, FE "Scrapie Prions: A Three-dimensional Model of an Infectious Fragment" on Internet www.ceberdyne.com/~tom/prion3d_theory.html

Jaynes, Julian *The Origin of Consciousness in the Breakdown of the Bicameral Mind.* Boston: Houghton Mifflin Company,1990

Johnson, Paul. *Modern Times: The World from the Twenties to the Eighties*. New York: Harper and Row, 1983

Jourdain, Robert. *Music, The Brain, and Ecstasy* New York: Avon Books, 1997

Jung, CG *Man and His Symbols*, New York: Dell Publishing Co., 1964

Kandel ER. Genes, synapses and long-term memory. *Journal of Neurological Sciences* 134:2, 1995

Karp,BP; Juliano DM, BermanKF, Weinberger,DR.Neurologic Outcome of Patients with Dorsolateral Prefrontal Leukotomy.. *J. Neuropsychiatry Clin Neurosci*. 1992l; 4(4): 415-21

Kauffman S, *At Home in the Universe: The Search of Laws of Self-Organization and Complexity* New York: Oxford University Press, 1995

Kauffman WJ,. Smarr LL *Supercomputing and the Transsformation of Science.* New York: Scientific American Library, 1993

Kinney HC, Korein J, Panigrahy A, Dikkes P, Goode R. Neuropathological Findings in the Brain of Karen Ann Quinlan--The Role of the Thalamus in the Persistent Vegetative State *New Engl J of Med* 330:May 26, 1994

John Keats, John: *Ode to a Grecian Urn*

Lane RJM. Recurrent Coital Amnesia. *J Neurol Neurosurg Psychiatry* 260 (letter) (1997)

Lash, JP *Helen And Teacher* New York: Delacorte Press,1980

Lewis, Dorothy Otnow. *Guilty by Reason of Insanity*: A Psychiatrist Explores the Minds of Killers. New York Ballentine Publishing Group, 1998

Lewis RWB. *The Jameses.* New York: Anchor Books, 1991

Lezak, Muriel Deutsch. *Neuropsychological Assessment (3rd Ed.)* New York: Oxford Univ Press, 1995

LeDoux JE, Iwata J, Cicchetti P, et al: Different Projections of the Central Amygdaloid Nucleus Mediate Autonomic and Behavioral Correlates of Conditioned Fear. *J Neurosci.* 8:2517-2529,1998

LeDoux, JE Emotion, Memory and the Brain. *Scientific American.* 270:32-39, 1994

LeVay, Simon. *Queer Science: The Use and Abuse of Research into Homosexuality.* Cambridge, MA: MIT Press, 1966

LeVay S. A difference in hypothalamic structure between heterosexual and homosexual men. Science. 1991:253:1034-7

Lhermitte F, Human Autonomy and the Frontal Lobes. Part II: Patient Behavior in Complex and Social Situations:The "Environmental Dependency Syndrome". Annals of Neurology 19: 335-343, 1986

Lhermitte F, Pillon B, Serdaru M Human Autonomy and the Frontal Lobes. Part I, Imitation and Utilization Behavior: A neuropsychological Study of 75 Patients. *Annals of Neurology* 1986;19: 326-334

Linduist S, Patino M, Glover J, Liu J-J, Protein Particles Similar to Those Suspected in "Mad Cow" Disease Found in Yeast Cells. *Science* August 2, 1996

Lindvall O, Widner H, Rehncrona S, Brundin P, Odin P, Gustavii B, Frackowiak R, Leenders KL, Sawle G, Rothwell JC et al. Transplantation of Fetal Dopamine Neurons in Parkinson's Disease: One Year Clinical and Neurophysiological Observations in Two Patients with Putaminal Implants. *Annals of Neurology* 31(2):155-65, 1992 Feb.

Loftus, Elisibeth F.Creating False Memories. *Scientific American* 277:70-75,1997 A

Luck, Georg *ARCANA MUNDI*, Baltimore: Johns Hopkins Univ. Press, 1985

Luria, AR. *The Mind of a Mnemonist.* New York: Avon, 1968

Luria AR, Tsvetkova LS and Futer DS Aphasia in a Composer. *J. Neurol. Sci*. 2:288-92 (1965)

McEwen BS, Alves SE, Bulloch K, Weiland NG Ovarian Steroids and the Brain:Implications for Cognition and Aging *Neurology* 48:S8-S15, 1997

Magarinos AM, Garcia Verdugo JM, McEwen BS. Chronic Stress Alters Synaptic Terminal Structure in the Hippocampus. *Proc. Natl. Acad. Sci. USA* 94:14002-14008, 1997

Mahowald, Mark W Synchrony, Sleep, Dreams, and Consciousness: Clues from K-complexes. Neurology 49(4):909-911,1997

Marler P, Hamilton WJ *Mechanisms of Animal Behavior* New York: John Wiley & Sons, 1966

Marks Cara, Marks James D. Phage Libraries -- A New Route to Clinically Useful Antibodies *New Engl. J. Med* 335(10):730-733 (9/5/96)

Mayr, Ernst. *Toward A New Philosophy of Biology: Observations of an Evolutionist.* Cambridge, MA: Harvard University Press, 1988

McConkey J (ed) "Memory and Creativity" in *The Anatomy of Memory* New York: Oxford University Press,1996

Mesulam, M.-Marsel. Frontal Cortex and Behavior. *Annals of Neurology* 1986;19(4): .320-325

Mesulam, M-Marsel. Large-Scale Neurocognitive Networks and Distributed Processing for Attention, Language, and Memory. *Annals of Neurology* 1990;28(5):597-613

Mesulam, MM *Principles of Behavioral Neurology* Philadelphia: F. A . Davis and company, 1985

Moses Maimonides: *The Guide of the Perplexed* Translated by Shlomo Pines Chicago: Univ of Chicago Press,1963

Miyashita YasushiHow the Brain Creates Imagery: Projection to Primary Visual Cortex. *Science* 268:1719-20, June 23 1995

Monrad-Krohn GH "Dysprosody or Altered "Melody of Language" *Brain* 70:405-415

Moore KL. *The Developing Human.* Philadelphia: WB Saunders Co.,1973

Moore, Patrick. *The New Atlas of the Universe* New York: Arch Cape Press, 1988

Moore, RV. Editorial: Vision Without Sight *New Engl J Medicine* 332 (1) p. 54-55 Jan 5, 1995

Morgane,P.J., Jacobs, M.S., and Galaburda, A. Evolutionary Aspects of Cortical Organization in the Dolphin Brain in: *Research on Dolphins* M.M. Bryden and Richard Harrison (eds.) Oxford:Clarendon Press, 1986 p. 92

Morris, M. R. (ed.)*Larousse ASTRONOMY* Phillippe de la Cotardiere (Ed. In Chief). New York: Facts on File Publications, 1986

Morrison P, Morrison P *Powers of Ten.* New York: Scientific American Library, 1982

Multi-Society Task Force on PVS. Medical Aspects of The Persistent Vegetative State (Second of Two Parts) *New Engl J of Med* 330: June 2, 1994

Murchie, Guy. *The Seven Mysteries of Life* Boston:Houghton Mifflin Company, 1978

Nauta, Walle, JH *Fundamental Neuroanatomy* New York: WH Freeman and Company, 1986

Netter, Frank Atlas THE CIBA COLLECTION OF MEDICAL ILLUSTRATIONS: Vol. 1: The Nervous System Summit, NJ: CIBA. 1972

Nicholls JF, Martin RA, Wallace BG *From Neuron to Brain* Sunderland, MA: Sinaurer Associates, Inc, 1992

Noton D, Stark L. Eye Movements and Visual Perception In: *Physiological Psychology*. San Francisco: WH Freeman & Co, 1971

*Origins of the Human Mind* Audiocassette Edward O. Wilson, Merlin Donald, Chales J. Lumsden, Philip Lieberman Audio-Scholar 1996 Mendacino, CA

Parmentier JC, Lynch RM. "The Physics and Physiology of Labyrinthine Stimulus and Response" in: *Dizziness and Balance Disorers: An Interdisciplinary Approach to Diagnosis, Treatment and Rehabilitation* I. Kauffman Arenberg (ed.) Amsterdam:Kulger Publications, 1993

Peele, Stanton "Send in the Clones" *Wall Street Journal* March 3, 1997, P. A18

Pelligrino, Charles *Return to Sodom and Gomorrah* New York: Avon Books, 1994

Penrose, Roger. *The Emperor's New Mind* New York:Oxford University Press, 1989

Penrose, Roger. *Shadows of the Mind* New York: Oxford University Press, 1994

Pfennig, David W. And Sheman, Paul W. "Kin Recognition" Scientific American 272:(6):98-103 June 1995

Pinner RW, Teutsch, SM, Simonsen L, et al.,Trends in Infectious Diseases Mortality in The United States" and Accompanying editorials. *Journal of the American Medical Association* 275(1996):189-193

Plato's Apology of Socrates

Plum F Posner JB, *The Diagnosis of Stupor and Coma. (*3rd Edition). Philadelphia: FA Davis Company, 1982

Plunkett, Matthew J. and Ellman, Jonathan A. Combinatorial Chemistry and New Drugs. *Scientific American* April 1997;276 (4):68-73

Pollack A. The Hologram Computers of Tomorrow. *New York Sunday Times* June 6, 1991

Raine, D.J. *Albert Einstein and Relativity* East Sussex, England: Wayland Publishers Ltd.,1975 .

Rakik, P. Mode of Cell Migration to the Superficial layers of the Fetal Monkey Neocortex. *J. Comp. Neurol.* 145:61-83, 1972

Rauscher FH, Shaw GN, Ky KN, Listening to Mozart Enhances Spatial-Temporal Reasoning. Towards a Neurophysiological Basis. *Neuroscience Letters* 1995; 185(1):44-47

Raviola E, Wiesel TN. The Neural Basis of Myopia *The Harvard Mahoney Neuroscience Institute Letter* ON THE BRAIN p.1-5 Summer 1995

Ridgway, S.H., Dolphin Brain Size In: *Research on Dolphins* M.M. Bryden and Richard Harrison (eds.) Oxfore: Clarendon Press, 1986 P. 59-70

Roberts, J.M. *A Concise History of the World.* New York: Oxford University Press, 1995

Ramachandran, VS; Cobb, S; Schwartz, BJ; Bloom, FE. Noninvasive Detection of Cerebral Plasticity in Adult Human Somatosensory Cortex. *Neuroreport.* 1994 Feb 24; 5(6):701-4

Ramachandran, VS and Rogers-Ramachandran, D Synesthesia Induced in Phantom Limbs with Mirrors. *Proc. Royal Society of London B Biol Sci* 1996 Apr 22; 263(1369): 377-86

Roberts JM *A Concise History of the World* New York:Oxford University Press, 1995

Rothschild, Michael. *Bionomics:Economy as Ecosystem* New York: Henry Holt and Company, 1990

Rothstein, E *Emblems of Mind.* New York:Times Books, 1995

Rowland, LP, *Merritt's Textbook of Neurology (Eighth Edition).* Philadelphia: Lea and Febiger, 1989

Sarnat, HB, Netsky, MG *Evolution of the Nervous System* New York: Oxford Univ. Press, 1974

Schacter, Daniel L.. Searching for Memory: The Brain, The Mind, and the Past. New York: BasicBooks, 1996

Scientific American "Off with Its Head!" in Technology and Business January 1998 278: p. 41

Searle, John R. *The Rediscovery of the Mind.* Cambridge, MA: Bradford Books MIT Press, 1992

Selkoe DJ Translating Cell Biology Into Therapeutic Advances in Alzheimer's Disease Nature 399 (Suppl) A23-A31, 1999

Sevush S, Jy W, Horstman LJ, Mao W, Kolodny l, Ahn YS Platelet Activation in Alzheimer Disease. *Arch Neurology* 55:530-535 (1998)

Shephard, GM *Neurobiology* 2nd Edition Oxford Univ Press New York, 1988

Silbersweig DA, Stem E, et al. A Functional Neuroanatomy of Hallucinations in Schizophrenia. *Nature* 378: 176-79,1995

Silk. J. *The Big Bang* W.H. Freeman & Co., San Francisco: WH Freeman and Company,1980

Spencer DD, Robbins RJ, Naftolin F, Marek KL, Vollmer T, Leranth C, Roth RH, Price LH, Gjedde A, Bunney Bs. Et al Unilateral Transplantation of Human Fetal Mesencephalic Tissue into the Caudate Nucleus of Patients with Parkinson's Disease. *New England Journal of Medicine* 327(22):1541-8, 1992,

Springer S, Deutsch G. *Left Brain Right Brain* San Franscisco: WH Freeman and Co., 1981

Stein, Gertrude. *Picasso* by New York: Dover Publications, 1984

Stevens, Anthony. *Private Myths: Dreams and Dreaming,* Cambridge, MA: Harvard University Press, 1995

Stockard, James J, Stockard, Janet K., Sharbrough, Frank W. Brainstem Auditory Evoked Potentials in Neurology: Methodology, Interpretation, Clinical Application In: *Electrodiagnosis in Clinical Neurology* Michael J. Aminoff (ed.) New York: Churchill Livingstone, 1980

Swaab DF, Hofman MA. An enlarged suprachiasmatic nucleus in homosexual men. Brain Res 1990, 537:141-8

Swedo SE, Leonard HL, Garvey MA et al. Pediatric AutoimmuneNeuropsychiatric Disorders Associated with Streptococcal Infections: Clinical Descriptions of the First 50 Cases. *Am J Pshychiatry* 155 (2):264-71, 1998

Tarnas R, *Passion of the Western Mind.* New York: Ballantine Books, 1991

Terborgh J *Diversity and the Tropical Rain Forest* New York: Scientific American Library WH Freeman and Company, 1992

Tipler, Frank *The Physics of Immortality* New York: Anchor Doubleday, 1994

Townsend ,Jeanette, www.medlib.med.utah.edu/webpath/tutorial/cns/cns.html

Vallenstein, Elliot S. *Great and Desperate Cures: The Rise and Decline of Psychosurgery and Other Radical Treatments for Mental Illness.* New York: Basic Books, 1986

Vines, AE & Rees, N *Plant and Animal Biology* Vol I New York: Pitman, 1968

Waite, Terry *Taken on Trust* New York: Harcourt, Brace & Co.,1993

Waitzkin, Fred. *Searching for Bobby Fischer:* The World of Chess Observed by the Father of a Child Prodigy New York: Random House, 1988

Walston J, Silver K, Bogardus C, Knowler WC, Celi FS, Auston S, Manning B, Strosberg AD, Stern MP, Raben N, et al. Time of Onset of Non-Insulin Dependent Diabetes Mellitus and Genetic Variation in the beta-3-adrenergic Receptor gene. *New Engl J of Med.*,1995; 333(6): 343-7

Waxman SG, Geschwind N. The Interictal Behavior Syndrome of Temporal Lobe Epilepsy. Arch Gen Psychiatry 1975;32:1580-1586

Weinberger, D.R. A Connectionist Approach to the Prefrontal Cortex. *J-Neuropsychiatry-Clin-Neurosci*ence 1993; 5(3): 241-53

Wertheim N. The Amusias Chapter 10 in Critchley M (ed.)*Music and the Brain*

Wilson CL, Walter E, Steeves TA *Botany,* 5th Ed New York: Holt, Rinehart and Winston, 1971

Wilson, Sarah *Matisse* Rizzoli International Publications, Inc. New York1992 "Dance I First Version Paris 1909 Oil on Canvas" Figure 56

Whitam FL , Diamond M , Martin J.: Homosexual orientation in twins: a report on 61 pairs and three triplet sets. *Arch Sex Behav*;22(3) (1993 Jun):187-206

Yang, TT; Gallen, CC; Ramachandran, VS; Cobb, S; Schwartz, BJ; Bloom, FE. Noninvasive Detection of Cerebral Plasticity in Adult Human Somatosensory Cortex. *Neuroreport*. 1994 Feb 24; 5(6):701-4

Yasushi Miyashita How the Brain Creates Imagery: Projection to Primary Visual Cortex. *Science* 268:1719-20, June 23 1995

Valenstein, ES *Great and Desperate Cures* New York: Basic Books, 1986

Vaughn D *Memory Power* Mechanicsburg, PA Dean Vaughn, 1981

Weinberg, Steven. Life in the Universe. *Scientific American* 271:(4) p.47 October 1994

Wells HG. *The Island of Doctor Moreau*

Widner H, Tetrud J, Rehncrona S, Snow B, Brundin P, Gustavii B, Bjorklund A, Linvail O, Langsotn JW Bilateral Mesencephalic Grafting in Two Patients with Parkinsonism Induced by 1-methyl-4-phenyl-1,2,3,6-tetrahydropyridine (MPTP). *New England Journal of Medicine* 327(22):1556-63, 1992 Nov 26

Wilson JM. Adenoviruses as Gene-Delivery Vehicles *New Engl J Med* 334:1184-1187 May 2, 1996

Wilson, EO *Sociobiology* The abridged Edition Cambridge, MA:The Belknap Press of Harvard University Press,1980

Wilson, EO *Consilience: The Unity of Knowledge* New York: Alfred A. Knopf, 1998

Wordsworth, William ODE: INTIMATIONS OF IMMORTALITY FROM RECOLLECTIONS OF EARLY CHILDHOOD 1770-1850 in: Williams O (Ed) *Immortal poems of the English Language* New York: Washington Square Press, 1952

Zametkin AJ, Liebenauer LL, Fitzgerald GA, King AC, Minkunas DV, Hercovitch P, Yamada EM, Cohen RM: Brain Metabolism in Teenagers with Attention-Deficit Hyperactivity Disorder. *Arch. Gen. Psychiatry* 1993;50:333-340

Zametkin AJ, Nordahl, TE, Gross M, King AC, Semple WE, Rumsey J, Hamberger S, Cohen RM: Cerebral Glucose Metabolism In Adults with Hyperactivity of Childhood Onset. *New Engl J Med* 1990;323:1361-66

# Endnotes

[1] Selkoe DJ Translating Cell Biology Into Therapeutic Advances in Alzheimer's Disease Nature 399 (Suppl) A23-A31, 1999

[2] Adkanas D, Engle WK Sporadic Inclusion-Body Myositis and Hereditary Inclusion-Body Myopathies:Diseases of Oxidative Stress and Aging? *Archives of Neurology* 55:922-928 (1998) See also Sevush S, Jy W, Horstman LJ, Mao W, Kolodny l, Ahn YS Platelet Activation in Alzheimer Disease. *Arch Neurology* 55:530-535 (1998)

[3] See Whitam FL , Diamond M , Martin J.: Homosexual orientation in twins: a report on 61 pairs and three triplet sets. *Arch Sex Behav*;22(3) (1993 Jun):187-206 also Peele, Stanton "Send in the Clones" *Wall Street Journal* March 3, 1997, P. A18

[4] See Walston J, Silver K, Bogardus C, Knowler WC, Celi FS, Auston S, Manning B, Strosberg AD, Stern MP, Raben N, et al. "Time of Onset of Non-Insulin Dependent Diabetes Mellitus and Genetic Variation in the beta-3-adrenergic Receptor gene. "*New Engl J of Med.*, 333(6) (1995 Aug 10): 343-7 Also *Scientific American* August 1996 "Gaining of Fat" by W.Wayt Gibbs Pima Indians of Arizona living on an American Diet high in fat have the highest incidence of diabetes anywhere in the world with one half developing it by age 35 and are an average of 57 pounds heavier than their Mexican relatives. They seem to have inherited an efficient method of energy handling under adverse conditions that has resulted in their being obese under better circumstances.

[5] Crick's support for this contention is that patients who have lesions in this general region of the brain, may experience either profound apathy known as abulia, and there are others whose hands seem to move, as if of their own free will, a phenomenon known as an "alien hand". The latter is an interesting condition in which the patient is uninvolved in his own hand's movement. He doesn't will it.

It just happens. As we shall see, the big picture is that this reasoning on the basis of lesions is erroneous. A brain lesion that results in a lack of function such as the cingulate gyrus that causes abulia or alien hand, merely shows that this region of the brain is necessary for the disturbed function, not that it is sufficient or responsible for the function. See Francis Crick THE ASTONISHING HYPOTHESIS: The Scientific Search for the Soul. Charles Scribner's Sons. New York © 1994

[6] His classic book was written with Jerome Posner,MD, *The Diagnosis of Stupor and Coma,* laid the basis for evaluation of alterations of awareness. Complex motor behaviors in patients with persistent vegetative state was presented in a lecture at the meeting for The Association for Research in Nervous and Mental Disease held in New York on December 5 & 6, 1997, "Isolated Psychophysiologic Activity in the Unconscious Brain."

[Φ] Under most circumstances, the persistent vegetative state results from widespread destruction of the cerebral cortex and is not due to blocking sensory input.

[7] "Neuropathological Findings in the Brain of Karen Ann Quinlan--The Role of the Thalamus in the Persistent Vegetative State" Kinney, Hannah C., Korein Julius, Panigrahy Ashok, Dikkes Pieter, Goode Robert New Engl J of Med 330:May 26, 1994 The article stresses the role of widespread thalamic destruction here but it is clear that her brain suffered much more widespread damage particularly the Cerebral Corices.

[Φ] The inability to recall an event does not mean that it might not affect a person in other ways that are difficult to quantify, for example, subliminally. However, atropine most probably blocks the initial entry point of memory engrams into to the temporal lobe. Therefore a memory is unlikely to affect a person subliminally or emotionally.

[8] From : Casey D, Blitt (Eds.): Monitoring in Anesthesia and Critical Care Medicine.Chap. 16 Grantham CD, Hameroff SR, Monitoring Anesthetic Depth pp 427-440 Chuchill Livingstone New York 1985

[9]Roger Penrose, *The Emperor's New Mind* (New York:Oxford University Press, 1989)382-384

[Φ] This close control of REM vs non-REM states occurs in elderly patients and those with Parkinson disease, particularly those treated with L-Dopa. Patients have fragmented sleep and possiibly REM sleep intrusions causing hallucinations. See Commella CL, Tanner CM, Ruzica RK. "Polysomnographic Sleep Measures in Parkinson's Disease Patients with Treatment-induced Hallucinations". *Annals of Neurolgy* 1193 (34):710-714

[10] For further discussion see Mahowald, Mark W Synchrony, Sleep, Dreams, and Consciousness: Clues from K-complexes. Neurology 49(4):909-911 1997

Ψ Technically recent work has shown that this waking electrical activity of the cortex, far from being desynchronized, is actually still synchronized by diffusely projecting rhythm making thalamic nuclei, which simply drive the cortex at faster frequencies.

[11] From Shepherd, Gordon NEUROBIOLOGY Second Edition Oxford Univ Press, New York 1988 Page 525.

[12]"Know, thou who studiest this Treatise: if you are of those who know the soul and its powers and have acquired true knowledge of everything as it really is, you already know that imagination exists in most living beings. As for the perfect animal, I mean the one endowed with a heart, the existence of imagination in it is clear. Accordingly, man is not distinguished by having imagination; and the act of imagination is not the act of the intellect but rather its contrary. For the intellect divides the composite things and differentiates their parts and makes abstractions of them, represents them to itself in their true reality and with their causes, and apprehends from one thing to very many notions, which differ for the intellect just as two human individuals differ in regard to their existence for the imagination. It is by means of the intellect the universal is differentiated from the individual, and no demonstration is true except by means of universals." Maimonides goes on to say that things apprehended by intellect are more remarkable by far than anything imaginable!! The truth goes well beyond our ability to imagine it. Who could imagine that with two persons living on opposite sides of the sphere which is our earth one should not fall off, yet this is in fact the case. Moses Maimonides: THE GUIDE OF THE PERPLEXED Translated by Shlomo Pines Univ of Chicago Press, Chicago Page 209-210

[13]Hamlet and Socrates treat sleep and death in the same vein. See in "To be or not to be" Soliliquy: To die-to sleep- No more; and by a sleep to say we end The heartache, and the thousand natural shocks That flesh is heir to. 'Tis a consummation Devoutly to be wish'd. To die-to sleep. To sleep-perchance to dream: ay there's the rub! For in that sleep of death what dreams may come When we have shuffled off this mortal coil, Must give us pause. ..." Act III, Scene I See also Plato: The apology of Socrates: "Let us reflect in another way, and we shall see that there is great reason to hope that death is a good, for one of two things: either death is a state of nothingness and utter unconsciousness, or, as men say, there is a change and migration of the soul from this world to another. Now if you suppose that there is no consciousness, but a sleep like the sleep of him who is undisturbed even by the sight of dreams, death will be an unspeakable gain. For if a person were to select the night in which his sleep was disturbed even by dreams, and were to compare with this the other days and nights of his life, and then were to tell us how many days and nights he had passed in the course of his

life better and more pleasantly than this one, I think that any man, I will not say a private man, but even the great king, will not find many such days or nights, when compared with the others. Now if death is like this, I say that to die is gain; for eternity is then only a single night...."

[14]See Gregory, Richard L (ED.), THE OXFORD COMPANION TO THE MIND "Dreams in Ancient Greece" p.203 Oxford Univ Press Oxford 1987

[15]See Luck, Georg *ARCANA MUNDI*, Johns Hopkins Univ. Press, Baltimore, 1985 pps 166-168. I was surprised to read in a biography THE JAMESES by R.W.B. Lewis Anchor Books, New York, 1991 that the great psychologist William James was a believer in necromancers in his own day, and seems to have placed some trust in a certain Mrs. Leonora Piper, a famous medium. Otherwise incisive and logical thinkers may not be immune to the influence of the occult. (See p. 491)

[16]"It is on such evidence that psychologists assume the existence of an unconscious psyche-though many scientists and philosophers deny its existence. They argue naively that such an assumption implies the existence of two "subjects," or (to put it in a common phrase) two personalities within the same individual. But this is exactly what it does imply-quite correctly. And it is one of the curses of modern man that many people suffer from this divided personality. It is by no means a pathological symptom; It is a normal fact that can be observed at any time and everywhere." I would agree but respectfully suggest that the word 'unconscious' seems to imply that the person has lost consciousness rather than entering a state of altered awareness. From CG Jung *MAN AND HIS SYMBOLS* Dell publishing, New York 1964, p.5

[17]This is fairly obvious for most of us but yet is of profound theoretical importance and frequently mentioned by both Freud and Jung See Sigmund Freud, *The Interpretation of Dreams* trans. James Strachey (New York :Avon Books, 1965) and C.G. Jung, *Man and His Symbols*, (New York: Dell Publishing Co. 1964)

[Φ] See, for example, Sally Springer, Georg Deutsch, *Left Brain Right Brain* (San Franscisco: WH Freeman and Co., 1981) Just one of a number of popular books on the subject written by psychologists.

[18] Richard L. Gregory and O.L. Zangwill, Eds., *The Oxford Companion to the Mind* (New York :Oxford University Press, 1987) 329

[19] See Dorothy Otnow Lewis (1998)*Guilty by Reason of Insanity* A Psychiatrist Explores the Minds of Killers Ballentine, New York

[20] See "Medical Aspects of The Persistent Vegetative State" (Second of Two Parts) by the Multi-Society Task Force on PVS New Engl J of Med Vol 330 June 2, 1994

[21] See Pinner RW, Teutsch, SM, Simonsen L, et al., "Trends in Infectious Diseases Mortality in The United States" and Accompanying editorials. *Journal of the American Medical Association* 275(1996):189-193

[22] Quote of David Liddle, Santa Fe Institute which supports nature-inspired research, Wall Street Journal, Tuesday January 16, 1996 P. 1 **"Back to Darwin** In Sunlight and Cells, Science Seeks Answers to High-Tech Puzzles by Gautam Naik

By Gautam Naik

[23] Figures taken from:Hickman Cleveland D. *Integrated Principles of Zoology (*Saint Louis :C.V. Mosby Company, 1970 ). 58 (mito) and Wilson, Carl, Loomis Walter E, Steeves, Taylor A. *Botany,* 5th Ed (New York: Holt, Rinehart and Winston, 1971).110 (choloroplast)

[24]Adapted from Vines, AE & Rees, N *Plant and Animal Biology* Vol I Pitman New York 1968 p. 187

[Ψ] Birdsong is obviously for male listeners as well, useful to mark out territory and discourage other males.

[25]See "Animal Sexuality" by David Crews *Scientific Amercan* 270 (Jan. 1994):108-114

[26] See SOCIOBIOLOGY by Edward O. Wilson The abridged Edition The Belknap Press of Harvard University Press Cambridge, MA 1980 p.225-228 and Ridgway, S.H., Dolphin Brain Size In: RESEARCH ON DOLPHINS M.M. Bryden and Richard Harrison (eds.) Clarendon Press, Oxford 1986 P. 59-70

[27] Morgane,P.J., Jacobs, M.S., and Galaburda, A. Evolutionary Aspects of Cortical Organization in the Dolphin Brain in: RESEARCH ON DOLPHINS M.M. Bryden and Richard Harrison (eds.) Clarendon Press, Oxford 1986 p. 92

[28] ibid p.94

[29]Churchland PS *NEUROPHILOSOPHY Toward a Unified Science of Mind/Brain* M.I.T. Press Cambridge, MA 1986 p.410

[τ] You might argue that the refractory period for a neuron, mentioned above is merely the inverse of the switching frequency of a Silicon element, but strictly speaking this isn't so. Just some of the arguments to be taken into consideration are, the relative state of excitation of the neuron, relative and absolute refractory periods and the influence of an enormous number of inputs from other nerve cells.

[30] See Rotaxanes: These organic molecules may serve as nanoswitches SCIENTIFIC AMERICAN August 1994 Volume 271 Number 2 Page 87, Also

Protein-Based Computers by Robert R. Birge SCIENTIFIC AMERICAN March 1995 Volume 272 Number 3 Pages 90-95

[31] Silicon, the basis of computer semiconductor structure and Carbon on which all biological molecules is based, belong to the same family of elements capable of forming four separate chemical bonds. Curiously old science fiction films often featured beings from outer space that had a Silicon based biology. Both elements seem somehow peculiarly suited for information storage and in my view combinations of Carbon and Silicon based information storage machines is inevitable.

[32] See Bill Gates THE ROAD AHEAD Viking Penguin, New York 1995.

Bill Gates provides one vision of the future, with every person being connected with literary, entertainment, financial and other information through information appliances.

[33] From Stockard, James J, Stockard, Janet K., Sharbrough, Frank W. Brainstem Auditory Evoked Potentials in Neurology: Methodology, Interpretation, Clinical Application IN: ELECTRODIAGNOSIS IN CLINICAL NEUROLOGY Michael J. Aminoff (ed.) Churchill Livingstone, New York 1980 P. 371

* It is estimated that there are 300,000 neurons under each square millimeter of visual cortex and about half this number beneath the same area elsewhere in the brain (Nauta, Walle, JH *Fundamental Neuroanatomy* WH Freeman and Company, New York 1986

# Peroxisomes, mitochondria, and Chloroplasts are all felt to have originated as invaders, remnants of parasitic or symbiotic organisms, in a theory popularized by biologist Lynn Margulis. This must have happened very early in evolution in cells ancestral to nearly every current animal and plant cell. Mitochondria and Chloroplasts contain some of their own genetic material. As generations pass some DNA of mitochondria and chloroplasts is passed into the nucleus of the cell. Peroxisomes must have been intracellular for so long, they no longer contain any DNA of their own and their structure is specified by the cell nucleus.

How these organelles might have gotten into the primeval cell is another interesting question. One theory is that the first anaerobic (non-oxygen using) cells were ameba like and just engulfed primitive organelle containing single celled organisms. At that time, billions of years ago there were probably no multicelled organisms yet, or at least none with differentiated cells. Primitive single celled parasites very likely abounded and was frequent invaders of other cells. These invading intra-cellular parasites, the forerunners of other intracellular invaders we know today as mycoplasma, Rickettsiae likely evolved into symbionts or commensals of the single cell organisms that evolved into today's plants and animals all of which contain the remnants of tiny parasites from billions of years ago.

Ψ Curare is the same as a lethal poison derived from plants shot from darts by South American Indians. It paralyzes muscles by blocking the effect of ACh on muscle membranes that can block respiratory muscle to cause death. Nerve gases result in the accumulation of ACh so you want to block it, but the problem is to determine how much might be a life-saving dose and what dose would cause a complete block and thus be lethal.

[34] See Dulbecco, Renato THE DESIGN OF LIFE Yale Univ. Press, 1987. p349Ackerman, Diane A NATURAL HISTORY OF THE SENSES Random House, New York 1990 , p169

[35] Guttmann, Ludwig, and Guttman, Laurie "Axonal Channelopathies: An evolving Concept in the pathogenesis of Peripheral Nerve Disorders." Neurology 1996;47:18-21

[36] See Laurie Garrett THE COMING PLAGUE: Newly Emerging Diseases in a World Out of Balance. Penguin Books, New York © 1994 especially pages 411-457

[37] See Marks Cara, Marks James D. "Phage Libraries -- A New Route to Clinically Useful Antibodies" NEJM 335(10):730-733 (9/5/96)

Φ New evidence points to the ends of chromosomes, the so-called telomeres, proteins that are telomerases as one factor, at least, that inscribes as specific number of possible doublings within a given cell line. Understanding of this process may allow manipulation that may make further doublings possible or conversely control certain malignant growths where defects in telomerase is expressed causing uncontrolled growth.

[38] There is a conflict as to whether this stream of consciousness which certainly seems real enough to most of us, is, in fact real, or an illusion, part of the characteristic way that events are recollected in our minds which is not at all an accurate picture compared to the way events actually occur. This is referred to as the "virtual machine" of the brain, the way the brain perceives the flux of time and reconstructs reality by Daniel Dennett see Scientific American Profile: Daniel C. Dennett 274(2) February 1996 p. 34 by Tim Beardsley and CONSCIOUSNESS EXPLAINED by Daniel C. Dennett Little, Brown and Company, Boston © 1991

Φ Consciousness and experience are linear just as computer processing in a microprocessor is linear. This linearity of consciousness makes our conception of time which we also tend to appreciate on a linear scale which may not be entirely accurate, by the way when we actually come to consider the effect of time on physical processes, especially those that are far different from the ordinary experience in our own macroscopic world.

[ℵ] See Wilson, James, M. Adenoviruses as Gene-Delivery Vehicles in New England Journal of Medicine 334:1184-1187 May 2, 1996. This is a good start but unfortunately does not cure the condition. Infected cells are removed by the patient's immune system which then prevents reinfection by the same virus. Somehow the gene (more precisely CFTR for the cystic fibrosis transmembrane conductance regulator) has to be administered so as to have a permanent effect.

[∅] This has been a theme in science fiction. One good example is The Island of Dr. Moreau by H.G. Wells in which the diabolical Dr. Moreau retreats to an island paradise in which he is all powerful creating hideous beings, part men and part beast, to whom he is a great God as creator and controller, and maker of the law. The story is a metaphor for the animal and human elements in human behavior.

[39] John G. Nichols, A. Robert Martin, Bruce C. Wallace, *FROM NEURON TO BRAIN A Cellular and Molecular Approach to the Function of the Nervous System,* 3rd Ed,. (Sunderland MA :Sinauer Asssociates, Inc,1992), 18

[40] Hawkins and Kandel In: *Neurobiology of Learning and Memory*, ed. G. Lynch, J.L. McGaugh, and N. M. Weinberger. (New York :Guilford Press, 1984)

[Φ] Antonio Damasio has refreshingly in his book, DECARTES ERROR argued that strict localization of consciousness within the brain, and even trying to find a organ of the will within the brain as Decartes tried to do (postulating that the Pineal gland performed this function) makes no logical sense. Damasio's view in capsule form is that body and brain work as a whole, to produce all the phenomena we speak of as being part of consciousness. As a neurologist, he speaks very little about anything that a biological knowledge can't explain. His message is that all function cannot be localized and is not localized only in the brain.

Also "...the latest finding in physiology suggests that *the mind* doesn't really dwell in the brain but travels the whole body on caravans of hormone or enzyme, busily making sense of the compound wonders we catalogue as touch, taste, smell, hearing, vision." From Diane Ackerman A NATURAL HISTORY OF THE SENSES © 1990 Random House, New York

[Ψ] from THE PORTRAIT OF AN ARTIST AS A YOUNG MAN. We rekindle old lamps long extinguished, obviously what we mean when we are inspired or in love. One of the deepest experiences is to rediscover and experience old emotions that have been put away. Of course, this process can be traumatic as well, as when fears are brought to the surface in psychotherapy, but for most of us uncovering a nearly dead emotion form our youth is a deep positive experience. In the stricter sense, these "lights" refer to cognitive and emotional abilities not experiences and emotions.

[41] Hannah Arendt *THE LIFE OF THE MIND* (San Diego : Harcourt Brace & Company, 1978), 129. Her quote is from Timaeus, 90c

See also discussion in Richard Tarnas *THE PASSION OF THE WESTERN MIND* (New York: Ballentine Books, 1991), 211-218

[Φ] Humans, by virtue of their ability to think, are held to be created in the image of God that is reflect the divine.

[42] See Berger RL 1990 "Nazi science -- the Dachau hypothermia experiments," *New Engl. J. Med.* 322 (20) (1990): 1435

[43] From: ODE: INTIMATIONS OF IMMORTALITY FROM RECOLLECTIONS OF EARLY CHILDHOOD 1770-1850 *Immortal poems of the English Language* Washington Square Press, New York (1952) p262

[44] Umberto Eco *Foucault's Pendulum* Ballantine Books New York (1989) Harcourt Brace Jovanovich, Inc.p.532

[45] Figure taken from NEUROPSYCHIATRY Barry S. Fogel, Randolph B. Schiffer &Stephen M. Rao (eds.) Williams and Wilkins, Baltimore ©1986, p.374

[46] Memory consolidation utilizes a molecular switch that turns on and off protein production, so-called CREB, cyclic AMP responsive element binding protein. CREB activates protein transcription inside the neuron. See Journal of Neurological Sciences Vol. 134, Nos. 1-2 December 1995 p.2 "Genes, synapses and long-term memory by Eric R. Kandel

[47] See Bach ME, Hawkins RD, Osman M, Kandel ER, Mayford M "Impairment of spatial but not contextual memory in CaMKII mutant mice with a selective loss of hippocampal LTP in the range of the theta frequency." Cell 1995 Jun 16;81(6):905-15 also Bartsch D, Ghirardi M, Skehel PA, Karl KA, Herder SP, Chen M, Bailey CH, Kandel ER "Aplysia CREB2 represses long-term facilitation: relief of repression converts transient facilitation into long-term functional and structural change." Cell 1995 Dec 15;83(6):979-92

[#] The work of Joseph E LeDoux and colleagues emphasizes a vigorous connectivity between the hippocampus and emotional and visceral areas of the brain, especially the amygdala. He suggests that conscious memory is laid down in parallel and simultaneous with emotional memory. See Emotion, Memory and the Brain by Joseph LeDoux Scientific American June 1994 p. 50-57

[48] Papez, JW, "A Proposed Mechanism of Emotion." *Arch Neurol Psychiatry* 38:725-43 (1937)

[49] Kluver H, and Bucy PC, "Psychic Blindness" and Other Symptoms following Bilateral Temporal Lobectomy in Rhesus Monkeys" *Am. J. Physiol.* 119:352-353 (1937)

[50] From: Gordon M. Shepherd *NEUROBIOLOGY* 2nd Edition. Oxford University Press New York 1988 page 575

[51] See Michael Waldholz "Panic Pathway: Study of Fear Shows Emotions Can Alter 'Wiring' of the Brain" The Wall Street Journal Wednesday Sept. 29, 1993 Scientists at the Yale VA Medical Center are working on the connection between traumatic experiences and memory in tests on laboratory animals and observations with patients with PTSD.

[52] Viktor Frankl THE DOCTOR AND THE SOUL From Psychoanalysis to Logotherapy (New York: Alfred A. Knopf, 1965) 73-74

[53]Serge Duckett, *The Pathology of the Aging Human Nervous System* Lea & Febiger, Philadelphia 1991, p130 Numerous pathological studies find decreased cholinergic output in widespread areas of brain and in the enzyme CAT, choline aceytyl tranferase that produces acetylcholine.

[ϕ] This is a lot like pavor nocturnus, panic disorder and post traumatic stress above, in which, high emotion, is amputated, occurs on its own and split from the inciting stimulus. Presumably, the successful therapist will re-establish the broken connection between the specific memory and the high emotion, at least in the latter two conditions.

[54] See Philip J. Hilts NY Times "In Research Scans, Telltale Sighs Sor False Memories from True" study Published in Neuron*, probably July or August 1996 See also Loftus, Elisibeth F. "Creating False Memories" *Scientific American* 277: Sept. 1997

[55] James McConkey makes this excellent point, equating memory and imagination, in his anthology THE ANATOMY OF MEMORY. He, in turn, attributes some of these ideas to the philosopher Mary Warnock. See "Memory and Creativity" in THE ANATOMY OF MEMORY James McConkey (ed) Oxford University Press, New York © 1996 P. 123-124

[56] See same anthology "Mathematical Creation" by Henri Poincare p,136-142.

[57] See Lane RJM. Recurrent Coital Amnesia. J Neurol Neurosurg Psychiatry 1997;260 (letter)

[58] Adapted from *THE TAO OF PHYSICS* by Fritjof Capra Bantoam Books Toronto © 1984 Page 224. Capra gives a lucid discussion of Feynman diagrams

which describe particle collisions and the creation and destruction of particles. The major point to be made in our discussion revolves about the reversibility of time on these physical scales.

[Ψ] In sleep we are dealing with a special and fascinating case of dreams with mental activity occurring even though the person is not awake. Sleep is special in that one is cut off from one's environment yet, not unconscious. The cerebral cortex is active, yet isolated, and the goings on in sleep capture the attention of the sleeping person. The sleeping brain, unlike the brain in deep coma, is active, especially in dream sleep, only it is cut off from the outside world. One example--a normal person in REM or dream sleep is unable to move and thus is cut off motorically unable to influence his environment.

[#] No one has been able to show that hypnosis or any chemical ("truth serum") enhances memory in any way. Facts obtained via hypnosis are quite often inaccurate and such recollections could have easy been gotten via simpler means. Hence the whole "science" of hypnotic memory enhancement, for example by police hypnosis experts and therapists is called into question. For example, how accurate or distorted are childhood memories of abuse brought out through hypnosis? Should this information be admitted in court? What about certain memories of Abduction or past lives as recounted by subjects. This is a whole vast literature in itself. See Frankel FH Discovering new memories in psychotherapy--childhood revisited, fantasy, or both? N Engl J Med 1995 Aug 31;333(9):591-4 John Mack *Abduction*

[59] From *Poems of Francis Thomson* Revised Edition, Appleton-Century-Crofts, New York, 1941 Rev. Terence L. Connolly (Ed.)

[Φ] Mnemosyne was the Greek Goddess of memory and the mother by Jupiter of the nine Muses. The daughter of Uranus and Ge, heaven and earth, she had a high station in the pantheon.

[60] See Memory Power by Dean Vaughn © 1981 Mechanicsburg, PA, apparently self-published. One of may similar memory systems.

[61] *Gardner, Howard *The Mind's New Science.* A History of the Cognitive Revolution Basic Books New York 1985 p.283 Presentation of views of Karl Pribram. See also Pribram, KH (1982) "Localization and Distribution of Function in the Brain" in: J. Orbach, ed., *Neuropsychology After Lashley.*

[62] See New York Sunday Times June 6, 1991, Andrew Pollack "The Hologram Computers of Tomorrow"

[63] Mayr, Earnst. *Toward A New Philosophy of Biology: Observations of an Evolutionist.* Harvard University Press, Cambridge MA © 1988 P. 16-17

[64] See Tipler, Frank J. PHYSICS AND IMMORTALITY Anchor Books Doubleday New York © 1994 Pages 21-35

[65] See Duke Richard C, Ojcius David M., Ding-E Young John "Cell Suicide in Health and Disease Scientific American 275 no 6: Dec. 1996 P 80-87

[66] For a fuller description see Ambros IM, Zellner A, Roald B et al. "Role of Ploidy, Chromosome 1p, and Schwann Cells in the Maturation of Neuroblastoma" NEJM 334 (23):1505-1511 6/6/96 also an Editorial by Brodeur G "Schwann Cells as Antineuroblasoma Agents" 1537-8 in same issue. Here I wish only to make the point that adjacent cells induce maturation and determine form.

[67] From Arey, Leslie Brainerd DEVELOPMENTAL ANATOMY revised 7th Edition WB Saunders Company Philadelphia 1974 P. 202

[68] ibid. previous figure

[69] Endlesss Forms Most Beautiful: The New Science of Evo Devo by Sean B. Carroll 2006, Norton and Company, New York

[ϕ] This is not to say that men evolved from frogs or birds, but share some common features. We do come down from ancient ancestors of these lineages. Mammals and birds perhaps diverged from a primitive reptilian form.

[70] See Tipler, Frank *The Physics of Immortality* Anchor Doubleday, New York © 1994

[71] See Stephen Jay Gould "The Evolution of Life on the Earth" Scientific American October 1994 "Life in the Universe" p. 85-91. That the advent of humankind is more accidental than inevitable and the theme that humans are by many criteria not really more advanced than other living forms but only a small constituent of the evolutionary theme is pervasive in Gould's books.

[72] See Stuart Kauffman AT HOME IN THE UNIVERSE Oxford University Press, New York © 1995 also Michael Behe DARWIN'S BLACK BOX Free Press, New York © 1996

[Φ] Very little is known about what drives many obvious asymmetries, placement of abdominal organs such as stomach, liver and spleen, the great vessels such as aorta, subclavian arteries, right and left heart, and subtle differences in the brain that determine language function and handedness or even what causes a curious condition, situs inversus in which the heart and great vessels are inverted as can be seen on a chest x-ray.

[73] From Arey, Leslie Brainerd DEVELOPMENTAL ANATOMY revised 7th Edition WB Saunders Company Philadelphia 1974 p454

[74] Figure from Malcolm B.Cartpenter CORE TEXT OF NEUROANATOMY Williams and Wilkins Company, Baltimore © 1972 P. 39

[75] From Arey, Leslie Brainerd *Developmental Anatomy* revised 7th Edition WB Saunders Company Philadelphia 1974. P480

[76] From Carpenter, Malcolm B. *Core Text of Neuroanatomy* Williams and Wilkins Co. Baltimore 1972 p. 80

[77] From Arey, Leslie Brainerd *Developmental Anatomy* revised 7th Edition WB Saunders Company Philadelphia 1974 p408

[Ψ] Part of the cerebellum may be pulled outside the skull into the upper neck or, even worse, the central canal of fluid or even ventricles might obstruct. The spinal fluid being constantly produced may be trapped and hydrocephalus can easily develop. A canal may form in the center of the cord that fills up with fluid causing numbness and paralysis in the arms and below the neck. This is rather fancifully termed Syringomyelia, after the maiden Syrinx who being pursued and overtaken by Pan, refused to give up her virginity, and was turned suddenly into a flute like hollow instrument made of reed.

[78] Diagram taken from John G. Nichols, A.Robert Martin, Bruce G. Wallace *From Neuron to Brain* 3rd Edition Sinauer Assiciates, Inc. Publishers, Sunderland MA © 1992, P. 345

[∅] Both Glia and neurons are derived from neuroectoderm and the cells are very similar in structure with elongated processes. Glia are not excitable cells but may very well serve functions that we traditionally reserve for neurons, including a mechanism for recording of memories in their own RNA and protein production, which is point of some scientific speculation.

[79] See Rakik, P. Mode of Cell Migration to the Superficial layers of the Fetal Monkey Neocortex. *J. Comp. Neurol.* 145:61-83, 1972

[80] From: M-Marsel Mesulam PRINCIPLES OF BEHAVIORAL NEUROLOGY F.A. Davis Company, Philadelphi © 1985 P. 13

[81] See GREAT AND DESPERATE CURES by Elliot S. Vallenstein 1986 Basic Books, New York.

[82] See The Mismeasure of Man, WW Norton and Co, New York., 1981

[Ψ] To complicate matters still further, there is information derived from a number of sources now, finding a slight negative correlation with intellectual and social ability and acceleration of gross motor milestones. Anna Freud (Freud A. and Burlingham D. Infants without Families. New York: Intern. Univ. Press, 1944 as cited in Child Development, Analysis and Synthesis, DiLeo, Joseph H. (M.D.) Brunner/Mazel, Publishers New York, 1977) was one of the first persons to find accelerated gross motor performance in children raised under adverse social conditions and its been found that children of parents of lower socio- economic status may attain gross motor milestones more rapidly.

[v] An alternative view is that the eye developed very early in evolution and that the rod and cone receptor cells, that contain portions very much like cilia, are actually derived from ciliated cells in the center of the neural tube. This view is probably accurate. The neural tube originally came from surface cells that ended up inside the animal as a streak invaginated to form a hollow tube, the precursor of our central nervous system (see "Beginnings"). At that point there were light sensitive cells and connecting neurons that crossed to the opposite muscular wall of the animal to control primitive movements in response to light changes. It was only later that parts of the tube close to the surface along with their connecting neurons that were external to light sensitive cells, actually formed a whole eye in evolution. This accounts for the ciliary origin of rods and cones whose structure is close to modified cilia, and for all the connections between them.

[83]See Sarnat, HB and Netsky, MG EVOLUTION OF THE NERVOUS SYSTEM Oxford Univ. Press, New York 1974 p. 150

[84] See Elio Raviola, and Torsten N. Wiesel, "The Neural Basis of Myopia" *The Harvard Mahoney Neuroscience Institute Letter* ON THE BRAIN p.1-5 Summer 1995

[85]From : Nicholls JF, Martin RA, Wallace BG FROM NEURON TO BRAIN Sinaurer Associates, Inc Sunderland, MA 1992 p. 565

[86]See: "Protein-Based Computers" by Robert R. Birge Scientific American 272(3):90-95. March 1995

[87]From Shephard, GM Neurobiology 2nd Edition p.339 Oxford Univ Press 1988 New York

[88]From : David Noton and Lawrence Stark "Eye Movements and Visual Perception" in Physiological Psychology WH Freeman & Co San Francisco 1971 p.265

[89] Picture courtesy of Vestibular Disorders Association Internet site.

[µ] Remarkably, the hair cells of the inner ear, both vestibular and auditory portions, and the retinal rods and cones all have a common origin as cilia invested sensory neurons.

[90]From Shephard, GM *Neurobiology* 2nd Edition Oxford Univ Press New York, 1988 P214

[91]From : Churchland, Patricia S. & Sejnowski, Terrence J. THE COMPUTATIONAL BRAIN The MIT Press, Cambridge, MA 1992 p.151

* Some might object to the concept of lower and higher animals. Evolution and embryology allow us to define these terms. Higher animals are whose brains (and other organs which pretty much evolve in parallel) evolved from lower forms. You can see this development in comparing brains and following evolution and also in the embryological development of the more advanced animal, which go through many of the same evolutionary stages. This is seen in the visual system where in lower animals most of the retinal input goes to the midbrain, whereas in more advanced animals especially man, visual input is sent to the thalamus on its way to the cerebral visual cortex, something far less developed in lower forms.

# There is a medial geniculate nucleus in the auditory pathway.

Ψ In some other animals that have eyes on the sides of their heads, fish, some rodents etc. mostly prey rather than predators, vision from the two eyes does not overlap very much. They lack stereoscopic vision. In these animals the optic nerves may cross to the opposite side completely which is termed total decussation. In the most extreme examples of stereoscopic animals with eyes in front of the head, man is an example, the optic nerves are hemidecussated. Because of how development (and evolution) works there is no such thing as an animal with two laterally placed eyes that is not decussated at all.

[92] Image courtesy http//hendrix.imm.dtu.dk/image/index.html

Ψ Simultanagosia, the inability to go to an object in a scene with your eyes, *optic apraxia,* the inability to reach without overshooting at an object in your visual field using your eyes as guide, so-called *optic ataxia,* constitute the three components of Balint's syndrome

[93]From: M-Marsel Mesulam *Principles of Behavioral Neurology* F. A . Davis and company, Philadelphia 1985 P. 277

[94]See Yasushi Miyashita "How the Brain Creates Imagery: Projection to Primary Visual Cortex. Science 268:1719-20, June 23 1995

ϕ In essence the ability to see and knowledge that one can see, and knowledge about what one is seeing are no longer congruent. Given the arguments above it is easy to see why this is so. The various visual functions reside in different modules and the language areas of the brain in the dominant hemisphere express conscious perception of the visual object. The obvious way to test whether someone is aware of seeing something by asking him. The incongruity between vision and expressed awareness is agnosopsia - that the subject is unaware he can see e.g. so-called blind sight, and gnosanopsia -the knowledge that one is blind. The latter is relevant to Anton's syndrome where a subject is blind but does not know it. Given that only the language areas of the brain express this "knowledge", it is theoretically possible that other regions of the brain are actually aware but are inarticulate. This knowledge is due to a disconnection with language areas in the left hemisphere. This discussion of the incongruity of sight and knowledge of sight was brought to my attention in a lecture of Antonio Damasio, at the 12/97 meeting of the American Association of Nervous and Mental Disease in New York.

[95] The retinal connections to the hypothalamus are called the retinothallamic tract. The suprachiasmatic nucleus in the hypothalamus most likely generates circadian rhythms and connects with the paraventricular nucleus, then to the sympathetic nervous system via the superior cervical ganglion, with the pineal gland, the endocrine gland that secretes melatonin, implicated in sleep-wake cycles. See editorial "Vision Without Sight by Robert V. Moore, New England Journal of Medicine 332 (1) p. 54-55 Jan 5, 1995 and in the same issue "Suppression of Melatonin Secretion in Some Blind Patients by Exposure to Bright Light by C.A. Czeisler and others pps. 6-11

[96] Ramachandran, VS and Rogers-Ramachandran, D Synesthesia Induced in Phantom Limbs with Mirrors. Proc. Royal Society of London B Biol Sci 1996 Apr 22; 263(1369): 377-86 also Yang, TT; Gallen, CC; Ramachandran, VS; Cobb, S; Schwartz, BJ; Bloom, FE. Noninvasive Detection of Cerebral Plasticity in Adult Human Somatosensory Cortex. Neuroreport. 1994 Feb 24; 5(6):701-4

[Ψ] Anosognosia: From gnosis=awareness, noso=disease, a=not hence unawareness of disease or deficit.

[Ψ] In its most extreme form deafferentiation is total loss of sensory contact with one's environment or at least unawareness of one's surroundings. Such extreme forms of non-contact are actually common in certain trances that insulate a person from environmental stimuli, in sleep and coma. See Chapter one for discussion of deafferntiation in relation to the persistent vegetative state

[97]From: *HELEN AND TEACHER* by Joseph P. Lash Delacorte Press1980 p. 542

[98]Edward Rothstein *EMBLEMS OF MIND* Times Books New York 1995 P. 12-13

[99]Fred Waitzkin SEARCHING FOR BOBBY FISCHER: The World of Chess Observed by the Father of a Child Prodigy Random House New York ©1988

[+] Everyone who loves music should see Leonard Bernstein's final performance of Beethoven's ninth symphony with the Berlin Philharmonic "Bernstein in Berlin", an astounding feat for both men (The Ninth was Beethoven's final public performance as well). You have the chance to see Bernstein, old and infirm, come alive with feeling that spans centuries. When you consider the historical implications as well, we have a deep statement healing and brotherhood as well. It's hard to believe a spirit such as Bernstein's or Beethoven's ever dies.

[100] Terry Waite TAKEN ON TRUST Harcourt, Brace & Co. New York 1993

[101] Albert Einstein "On The Generalized Theory of Gravitation" THE LAUREATES ANTHOLOGY 1990 Scientific American p. 1

[102] Patrick Moore THE NEW ATLAS OF THE UNIVERSE Arch Cape Press, New York 1988 p.194 quasar "20051-279 is thought to be about 13,000 light years away." But estimates of quasar distances are made on the basis of Doppler red shifts, in other words on the idea that since the Big Bang the speed that a very distant object is moving away from us is proportionate to the distance of the object and these estimates may be totally off.

[103] Herein we may find the answer to the question: Why was our universe made in just the way we find it? Why is it not slightly different than it is? Just to tease us, just so that there will always be something just beyond the limits or our understanding, in other words things are just as they are, because we, sapient beings, inhabit the universe. Another way of stating what has been called "The anthropic cosmological principle."

[104] "The most important fundamental laws and facts of physical science have all been discovered, and these laws are so firmly established that the possibility of their ever being supplanted in consequence of new discoveries is extremely remote." As quoted in *The Seven Mysteries of Life* by Guy Murchie Houghton Mifflin Company, Boston 1978, p 573

[Ψ] PET=Positron Emission tomography, SPECT=Single Photon emission computerized tomography, MRI=Magnetic Resonance Imagery. These are all imaging tests that allow researchers to monitor activity, mostly in the form of blood flow and metabolism of various brain regions as tasks are performed. It is not important to know exactly what they are or how these tests work here for purposes of discussion, only that they are tools for gauging the activity of parts of the brain.

[105] For the general concept of looking at the known scales of objects, illuminating to me was *POWERS OF TEN* by Philip Morrison and Phylis Morrison and the Office of Charles and Ray Eames Scientific American Library New York , 1982, also for the figures on the chart

[Ψ] There is some debate as to whether there may be certain exceptions to this rule as when certain events known to be outside each other's event horizons are linked in a certain predetermined way. This may possibly happen, for example with particle decay where the end products are known. In General relativity these space-time axes are deformed by gravity so that the cones so represented may actually tilt, that is lose their verticality as represented here, but these matters will not affect my arguments.

[106]These considerations arise from the discussion of consciousness of time as presented in SHADOWS OF THE MIND by Roger Penrose Oxford Univ. Press, New York, 1994, especially as presented on p.384. "According to general relativity, 'time' is merely a particular choice of coordinate in the description of the location of a space-time event. Thee is nothing in the physicists' space-time descriptions that singles out 'time' as something that 'flows'." If so, then physical or mathematical descriptions of time are inadequate, but another way of looking at all of this of course, is that as in many cases, human perception or intuition vis à vis time may be inaccurate and the scientific or more rigorous treatment of time may be correct.

[107]Taken from: Joseph Silk THE BIG BANG W.H. Freeman & Co., San Francisco 1980 Figure 5.4 page 86

[108] Bever, TG & Chiarillo, RJ Cerebral Dominance in Musicians and Non-Musicians Science 185: 137-39 (1974)

[109] See "The Amusias" by N. Wertheim Chapter 10 in Critchley (ed.)MUSIC AND THE BRAIN also A.R. Luria, L.S. Tsvetkova and D.S. Futer "Aphasia in a Composer" J. Neurol. Sci. 2:288-92 (1965) The latter is a case report regarding the Russian composer V.G. Shebalin who had a stroke involving the dominant hemisphere with mostly a receptive aphasia yet continued as an active composer. Maurice Ravel had a non-specified degenerative disease with a receptive aphasia indicating Left hemishpere dysfunction yet was unable to function musically. However chance are in Ravel's case that much more than his left hemisphere was affected see Alajouanine, T, Aphasia and Artistic Realization. Brain 71:228-241 (1948)

[Ψ] The melodic structure of language may be altered too in what are called "dysprosodias" or defects in the prosidy of speech usually produced in right hemishpere lesions. See G.H. Monrad-Krohn "Dysprosody or Altered "Melody of Language" Brain 70:405-415 Here too there are receptive and expressive dysprosodias

[110]Jaynes, Julian THE ORIGIN OF CONSCIOUSNESS IN THE BREAKDOWN OF THE BICAMERAL MIND. Houghton Mifflin Company, Boston 1990 pages 29-30 Part of his example is drawn from Donald Hebb: "The Mind's Eye" Psychology Today 1961

[111] J. M. Roberts A CONCISE HISTORY OF THE WORLD Oxford University Press, New York 1995 p. 20

[112]See THE FIRST HUMANS Göran Burenhult, Gen Editor, Harper San Francisco "The Rise of Art" by same author P. 97-122

[113]From:THE FIRST HUMANS Göran Burenhult (Ed.) Harper San Francisco New York 1993 P.116

[114]ibid. page103

[115]From *Picasso* by Gertrude Stein Dover Publications, New York 1984 P. 46

Φ "When old age shall this generation waste,

Thou shalt remain, in midst of other woe

Than ours, a friend to man, to whom thou say'st

Beauty is truth, truth beauty, -that is all

Ye know on earth, and all ye need to know."

-John Keats: *Ode to a Grecian Urn*

# Earlier this century these lobe-finned fish that walked onto dry land, whose ancestors were the forebears of amphibians, reptiles, birds and mammals, using their front ventral fins to walk onto dry land, were found alive in waters off Africa. Before this these ungainly creatures were thought to be extinct. The Devonian period was the age of fish in which the first vertebrates, modified fish, crawled upon the land.

# It can be argued that we still have no idea of the vastness of the cosmos to this day, though our concept is considerably larger than in the past. Arguments rage about an "open" or "closed" universe, parallel universes and a host of other issues.

[116]Detail of head taken from: FREUD: Character and Culture "The Moses of Michelangelo" Edited by Philip Rieff, Collier Books, New York 1963. Freud interprets the Statue in light of the Biblical text noting details about how the tablets of the law are held. Freud maintains that Moses is about to break the tablets. In describing other portions of the statue in great detail, Freud seems to miss the horns of light. But according to the Biblical text this transformation happened *after* the first set of tablets had been broken with the second giving of the law, which totally negates Freud's interpretation. Note also the Christian tendency also to misinterpret the Mosaic transformation ignoring the shining of his face and mistaking the literal but figurative word for "horns" with what it really should be, "beams" of light.

[117] Hannah Arendt THE LIFE OF THE MIND Harcourt, Brace Jovanovich, New York © 1978 p. 111

[118] Viktor Frankl THE DOCTOR AND THE SOUL From Psychoanalysis to Logotherapy (New York: Alfred A. Knopf, 1965) 73-74

[119] Blum, K, Cull, J, Braverman, E., Comings, DE. (1996) Reward Deficiency Syndrome. *American Scientist* , 84,2,132-145

Blum K,Noble EP,Sheridan PJ, et al., (1990) Association of Human Dopamine D2 receptor gene in Alcoholism. *Journal of The American Medical Association* 263(15):2055-2060

Blum K, Sheridan, PJ, Wood RC, Braveman ER, Chen T, Comings DE. (1995) Dopamine $D_2$ Receptor Gene Variants: Association and Linkage Studies in Impulsive-Addictive-Compulsive Behavior. *Pharmacogenetics.* 5:121-141

Bolos, AM, Dean M, Lucas-Derse S. Et al., (1990) Population and Pedigree Studies Reveal a Lack of Association Between the Dopamine D2 Receptor Gene and Alcoholism. *Journal of the American Medical Association* 264:3156-3160

[120] Comings, DE (1998) The Molecular Genetics of Pathological Gambling *CNS Spectrums* 3(6), 20-37

[121] David Gelernter, *1939: The lost World of the Fair* (New York: The Free Press, 1995)

[*]Wagner was the most remarkable combination of the sublime and ridiculous. His music, which is undeniably wonderful, has an almost diabolical appeal that lays the basis for a whole line of composers who came after him, from Mahler and Bruckner to Berg. But many of his ideas, steeped in his time and culture, today appear totally ridiculous, especially those expressed in *Parsifal,* his last opera. *Parsifal* is concerned with the purity of blood lines that have been polluted, especially by mongrel Jews, which is further weakened by consuming meat, of all things. Blood ultimately needs to be purified through the Grail, chalice of the blood of Christ. Such overt racism was basic to the German psyche of the romantic age, and would be laughable, if not for its terrible effect.

[122]See Cases O, Seif I, Grimsby J, Gaspar P, Chen K, Pournin S, Muller U, Aguet M, Babinet C, Shih JC, De Maeyer E, "Aggressive Behavior and Altered Amounts of Brain Serotonin and Norepinephrine in Mice Lacking MAOA", *Science* 268: 1763-66, June 23, 1995, Also Brunner HG, Nelen M, Breakefield HH, Ropers BA, Van Oost A, *Science* 262:578 (1993) (Article on Dutch Kindred).

[123]See Pfennig, David W. And Sheman, Paul W. "Kin Recognition" Scientific American 272:(6):98-103 June 1995

[Φ] Most genetic defects involving production of an abnormal protein or enzyme are recessive. If you inherit a single dose (gene) from only one parent you will have more than enough of the normal enzyme to get by. But if you inherit a double dose, one gene from each parent, and thus no normal protein, it could be lethal. Tay-Sachs, Gaucher's disease, Cystic fibrosis are but some of a number of examples. If your parents are closely related, it increases your chances enormously of your inheriting a double dose of a defective gene, which would not be a problem were you just to inherit one copy from either parent. By some estimates, each of us on average carries two lethal recessive traits. If two first degree relatives were to mate, the chances of inheriting a double dose of a lethal gene would be somewhat less than one in four.

The classic example of a defective recessive trait that survives in great numbers in the population is the sickle cell trait. Here the heterozygote actually has an advantage. At least in malaria endemic areas the heterozygote is more resistant to the malaria plasmodium which spends part of its life cycle in red blood corpuscles. The homozygote, who inherits two doses of the defective sickle cell gene from each of his two parents, has an essentially lethal, extremely severe, disease. The sickle cell gene is an oddity surviving in a large population (mostly black) by virtue of the advantage conferred on the heterozygote. Other lethal recessive traits may survive in populations because the heterozygote confers an advantage for infectious disease. Tay-Sachs heterozygotes may more easily survive tuberculosis. Cystic fibrosis heterozygotes may have some advantage in certain forms of infectious diarrhea or cholera.

‡ I am indebted to the excellent discussion of these epistomological issues to be found in THE PASSION OF THE WESTERN MIND by Richard Tarnas Ballantine Books, New York, 1991

[124] See AT HOME IN THE UNIVERSE The Search of Laws of Self-Organization and Complexity by Stuart Kauffman Oxford University Press 1995

ϕ With a little care we could make predictions today, and it should be put to better use. Past behavior better uncovers proclivities than a psychological test. Unusual sexual interests, and behaviors predict future occurrences ranging from pedophilia to spousal abuse. Criminals do not tend to perform acts that are different from their previous repertoire of behaviors, so that past acts and the frequency of such acts, need to be carefully evaluated. A brutal murder often is not a first time offense. A perpetrator of crime would be expected to have an accelerated pattern of violence reflected in previous acts. The popular case of OJ Simpson neatly illustrates this point. That he should suddenly stab his wife out of the blue is not expected, but if he on the other hand, had a history of physical abuse of his wife that would make his violent crime more plausible. Fully realizing that you can't punish persons for acts that they have not yet committed, certain acts raise the risk that violent crimes will be performed and hence many if not most violent acts are preventable. To pre-emptively imprison or even mark persons who commit violent acts may violate civil rights. On the other hand, past behavior should not be ignored in a court of law. An individual's past is data that needs to be exploited in courts of law. This may prevent crimes and other tragic consequences. Crime ruins the life of the perpetrator as well as the victim. Some nascent criminals may be helped early on by treatment. If not, they should certainly be prevented from performing violent or sexually deviant acts on unsuspecting victims!

[125] For a classic discussion on the historical misuse and abuse of IQ testing and other related matters see Steven Jay Gould THE MISMEASURE OF MAN W.W. Norton and Company New York 1981. See also Carl N. Degler IN SEARCH OF HUMAN NATURE Oxford Univ. Press, New York 1991

[126]John Horgan "Get Smart, Take a Test" From *Science and the Citizen* , Scientific American Vol 273,(5) November1995 page12-14

[127] I am not writing here about psychological tests. I am merely trying to create some superficial familiarity to raise certain points, yet certain features of these exams are worth mentioning. The Wechsler, tests for components of intellect which are surprisingly straightforward including among others, vocabulary, memory, pattern recognition, digit symbols and others, the MMPI, The Minnesota Multiphasic Personality Inventory was developed at the Mayo Clinic and is composed of a large number of scales quantified by answers on a 600+ question true false format, scales consisting of such things as neuroticism, schizophrenia, depression and even truthfulness and consistency

[128]I am indebted for a lot of this discussion to D. Frank Benson *THE NEUROLOGY OF THINKING*, Oxford Univ. Press, New York 1994 p.198-207

[129] See Pamela Y. Blake MD, Jonathan H. Pincus, MD, Cary Buckner, MD "Neurologic Abnormalities in Murderers" NEUROLOGY 1995;45:1641-1647. The major thesis of the article which looked at individuals awaiting trial for murder is that Murderers had a high incidence of frontal lobe dysfunction and childhood abuse as well as a paranoid style of analysis. See also Dorothy Otnow Lewis *Guilty by Reason of Insanity*: A Psychiatrist Explores the Minds of Killers. Ballentine Publishing Group, New York 1998 Dr Lewis did many evaluations of prisoners with Dr. Pincus and comes to firm conclusions about necessary and sufficient conditions for violence.

[130] Jonathan M. Borwein and Peter B. Borwein "Ramanujan and PI" Scientific American 1989 p.112-117

[131] See D.J. Raine ALBERT EINSTEIN AND RELATIVITY Wayland Publishers Ltd. East Sussex, England © 1975 which describes the familiar story of Einstein's not very successful academic career.

[132] M-Marsel Mesulam Large-Scale Neurocognitive Networks and Distributed Processing for Attention, Language, and Memory Annals of Neurology 28(5):597-613 (1990) Of memory Mesulam says," After a critical gestation period, routinely used information becomes so massively distributed (i.e. consolidated) that it no longer requires the limbic system for retrieval."

[ϕ] See "Vision" Chapter 3 for a description of the entire visual attention process. This illustrates how the generally receptive parieto-occipital area cooperates with the motor frontal area to direct attention, throwing the central vision of the eyes, "foveating" an object of consequence. For example nothing catches visual attention better than an object moving in the peripheral vision of an eye. The visual area, composed mostly of imprecise-seeing rods, is sensitive to movement. Reflex mechanisms force the eye to move so as to place the moving object into the precise-seeing fovea of the eye.

[Ψ] Many other areas, including the basal ganglia and cerebellum that coordinate motor output and the primary visual and proprioceptive systems on the sensory side are part of this attentional network as well.

[133] See Zametkin AJ, Nordahl, TE, Gross M, King AC, Semple WE, Rumsey J, Hamberger S, Cohen RM: Cerebral Glucose Metabolism In Adults with Hyperactivity of Childhood Onset NEJM 323:1361-66 (1990) also Zametkin AJ, Liebenauer LL, Fitzgerald GA, King AC, Minkunas DV, Hercovitch P, Yamada EM, Cohen RM: Brain Metabolism in Teenagers with Attention-Deficit Hyperactivity Disorder Arch. Gen. Psychiatry 50:333-340 (1993)

[Ψ] from fascicle a little bundle or sheaf. A Fasciculus is a bundle of axons connecting one brain region to another.

[134] Neurologic outcome of patients with dorsolateral prefrontal leukotomy. Karp-BP; Juliano-DM; Berman-KF; Weinberger-DR J-Neuropsychiatry-Clin-Neurosci. 1992 Fall; 4(4): 415-21 There is general agreement that prefrontal leukotomy has little measurable effect on most psychological objective tests.

[Φ] Considering the frontal lobotomy as a type of lesion in the brain it is possible to conclude that in lesioning a part of the frontal lobe and turning a formerly aggressive patient into a passive individual, that somehow aggressively must be the function of the lessened frontal lobe area. If you were to draw this conclusion you would be wrong as we shall see, yet such logic is not unlike that applied to many other lesioning data.

[135]John R. Searle THE REDISCOVERY OF THE MIND , Bradford Books 1992 MIT Press

[136]This is an oft quoted passage description of J Harlow taken from Benson THE NEUROLOGY OF THINKING Oxford Univ Press, NY 1994 p 211 but this is such a celebrated case that this passage is multiply quoted in many texts because of its eloquent description.

[137] A good part of this immediate discussion is taken from a June 1 1990 lecture NEUROPSYCHIATRY at Univ. Pittsburgh given by Elkhnan Goldberg on the frontal lobe.

[138]Lhermitte F, Pillon B, Serdaru M. Human Autonomy and the Frontal Lobes: Part I: Imitation and Utilization Behavior: A Neuropsychological Study of 75 Patients. *Annals of Neurol.* 19:326-34, 1986

[139] A connectionist approach to the prefrontal cortex. Weinberger-DR J-Neuropsychiatry-Clin-Neurosci. 1993 Summer; 5(3): 241-53

[140] Lhermitte F, Human Autonomy and the Frontal Lobes. Part II: Patient Behavior in Complex and Social Situations:The "Environmental Dependency Syndrome". Annals of Neurology 19: 335-343, 1986 also an the excellent discussion by M.-Marsel Mesulam Frontal Cortex and Behavior Annals of Neurology 19(4) April 1986 p.320-325

[141] Bear, DM, Fedio P Quantitative Analysis of Interictal Behavior in Temporal Lobe Epilepsy Arch. Neurology 1977;34:453-467 also Waxman SG, Geschwind N. The Interictal Behavior Syndrome of Temporal Lobe Epilepsy. Arch Gen Psychiatry 1975;32:1580-1586

[142] Rauscher FH, Shaw GN, Ky KN, Listening to Mozart Enhances Spatial-Temporal Reasoning Towards a Neurophysiological Basis. Neuroscience Letters 185(1):44-47 (1995)

[143] See Paul Johnson MODERN TIMES

[144] See for example Silbersweig DA, Stem E, et al. A Functional Neuroanatomy of Hallucinations in Schizophrenia. *Nature* 378: 176-79,1995

[145] Swedo SE, Leonard HL, Garvey MA et al. Pediatric AutoimmuneNeuropsychiatric Disorders Associated with Streptococcal Infections: Clinical Descriptions of the First 50 Cases. *Am J Pshychiatry* 155 (2):264-71, 1998

[#] I don't mean to imply that atheists or agnostics are not ethical. In fact, among scientists and many others who don't practice religion, ethics strikingly similar to those of Western religions are highly prized and rigorously pursued. People who never set foot in church, live their lives according to basic principles of fair-dealing, avoidance of deceit, monogamy, incest taboos, and family obligations, benevolence, proscriptions against murder and violence etc. Indeed, it is striking how many pious individuals behave unethically. Violent acts and murders are committed by religious persons in our own day as they have throughout history. Moral and ethical behavior is only slightly correlated with religious practice. Religionists tend to rationalize despicable behavior by invoking "higher" principles, especially belief in an other world or afterlife which allows them to be less focused on ethics and the welfare of others in their temporal mundane world. Autocrats place principle above simple values and human welfare in much the same way, whether they happen to be "religious" or not. This other worldism (or in the case of the dictator attachment to abstract ideas (namely preservation of their own rule)) is fundamentally the same in that belief in some higher power or principle is justifies war, slavery, sadistic killings, immolation, and all manner of unethical conduct. On the other hand, scientists and non-religious ethical persons may not realize just how indebted they are to religion and how much they have been influenced in their ethical behavior, by religious ethical precepts as laid down over thousands of years. Most of these ideas are by now considered to be extremely basic, but are not at all obvious or inevitable given by our animal origin and background even considering our level of civilization. Religious teachings may have a civilizing influence especially Judeo-Christian ethics by which I mean they are most successful in converting persons from a savage almost hunter-gatherer type of existence into social and socially conscious beings.

[Φ] In humans the sexual capacity, as compared with other animals is highly embellished especially among those obsessed with it. It's been said that male sexual organs both testes and penis, are quite enlarged in comparison with our nearest primate relatives. Resulting increase in numbers of copulations perhaps figures into our advanced social relationships. In this area increased capacity for cerebration must figure even more prominently. Legendary sexual figures from Don Juan to the Marquis de Sade have a genius and proclivity expressed in their varied and developed tastes, as well as their tendency to wax eloquent and lure us in with their lurid and highly developed descriptions far beyond the imaginings of the ordinary person. The same holds for gastronomes. Obsessed with highly developed taste for foods that goes far beyond normal capacities, they tend of course to be fat. In a sense these persons have hypertrophied abilities, and as such they have much to teach the rest of us. The facility that these persons have is more than a mere neurological deficit. Actually a proclivity or genius is the other side, the obverse of a deficit, which for which the biologist and neurologist have an inadequate vocabulary, store of concepts.

[φ] It should be understood that all of these examples are caricatures and stereotypes used for purposes of illustration and argument only. In no way do I wish to place US Marines, whom I admire, in a pejorative light, as adventurous brigands with little frontal lobe function. In fact, a marine should be thought of in opposite way, as a highly disciplined planner, with supreme loyalty to high purpose, his country and his fellow marines. In the end, though for our purposes here, the soldier is a doer as opposed to a ruminator. He may be willing to assume danger because he either considers it less or relishes the idea of living on the brink of death as opposed to the common notion of being able with courage to overcome fear. It is a fact, that some persons just have more, some less, fear (which is related to other forms of inhibition) to begin with. In any case I wish to use this example merely to make a point that has nothing to do with a profile of any group mentioned here.

[146] For further discussion of the topic of dream facilitation of associations see Hartmann, Ernest A. OUTLINE FOR A THEORY ON THE NATURE, AND FUNCTIONS OF DREAMING in Dreaming vol. 6, No. 2 1966

[147] See also Elisibeth F. Loftus "Creating False Memories" *Scientific American* 277:70-75 September 1997 A whole literature on false childhood memories exists. See also Ceci S.J. Bruck M. Suggestibility of the Child Witness A historical Review and Synthesis *Psychological Bulletin* 113(3):403-39, 1993 May.

Ceci SJ, Loftus EF, Leichtman MD Bruck M. The Possible Role of Source Misattributions in the Creation of False Beliefs Among Preschoolers. *International Journal of Clinical & Experimental Hypnosis.* 42(4):304-320, 1994 Oct.

Brenneis CB. Belief and Suggestion in the Recovery of Memories of Childhood Sexual Abuse (Review) *Journal of the American Psychoanalytic Association.* 42(4):1027-53,1994.

*Searching for Memory: The Brain, The Mind, and the Past.* Daniel L. Schacter. BasicBooks, 1996

[ψ] Animals and humans must make a living in order to survive. The analogies between biological and economic adaptation are obvious and extensive. Just one comparison between biological and mental adaptation is that this adaptation thought or idea occurs ("In the beginning was the word") first, and only then is empirically applied in a given situation. See Michael Rothschild *BIONOMICS Economy as Ecosystem* Henry Holt and Company New York © 1990 for extensive discussion of these and related ideas.

[148] See Matthew J. Plunkett and Jonathan A. Ellman Combinatorial Chemistry and New Drugs SCIENTIFIC AMERICAN April 1997 Volume 276 Number 4 Pages 68-73

[∅] The round structure of Benzene a ring of six Carbon atoms as opposed to the linear structure of an ordinary chain of Carbon has often be compared with the *Uroborus* an ancient symbol of a serpent bent in a circle and biting its own tail symbolic of the eternal cycle of nature. See *Private Myths* by Anthony Stevens (above).

[149] See "Update on the Neurology of Sleep" Course given at the Annual Meeting of the American Academy of Neurology 1997 Syllabus 146 pps67-71.

[150] Anthony Stevens *Private Myths: Dreams and Dreaming,* Harvard University Press © 1995 p.324

[151] Mayr, Ernst. *Toward A New Philosophy of Biology: Observations of an Evolutionist.* Harvard University Press, Cambridge MA © 1988 P. 16-17

[φ] I don't deny that many other organisms without advanced consciousness, bacteria and insects especially, are giving us a real run for our money.

[152] Goldhagen, Daniel: *Hitler's Willing Executioners:* Ordinary Germans and the Holocaust. Alfred Knopf, New York 1996

[153] See Gershom Sholem *Kabbalhah* Dorset Press, New York © 1974 pp. 351-355

[154]From: Steven Weinberg, LIFE IN THE UNIVERSE Scientific American 271:(4) p.47 October 1994

[155]See, for example, Charles Pelligrino Reurn to Sodom and Gomorrah Avon Books New York New York © 1994 p 90

[156] For a fuller discussion of this topic see Michael Rothschild (1990) *Bionomics: Economy as Ecosystem* Henry Holt and Company NewYork

[157] John Terborgh *Diversity and the Tropical Rain Forest* Scientific American Library, W.Hl Freeman and Company, New York © 1992 P. 71

[158] John Terborgh *Diversity and the Tropical Rain Forest* Scientific American Library, W.Hl Freeman and Company, New York © 1992 P. 71

[159] Jared Diamond *The Third Chimpanzee* Harper Collins New York © 1992

Ψ The Anthropic Cosmoligical Principle is idea that the universe would have to be just exactly as it is in order to accommodate an intellectual being capable of thinking of and writing about it. If things were even slightly different, the universe would not have humankind and thus would have no being able to understand and think about it.

Ψ In the case of Carbon, these eight electrons will fill shell 2, its outer shell, achieving perfect happiness, if such exists. For silicon its outer shell (3) requires 18 electrons, but a smaller part of the shell three, an orbital, will then be filled when the third shell has 8 electrons. The third shell has orbitals termed s, and p and so on. The s orbital always has 2 electrons, the p, 6 so that for Silicon 8 electrons in the outer shell will at least fill its s and p orbitals. For a given shell the number of ideal orbitals or sub orbits equals the number of that shell, 1 for 1, 2 for 2 etc.

[160] Much of this discussion is abstracted from William J Kauffman and Larry L. Smarr *Supercomputing and the Transsformation of Science* Scientific American Library, New York © 1993

[161] I refer here, of course, to Carl Sagan's 1985 novel.

[162] See Widner H, Tetrud J, Rehncrona S, Snow B, Brundin P, Gustavii B, Bjorklund A, Linvail O, Langsotn JW Bilateral Mesencephalic Grafting in Two Patients with Parkinsonism Induced by 1-methyl-4-phenyl-1,2,3,6-tetrahydropyridine (MPTP). *New England Journal of Medicine* 327(22):1556-63, 1992 Nov 26 also Spencer DD, Robbins RJ, Naftolin F, Marek KL, Vollmer T, Leranth C, Roth RH, Price LH, Gjedde A, Bunney Bs. Et al Unilateral Transplantation of Human Fetal Mesencephalic Tissue into the Caudate Nucleus of

Patients with Parkinson's Disease. *New England Journal of Medicine* 327(22):1541-8, 1992, Lindvall O, Widner H, Rehncrona S, Brundin P, Odin P, Gustavii B, Frackowiak R, Leenders KL, Sawle G, Rothwell JC et al. Transplantation of Fetal Dopamine Neurons in Parkinson's Disease: One Year Clinical and Neurophysiological Observations in Two Patients with Putaminal Implants. *Annals of Neurology* 31(2):155-65, 1992 Feb.

[Φ] A defibrillator senses ventricular fibrillation an electrical condition in the heart that is most often lethal in which the ventricle stops pumping blood and instead fibrillates. The wall of the heart just wiggles but does not pump. Ventricular fibrillation is one of the most common causes of sudden death, and if a person can be prevented from going into ventricular fibrillation then sometimes point out jokingly, he cannot die.

[φ] For most of us, if we "know" a person, we would like to think, we can predict, within reasonable degree of certainty what he or she might do from one moment to the next. Unpredictable persons are usually felt to be mentally ill, quixotic, or at the least willful but as a general rule we all know some of us are more excitable, some calm, more or less anxious, rational, irrational etc. For philosophers the ages-old philosophic debate between determinists and non-determinists is that for determinists, having all this data, if you know the state of a person at $t_0$ you will then always know exactly their status at $t_{0+1}$, this provided you knew also about all stimuli. The non-determinist holds that human will intervenes so that behavior at the next moment in time can't entirely be predicted.

[ψ] It may well be that this device will not be made of Silicon but Gallium or some other data storing substance. These considerations are merely for the purpose of argument.

[163] © 1996 Dr. John B. Chittick, used by permission

[Φ] for Cytosine, Adenine, Guanine used as an example here, the repeat that causes the Huntington's disease residing on the fourth chromosome and, as it turns out a majority of neurological diseases due to this particular repeating strand, the codon for the excitatory amino acid glutamate. Thus adding extra CAG codons will translate into the resulting protein into a polyglutamate strand.

[164] See Haywood, Anne M., Transmissible Spongiform Encephalopathies New England Journal of Medicine 337,1821-28, Dec. 18, 1997 also Huang, Z, Pruisiner SB, Cohen, FE "Scrapie Prions: A Three-dimensional Model of an Infectious Fragment" on Internet www.ceberdyne.com/~tom/prion3d_theory.html

[165] See Scientific American "Off with Its Head!" in Technology and Business January 1998 278: p. 41

[166] Picture courtesy of Scientific American. News and Analysis: "Off With Its Head" January 1998 (Vol 278) p.41

[167] See DeArmond SJ, Mobley WC, DeMott DL, Barry RA, Beckstead JH, Prusiner SB. Changes in the Localization of Brain Prion Proteins During Scrapie Infection Neurology 50: 1271-80, 1998

[168] see Linduist S, Patino M, Glover J, Liu J-J, "Protein Particles Similar to Those Suspected in "Mad Cow" Disease Found in Yeast Cells. Science August 2, 1996

[Ψ] One could think of some circumstances where polygyny would be adaptive and perhaps these are where it got it start in certain cultures. In a warring society which allowed only for the survival of few males, polygyny might make sense. The classic example of a polygynous arrangement is the Harem. Here, polygyny occurs only because of economic and societal dominance of certain males. Polygyny occurs in America as well. Certain males have multiple relationships and sire large number of offspring. In certain strata of our society adulterous relationships with women is also quite popular. This is quite advantageous from a biological perspective, particularly if the cuckolded husband is unaware he is supporting another man's child. One would think a society tolerating mating with multiple females would be less able to compete (see below) because less of the father's attention would inevitably accrue his children. Lacking certain knowledge about paternity would also increase the likelihood of incestuous or near-relative matings.

[169] Adapted from Wilson, Sarah *Matisse* Rizzoli International Publications, Inc. New York1992 "Dance I First Version Paris 1909 Oil on Canvas" Figure 56

CPSIA information can be obtained
at www.ICGtesting.com
Printed in the USA
LVOW11s1503010917
547247LV00001B/81/P

9 781432 754013